Modeling Multiphase Materials Processes

Manabu Iguchi • Olusegun J. Ilegbusi

# Modeling Multiphase Materials Processes

Gas-Liquid Systems

Springer

Manabu Iguchi
Graduate School of Engineering
Division of Materials Science
& Engineering
Hokkaido University
Sapporo, Hokkaido 060-8628, Japan
gaku@eng.hokudai.ac.jp

Olusegun J. Ilegbusi
Department of Mechanical, Materials
& Aerospace Engineering
University of Central Florida
Central Florida Blvd. 4000
Orlando, Fl 32816, USA
ilegbusi@mail.ucf.edu

ISBN 978-1-4419-7478-5    e-ISBN 978-1-4419-7479-2
DOI 10.1007/978-1-4419-7479-2
Springer New York Dordrecht Heidelberg London

Library of Congress Control Number: 2010938939



Printed on acid-free paper

Springer is part of Springer Science+Business Media (www.springer.com)

# Preface

Multiphase flow phenomena abound in many materials processing operations, including gas-stirred ladles, blast furnaces, and the like. This book focuses primarily on systems in which liquid and gaseous phases coexist. In general, such systems are complex and difficult to characterize. The phases can be continuous or dispersed, with mass, momentum, and energy exchange across the interfaces. For the past few years, there has been a heavy focus on relating the interfacial fluxes to measurable macroscopic properties of the flow so as to provide a fundamental understanding of the mixing process. Significant advances have also been made in the application of computational and experimental techniques to multiphase transport phenomena in materials processing operations. This book presents the application of these modeling tools to dispersed gas–liquid metallurgical systems.

The book is an outgrowth of recent research activities of the authors in the modeling of multiphase flow phenomena in materials processes. It emphasizes the synergistic relationship that should exist between physical and mathematical modeling techniques through a variety of validation examples. The primary intended audience is materials researchers in industry and academia faced with practical problems of modeling multiphase flow phenomena in metallurgical reactors. The book can also be used as a secondary reference for an upper-level multiphase flow course, and serve as a research monograph for graduate students. The typical reader would have had previous exposure to the scientific basis of physical and mathematical modeling techniques.

Several new features are included in the book beyond the traditional texts on multiphase flow. For example, the conflicting roles of hydrodynamic instability in materials processing are discussed. Numerical methods are presented for modeling the complete continuous casting operation, including gas–liquid and solid–liquid multiphase systems, with examples. Nanoscale and microscale multiphase phenomena are reviewed, bringing into focus recent advances in this emerging field. The book is organized into 11 chapters as follows:

Chapter 1 provides a general overview and introduction of the principles and techniques of physical and mathematical modeling discussed in the book. It provides the rationale for modeling two-phase flow in gas-agitated reactors of materials processes. Chapter 2 presents the turbulence structure of two-phase jets and the impact on the mixing and chemical reaction rates in materials reactors agitated by gas injection. Chapter 3 describes the principle of the Coanda effect, including the diversion of bubble path near the vessel walls as well as the merging of bubbles generated from multiple sources. The implication is also discussed of these phenomena on the mixing intensity and mixing time in the reactor. Chapter 4 focuses on the interaction between bubbles and walls, including dynamic wettability associated with the advancing contact angle, and the effect on bubble shape and size. Experimental results are also provided for bubble and liquid characteristics in systems with walls of varying wettability.

Chapter 5 discusses swirling flow, the underlying hydrodynamic instability, and the impact on the mixing process. Chapter 6 describes the interaction between molten metal and top slag in a gas-agitated reactor, and its effect on mass transfer mechanism and the mixing phenomena. Chapter 7 focuses on surface control of the reactor and the effect on mixing time. Chapter 8 discusses mold powder entrapment in molten steel in continuous casting molds and the impact on product quality. Results from flow visualization and particle image velocimetry (PIV) techniques are presented to elucidate the underlying mechanism.

Chapter 9 describes the mathematical techniques for modeling two-phase gas–liquid multiphase systems. A variety of methods are assessed with the aid of examples comparing predictions with the experimental data. Chapter 10 presents a detailed description of the application of numerical techniques to continuous casting systems in general, including alloy solidification. Finally, Chapter 11 reviews nanoscale and microscale multiphase phenomena in materials processing. The important global and local characteristics relevant to understanding and modeling of such phenomena are described, and examples are presented on fluid flow and heat transfer in such systems. There is an Appendix that summarizes the transport data for common fluids, including air, water, argon, helium, hydrogen, etc.

The modeling of multiphase transport in materials processes is multidisciplinary, involving concepts and principles not often associated completely with one particular field of study. We have therefore ensured that this book is self-sustaining through the variety and breath of topics covered. The text could therefore serve as a useful guide to the types of questions that should arise in the modeling of gas-agitated metallurgical processes and how to seek relevant answers.

Many people deserve thanks for their part in this effort. MI gratefully acknowledges the useful advice given by Dr. Zen-ichiro Morita, Professor Emeritus of Osaka University. Dr. Mahmut Mat of Nidge University deserves special thanks for his contribution to the mathematical modeling effort. We appreciate the mentorship of the late Professor Julian Szekely of the Massachusetts Institute of Technology (MIT) as well as his pioneering role in the field of mathematical modeling of materials processing operations, and gas-agitated systems in particular. We acknowledge the contributions of our graduate students who have participated in our multiphase flow research over the years. Mr. Hirotoshi Kawabata of Osaka University deserves special mention. Finally, we say a big thank you to our families for their support and understanding during the preparation of this book.

Sapporo, Japan — Manabu Iguchi
Orlando, FL — Olusegun J. Ilegbusi

# Contents

# Chapter 1
# Introduction

## 1.1 Introductory Remarks

The flow pattern and local turbulence structures largely determine the performance characteristics of metallurgical operations. The transport phenomena in these processes typically have the following basic features:

- Flows are complex and multiphase with strong turbulence.
- Fluids are usually opaque with very high temperature. For example, the melting point of molten metal is typically above 1,000 °C.
- In most cases, fluid flows are coupled with heat and mass transfer.
- Fluid flows are usually accompanied by metallurgical reactions.
- The walls of the reactors and pipelines are usually poorly wetted by molten metal.

The variety of multiphase flows encountered in metallurgical processes [1–6] include; gas–liquid, liquid–liquid, gas–solid, solid–liquid, and gas–liquid–solid systems. This book focuses on gas–liquid two-phase flows, being the most popular and relevant to current metallurgical processes. Due to their complexity, the exact equations governing gas–liquid two-phase flows are not known except for very limited situations. A typical example is one in which dispersed bubbles in a liquid are spherical and uniformly distributed since the interface between the two phases is deformable and the gas phase is compressible. It is therefore imperative to develop a set of approximate or simplified governing equations to describe the flow field.

The measurement of fluid flow characteristics in real processes is quite difficult due to the opacity and high temperature characteristics. In addition, model experiments using molten metals as in the real process are often cumbersome. The situation is further complicated when heat and mass transfer, and chemical reactions occur simultaneously. Under these severe and complex conditions, it is very difficult to carry out experimental investigations on the real processes in order to improve efficiency or to develop novel processes.

The poor wettability at the metal/solid interface implies that the boundary conditions on solid walls immersed in molten metal–gas two-phase flows would be different from the relatively well-known conditions for wetted interface. The heat and mass transfer boundary conditions would be similarly different. In addition,

M. Iguchi and O.J. Ilegbusi, *Modeling Multiphase Materials Processes: Gas-Liquid Systems*, DOI 10.1007/978-1-4419-7479-2_1,

any presumption regarding the true situation cannot be readily verified through experimentation on real processes.

In light of the above, we rely on model experiments and theory to provide an improved understanding of fluid flow phenomena in metallurgical processes. In this chapter, the modeling of gas–liquid two-phase flows is classified into two broad categories. Some basic physical situations of relevance in multiphase metallurgical processes are then introduced. Finally, the nature of individual multiphase flows in real processes is discussed.

## 1.2 Classification of Models

### 1.2.1 Physical Modeling

Modeling can generally be classified into two categories: physical and mathematical [1,4,6]. The primary objective of physical modeling is usually to represent a system to be modeled by changing the materials to be handled, and also the scale of the operation, so as to achieve realistic representation of the process [4]. If the governing differential equations for the fluid flow, heat and mass transfer are known, representative scales are introduced for time, length, mass and so on to non-dimensionalize the equations. As a result, some groups of dimensionless parameters are obtained. Physical models are typically designed so that the fluid flow, heat and mass transfer phenomena are readily measured using fluids that are different from the real fluids. On the other hand, in complex systems with limited information in the governing equations, the Bukingham's $\Pi$ theorem [7] is useful for identifying the relevant dimensionless parameters.

Physical modeling can be classified into the following three stages in order of descending rigor [1]:

(a) Results obtained from a physical model can be translated directly to quantitatively describe the behavior of the real system. Models in slightly smaller scale than the prototype are usually classified into this category.
(b) A physical model experiment is performed to verify the applicability of certain mathematical models. The same fluid as in the real system is usually employed in such experiments.
(c) A physical model is constructed to obtain qualitative information on a system. Water is usually used to simulate molten metal. This is because the kinematic viscosity of water is approximately equal to that of molten metal. The results based on water models are strictly applicable to molten metal flows inside the bath. Care must therefore be taken before applying the results to the estimation of interfacial phenomena for which the physical properties such as surface tension play an important role.

A variety of examples of the second and third categories are given in this book, and the key features of physical modeling are examined.

### 1.2.2 Mathematical Modeling

Mathematical modeling involves the use of a set of algebraic or differential equations to describe the fluid flow and associated heat and mass transfer in the real processes [1,4,6]. The models are based on the fundamental laws of physics, chemistry, fluid, and solid mechanics. Due to recent rapid advances in computer technology and developments in numerical calculation techniques, mathematical modeling is now widely applied in engineering and science practice. The techniques of mathematical modeling and their applications to metallurgical processes are given in detail in Chaps. 9 and 10. Additional information on mathematical modeling principles is also available in other texts [1,4,6].

In physical as well as mathematical modeling of transport phenomena, it is important to consider the existence or otherwise in the real process, any possible flow transition from laminar to turbulent flow, natural convection to forced convection, or subsonic to supersonic flow. This is because of the significant impact that such transitions may have on the process.

## 1.3 General Strategy for Modeling Two-Phase Phenomena

In physical modeling, it is essential to choose the most influential dimensionless parameters in the flow field. These parameters can be derived directly from the governing equations or determined by using the Buckingham's $\Pi$ theorem. Each parameter should then be varied over the range typical of the real process. In complex processes, a number of dimensionless parameters are obtained. It is however practically impossible to cover the complete range of all these parameters. Therefore a decision has to be made as to which are the most dominant. Fortunately, if the flow is turbulent, relaxation of similitude such as the Reynolds number similitude is allowed in many situations, thus significantly reducing the number of dimensionless parameters.

In mathematical modeling the exact governing equations for gas–liquid two-phase flows are not well known. Constructing a set of governing equations, which can properly approximate the real fluid flow situations is therefore crucial. Although the level of approximation employed depends on individual modeler's experience and insight into the phenomena, the fundamental procedure in mathematical modeling is common to a certain extent even if the processes differ from one another. The same is true in physical modeling.

## 1.4 Basic Physical Situations of Relevance in Gas–Liquid Processes

### *1.4.1 Gas–Liquid Two-Phase Flows in Cylindrical Bath*

#### 1.4.1.1 Bubbling and Jetting

Two types of bubble dispersion patterns are known to exist in a bath agitated by gas injection through a submerged single-hole nozzle (see Fig. 1.1) [5, 8, 9]. In a relatively low gas flow rate regime, discrete bubbles are formed successively at the nozzle exit, some of which break up into smaller bubbles. On the contrary, a continuous bubble column like a cone is formed above the nozzle exit and small bubbles are generated at a gas/liquid interface due to high shearing stress. The former is called bubbling and the latter is named jetting. The transition from bubbling to jetting is of practical importance, because it is closely associated with mixing in the bath and the wear of the nozzle material [5].

Mori and Sano stated, based on their experimental results, that the transition from bubbling to jetting occurs at a Mach number of unity [9]. In other words, the transition occurs in the transonic region [10]. Recently, Chen and Richter [11] have developed an instability theory to explain the transition from bubbling to jetting by studying a circular compressible jet in a liquid. This theory indicates that small disturbances grow immediately in any position without limitation in the subsonic

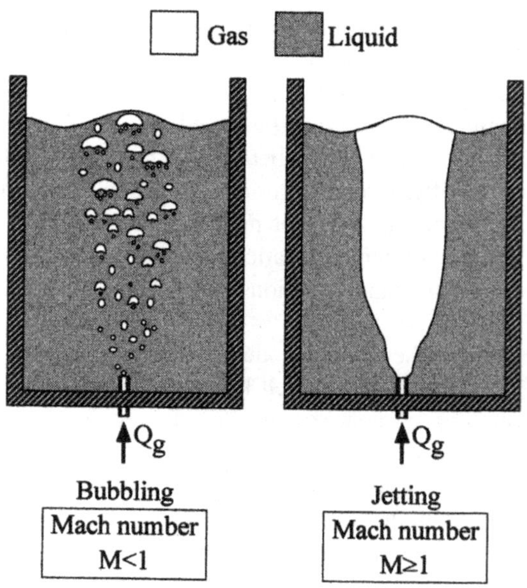


**Fig. 1.1** Bubbling and jetting

region, causing the jet to break up and bubbling to occur. In the supersonic region, small disturbances ultimately diminish with time, and the gas jet remains stable. Consequently, bubbling exists in the subsonic region and the transition occurs in the transonic region, which is consistent with the experimental findings of Mori and Sano [9].

#### 1.4.1.2 Characteristic Parameters

(i) Bubble characteristics

The dynamic behavior of bubbles rising in a bath is quantitatively characterized by the following parameters [12–15]:

(a) Gas holdup, $\alpha$.
(b) Bubble frequency, $f_B$. This is the number of bubbles passing through a monitored position in 1 s.
(c) Mean bubble rising velocity, $u_B$.
(d) Mean bubble chord length $L_B$.

Gas holdup and bubble frequency can be measured with a single-needle electroresistivity probe, while a two-needle probe is required for mean bubble rising velocity and mean bubble chord length.

Bubbles are usually not perfectly spherical in shape. Accordingly, the mean chord length is used to represent the bubble size unless the bubble shape is radically different from spheroidal. For definition, the equivalent spherical bubble has the diameter $d_B$. The mean bubble chord length $L_B$ is given by [12]:

$$L_B = d_B/1.5 \tag{1.1}$$

Figure 1.2 shows a schematic of the output signals of a two-needle electroresistivity probe. The aforementioned four quantities can be determined from the following relations:

$$\alpha = 100\Sigma t_i/T \tag{1.2}$$
$$f_B = \Sigma n_i/T = N/T \tag{1.3}$$
$$u_B = \Sigma(L_p/\Delta t_i)/N \tag{1.4}$$
$$L_B = \Sigma(L_p/\Delta t_i)t_i/N \tag{1.5}$$

where $t_i$ is the time duration of the $i$th bubble signal, $T$ is the measurement duration, $N$ is the total number of bubbles, $L_p$ is the vertical distance between the two electrode needles, and $\Delta t_i$ is the time delay for the $i$th bubble to pass through the two needles.

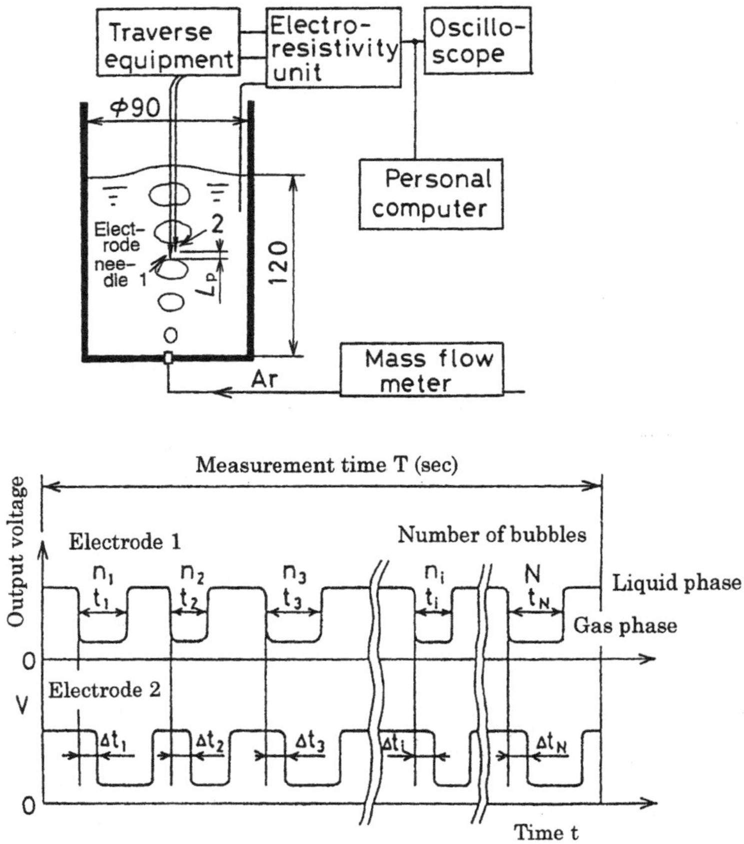


**Fig. 1.2** Output signals of two-needle electroresistivity probe

The units of $\alpha$, $f_B$, $u_B$, and $L_B$ are %, Hz, m/s, and m, respectively. In actual experiments some additional conditions are usually introduced in order to remove noise and enhance the measurement accuracy as much as possible [14].

(f) Bubble shape

It is necessary to determine the shape of bubbles for the evaluation of the total interfacial area between the bubbles and molten metal in the bath. The total interfacial area is closely associated with metallurgical reaction rates, and hence, the efficiency of the process. The shape of a single bubble rising in a still, transparent liquid bath has been extensively investigated by many researchers, and the results are summarized in a map by Clift et al. [16]. The shape of bubbles successively rising in a molten metal bath has been observed with an X-ray fluoroscope [17] and a multi-needle electroresistivity probe [18, 19]. Iguchi et al. [19] classified the bubble shapes into five categories and drew a map in terms of a modified Reynolds number $Re_m$ and a modified Weber number $We_m$, as shown in Fig. 1.3. These two dimensionless parameters, $Re_m$ and $We_m$, are defined as follows:

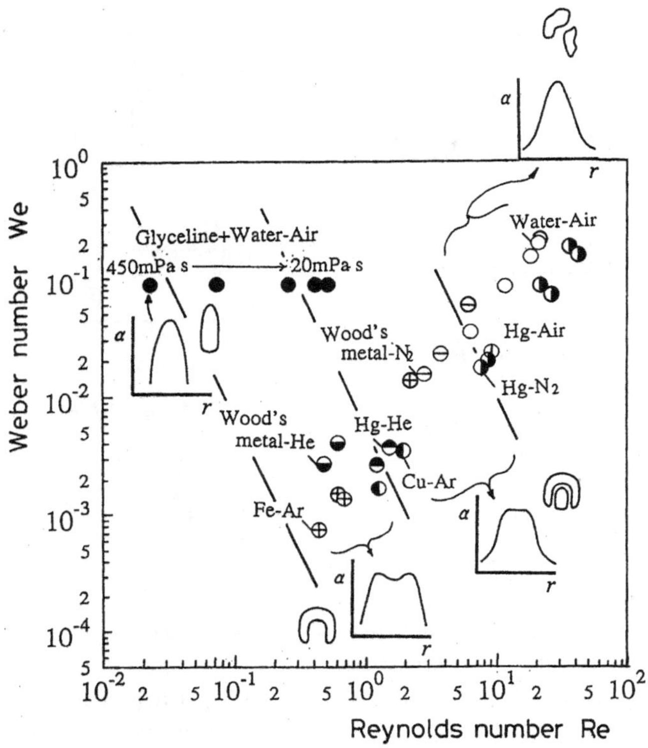


**Fig. 1.3** Classification of the shape of successively rising bubbles

$$Re_m = \rho_g d_B u_B / \mu_L \tag{1.6}$$

$$We_m = \rho_g d_B u_B^2 / \sigma \tag{1.7}$$

where $\rho_g$ is the gas density, $d_B$ is the mean bubble diameter, $u_B$ is the mean bubble rising velocity, $\mu_L$ is the viscosity of liquid, and $\sigma$ is the surface tension of liquid.

The shape of successively rising bubbles in molten metal baths is closely related to the horizontal distribution of gas holdup, as shown in Fig. 1.3. Thus the shape can be predicted from the results of gas holdup measurements.

(ii) Liquid flow characteristics

Figure 1.4 shows a schematic of the output signals of a two-channel laser Doppler velocimeter used for measuring the axial and radial velocity components, $u$ and $v$, of water flow in a bottom blown bath. The main quantities characterizing the water flow can be determined from the following relations [20]:

(a) Mean velocity components

$$\bar{u} = \Sigma u_i / N \tag{1.8}$$

$$\bar{v} = \Sigma v_i / N \tag{1.9}$$

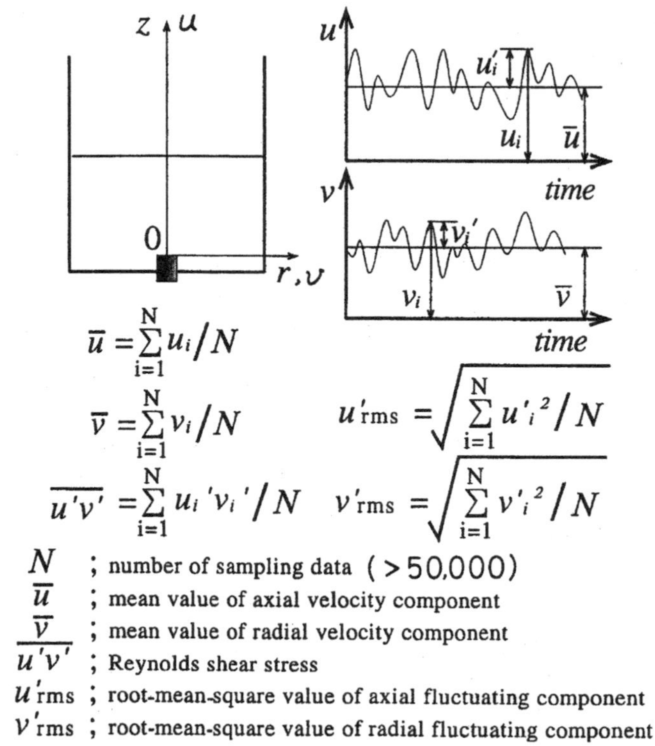


**Fig. 1.4** Output signals of two-channel laser Doppler velocimeter

where subscript $i$ denotes the $i$th digitized datum and $N$ is the number of velocity data.

(b) Root-mean-square (rms) values of turbulence components

$$u'_{\mathrm{rms}} = [\Sigma(u_i - \bar{u})^2/N]^{1/2} \tag{1.10}$$

$$v'_{\mathrm{rms}} = [\Sigma(v_i - \bar{v})^2/N]^{1/2} \tag{1.11}$$

(c) Turbulence intensity

$$\mathrm{Tu} = u'_{\mathrm{rms}}/u \tag{1.12}$$

(d) Reynolds shear stress

$$\overline{u'v'} = [\Sigma(u_{\mathrm{i}} - \bar{u})(v_{\mathrm{i}} - \bar{v})/N]^{1/21} \tag{1.13}$$

(e) Skewness factor $S$ and flatness factor $F$

$$S_{\mathrm{u}} = \Sigma(u_{\mathrm{i}} - \bar{u})^3/({u'_{\mathrm{rms}}}^3 N) \tag{1.14}$$

$$S_{\mathrm{v}} = \Sigma(v_{\mathrm{i}} - \bar{v})^3/({v'_{\mathrm{rms}}}^3 N) \tag{1.15}$$

$$F_u = \Sigma(u_i - \bar{u})^4/({u'_{rms}}^4 N) \tag{1.16}$$

$$F_v = \Sigma(v_i - \bar{v})^4/({v'_{rms}}^4 N) \tag{1.17}$$

These skewness factors are widely used to specify the higher-order turbulence components, because the turbulence intensity in gas–liquid two-phase flows in the baths of metallurgical processes is much higher than those of single-phase pipe flows (Tu $< 15\%$) and single-phase jets (Tu $< 30\%$) [21,22]. It is known that $S = 0$ and $F = 3$ for a Gaussian (normal distribution) error curve.

(iii) Mixing time

The concept of mixing time is extensively used to quantify the intensity of mixing in the baths of metallurgical refining processes [23–25]. Figure 1.5 shows a schematic illustration of the output voltage of an electrical conductivity probe immersed in a bath agitated by gas injection. The voltage begins to increase immediately when the tracer is injected onto the bath surface and finally reaches a constant value. The mixing time is defined as the time interval required for the output voltage of the electrical conductivity probe to settle to within $\pm 5\%$ deviation about the final value $V_f$. This definition is referred to as the 95% mixing criterion. Since the pioneering work of Nakanishi et al. [26], remarkable progress has been made on the determination of mixing time. Previous empirical relations for the mixing time have been reviewed by Mazumdar and Guthrie [23,24].

In model experiments using molten steel, Cu is often chosen as a tracer, but the sampling is made discretely to determine mixing time. It is desirable to develop a simplified method capable of measuring the concentration of tracer in a molten metal bath continuously. Such a method would also be useful for judging whether or not fine powder disperses uniformly in the molten metal bath according to the established powder injection metallurgy.

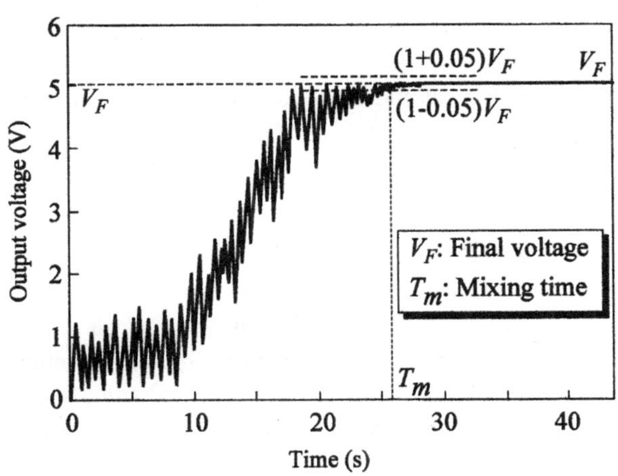


**Fig. 1.5** Determination method of mixing time

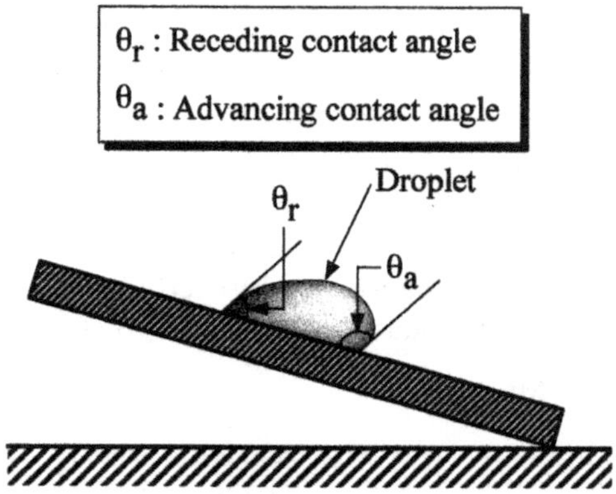


**Fig. 1.6** Inclined plate method for the determination of advancing and receding Contact angles, $\theta_a$ and $\theta_r$

(iv) Wettability

The wettability of a plate is quantitatively evaluated in terms of the advancing contact angle $\theta_a$. When $\theta_a$ of a plate is less than 90°, the plate is classified as being of good wettability or wetted, while a plate of $\theta_a$ ranging from 90 to 180° is regarded as having poor wettability. Both the advancing and receding ($\theta_r$) contact angles can be determined by the inclined plate method, as shown in Fig. 1.6 [27–29].

When a mixture of gas and liquid impinges on a wall or flows along the wall, the gas preferably attaches to the surface. The flow in the vicinity of the wall therefore strongly depends on the wettability of the wall. Consequently, heat and mass transfer from the wall is also impacted by the wettability.

(v) Heat and mass transfer

Heat transfer between injected gas and molten metal near the nozzle exit is closely associated with the expansion of the gas, formation of thermal accretion around a bottom nozzle, and the so-called back attack [9, 30, 31]. Furthermore, mass transfer between bubbles and molten metal is essential for the efficiency of refining processes [32, 33]. A discussion on these subjects is beyond the scope of this book and interested readers should refer to other specialized texts, for example, Ilegbusi et al. [6].

### *1.4.2 Gas–Liquid Two-Phase Flows in Pipes*

Investigations on gas–liquid two-phase flows in pipes of good wettability started approximately 70 years ago mainly in mechanical and atomic energy engineering [34–39]. Since then many experimental and theoretical studies have been carried out on this subject. Some of the findings of relevance to metallurgical operations will be briefly reviewed here.

#### 1.4.2.1 Vertical Pipes

(i) Flow regimes

Focus will be on gas–liquid two-phase flows in which both fluids move vertically upwards. As shown in Fig. 1.7 the flow regimes in a vertical pipe of good wettability can be classified into the following seven categories [36]:

(a) Bubbly flow
(b) Slug flow
(c) Froth flow or Churn flow
(d) Annular-mist flow
(e) Mist flow
(f) Slug-Annular flow
(g) Bubbly-Annular flow

The nature of flow in each flow regime has been described in great detail in other texts [36, 37]. Many researchers have also given the flow regime maps. The boundaries between these regimes are typically expressed as functions of the superficial velocities of gas and liquid, $j_g$ and $j_L$, defined as,

$$j_g = 4Q_g/(\pi D^2) \tag{1.18}$$
$$j_L = 4Q_L/(\pi D^2) \tag{1.19}$$

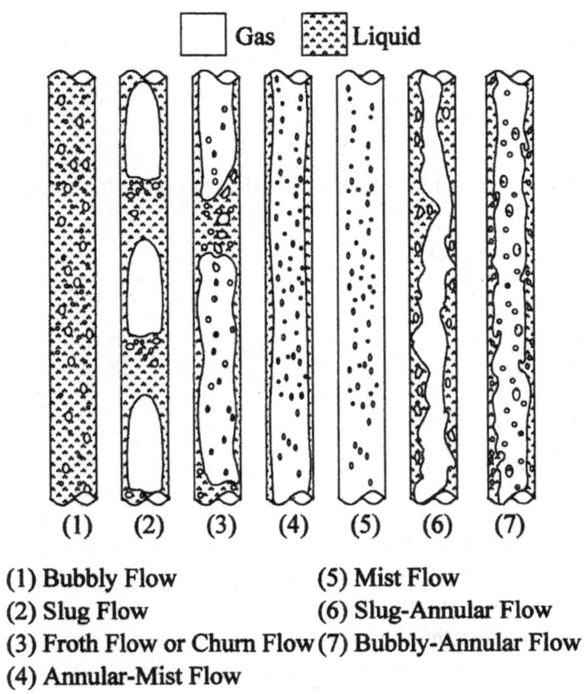


**Fig. 1.7** Flow regimes in vertical pipe of good wettability

in which $D$ is the pipe diameter, and $Q_g$ and $Q_L$ are the gas flow rate and liquid flow rate, respectively. The boundaries, however, are different depending on the researchers.

(ii) Bubble and liquid flow characteristics

Data on the quantities given in (1.2)–(1.5) and (1.8)–(1.12), as well as the pressure losses, and heat and mass transfer have been accumulated by numerous researchers [36, 37]. Mathematical models and empirical relations have also been proposed for the quantities in mechanical, chemical, and atomic energy engineering. Most of these investigations, however, considered pipes of good wettability.

Gas–liquid two-phase pipe flows are also prevalent in metallurgical processes. A representative example is molten steel–argon two-phase flow in the immersion nozzle in continuous casting [40–43]. However, unlike situations for which correlations have been established, the pipes of metallurgical systems are poorly wetted by the molten metal. In addition, studies on molten metal–gas two-phase pipe flows are rather limited. This book describes the flow regimes and the bubble and liquid flow characteristics in pipes of poor wettability typical of materials engineering operations in order to elucidate the difference between the behavior of bubbles passing through pipes of varying wettability. The flow regime map illustrated in Fig. 1.7 is not applicable to pipes of poor wettability when the superficial velocity of liquid is less than a certain critical value due to the attachment of bubbles to the pipe wall [44].

#### 1.4.2.2 Horizontal Pipe

(i) Flow regimes

The flow regimes in horizontal pipes of good wettability have been classified into the following (see Fig. 1.8) [36]:

(a) Stratified flow
(b) Wavy flow
(c) Bubbly flow
(d) Plug flow
(e) Slug flow
(f) Froth flow
(g) Annular Mist flow
(h) Mist flow

A number of researchers have also classified gas–liquid two-phase flows in horizontal pipes of good wettability based on other criteria [37]. Unfortunately, there is yet no consensus among these various investigations. Although considerable effort has been devoted to understanding the flow regimes in horizontal pipes of good wettability, investigations on gas–liquid two-phase flows in pipes of poor wettability are quite limited.

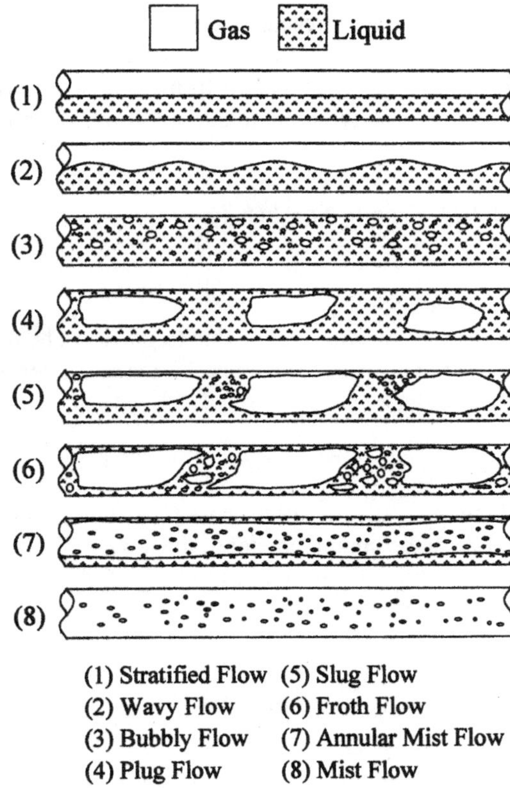


**Fig. 1.8** Flow regimes in horizontal pipe of good wettability

(ii) Bubble and liquid flow characteristics

These characteristics as well as pressure losses and heat and mass transfer have been presented in previous monographs for pipes of good wettability [34–39]. Similar investigations on poorly wetted horizontal pipes are very limited [45, 46].

### *1.4.3 Dimensionless Parameters*

The basic dimensionless parameters encountered in metallurgical engineering are briefly described below. Additional dimensionless parameters are listed in Appendix 1 together with their physical meanings.

#### 1.4.3.1 Reynolds Number, *Re*

$$Re = vL/\nu\,(\text{inertial force}/\text{viscous force}) \tag{1.20}$$

where $v$ is the representative velocity scale, $L$ is the representative length scale, and $\nu$ is the kinematic viscosity of the fluid. This dimensionless parameter is useful for describing the transition from laminar to turbulent flow as well as the reverse transition from turbulent to laminar flow (i.e., relaminarization).

#### 1.4.3.2 Froude Number, *Fr*

$$Fr = v/[gL]^{1/2} \text{ (inertial force/buoyancy force)} \tag{1.21}$$

where $g$ is the acceleration due to gravity. The following modified Froude number, $Fr_{\mathrm{m}}$, is widely used to characterize the movement of bubbles generated with a single-hole nozzle:

$$Fr_{\mathrm{m}} = \rho_{\mathrm{g}} Q_{\mathrm{g}}^2/[\rho_{\mathrm{L}} g d_{\mathrm{ni}}^5] \tag{1.22}$$

where, $\rho_{\mathrm{g}}$ is the density of gas, $Q_{\mathrm{g}}$ is the gas flow rate, $\rho_{\mathrm{L}}$ is the density of liquid, and $d_{\mathrm{ni}}$ is the inner diameter of the nozzle. This parameter represents the ratio of the inertial force of injected gas to the buoyancy force acting on bubbles generated at the nozzle exit.

#### 1.4.3.3 Weber Number, *We*

$$We = \rho L v^2/\sigma \text{(inertial force/force due to surface tension)} \tag{1.23}$$

where, $\rho$ is the density of fluid and $\sigma$ is the surface tension. In metallurgical engineering systems, the Weber number is an important parameter because the surface tension of molten metal is usually much higher than that of fluids such as water often used in mechanical and chemical engineering practice.

#### 1.4.3.4 Mach Number, *M*

$$M = v/c \text{(inertial force/elastic force)} \tag{1.24}$$

where $c$ is the speed of sound. The Mach number specifies the effect of fluid compressibility on the flow. The definition for transition from bubbling to jetting can be expressed by the following relation,

$$M = 1 \tag{1.25}$$

$$v = Q_{\mathrm{g}}/(\pi d_{\mathrm{ni}}^2/4) \tag{1.26}$$

where $v$ denotes the cross-sectional mean velocity of gas in the nozzle, $Q_{\mathrm{g}}$ is the gas flow rate, and $d_{\mathrm{ni}}$ is the inner diameter of the nozzle.

#### 1.4.3.5 Strouhal Number, *St*

$$St = fL/v \tag{1.27}$$

This parameter represents the ratio of inertial force due to temporal acceleration of fluid motion to inertial force due to spatial acceleration. It is conveniently used to describe the shedding of vortices from solid bodies immersed in fluids.

## 1.5 Closing Remarks

As briefly introduced in the preceding sections, gas–liquid two-phase flows find wide applications in current metallurgical processes. This book, however, considers only flow fields such as those in baths agitated by gas injection, in the continuous casting molds, and in vertical pipes. There is only limited information even on these fundamental flow fields. Although gas–liquid two-phase flows are generally complex and difficult to measure, additional comprehensive investigations are needed to provide a better understanding of current metallurgical processes as well as useful information on the design of novel processes for high efficiency. In particular, the book is meant to provoke researchers to confront the difficulties and open a new horizon in metallurgical engineering practice.

## References

1. Szekely J (1979) Fluid flow phenomena in metals processing. Academic, New York
2. ISIJ (1984) Recent development in steelmaking technologies using gas injection. In: Proc. 100th and 101st Nishiyama memorial lecture, ISIJ, Tokyo
3. Szekely J, Carlsson G, Helle L (1988) Ladle metallurgy. Springer, New York
4. Szekely J, Ilegbusi OJ (1989) The physical and mathematical modeling of tundish operations. Springer, Berlin
5. Sahai Y, St. Pierre GR (1992) Advances in transport processes in metallurgical systems. Elsevier, Amsterdam, pp 260–263
6. Ilegbusi OJ, Iguchi M, Wahnshiedler W (1999) Mathematical and physical modeling in materials processing operations. Chapman & Hall/CRC, Boca Raton
7. Perry RH, Chilton CH, Kirkpatrick SD (1963) Chemical engineer's handbook. McGraw-Hill, New York
8. Ozawa Y, Mori K, Sano M (1981) Behavior of injected gas observed at the exit of a submerged orifice in liquid metal. Tetsu-to-Hagane 67:2655–2664
9. Mori K, Sano M (1981) Process kinetics in injection metallurgy. Testsu-to-Hagane 67: 672–695
10. Ruzika MC, Drahos J, Zahradnik J, Thomas NH (1997) Intermittent transition from bubbling to jetting regime in gas-liquid two phase flows. Int J Multiphase Flow 23(4):671–682
11. Chen K, Richter HJ (1997) Instability analysis of the transition from bubbling to jetting in a gas injected into a liquid. Int J Multiphase Flow 23(4):699–712
12. Kawakami M, Hosono S, Takahashi K, Ito K (1992) Bubble dispersion phenomena in water, mercury, molten iron and molten copper baths. Tetsu-to-Hagane 78:267–274

13. Tacke TH, Schubert HG, Weber DJ, Schwerdfeger K (1985) Characteristics of round vertical gas bubble jets. Metall Trans B 16:263
14. Castillejos AH, Brimacombe JK (1987) Measurement of physical characteristics of bubbles in gas-liquid plumes: Part II. Local properties of turbulent air-water plumes in vertically injected jets. Metall Trans B 18:659–671
15. Iguchi M, Kawabata H, Nakajima K, Morita Z (1995) Measurement of bubble characteristics in a molten iron bath at 1600°C using an electroresistivity probe. Metall Mater Trans B 26:67–74
16. Clift R, Grace JR, Weber ME (1978) Bubbles, drops, and particles. Academic, New York
17. Iguchi M, Chihara T, Takanashi N, Ogawa Y, Tokumitsu N, Morita Z (1995) X-ray fluoroscopic observation of bubble characteristics in a molten iron bath. ISIJ Int 35:1354–1361
18. Iguchi M, Nakatani T, Kawabata H (1997) Development of a multineedle electroresistivity probe for measuring bubble characteristics in molten metal baths. Metall Mater Trans B 28:409–416
19. Iguchi M, Nakatani T, Tokunaga H (1997) The shape of bubbles rising near the nozzle exit in molten metal baths. Metall Mater Trans B 28:417–423
20. Iguchi M, Ueda H, Uemura T (1995) Bubble and liquid flow characteristics in a vertical bubbling jet. Int J Multiphase Flow 21:861–873
21. Schlichting H (1979) Boundary layer theory, 7th edn (trans: Kestin J). Mcgraw-Hill, New York
22. Hinze JO (1975) Turbulence, 2nd edn. McGraw-Hill, New York
23. Mazumdar D, Guthrie RIL (1995) The physical and mathematical modelling of gas stirred ladle systems. ISIJ Int 35:1–20
24. Mazumdar D, Guthrie RIL (1994) ISS Trans 21:89–96
25. Iguchi M, Nakamura K, Tsujino R (1998) Mixing time and fluid flow phenomena in liquids of varying kinematic viscosities agitated by bottom gas injection. Metall Mater Trans B 29:569–575
26. Nakanishi K, Fujii T, Szekely J (1975) Ironmaking Steelmaking 2–3:193–197
27. MacDougal G, Ockrent C (1942) Surface energy relations in liquid/solid systems. I. The adhesion of liquids to solids and a new method of determining the surface tension of liquids. Proc R Soc Lond A 180:151–173
28. Shoji M, Zhang XY (1992) Study of contact angle hysteresis: In relation to boiling surface wettability. Trans Jpn Soc Mech Eng B 58:1853–1859
29. Fukai J, Shibata Y, Yamamoto T, Miyatake O, Poulikakos D, Megaridis CM, Zhao Z (1995) Wetting effects on the spreading of a liquid droplet colliding with a flat surface: Experiment and modeling. Phys Fluids 7:236
30. Ishibashi M, Shiraishi K, Yamamoto S, Shimada M (1975) Tetsu-to-Hagane 61:S. 111
31. Ishibashi M, Yamamoto S (1979) Tetsu-to-Hagane 65:A. 133
32. Kitamura S, Miyamoto K, Tsujino R (1994) The Evaluation of gas-liquid reaction rate at bath surface by the gas adsorption and desorption tests. Tetsu-to-Hagane 80:101–106
33. Kitamura S, Yano M, Harashima K, Tsutsumi N (1994) Decarburization model for vacuum degasser. Tetsu-to-Hagane 80:213–218
34. Wallis GB (1969) One-dimensional two-phase flow. McGraw-Hill, New York
35. Azbel D (1981) Two-phase flows in chemical engineering. Cambridge University Press, Cambridge, p 19
36. Akagawa K (1974) Gas-liquid two-phase flow. Corona Pub. Co. Ltd., Tokyo
37. Hetsroni G (1982) Handbook of multiphase systems. Hemisphere, Washington
38. JSME (1989) Handbook of gas-liquid two-phase flow technology. Corona Pub. Co. Ltd., Tokyo
39. Ueda T (1989) Gas-liquid two-phase flow-fluid flow and heat transfer. Yokendo Pub. Co. Ltd., Tokyo, pp 193–199
40. ISIJ (1996) History of steel continuous casting technology in Japan. ISIJ, Tokyo
41. Yokoya S, Takagi S, Souma H, Iguchi M, Asako Y, Hara S (1998) Removal of inclusion through bubble curtain created by swirl motion in submerged entry nozzle. ISIJ Int 38:1086–1092
42. JSPS (1994) 19th Committee: recent developments in investigations on the removal of nonmetallic inclusions. JSPS, Keyo, Tokyo

43. Zhe W, Mukai K, Matsuoka K (1997) Water model experiment for the behaviors of bubbles and liquid flow on the inside of the nozzle and mold of continuous casting process. CAMP ISIJ 10:68–71
44. Iguchi M, Terauchi Y (2001) Boundaries among bubbly and slug flow regimes in air–water two-phase flows in vertical pipe of poor wettability. Int J Multiphase Flow 27:729–735
45. Watanabe K (2000) Jpn J Multiphase Flow 14(1):208–211
46. Terauchi Y, Iguchi M, Kosaka H, Yokoya S, Hara S (1999) Wettability effect on the flow pattern of air-water two-phase flows in a vertical circular pipe. Tetsu-to-Hagane 85:645–651

# Chapter 2
# Turbulence Structure of Two-Phase Jets

## 2.1 Mean Flow Characteristics

### *2.1.1 Introduction*

Gas injection techniques have been extensively used for current steelmaking processes, such as converters, ladles, RH degassing processes, and other metals refining processes. In order to enhance the efficiency of these processes, precise information on the behavior of rising bubbles in a molten metal bath and resultant molten metal flow induced by the bubbles is required. Since it is difficult to study bubble and liquid flow characteristics using actual processes, many experimental and theoretical model investigations have been carried out. Details of the previous investigations should be referred to, for example, a review article by Mazumdar and Guthrie [1].

Concerning the bubble characteristics specified by gas holdup $\alpha$, bubble frequency $f_B$, mean bubble rising velocity $u_B$, and mean bubble chord length $L_B$ in a high temperature molten metal bath agitated by bottom gas injection, hot model experiments have been carried out for molten copper and molten iron baths using a newly developed two-needle electro-resistivity probe [2–5]. In the axial region far from the nozzle exit, experimental results of $\alpha$, $f_B$, $u_B$, and $L_B$ for iron–Ar systems at 1,250 and 1,600°C and a copper–Ar system at 1,250°C were found to be predicted satisfactorily by empirical correlations derived originally for a water–air and a mercury–air system [6, 7] as long as the radial distributions of gas holdup and bubble frequency follow Gaussian distributions.

Compared with the bubble characteristics, the information on the liquid flow characteristics specified by the axial and radial mean velocities, $u$ and $v$, the root-mean-square values of the axial and radial turbulence components, $u'_{rms}$ and $v'_{rms}$, the Reynolds shear stress $\overline{u'v'}$, and higher correlations of turbulence components, such as the skewness and flatness factors, are limited except for a water–air system [8–12].

In general, precise measurements of the liquid flow characteristics in a molten metal bath at high temperatures are very difficult. At present, the mean velocity can be measured under limited conditions by using reaction probes [13–16] and Karman

M. Iguchi and O.J. Ilegbusi, *Modeling Multiphase Materials Processes: Gas-Liquid Systems*, DOI 10.1007/978-1-4419-7479-2_2,

vortex probes [17, 18] even if the bath temperature is higher than 1,000°C, whereas the measurement of turbulence is practically impossible in this temperature range.

Numerical simulation techniques therefore have acted as a bridge between water models and actual processes [19,20]. The aforementioned liquid flow characteristics in a molten metal bath agitated by gas injection were predicted numerically by using turbulence models, such as the $k - \varepsilon$ model, developed originally for single-phase flows [19, 20]. Numerical results thus obtained, however, have not received reliable experimental confirmation even for mercury and Wood's metal flows.

In this section, a description is provided of the bubble and liquid flow characteristics measured in a molten Wood's metal bath stirred by bottom helium gas injection. The aim is to provide experimental data for judging the adequacy of numerical results and to examine whether the results of liquid flow characteristics obtained for a water–air system are useful to predict molten metal flows.

A magnet probe, specifically the Vives probe, [21, 22] was used to measure the mean velocity and turbulence components of molten Wood's metal flow at around 100°C. Helium gas was chosen to simulate the density ratio between molten steel and Ar.

## 2.1.2 Experiment

### 2.1.2.1 Bubble characteristics

Figure 2.1 shows a schematic diagram of the experimental apparatus. The cylindrical vessel made of stainless steel had an inner diameter $D$ of 20.4 cm and a height $H$ of 30.0 cm. Wood's metal (commercial name: U-alloy 70, melting temperature: 70°C) was melted in it to a depth, $H_L$, of 15.0 cm. The bath temperature was kept at 105 ± 1°C using a thermo-controller. The physical properties of the molten Wood's metal are as follows: density $\rho_s = 9.56\,\mathrm{g/cm^3}$, kinematic viscosity $\nu_s = 0.341\,\mathrm{mm^2/s}$, and surface tension $\sigma = 460\,\mathrm{mN/m}$.

Helium gas was preheated up to 105 ± 5°C with a heat exchanger installed upstream of the vessel and injected at a flow rate $Q_g$ of 60 $\mathrm{Ncm^3/s}$ through a centered single-hole bottom nozzle with an inner diameter, $d_{ni}$, of 0.5 mm. Therefore, it is not necessary to consider the expansion of gas at the nozzle exit due to temperature difference between the bath and the injected gas. The density ratio of the molten Wood's metal to the Helium gas is the same order of magnitude as that of steel to Ar. The nozzle was made of fluororesin (commercial name: Teflon). The origin of the axial and radial coordinates, $z$ and $r$, were placed at the center of the nozzle exit. The volumetric gas flow rate at the nozzle exit was 67 $\mathrm{cm^3/s}$. Some additional measurements were made for $Q_g = 95\,\mathrm{Ncm^3/s}$.

Gas holdup $\alpha$, bubble frequency $f_B$, mean bubble rising velocity $u_B$, and mean bubble chord length $L_B$ were measured using a two-needle electroresistivity probe (not shown in Fig. 2.1). The vertical distance between the tips of the two electrode needles was set at 2 mm. The output signals of the electroresistivity probe system

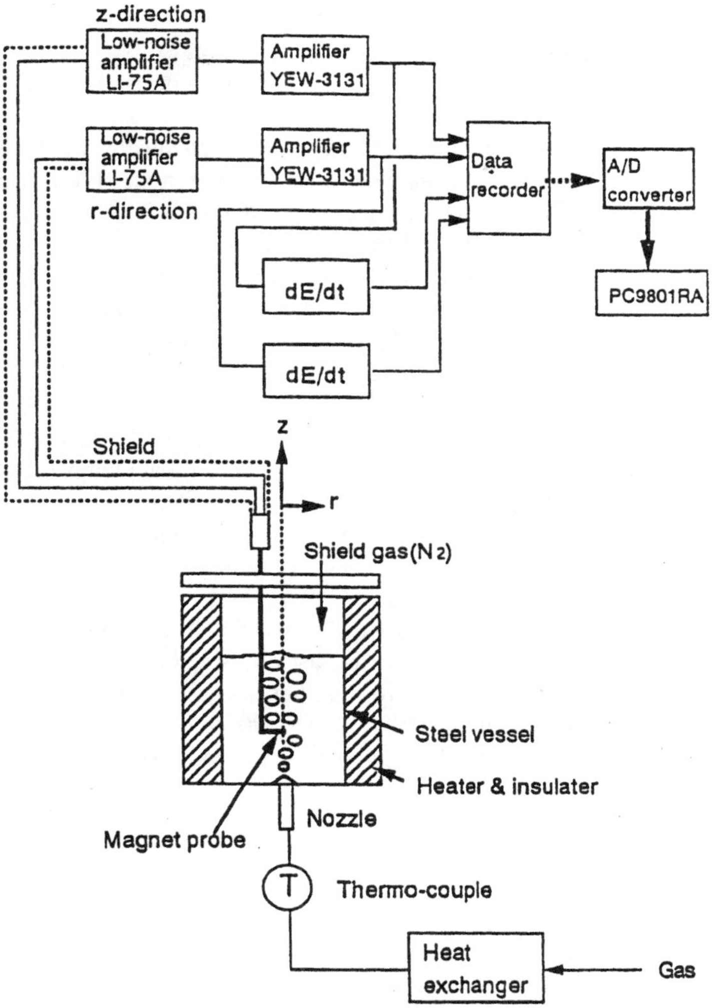


**Fig. 2.1** Experimental apparatus

were A/D converted and then processed on a personal computer to determine these quantities. Further details of the measurement method should be referred to a previous paper [6, 7]. A special consideration was provided to prevent oscillations of the electrode needles induced by quasi-periodical arrival of bubbles. The measurement errors were approximately ±3% for $\alpha$ and $f_B$ and ±5% for $u_B$ and $L_B$.

Electroresistivity probe measurements were carried out on the centerline of the bubbling jet at equal intervals of 1 cm. Also, the radial distributions of the bubble characteristics were measured at four representative axial positions of $z = 1.5, 4.0, 7.0$, and 10.0 cm. The experimental results were compared with empirical relations proposed previously by Iguchi et al. [6, 7] for an arbitrary combination of metal and gas and with those proposed by Xie et al. [23] for a molten Wood's metal–$N_2$ system.

### 2.1.2.2 Liquid Flow Characteristics

The axial and radial velocity components, *u* and *v*, were measured by means of a two-channel magnet probe. The magnet probe works on the basis of Faraday's law [21, 22]. According to Weissenfluh, [22] if an electric conductor moves through a magnetic field, the conductor has an electromotive force induced in it which is in a direction normal to the magnetic field and the direction of motion. The electromotive force generates an electric field which is nearly proportional to the intensity of the magnetic field and to the velocity of the conductor. Therefore, it is possible to measure the velocity of the conductor by detecting the electromotive force. Further details of the working principle of the magnet probe are found in the articles of Ricou and Vives [21] and Weissenfluh [22].

The magnet probe is applicable to molten metal flows at a temperature lower than the Curie point of the permanent magnet. Previous magnet probe measurements for a Wood' metal bath were carried out by Xie et al. [23] at bath temperatures lower than 150°C. The magnet probe used by Xie et al. cannot detect the axial and radial velocity components simultaneously. Accordingly, data on the Reynolds shear stress and higher order turbulence correlations, such as the skewness and flatness factors, have not been published.

An outline of the magnet probe described here is illustrated in Fig. 2.2. The original output voltage of the magnet probe is basically small (approximately 50 $\mu$V) under normal conditions, and hence, two amplifiers are necessary (see Fig. 2.1). The original voltage was amplified by 10,000 (100 × 100) times and then recorded on a data recorder. Signals originating from bubbles were removed using a differentiating

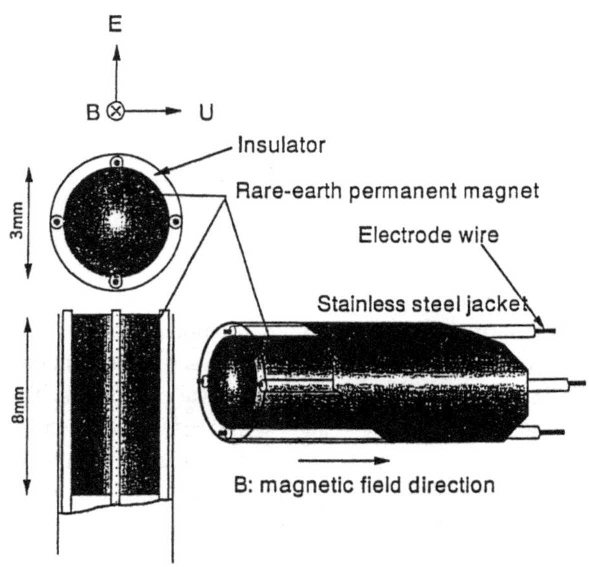


**Fig. 2.2** Schematic of the present magnet probe

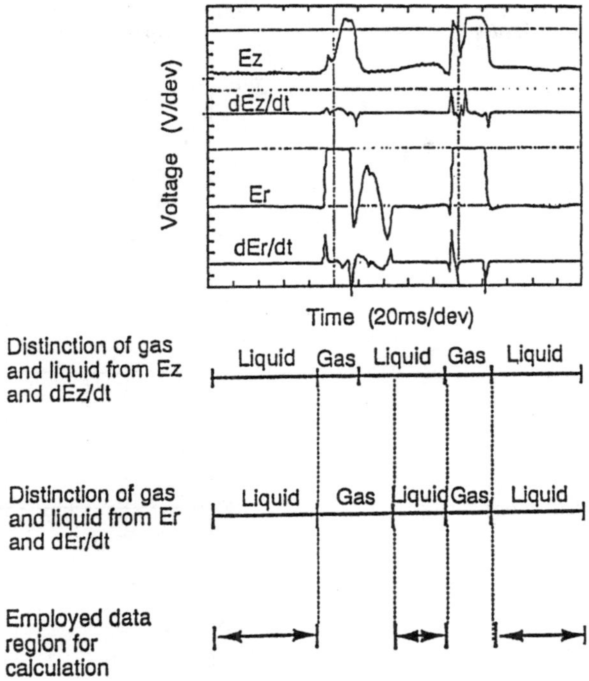


**Fig. 2.3** Discrimination of gas and liquid signals

circuit and a personal computer in a manner shown in Fig. 2.3, where Ez and Er denote the output voltage values for the axial and radial velocity components, respectively, and $t$ is time.

The measurement of the radial velocity component, $v$, is relatively more difficult than the axial velocity component, $u$, because the magnitude of $v$ is much smaller than that of $u$. The errors in measuring $u$ are estimated to be $\pm 5\%$, while those for $v$ are $\pm 10\%$. The time required for measuring the Reynolds shear stress is basically quite long. Thus, the measurement at only one axial position $z = 4\,\mathrm{cm}$ was done.

## *2.1.3 Experimental Results*

### 2.1.3.1 Bubble Characteristics

Axial distribution

Figures 2.4–2.7 show measured values of gas holdup, $\alpha_{cl}$, bubble frequency, $f_{B,cl}$, mean bubble rising velocity, $u_{B,cl}$, and mean bubble chord length, $L_{B,cl}$, where the subscript cl designates the value on the centerline of the bath. In Fig. 2.4, the

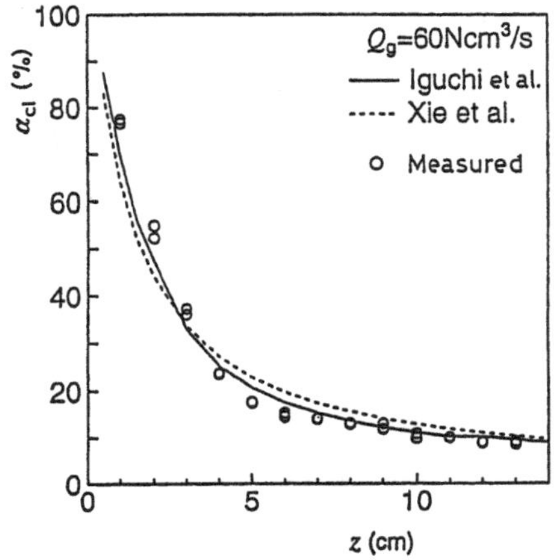


**Fig. 2.4** Axial distributions of gas holdup on the centerline

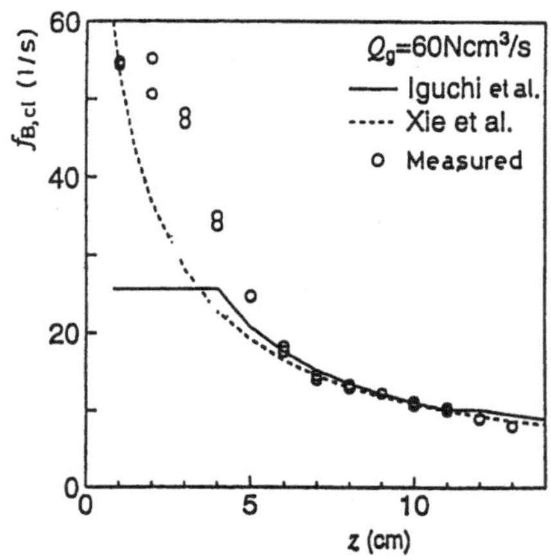


**Fig. 2.5** Axial distributions of bubble frequency on the centerline

measured $\alpha_{cl}$ values are approximated satisfactorily by two empirical equations [6, 7, 23]. The same degree of agreement between the measured $f_{B,cl}$ values and two empirical equations can be seen for $z > 5$ cm, as shown in Fig. 2.5. The axial distribution of the mean bubble rising velocity on the centerline, $u_{B,cl}$, shown in Fig. 2.6, is well correlated by the empirical relation of Xie et al. [23] in the axial region of $z > 5$ cm. However, it is not reasonable to judge the adequacy of the empirical correlations by focusing solely on the centerline values, as can be seen from comparisons between the empirical correlations and measured radial distributions of the bubble characteristics shown in a subsequent section.

In Fig. 2.7, the measured $L_{B,cl}$ distribution for $z > 5$ cm is also approximated satisfactorily by the empirical correlation of Xie et al. [23] The measured $L_{B,cl}$ value obtained at $z = 1$ cm is much larger than 2 cm, implying that the bubble generation region extends far beyond $z = 2$ cm. Consequently, the measured $u_{B,cl}$ value in the close vicinity of the nozzle exit does not present the real bubble

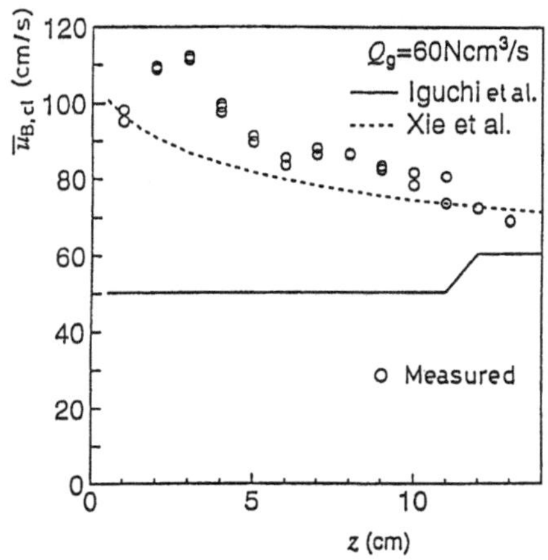


**Fig. 2.6** Axial distributions of mean bubble rising velocity on the centerline

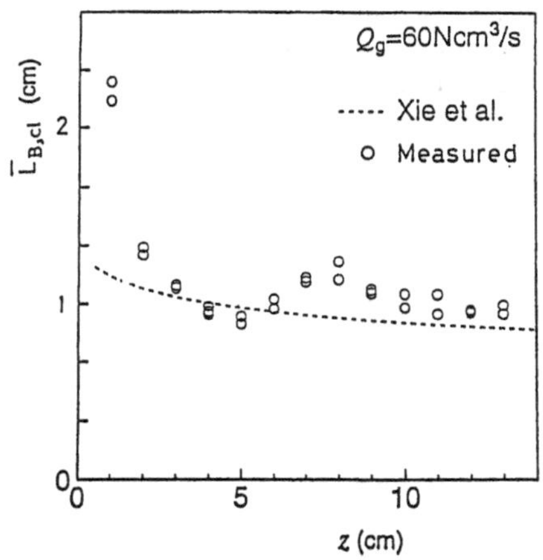


**Fig. 2.7** Axial distributions of mean bubble chord length on the centerline

rising velocity of a bubble but includes the contribution of the expansion velocity of the bubble. A peak appearing at around $z = 2.5$ cm in the axial distribution of $u_{B,cl}$ shown in Fig. 2.6 may be considered to be related to such an expansion velocity.

Radial distribution

The radial distributions of gas holdup, $\alpha$, measured by Iguchi et al. [24] at four representative axial positions are shown in Fig. 2.8. The $\alpha$ distribution at $z = 1.5$ cm is very steep near the centerline of the bubbling jet. Two peaks appeared in the $\alpha$ distributions measured at $z = 4.0$ and $7.0$ cm. These two peaks are not caused by the swirl motion of a bubbling jet because no swirl motion was observed under the

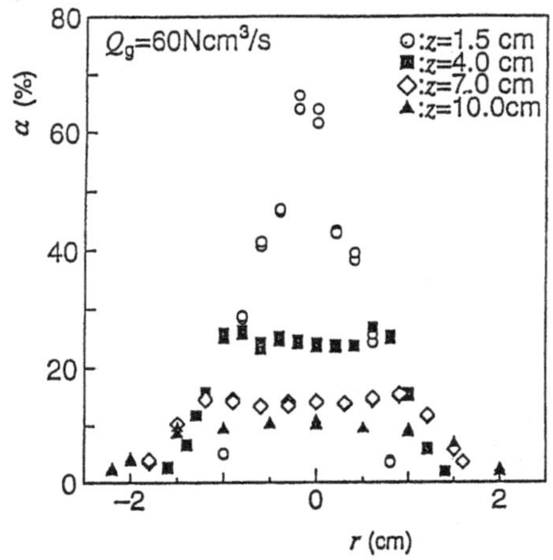


Fig. 2.8 Radial distributions of gas holdup

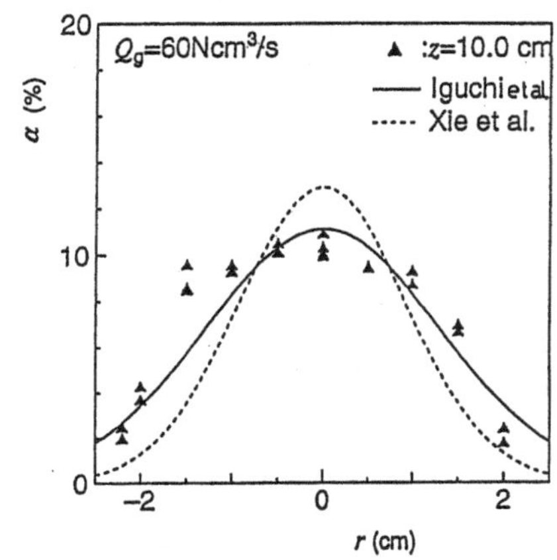


Fig. 2.9 Radial distributions of gas holdup at $z = 10\,\mathrm{cm}$

experimental conditions considered. They are closely associated with the shape of bubbles like a bell commonly seen in oriental temples [25].

Although any evidence is not given, agreement between the measured $\alpha$ values and the empirical correlations is not good at $z = 4.0$ and $7.0\,\mathrm{cm}$ because these two correlations assume Gaussian distributions. The measured radial distribution of $\alpha$ approached a Gaussian distribution as z increased further (see α distribution at $z = 10\,\mathrm{cm}$). As a result, the measured $\alpha$ values at $z = 10\,\mathrm{cm}$ were approximated adequately by the empirical correlation by Iguchi et al. [6, 7], as demonstrated in Fig. 2.9.

In Fig. 2.10, showing the radial distributions of bubble frequency $f_B$, a plateau appeared in the distributions at $z = 4.0$ and $7.0\,\mathrm{cm}$. This is also attributed to the formation of bell-shaped bubbles [25]. The radial $f_B$ distribution measured at

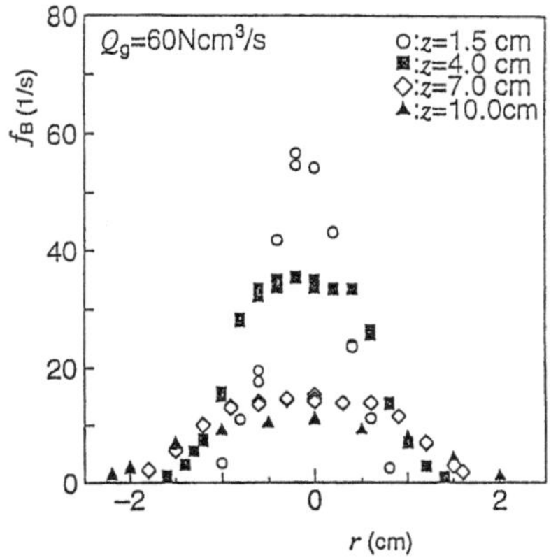


**Fig. 2.10** Radial distributions of bubble frequency

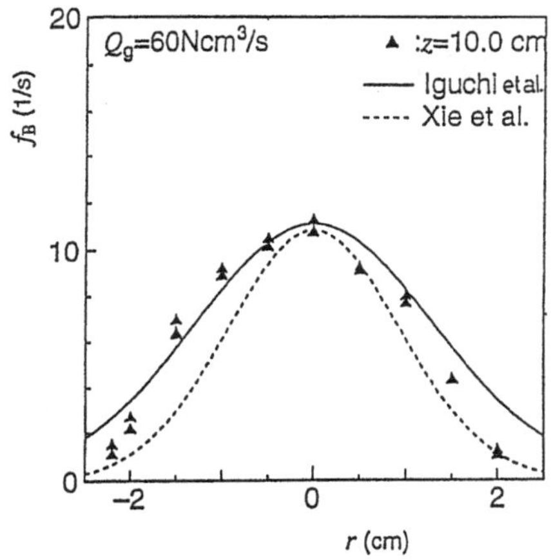


**Fig. 2.11** Radial distributions of bubble frequency at $z = 10$ cm

$z = 10.0$ cm was also predicted by the two empirical correlations (see Fig. 2.11) to within approximately the same accuracy.

Figures 2.12 and 2.13 show the radial distributions of mean bubble rising velocity $u_B$ and mean bubble chord length $L_B$, respectively. The measured values of $u_B$ and $L_B$ at $z = 1.5$ cm were not plotted. The relatively large $u_B$ values near the centerline ($r \cong 0$) at $z = 4$ and 7 cm may be caused by the vertical elongation of bell-shaped bubbles [25]. The empirical correlation of $u_B$ proposed by Iguchi et al. [6, 7] can predict the measured values only in the outer region of the bubbling jet. At every axial position, the measured $L_B$ value is relatively small near the centerline, being caused by thinning of the bell-shaped bubbles. Any empirical correlations for the radial distributions of $u_B$ and $L_B$ were not proposed by Xie et al.

A reexamination of previous experimental results for an Fe-Ar system, [2, 3, 5] a Cu-Ar system [4], and a Wood's metal–$N_2$ system [26] reveals that the radial distributions of $\alpha$, $f_B$, $u_B$, and $L_B$ are similar to those mentioned above near the nozzle exit, although each radial distribution is not so distinguished compared with the results presented in Figs. 2.12 and 2.13.

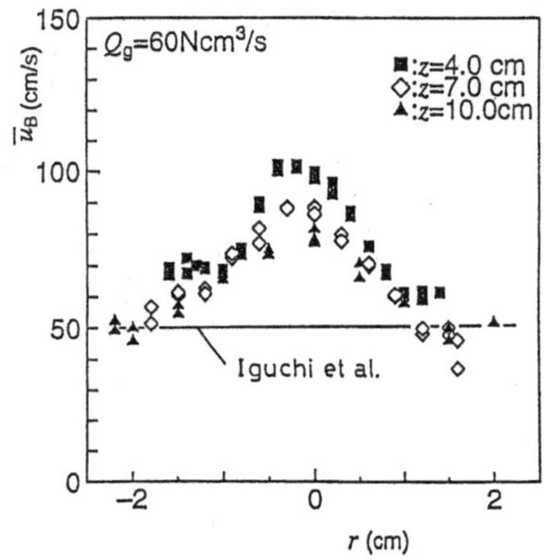


Fig. 2.12 Radial distributions of mean bubble rising velocity

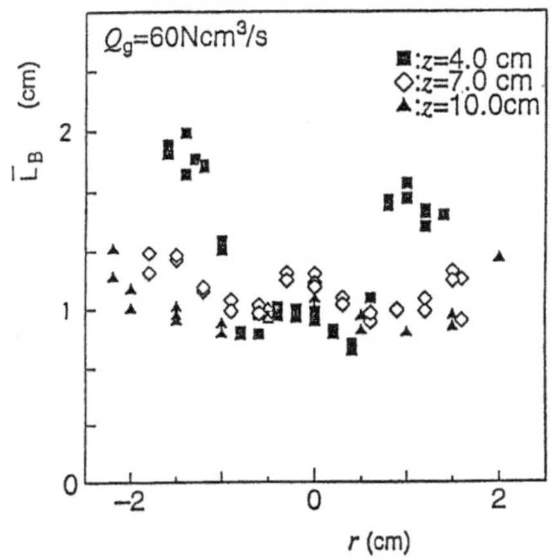


Fig. 2.13 Radial distributions of mean bubble chord length

Iguchi et al. [6, 7] did not observe two peaks in the radial distribution of $\alpha$ for a water–air system and a mercury–air system. For the water–air system, this might be because even if bell-shaped bubbles are generated, they are likely to disintegrate into smaller bubbles in complex shapes near the nozzle exit due to low surface tension and high turbulent shear stress of liquid flow.

Two peaks did not appear in $\alpha$ distributions for the mercury–air system, either. The surface tension of mercury is nearly equal to that of the Wood's metal, but the density ratio $\rho_L/\rho_g$ for the mercury–air system is much lower than that of the Wood's metal–He system, where the subscripts L and g designate liquid and gas, respectively. Consequently, the density ratio $\rho_L/\rho_g$ seems to be another candidate responsible for the difference in the bubble characteristics near the nozzle exit [27, 28].

### 2.1.3.2 Liquid Flow Characteristics

Axial Distribution

Figure 2.14 shows the axial mean velocity on the centerline of the bath, $u_{cl}$, and the root-mean-square values of the axial and radial turbulence components, $u'_{rms,cl}$ and $v'_{rms,cl}$, against the axial distance, $z$. The measured $\overline{u_{cl}}$ value remains almost unchanged in the axial direction and approaches the following empirical correlations originally proposed for a water–air system [9]:

$$\overline{u_{cl}} = 2.3u_r Fr_m^{0.036}(0.1 < P < 0.5) \tag{2.1}$$

$$= 1.2u_r P^{-0.28}(0.02 < P < 0.1) \tag{2.2}$$

$$Fr_m = \rho_g Q_g^2/(g d_{ni}^5 \rho_L) \tag{2.3}$$

$$u_r = (gQ_g/z)^{1/3} \tag{2.4}$$

$$P = [Q_g^2/(gz^5)]^{1/5} \tag{2.5}$$

where $u_r$ is a representative velocity (cm/s), $Fr_m$ is the modified Froude number, $g$ is the acceleration due to gravity (cm/s$^2$), and $P$ is a dimensionless parameter [9]. The units of $u_{cl}$, $\rho_g$, $Q_g$, $\rho_L$, $d_{ni}$, and $z$ are cm/s, g/cm$^3$, cm$^3$/s, g/cm$^3$, cm, and cm, respectively.

The axial turbulence intensity on the centerline, $u'_{rms,cl}/\overline{u}_{cl}$, is about 0.55 in the axial region except near the nozzle exit. The radial turbulence intensity on the centerline, $v'_{rms,cl}/\overline{u}_{cl}$, is also approximately 0.55 for the Wood's metal–He system. It is known that $u'_{rms,cl}/\overline{u}_{cl} = 0.5$ and $v'_{rms,cl}/\overline{u}_{cl} = 0.4$ for a water–air system. The measured values of $u'_{rms,cl}/\overline{u}_{cl}$ are almost the same for the two systems, whereas those of $v'_{rms,cl}/\overline{u}_{cl}$ are significantly different. This difference seems to be due mainly to the difference in the shapes of bubbles that influence the turbulence production in the wake of the bubbles.

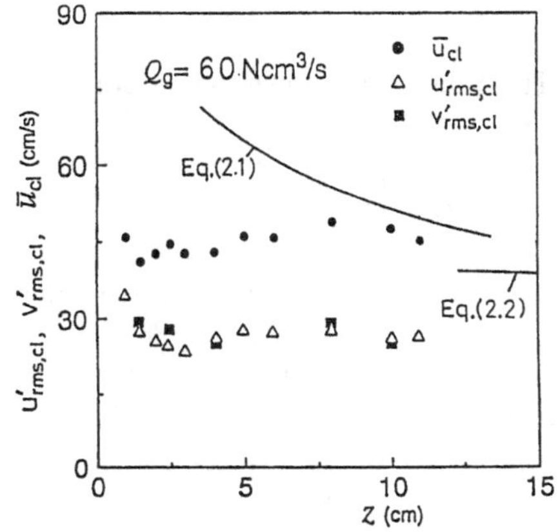


**Fig. 2.14** Axial mean velocity and the rms values of turbulence components on the centerline as functions of axial distance

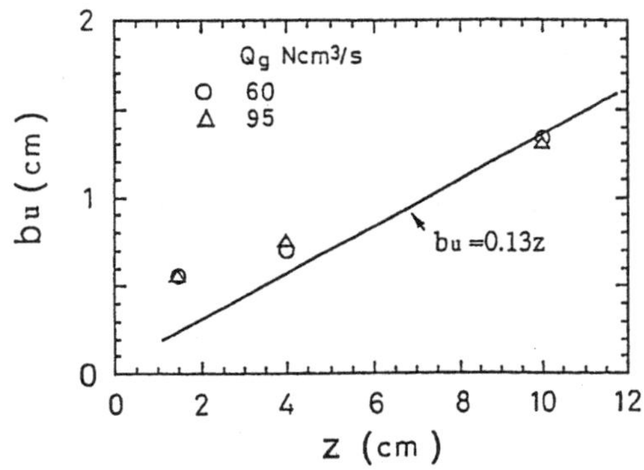


**Fig. 2.15** Radial distributions of axial mean velocity

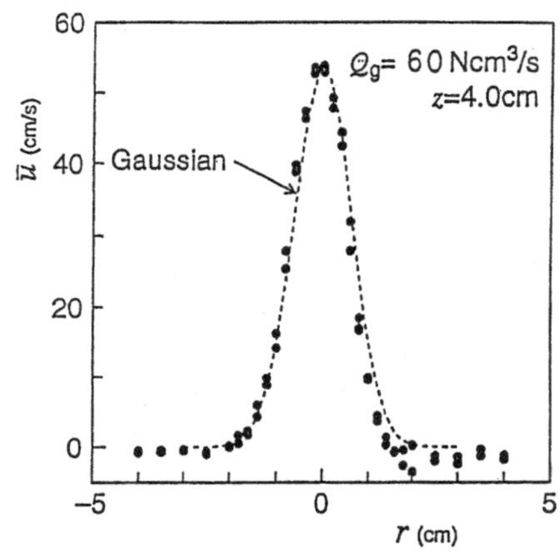


**Fig. 2.16** Comparison of radial distributions of $\overline{u}$ between Wood's metal–He and water–air systems

The measured values of the half-value radius, $b_u$, of the radial distribution of $\overline{u}$ are shown in Fig. 2.15. The $b_u$ value approaches the following empirical correlation as $z$ increases:

$$b_u = 0.13z \tag{2.6}$$

This relation is known to hold for a water–air system [11].

## Radial Distribution

The radial distribution of the axial mean velocity $\overline{u}$ measured at $z = 4$ cm is shown in Fig. 2.16. In the bubbling jet region ($r/b_u < 1.5$) where almost all rising bubbles are present, the radial $\overline{u}$ distribution is approximated satisfactorily by a Gaussian distribution marked by the broken line, although the radial $\alpha$ distribution does not follow a Gaussian distribution, as shown in Fig. 2.7. Measured $\overline{u}$ values for $z =$ 10 cm are also approximated by a Gaussian distribution.

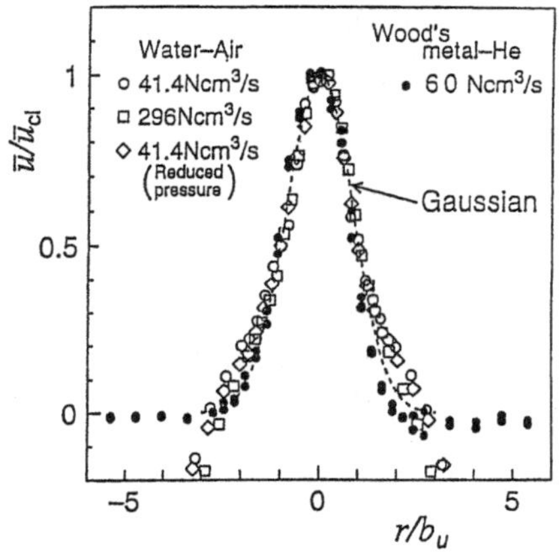


**Fig. 2.17** Relation between half-value radius and axial distance

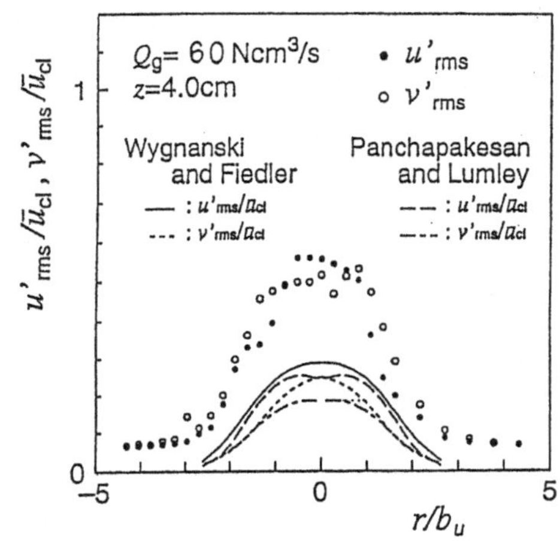


**Fig. 2.18** Radial distributions of the rms values of axial and radial turbulence components

The measured $\overline{u}$ values shown in Fig. 2.16 and those obtained at an other measurement position ($z = 10\,\text{cm}$) are nondimensionalized by $\overline{u}_{\text{cl}}$ and replotted against $r/b_{\text{u}}$ in Fig. 2.17. The measured values of $\overline{u}/\overline{u}_{\text{cl}}$ for a water–air system are also included. The radial distributions of $\overline{u}/\overline{u}_{\text{cl}}$ for the two systems agree with each other for $r/b_{\text{u}} < 1.5$.

Figure 2.18 compares the radial distributions of the rms values of the axial and radial turbulence components, $u'_{\text{rms}}$ and $v'_{\text{rms}}$, measured at $z = 4\,\text{cm}$ in the Wood's metal bath with those for single-phase free jets [29, 30]. Due to additional turbulence production in the wake of bubbles, the measured $u'_{\text{rms}}/\overline{u}_{\text{cl}}$ and $v'_{\text{rms}}/\overline{u}_{\text{cl}}$ values for the bubbling jet in the Wood's metal bath are much larger than those for the single-phase free jets [11]. The measured $u'_{\text{rms}}/\overline{u}_{\text{cl}}$ and $v'_{\text{rms}}/\overline{u}_{\text{cl}}$ values for the Wood's metal–He system are nearly the same both in the bubbling jet and recirculation regions. The latter region is located outside the former.

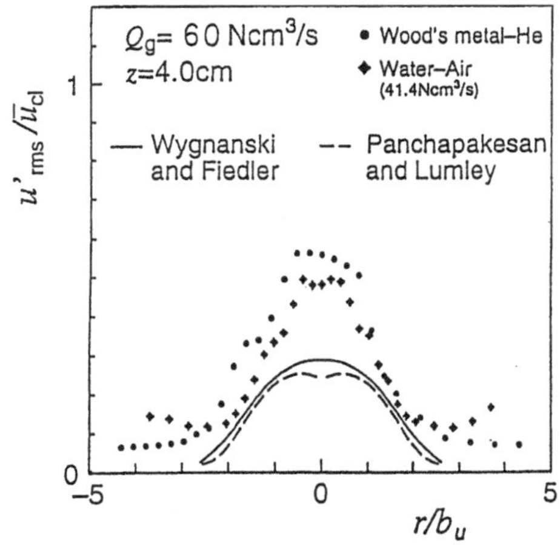


**Fig. 2.19** Radial distributions of the rms values of axial turbulence component

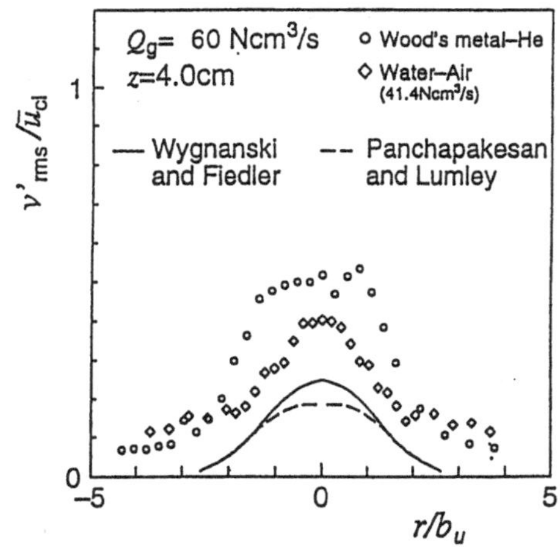


**Fig. 2.20** Radial distributions of the rms values of radial turbulence component

The measured rms values of the axial and radial turbulence components shown in Fig. 2.18 are replotted in Figs. 2.19 and 2.20, respectively. The values for a water–air system are also shown in these figures [10, 11]. In Fig. 2.19, agreement between the measured $u'_{\rm rms}/\overline{u}_{\rm cl}$ values for the Wood's metal–He system and the water–air system is quite good in the bubbling jet region ($r/b_{\rm u} < 1.5$). Meanwhile, in Fig. 2.20, the Wood's metal–He system exhibits a relatively large $v'_{\rm rms}/\overline{u}_{\rm cl}$ value compared with the water–air system. Such a difference seems to be caused mainly by the difference in the turbulence production in the wake of the bubbles [25].

In the recirculation region, the measured values of $u'_{\rm rms}/\overline{u}_{\rm cl}$ and $v'_{\rm rms}/\overline{u}_{\rm cl}$ for the Wood's metal–He system are the same, as observed for a water–air system [8, 11, 12].

The radial distributions of the Reynolds shear stress $\overline{u'v'}$ are shown in Fig. 2.21. In the bubbling jet region, $r/b_{\rm u} < 1.5$, the values for the Wood's metal–He system

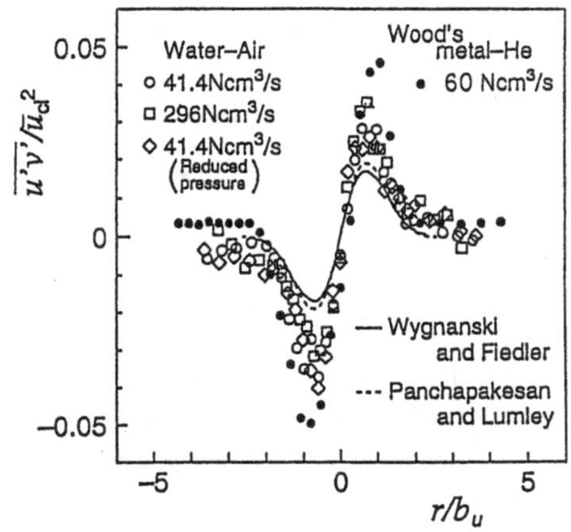


**Fig. 2.21** Radial distributions of the Reynolds shear stress

are much larger than those for a single-phase free jet because of the additional turbulence production in the wake of bubbles. Furthermore, the measured values of $\overline{u'v'}$ for the Wood's metal–He system are slightly larger than those for a water–air system. As mentioned above, this difference seems also to be attributable to the shape of bubbles.

## 2.2 Conditional Sampling

### *2.2.1 Introductory Remarks*

Many experimental studies have been carried out to investigate the turbulence characteristics in gas–liquid and gas–solid two-phase jets under various flow conditions [31–35]. In single-phase jets, turbulence is produced mainly by entrainment of surrounding fluid into the jets, [29, 30] while in two-phase jets, turbulence production takes place also in the wake of bubbles or solid particles in addition to turbulence production due to the entrainment typical of single-phase jets. The size and shape of bubbles or solid particles are considered to affect the structure of turbulence [36, 37]. The effects of the size of bubbles or solid particles on the modulation of turbulence characteristics have been reported by many researchers [35–38]. However, the information on the effect of bubble shape is quite limited.

As shown in the previous section, the shape of bubbles was found to affect preferably the modulation of radial (horizontal) turbulence component [24]. However, the mechanism of turbulence production was not mentioned. In this section, the application is described of the four-quadrant classification method, [39] one of the most popular conditional sampling methods, adopted to elucidate the coherent structure of turbulence or the mechanism of turbulence production in a He–Wood's metal bubbling jet.

### 2.2.2 Experimental Apparatus and Procedure

A detailed description of the apparatus and experimental method for molten Wood's metal flow has been provided in the preceding section.

### 2.2.3 Shape and Size of Helium Bubble

Before discussing the turbulence structure in a bubbling jet in detail, it is worth describing first the shape and size of bubbles. The radial distribution of gas holdup in a vertical bubbling jet is widely known to obey a Gaussian distribution when bubbles can be classified as the so-called wobbling type, as schematically shown in Fig. 2.22 [40]. The axial mean velocity of liquid flow also follows a Gaussian distribution in the presence of wobbling bubbles [39]. Under the experimental condition being described here, the radial distribution of gas holdup has a plateau near the centerline of the bubbling jet and exhibits a mild peak outside the plateau. This somewhat curious distribution of gas holdup can be explained as follows.

The cross-section of bubbles was observed using a multi-needle electroresistivity probe [25]. The bubbles have a cross-section illustrated in Fig. 2.23, and could be classified into the well-known skirted type. This kind of bubble appears when the following modified Weber number, $We_{\mathrm{m}}$, and the Reynolds number, $Re_{\mathrm{m}}$, are relatively small [41].

$$We_{\mathrm{m}} = (\rho_{\mathrm{g}}/\rho_{\mathrm{L}})\rho_{\mathrm{L}} d_{\mathrm{B}} u_{\mathrm{B}}{}^{2}/\sigma, \tag{2.7}$$

$$Re_{\mathrm{m}} = (\rho_{\mathrm{g}}/\rho_{\mathrm{L}}) d_{\mathrm{B}} u_{\mathrm{B}}/\nu_{\mathrm{L}}, \tag{2.8}$$

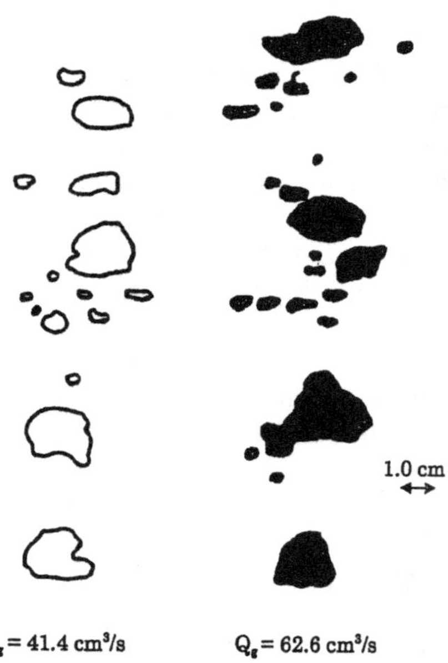


**Fig. 2.22** Shape of bubbles visualized with a high-speed video camera in air–water system

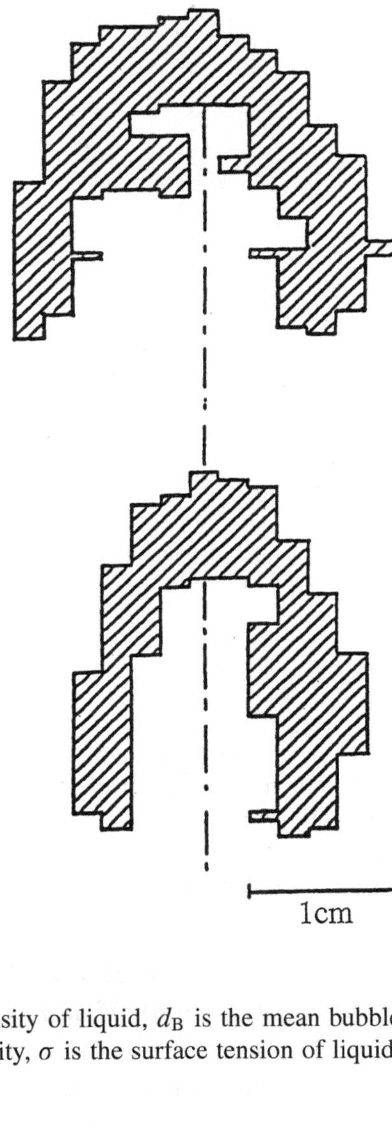


**Fig. 2.23** Measured bubble shape at $z = 0.050$ m in He–Wood's metal bubbling jet

where $\rho_g$ is the density of gas, $\rho_L$ is the density of liquid, $d_B$ is the mean bubble diameter, $u_B$ is the mean bubble rising velocity, $\sigma$ is the surface tension of liquid, and $\nu_L$ is the kinematic viscosity of liquid.

### *2.2.4 Four-Quadrant Classification Method*

Turbulent motions in single-phase boundary layers, pipe flows, and jets are not completely random but have coherent or ordered structure [42–45]. According to the four-quadrant classification method, the turbulent motions are grouped into four distinct categories, as illustrated in Fig. 2.24. In this figure, "ejection" denotes a higher momentum fluid motion directed outward, "outward interaction" denotes a lower momentum fluid motion directed outward, "sweep" denotes a lower momentum

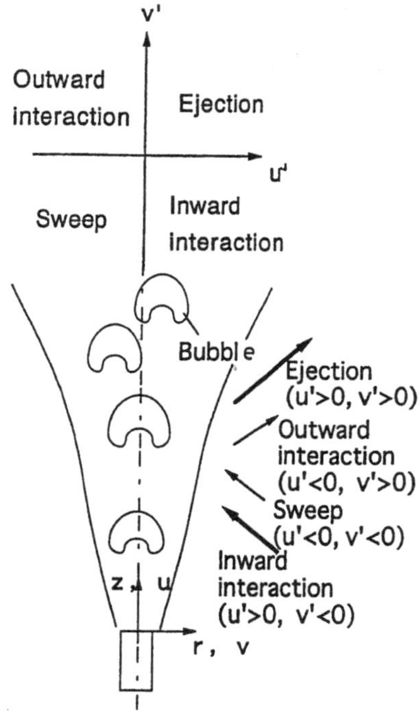


**Fig. 2.24** Classification of turbulent motions

fluid motion directed inward and "inward interaction" denotes a higher momentum fluid motion directed inward. The frequency of occurrence of these four turbulent motions, and their contributions to the turbulence kinetic energies and to the Reynolds shear stress, are discussed subsequently by comparing them with those for the air–water bubbling jet [39, 45].

### *2.2.5 Experimental Results Based on Four-Quadrant Classification Method*

Figure 2.25 shows the occurrence frequencies of the four turbulent motions in the He–Wood's metal bubbling jet together with those in an air–water bubbling jet [25]. In the region near the center line ($r = 0$), the occurrence frequency of each turbulent motion is approximately 0.25, and hence, there is no coherent structure there. In a radial region of $0 < r/b_u < 1.0$, the occurrence frequency of sweep is the highest, followed by the ejection. The occurrence frequency of inward interaction is the lowest. The order of the first two motions is the same as that for an air–water bubbling jet [39]. As $r/b_u$ becomes greater than approximately 1.0, around which the peak of gas holdup appeared, the occurrence frequency of sweep decreases drastically, while that of inward interaction increases. Consequently, the coherent structure for $r/b_u > 1.0$ in the He–Wood's metal bubbling jet differs significantly from that in the air–water bubbling jet.

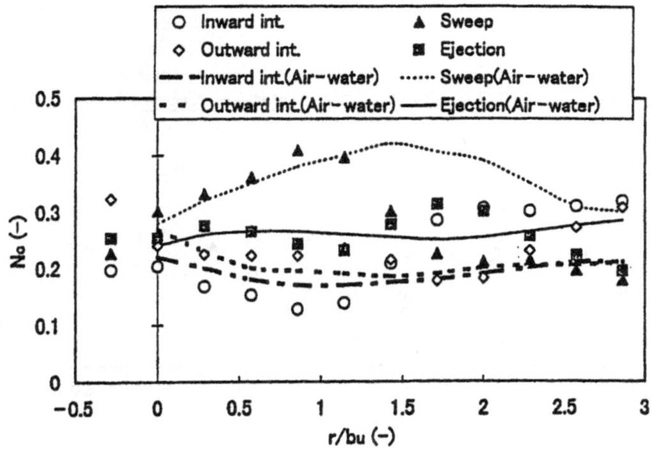


**Fig. 2.25** Relative appearance frequency Nc

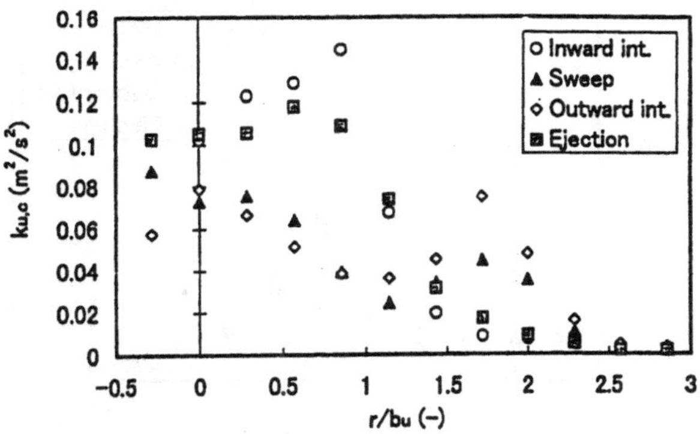


**Fig. 2.26** Radial distributions of the axial turbulence kinetic energy for each turbulent motion

Figure 2.26 presents the axial turbulence kinetic energy for each turbulent motion, $k_{\mathrm{u,c}}$, calculated from the axial turbulence component $u'$, thus,

$$k_{\mathrm{u,c}} = [\Sigma(u_i - u)^2]_{\mathrm{c}}/N_{\mathrm{c}}, \tag{2.9}$$

where the subscripts c and i designate each turbulent motion and the $i$th digitized datum, respectively. The contributions of ejection and inward interaction to the axial turbulence kinetic energy are dominant for $r/b_{\mathrm{u}} < 1.0$. Although the appearance frequency of inward interaction is the lowest for $r/b_{\mathrm{u}} < 1.0$, its contribution to the axial turbulence kinetic energy is the largest. For $r/b_{\mathrm{u}} > 1.0$, the sweep and outward interaction play an important role on the axial turbulence kinetic energy.

Figure 2.27 compares the measured data on the contribution of each turbulent motion to the total turbulence kinetic energy in the axial direction with those in an

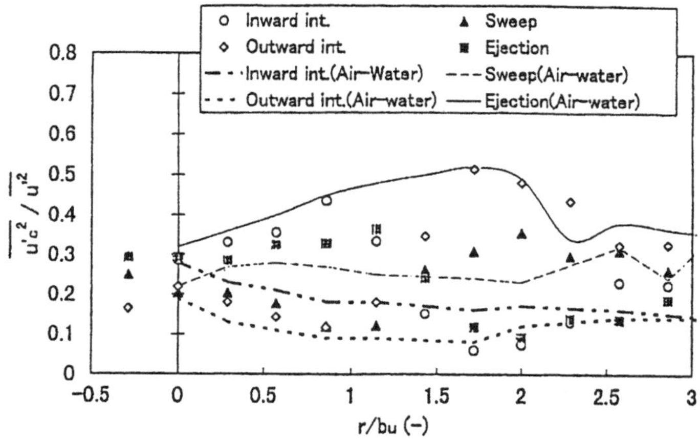


**Fig. 2.27** Contributions of four turbulent motions to the axial turbulence kinetic energy

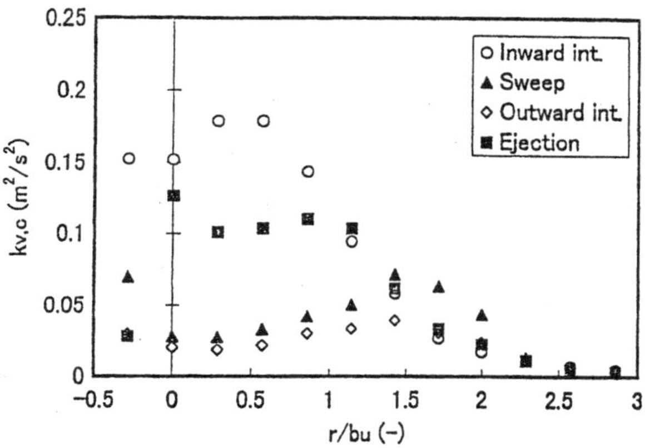


**Fig. 2.28** Radial distributions of the radial turbulence kinetic energy for each turbulent motion

air–water bubbling jet [45]. The contribution of ejection is most dominant everywhere in the air–water bubbling jet, while in the Wood's metal–He bubbling jet, the contribution of inward interaction is most dominant for $r/b_u < 1.0$ and that of outward interaction is so for $r/b_u > 1.0$. These results imply that the wake behind bubbles play an essential role on turbulent production in bubbling jets because the behavior of wake is closely associated with the shape and size of bubbles.

Figure 2.28 illustrates the radial turbulence kinetic energy for each turbulent motion. The contribution of inward interaction is particularly large for $r/b_u < 1.0$ in spite of its lowest appearance frequency just like the axial turbulence kinetic energy. Meanwhile, in this radial region ($r/b_u < 1.0$), the contributions of sweep and outward interaction are hardly different from each other.

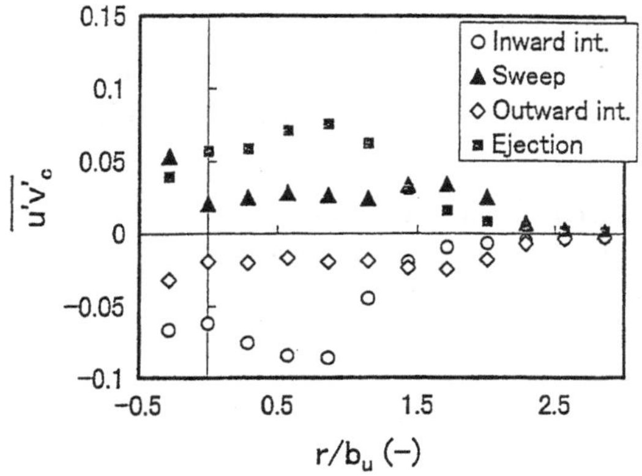


**Fig. 2.29** Radial distributions of Reynolds shear stress for each turbulent motion

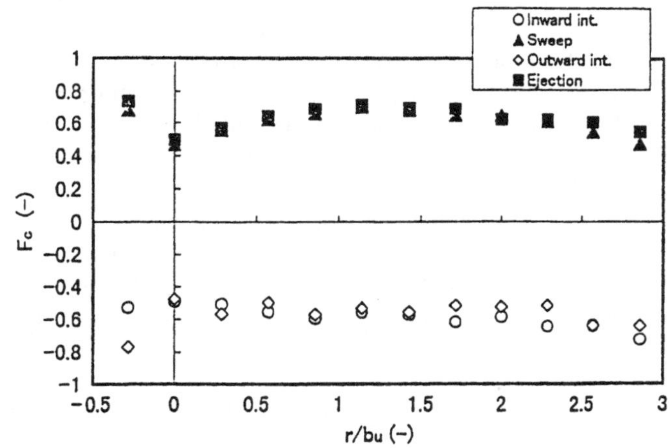


**Fig. 2.30** Correlation factor for each turbulent motion

Figure 2.29 shows the radial distribution of the Reynolds shear stress for each turbulent motion. The absolute values for the inward interaction and ejection are large, as suggested from the turbulence kinetic energies.

Figure 2.30 shows the measured values of the correlation factor, $F_c$, defined by

$$F_c = \Sigma[(u_i - \overline{u})(v_i - \overline{v})]_c / [(\Sigma(u_i - \overline{u})^2)_c(\Sigma(v_i - \overline{v})^2)_c]^{1/2} \tag{2.10}$$

The measured values of $F_c$ for each turbulent motion remain almost unchanged in the radial direction, but in a strict sense, those for ejection and sweep changed trend around $r/b_u = 1.0$. This implies that the He–Wood's metal bubbling jet has two large-scale coherent structures, the boundary being located around $r/b_u = 1.0$. This result is consistent with the above-mentioned findings on the appearance frequency and the contributions of each turbulent motion to the turbulence kinetic energies and the Reynolds shear stress.

## 2.3 Summary

### 2.3.1 Mean Flow Characteristics

The bubble and liquid flow characteristics in a molten Wood's metal bath agitated by bottom Helium gas injection measured using a two-needle electroresistivity probe and a magnet probe, respectively, can be summarized as follows:

#### 2.3.1.1 Bubble Characteristics

(a) The axial distributions of gas holdup and bubble frequency on the centerline, $\alpha_{\mathrm{cl}}$ and $f_{\mathrm{B,cl}}$, for $z > 5\,\mathrm{cm}$ are approximated satisfactorily by two empirical correlations [6, 7, 23]. The empirical correlations of Xie et al. [23] for the mean bubble rising velocity and the mean bubble chord length on the centerline, $u_{\mathrm{B,cl}}$ and $L_{\mathrm{B,cl}}$, also approximate adequately the measurement for $z > 5\,\mathrm{cm}$.
(b) The radial distributions of gas holdup $\alpha$ and bubble frequency $f_{\mathrm{B}}$ measured at the axial position of $z = 1.5\,\mathrm{cm}$ are very steep near the centerline of a bubbling jet. Two peaks appear in the $\alpha$ distributions at $z = 4.0$ and $7.0\,\mathrm{cm}$. The two-peak distribution is closely associated with bell-shaped bubbles. Further downstream of the nozzle exit ($z = 10\,\mathrm{cm}$), the disintegration of bell-shaped bubbles occurs and the distributions of $\alpha$ and $f_{\mathrm{B}}$ are nearly Gaussian. The empirical distributions of $\alpha$ and $f_{\mathrm{B}}$ [6, 7, 23] can approximate the measured radial distributions only at $z = 10\,\mathrm{cm}$.

The mean bubble rising velocity $u_{\mathrm{B}}$ and mean bubble chord length $L_{\mathrm{B}}$ measured at $z = 1.5$, 4.0, and 7.0 cm, exhibit a complex behavior in the radial direction. These radial distributions are also closely related to the shape of bubbles. An empirical correlation of $u_{\mathrm{B}}$ is proposed that could approximate the measurements only at the outer region of a bubbling jet.

#### 2.3.1.2 Liquid Flow Characteristics

(a) In the axial region where the measured radial distribution of gas holdup $\alpha$ follows a Gaussian distribution, the axial mean velocity on the centerline, $\overline{u}_{\mathrm{cl}}$, and the half-value radius of $\overline{u}$, denoted by $b_{\mathrm{u}}$, are well correlated by empirical correlations derived originally for a water–air system. The axial turbulence intensity on the centerline, $u'_{\mathrm{rms,cl}}/\overline{u}_{\mathrm{cl}}$, is approximately 0.55 except near the nozzle exit. This value is nearly the same as that for the water–air system. The radial turbulence intensity on the centerline, $v'_{\mathrm{rms,cl}}/\overline{u}_{\mathrm{cl}}$, is also approximately 0.55 for the Wood's metal–He system, while it is about 0.4 for the water–air system. This difference seems to be due to the fact that the turbulence production in the wake of bubbles is dependent on the shape of the bubbles.

(b) In the bubbling jet region ($r/b_u < 1.5$), the radial $\overline{u}$ distributions measured at axial positions except near the nozzle exit are Gaussian, as observed for a water–air system. The radial distributions of the axial turbulence intensity, $u'_{rms}/\overline{u}_{cl}$, measured at $z = 4$ and 10 cm also compare well with those for the water–air system. On the other hand, the radial distributions of the radial turbulence intensity, $v'_{rms}/\overline{u}_{cl}$, are larger for the Wood's metal–He system than for the water–air system. As a result, the radial distributions of the Reynolds shear stress $\overline{u'v'}/\overline{u}_{cl}{}^2$ in the two systems are different from each other. This difference seems also to be attributable to the shape of bubbles.

In the recirculation region, the measured values of $u'_{rms}/\overline{u}_{cl}$ and $v'_{rms}/\overline{u}_{cl}$ are nearly equal as in the case of a water–air system.

The experimental results suggest that empirical correlations of the bubble characteristics and the axial mean velocity and turbulence components of liquid flow, derived from cold model experiments, are applicable to actual refining processes stirred by bottom gas injection when the radial distributions of gas holdup and bubble frequency follow Gaussian distributions. These distributions appear to be a result of the disintegration of rising bubbles due to highly turbulent liquid motion in the bath.

### *2.3.2 Conditional Sampling*

A description is provided of the use of the 4-quadrant classification method, for measuring the coherent structure of turbulence in a He–Wood's metal bubbling jet. The results were compared with those obtained in an air–water bubbling jet. The main findings can be summarized as follows.

There exist two different types of coherent structures with respect to the radial distance in a He–Wood's metal bubbling jet. The boundary between the two types is located around $r/b_u = 1.0$. This radial position nearly corresponds to the outer edge of the bubble dispersion region. Most He bubbles rise in the radial region of $r/b_u < 1.0$. Turbulence is produced mainly by inward interaction for $r/b_u < 1.0$, and by outward interaction for $r/b_u > 1.0$.

There exists only one coherent structure of turbulence in an air–water bubbling jet. Turbulence is produced mainly by ejection everywhere in the bubbling jet. The difference in the turbulence structures in the two types of bubbling jets is attributable to the shape and size of bubbles. Bubbles in the He–Wood's metal bubbling jet are of the skirted type, while those in an air–water bubbling jet adopted for comparison can be classified into the wobbling type.

## References

1. Mazumdar D, Guthrie RIL (1995) The Physical and mathematical modelling of gas stirred ladle systems. ISIJ Int 35:1–20
2. Iguchi M, Kawabata H, Itoh Y, Nakajima K, Morita Z (1994) Continuous measurements of bubble characteristics in a molten iron bath with Ar gas bubbling. Tetsu-to-Hagane 80: 365–370
3. Iguchi M, Kawabata H, Nakajima K, Morita Z (1995) Measurement of bubble characteristics in a molten iron bath at 1600°C using an electroresistivity probe. Metall Mater Trans B 26B: 67–74
4. Iguchi M, Kawabata H, Nakajima K, Morita Z (1996) Bubble characteristics in molten copper bath with gas injection. Trans Jpn Soc Mech Eng 62–593:79–84
5. Iguchi M, Kawabata H, Itoh Y, Nakajima K, Morita Z (1996) Continuous measurements of bubble characteristics in a molten iron bath with Ar gas injection. ISIJ Int 34:980–985
6. Iguchi M, Demoto Y, Sugawara N, Morita Z (1992) Behavior of Hg-air vertical bubbling jet in a cylindrical vessel. Tetsu-to-Hagane 78:407–414
7. Iguchi M, Demoto Y, Sugawara N, Morita Z (1992) Bubble behavior in Hg-air vertical bubbling jets in a cylindrical vessel. ISIJ Int 32:998–1005
8. Iguchi M, Takeuchi H, Morita Z (1990) The flow field in air-water vertical bubbling jets in a cylindrical vessel. Tetsu-to-Hagane 76:699–706
9. Iguchi M, Kondoh T, Morita Z, Nakajima K, Hanazaki K, Uemura T, Yamamoto F (1995) Velocity and turbulence measurements in a cylindrical bath subject to centric bottom gas injection. Metall Mater Trans B 26:241–247
10. Sheng YY, Irons GA (1992) Measurements of the internal structure of gas-liquid plumes. Metall Mater Trans B 23:779–788
11. Iguchi M, Ueda H, Uemura T (1995) Bubble and liquid flow characteristics in a vertical bubbling jet. Int J Multiphase Flow 21:861–873
12. Iguchi M, Takeuchi H, Morita Z (1991) The flow field in air-water vertical bubbling jets in a cylindrical vessel. ISIJ Int 31:246–253
13. Hsiao TC, Lehner T, Kjellberg B (1978) Fluid flow in ladles – experimental results. Scand J Metal 9:105–110
14. Iguchi M, Kawabata H, Demoto Y, Morita Z (1991) Cold model experiment for developing a new velocimeter applicable to molten metal. Tetsu-to-Hagane 79:1046–1052
15. Mikrovas AC, Argyropoulos SA (1993) Measurement of velocity in high-temperature liquid metals. Metall Mater Trans B 24:1009–1022
16. Iguchi M, Kawabata H, Demoto Y, Morita Z (1994) Cold model experiments for developing a new velocimeter applicable to molten metal. ISIJ Int 34:461–467
17. Iguchi M, Takeuchi M, Kawabata H, Ebina K, Morita Z (1994) Development of a Kármán vortex probe for measuring the velocity of molten metal flow. Trans JIM 35:716–722
18. Iguchi M, Kawabata H, Ogura T, Hayashi A, Terauchi Y (1996) A new prove for directly measuring flow velocity in the continuous casting mold, Proc first int steelmaking conference, Chiba, Japan, ISIJ, p 40
19. Sawada I, Tani M, Szekely J, Ilegbusi OJ (1991) Recent developments and possibilities of computational fluid dynamics in materials processing. Tetsu-to-Hagane 77:1234–1242
20. Sahai Y, St. Pierre GR (1992) Advances in transport processes in metallurgical systems. Elsevier, Amsterdam
21. Ricou R, Vives C (1982) Local velocity and mass transfer measurements in molten metals using an incorporated magnet probe. Int J Heat Mass Transf 25–10:1579–1588
22. von Weissenfluh T (1985) Probes for local velocity and temperature measurements in liquid metal flow. Int J Heat Mass Transf 28–8:1563–1574
23. Xie Y, Orsten S, Oeters F (1992) Behaviour of bubbles at gas blowing into liquid wood's metal. ISIJ Int 32:66–75
24. Iguchi M, Tokunaga H, Tatemichi H (1997) Bubble and liquid flow characteristics in a wood's metal bath stirred by bottom helium gas injection. Metall Mater Trans B 28B:1053–1061

25. Iguchi M, Nakatani T, Kawabata H (1997) The shape of bubbles rising near the nozzle exit in molten metal baths. Metall Mater Trans B 28B:417–423
26. Kondoh H (1993) Bachelor's dissertation. Fac Eng Osaka Univ
27. Iguchi M, Takanashi N, Ogawa Y, Tokumitsu N, Morita Z (1994) X-ray fluoroscopic observations of bubble characteristics in a molten iron bath. Tetsu-to-Hagane 80–7:515–520
28. Iguchi M, Chihara T, Takanashi N, Ogawa Y, Tokumitsu N, Morita Z (1995) X-ray fluoroscopic observation of bubble characteristics in a molten iron bath. ISIJ Int 35–11:1354–1361
29. Wygnanski I, Fiedler H (1969) Some measurements in the self-preserving jet. J Fluid Mech 38–3:577–612
30. Panchapakesan NR, Lumley JL (1993) Turbulence measurements in axisymmetric jets of air and helium. Part 1. Air jet. J Fluid Mech 246:197–223
31. Theofanous TG, Sullivan J (1982) Turbulence in two-phase dispersed flows. J Fluid Mech 116:343–362
32. Mostafa AA (1992) Turbulent diffusion of heavy-particles in turbulent jets. Trans ASME J Fluids Eng 114:667–671
33. Tsuji Y, Morikawa Y (1982) LDV measurements of an air–solid two-phase flow in a horizontal pipe. J Fluid Mech 120:385–409
34. Lance M, Bataille J (1991) Turbulence in the liquid phase of a uniform bubbly air–water flow. J Fluid Mech 222:95–118
35. Longmire EK, Eaton JK (1992) Structure of a particle-laden round jet. J Fluid Mech 236: 217–257
36. Gore RA, Crowe CT (1989) Effect of particle size on modulating turbulent intensity. Int J Multiphase Flow 2:279–285
37. Hetsroni G (1989) Particles-turbulence interaction. Int J Multiphase Flow 5:735–746
38. Iguchi M, Okita K, Nakatani T, Kasai N (1997) Structure of turbulent round bubbling jet generated by premixed gas and liquid injection. Int J Multiphase Flow 23:249–262
39. Iguchi M, Nakatani T, Ueda H (1997) Model study of turbulence structure in a bottom blown bath with top slag using conditional sampling. Metall Mater Trans B 28B:87–94
40. Iguchi M, Nozaawa K, Tomida H, Morita Z (1992) Bubble characteristics in the buoyancy region of a vertical bubbling jet. ISIJ Int 32:747–753
41. Iguchi M, Nakatani T, Tokunaga H (1997) The shape of bubbles rising near the nozzle exit in molten metal baths. Metall Mater Trans B 28B:417–423
42. Brodkey RS, Wallace JM, Eckelmann H (1974) Some properties of truncated turbulence signals in bounded shear flows. J Fluid Mech 63:209–224
43. Kim J, Moin P, Moser P (1983) Turbulence statistics in fully developed channel flow at low Reynolds number. J Fluid Mech 177:133–166
44. Iguchi M (1988) The structure of turbulence in pulsatile pipe flow accompanied by relaminarization. JSME Int J 31:623–631
45. Iguchi M, Nakatani T, Okita K, Yamamoto F, Morita Z (1996) Turbulence in reactors agitated by bottom gas injection. ISIJ Int Suppl 36:38–41

# Chapter 3
# The Coanda Effect

## 3.1 General Features

### *3.1.1 Overview*

Attachment of flow to a wall or walls placed around the flow is usually referred to as the Coanda effect [1]. According to Kadosch [1, 2], this effect was first reported by Young in 1800. Since then many investigations have been made on its impact in a variety of flow fields [3–6]. In particular, the application to flow control in pneumatics and hydraulic circuits has been extensively investigated [3, 6]. Kirshner [1] identified the following effects on wall attachment in fluid amplifiers.

- Changing position of walls
- Diverging walls
- Single-sided case
- Mean flow and fluctuation
- Secondary nozzles
- Pressure distribution
- Wall height and aspect ratio

Wall attachment has also been effectively used in combustion, sizing of particles, air conditioning, and so on [4–6].

In addition to the wall attachment mentioned above, the Coanda effect emerges when single-phase jets are generated using dual nozzles placed a short distance apart [7, 8]. The effect also appears when two bubbling jets rise side by side. The two bubbling jets pull each other and merge into a larger scale single bubbling jet.

Gas injection through multi-nozzles is intended to enhance the mixing and chemical reaction rates in the baths of metallurgical processes [9–13]. The merging phenomena of bubbling jets may not be effective or suitable for these purposes. In this chapter, both the wall attachment of a bubbling jet and interaction of two bubbling jets will be discussed. Concerning the wall attachment of a vertical bubbling jet to a vertical wall, the attachment length and the bubble and liquid flow characteristics will be described. For two bubbling jet interactions, a detailed description is

M. Iguchi and O.J. Ilegbusi, *Modeling Multiphase Materials Processes: Gas-Liquid Systems*, DOI 10.1007/978-1-4419-7479-2_3,

provided of the critical condition for two jets to merge, the merging length defined as the distance from the nozzle exit to the merging position, and the bubble and liquid flow characteristics in the bubbling jets.

### 3.1.2 Mechanism of Coanda Effect

The mechanism of the Coanda effect will be qualitatively described with reference to Fig. 3.1. A bubbling jet generated vertically upwards through a single nozzle rises, entraining the surrounding liquid, as shown in Fig. 3.1a, and as a result, spreads

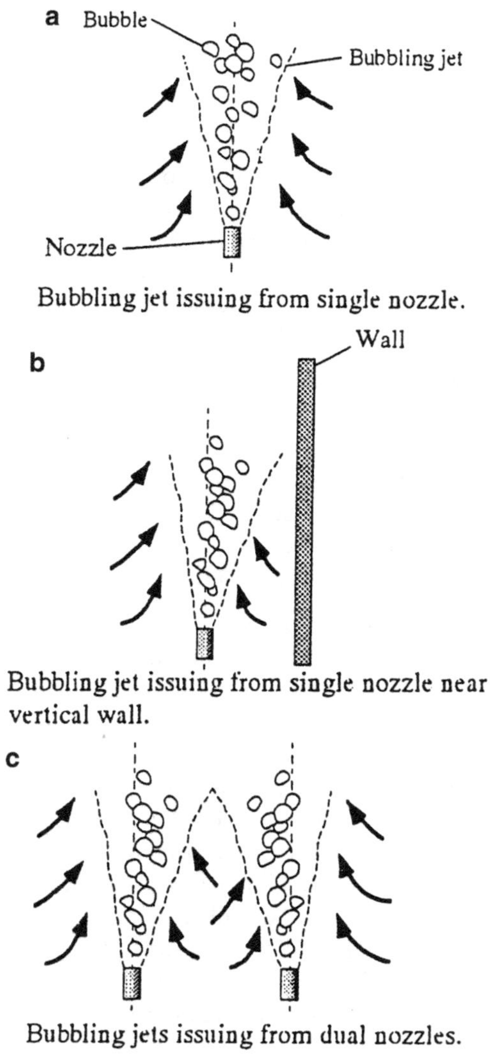


**Fig. 3.1** Explanation of Coanda effect

in the horizontal direction. If there is no wall or another jet around the bubbling jet, the pressure around it is uniform at any vertical position because the fluid is uniformly entrained into the jet from all directions. Accordingly, the bubbling jet ascends vertically upwards.

If there is a vertical wall nearby (see Fig. 3.1b), the pressure in the fluid between the jet and the wall decreases as the outer edge of the jet approaches the wall. That is, the pressure of fluid between the jet and the wall becomes low compared with the pressure of fluid on the opposite side of the jet as the jet develops in the vertical direction. Due to this pressure difference, the jet is pulled towards the wall and subsequently attaches to it. A similar phenomenon occurs for a bubbling jet rising in the vicinity of another, as shown in Fig. 3.1c.

## 3.2 Wall Interaction in Metallurgical Reactor

### *3.2.1 Bubble Characteristics*

The Coanda effect will occur in real metallurgical reactors agitated by gas injection through a bottom nozzle or a top lance placed near the side wall of the reactors. The attachment of a bubbling jet to the side wall of the reactors may cause undesirable events. For example, the intensity of mixing in the reactors may be reduced and erosion of the refractory may become serious. The main purpose of this section therefore is to describe the behavior of an air–water bubbling jet subjected to the Coanda effect. Bubble frequency, gas holdup, mean bubble rising velocity, and mean bubble chord length measured with a two-needle electroresistivity probe in the vertical region extending from the nozzle exit to the bath surface are presented. Particular attention is paid to the vertical and horizontal distributions of these quantities near the side wall.

#### 3.2.1.1 Experimental Apparatus and Procedure

Experimental Apparatus

Figure 3.2 shows a schematic of the experimental apparatus. The cylindrical vessel made of transparent acrylic resin has an inner diameter $D$ of 0.200 m and a height $H$ of 0.400 m. This vessel is enclosed with another vessel with a square cross-section, and water is used to fill the gap between the two vessels in order to reduce parallax effect as much as possible. Air is injected vertically upward into a water bath of a depth of 0.350 m through a single-hole nozzle at the exit of a glass top lance. The inner diameter of the nozzle, $d_{ni}$, is varied from $1.3 \times 10^{-3}$ to $3.8 \times 10^{-3}$ m. The center of the nozzle exit is located on the horizontal plane $5 \times 10^{-2}$ m above the bottom wall.

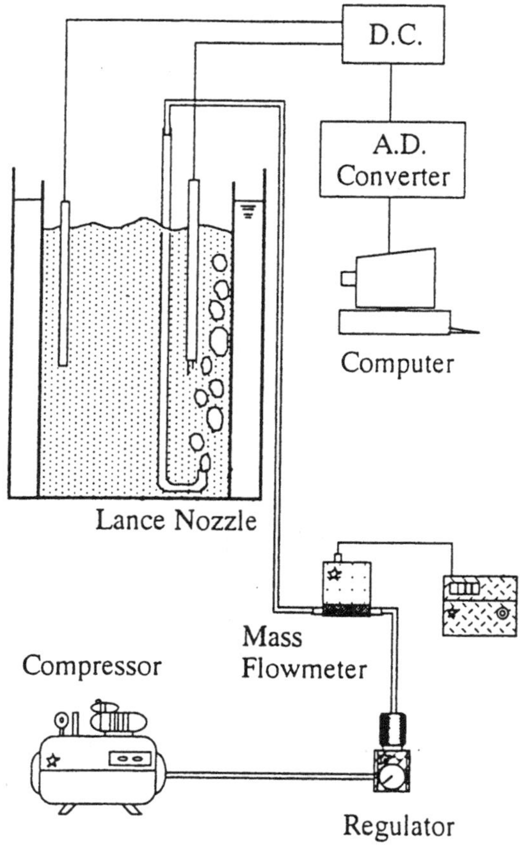


**Fig. 3.2** Experimental apparatus

The origin of the cylindrical coordinate system is placed at the center of the bath, as shown in Fig. 3.3. The horizontal distance from the side wall is designated by $\eta(=R-r)$. The distance from the side wall to the nozzle exit, $\eta_n$, is varied from $1.75\times10^{-2}$ to $3.5\times10^{-2}$ m. The horizontal position at which a peak appears in the gas holdup distribution is designated by $\eta_{\alpha,\max}$ and the half-value width by $\eta_{\alpha,\max/2,}$. These quantities are introduced to represent the horizontal extent of the bubble dispersion region. In the same manner, the peak position and half-value width of the axial mean velocity $\bar{u}$ are defined. These two representative scales will be discussed in a later section. The attachment length $L_a$ is defined as the vertical distance from the nozzle exit to the position at which bubbles attach to the side wall.

Measurement of Attachment Length

The attachment length $L_a$ was determined from a picture taken with a still camera of a bubbling jet subjected to the Coanda effect.

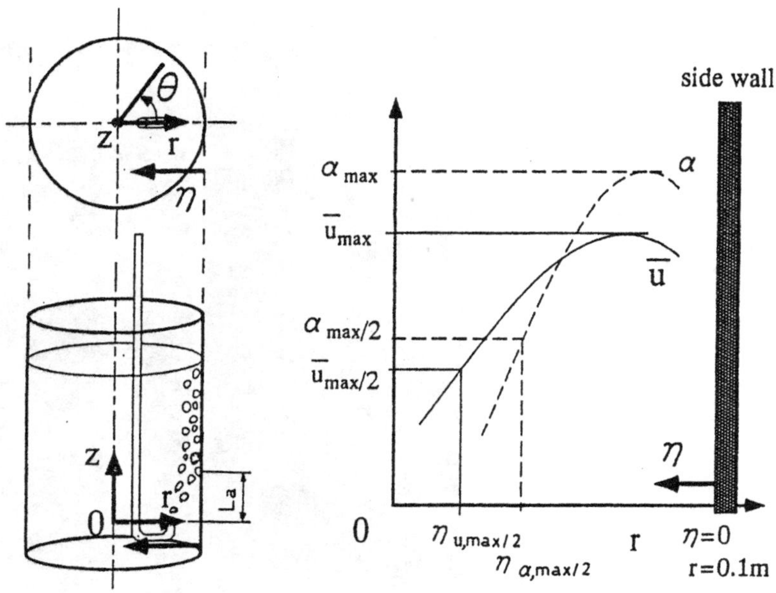


**Fig. 3.3** Coordinate system

Measurement of Bubble Characteristics

The bubble characteristics represented by the bubble frequency, gas holdup, mean bubble rising velocity, and mean bubble chord length were measured at $z = 0.050$, 0.100, 0.150, and 0.190 m with a two-needle electroresistivity probe [14–20]. The inner and outer diameters of the nozzle were $2.0 \times 10^{-3}$ and $4.0 \times 10^{-3}$ m, respectively, and the distance $\eta_n$ was $2 \times 10^{-2}$ m. The gas flow rate $Q_g$ was $41.4 \times 10^{-6}$, $100 \times 10^{-6}$, or $293 \times 10^{-6}$ m$^3$/s. Although the measurements were carried out in the $r$, $\theta$, and $z$ directions, the results obtained on the $r - z$ plane ($\theta = 0$) will be primarily presented to discuss the Coanda effect on an air–water bubbling jet rising near the side wall of a cylindrical vessel.

#### 3.2.1.2 Experimental Results

Attachment Length

Figure 3.4 shows a photograph of a bubbling jet attaching to the side wall. When the nozzle is far from the side wall, the bubbling jet thus generated rises vertically upward. Accordingly, the attachment of the bubbling jet shown in Fig. 3.4 can be attributed to the Coanda effect.

The measured values of the attachment length $L_a$ by $\eta_n$ are shown in Fig. 3.5. The $L_a/\eta_n$ values are independent of the inner diameter of the nozzle, $d_{ni}$, but

**Fig. 3.4** Photograph showing bubbling jet subjected to Coanda effect

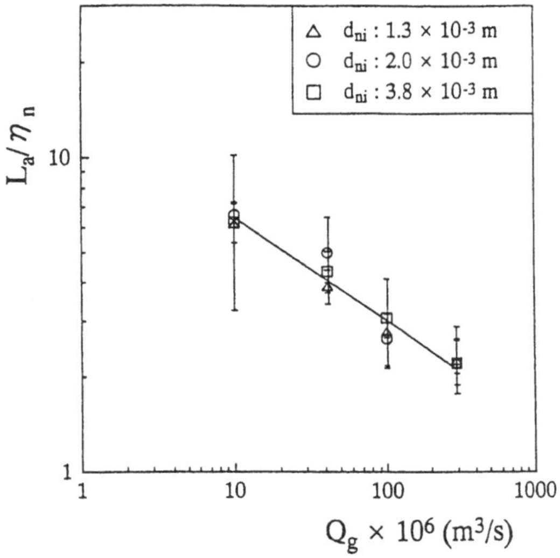


**Fig. 3.5** Relation between attachment length and gas flow rate

dependent on the gas flow rate $Q_g$. This implies that the attachment length is nearly independent of the inertia force of injected gas, while buoyancy force exerts primary influence on the length. From single-phase jet flow study [5], the attachment length is nearly independent of the inertia force of injected gas as long as $\eta_n/d_{ni} > 10$.

**Fig. 3.6** Bubble frequency near the side wall

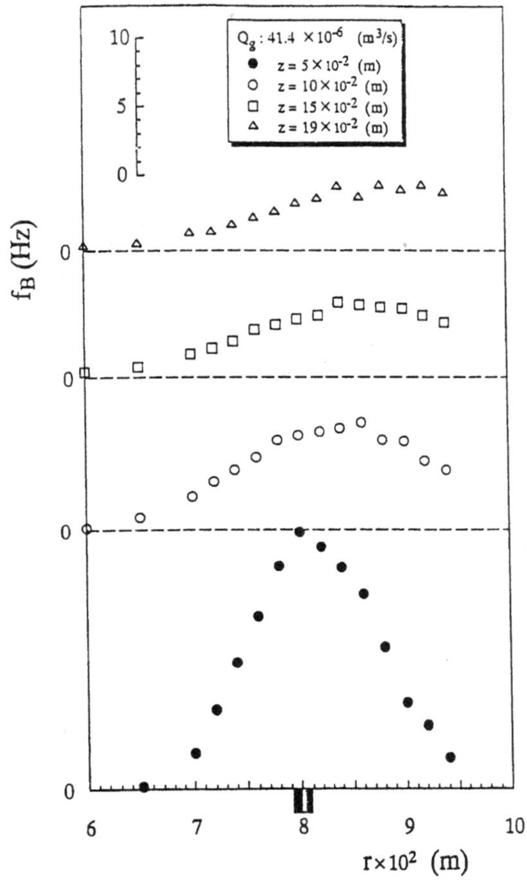


The results for the air–water bubbling jets in Fig. 3.5 can be represented by the empirical relation.

$$L_a/\eta_n = 0.14\, Q_g^{-1/3} \tag{3.1}$$

with $Q_g$ expressed in $m^3/s$. This relation is valid for $30 > \eta_n/d_{ni} > 5$.

### Horizontal Distribution of Bubble Characteristics

The horizontal distribution of bubble frequency $f_B$ and gas holdup $\alpha$, measured at the aforementioned four vertical positions, are shown in Figs. 3.6 and 3.7, respectively. The horizontal position at which a peak appears in the $f_B$ distribution moves towards the side wall as $z$ increases. The same peak-shift can be seen in the horizontal distribution of gas holdup $\alpha$.

Both the measured values of mean bubble rising velocity $u_B$ and mean bubble chord length $L_B$ are approximately uniform in the horizontal direction, as illustrated

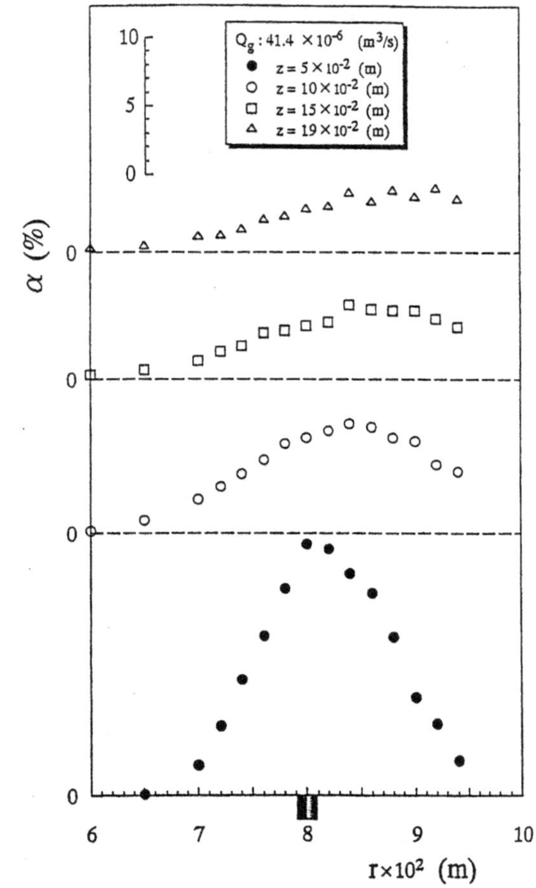


**Fig. 3.7** Gas holdup near the side wall

in Figs. 3.8 and 3.9, respectively. In addition, changes in $u_B$ and $L_B$ in the vertical direction are relatively small, indicating that bubbles rise without any significant disintegration or coalescence.

Axial Variation of Maximum Values of Bubble Characteristics

The maximum value of bubble frequency, $f_{Bmax}$, is plotted against the vertical distance $z$ for the three gas flow rate values in Fig. 3.10. The solid and broken lines designate $f_{B,max}$ values measured for vertical bubbling jets in the absence of the Coanda effect [21]. Agreement between the measurement and the vertical bubbling jet data is quite good for the two gas flow rates $Q_g = 41.4 \times 10^{-6}$ and $293 \times 10^{-6}\,m^3/s$ except near the bath surface. Data on $f_{B,max}$ for the vertical bubbling jet in the absence of the Coanda effect are not available for $Q_g = 100 \times 10^{-6}\,m^3/s$.

The measured maximum values of the gas holdup, $\alpha_{max}$, are plotted in Fig. 3.11. The trend follows that for $f_{B,max}$ described above.

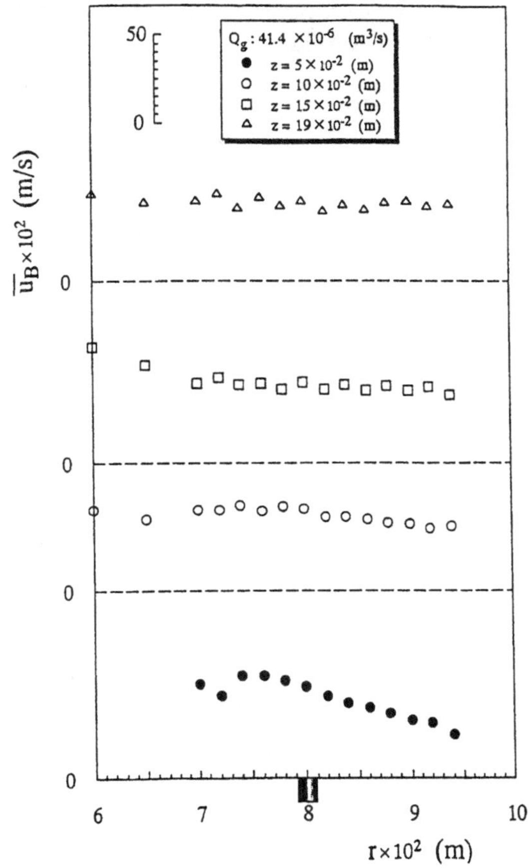


**Fig. 3.8** Mean bubble rising velocity near the side wall

The maximum values of mean bubble rising velocity, $u_{B,max}$, and mean bubble chord length, $L_{B,max}$, are shown in Figs. 3.12 and 3.13, respectively. In each figure the measurement for $Q_g = 41.4 \times 10^{-6}$ and $293 \times 10^{-6}$ $m^3/s$ can be approximated by the data obtained for the vertical bubbling jet that is free from the Coanda effect.

The experimental results shown in Figs. 3.6–3.13 collectively suggest that the maximum values of the bubble characteristics in the bubbling jet rising very near the side wall of the vessel are not significantly affected by the wall although the jet is pulled towards the wall and finally rises along it.

### Horizontal Spread of Bubble Dispersion Region

Figure 3.14 shows the characteristic length $\eta_{fB,max/2}$, defined in the same manner as $\eta_{\alpha,max/2}$ illustrated in Fig. 3.3. These length scales were introduced to represent the spread of a bubbling jet rising along the side wall of a cylindrical vessel. It is evident that the dependence of $\eta_{fB,max/2}$ on the air flow rate $Q_g$ is weak and $\eta_{fB,max/2}$

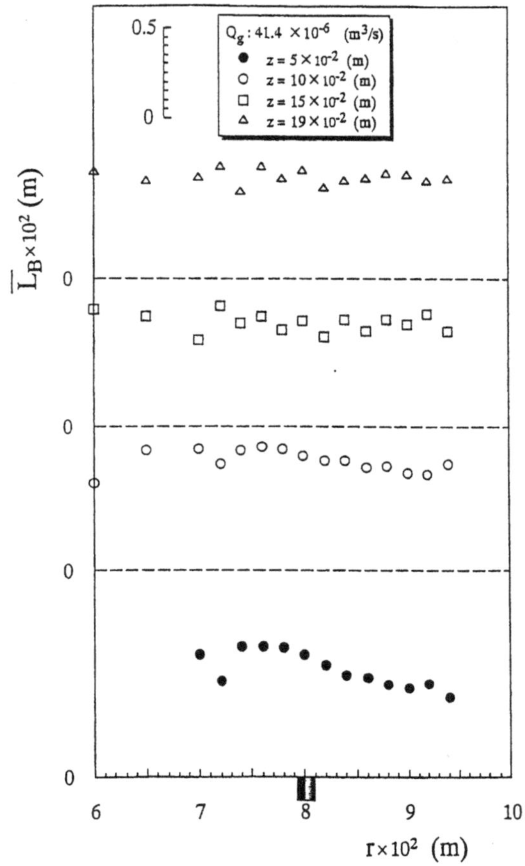


**Fig. 3.9** Mean bubble chord length near the side wall

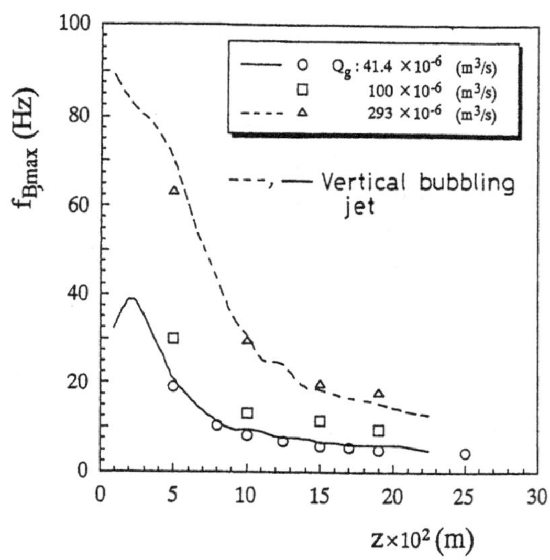


**Fig. 3.10** Maximum value of bubble frequency

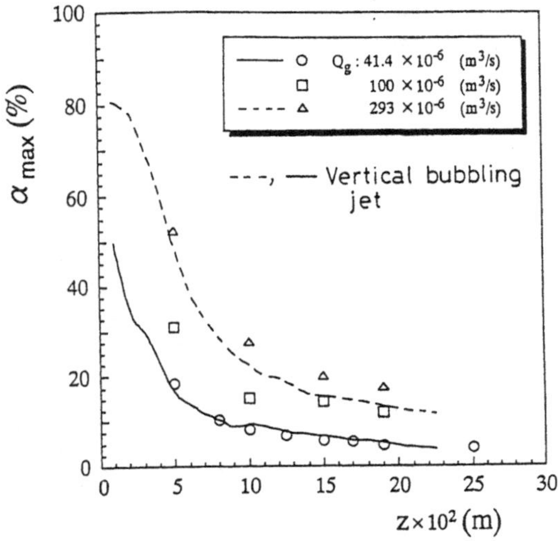


**Fig. 3.11** Maximum value of gas holdup

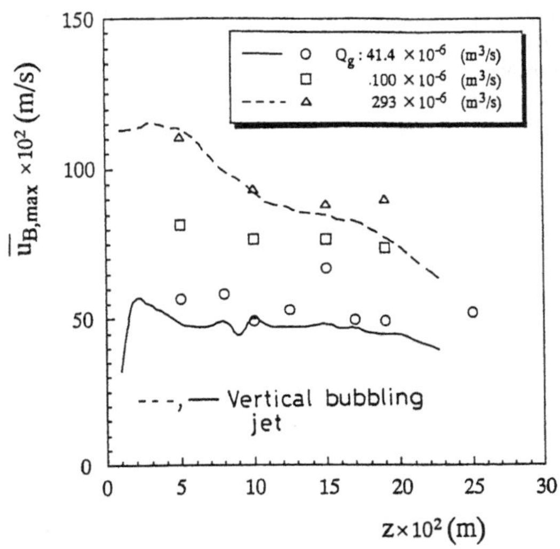


**Fig. 3.12** Maximum value of mean bubble rising velocity

remains nearly constant in the vertical direction. Allowing for mass balance of gas, it is curious that the $\alpha_{max}$ and $\bar{u}_{B,max}$ values for the bubbling jets with and without the Coanda effect are approximately the same in Figs. 3.10 and 3.12 in spite of the difference in the measured half widths. Mass balance of gas, however, must be considered after the bubble dispersion in the tangential ($\theta$) direction is known. So far this issue has not been addressed.

In the vicinity of the nozzle exit, $\eta_{fB,max/2}$ is approximated by

$$\eta_{fB,max/2} = \eta_n + b_f, \tag{3.2}$$

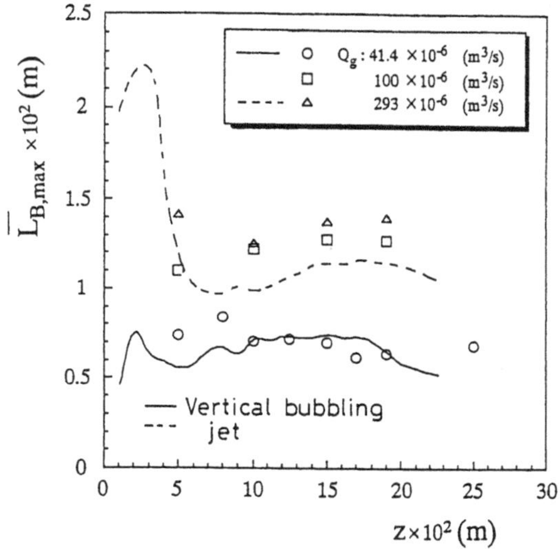


**Fig. 3.13** Maximum value of mean bubble chord length

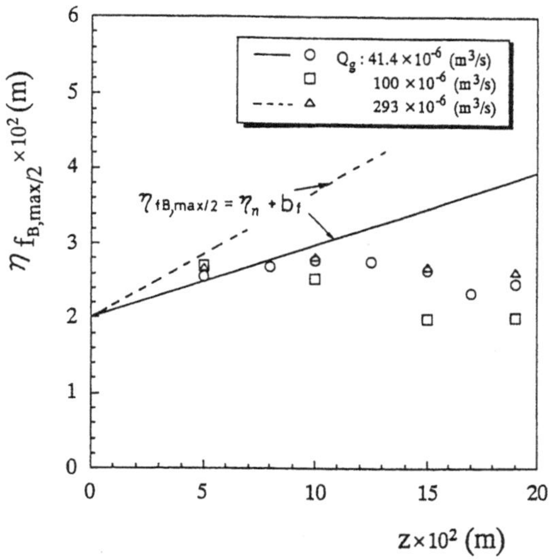


**Fig. 3.14** Relation between $\eta_{f\mathrm{B,max}/2}$ and vertical distance $z$

where $b_\mathrm{f}$ is the half-value radius of the horizontal distribution of the bubble frequency in the vertical bubbling jet, which is not subject to the Coanda effect [19].

The parameter $\eta_{\alpha,\max/2}$ follows the same relation, as shown in Fig. 3.15. It is interesting that the measured values of $\eta_{f\mathrm{B,max}/2}$ compare well with those of $\eta_{\alpha,\max/2}$.

### Distribution of Bubble Characteristics in the Horizontal Plane

The measured bubble frequency $f_\mathrm{B}$ for $Q_\mathrm{g} = 41.4 \times 10^{-6}\,\mathrm{m^3/s}$ are nondimensionalized by the maximum value, $f_{\mathrm{B,max}}$, and plotted against the dimensionless, hori-

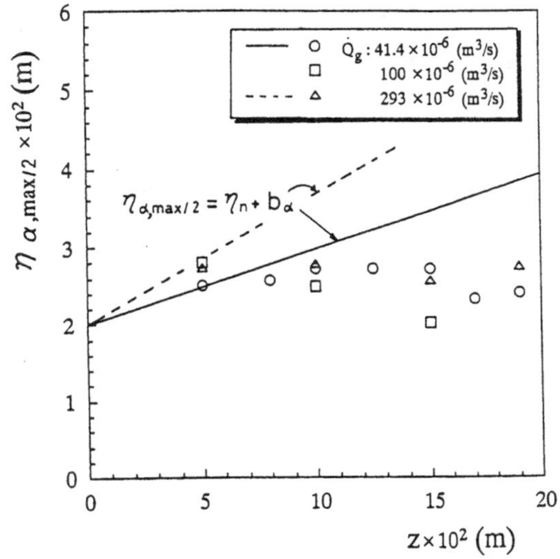


**Fig. 3.15** Relation between $\eta_{\alpha,\max/2}$ and vertical distance $z$

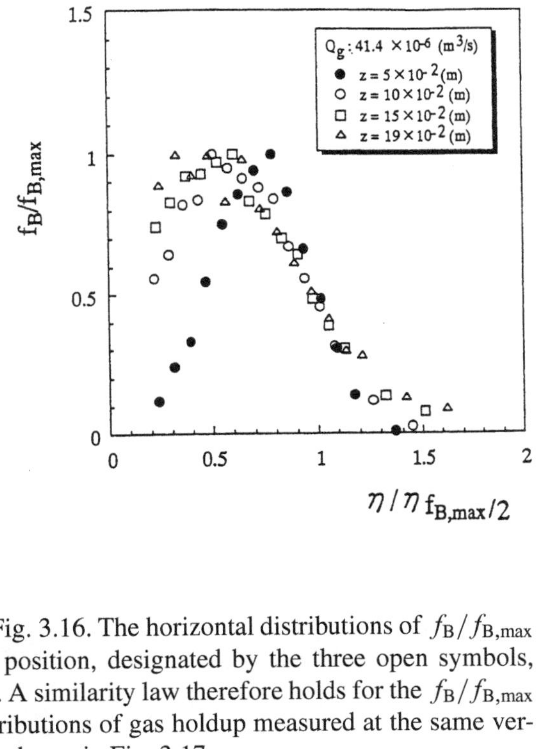


**Fig. 3.16** Horizontal distributions of bubble frequency for $Q_g = 41.4 \times 10^{-6}\ \text{m}^3/\text{s}$

zontal distance, $\eta/\eta_{f\text{B},\max/2}$, in Fig. 3.16. The horizontal distributions of $f_\text{B}/f_{\text{B},\max}$ measured above the attachment position, designated by the three open symbols, agree favorably with one another. A similarity law therefore holds for the $f_\text{B}/f_{\text{B},\max}$ distributions. The horizontal distributions of gas holdup measured at the same vertical positions are also similar, as shown in Fig. 3.17.

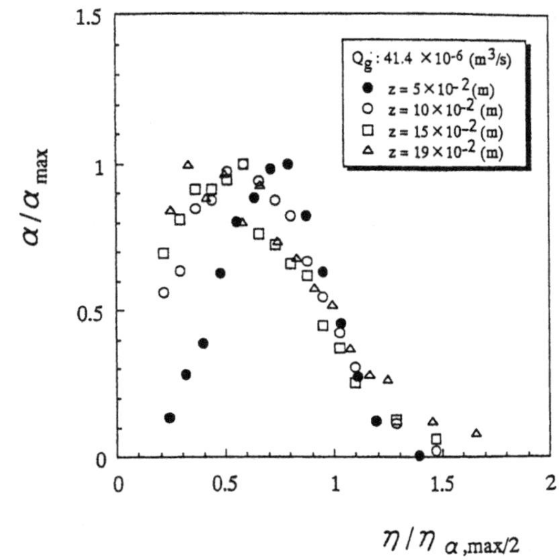


**Fig. 3.17** Horizontal distributions of gas holdup for $Q_g = 41.4 \times 10^{-6}\,\mathrm{m^3/s}$

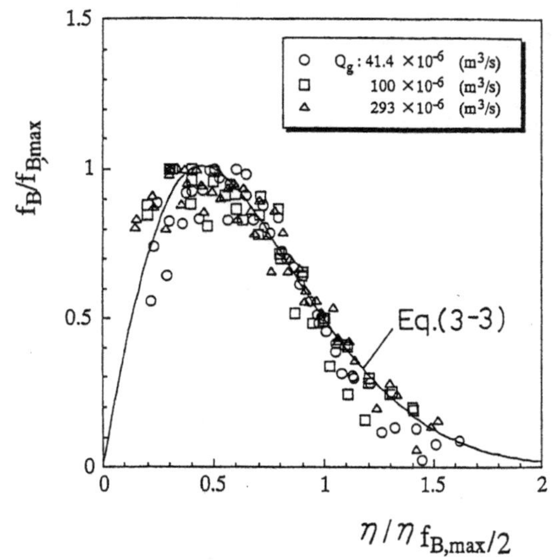


**Fig. 3.18** Similar distribution of bubble frequency

The $f_B/f_{B,\max}$ distribution for $Q_g = 100 \times 10^{-6}$ and $293 \times 10^{-6}\,\mathrm{m^3/s}$ also are similar, as demonstrated in Fig. 3.18. The solid line is drawn to pass through the mean of the measured values as close as possible. The distribution can be expressed by

$$f_B/f_{B,\max} = 1.52\exp(-1.1\eta_f'^2)\tan h(2.64\eta_f'), \tag{3.3}$$

$$\eta_f' = \eta/\eta_{f\mathrm{B},\max/2}. \tag{3.4}$$

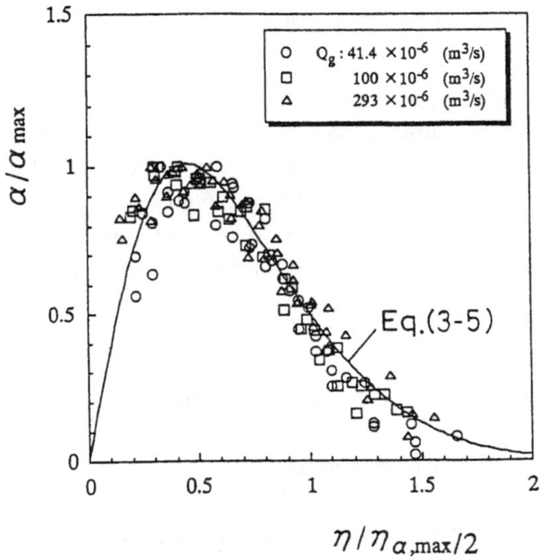


**Fig. 3.19** Similar distribution of gas holdup

In addition, Fig. 3.19 illustrates that the same functional relationship as above is valid for the $\alpha/\alpha_{max}$ distributions. Accordingly, $\alpha/\alpha_{max}$ can be expressed by,

$$\alpha/\alpha_{max} = 1.52\exp(-1.1\eta_\alpha'^2)\tan h(2.64\eta_\alpha'), \tag{3.5}$$

$$\eta_\alpha{}' = \eta/\eta_{\alpha,max/2}. \tag{3.6}$$

#### 3.2.1.3 Summary on Bubble Characteristics

(a) An air–water bubbling jet generated vertically upward near the side wall of a cylindrical vessel is pulled towards the side wall and attaches to it through the Coanda effect. The vertical distance from the nozzle exit to the attachment position, defined as the attachment length $L_a$, is nearly independent of the inner diameter of the nozzle, $d_{ni}$, but dependent on the air flow rate, $Q_g$, and the horizontal distance between the nozzle exit and the side wall, $\eta_n$. An empirical relation, (3.1), is given for $L_a/\eta_n$.

(b) The maximum values of the bubble frequency, gas holdup, mean bubble rising velocity, and mean bubble chord length are hardly influenced by the side wall of the vessel and agree with the experimental data for a vertical bubbling jet free from the Coanda effect.

(c) The horizontal extent of bubble dispersion region, represented by $\eta_{f B,max/2}$ or $\eta_{\alpha,max/2}$, is essentially independent of the air flow rate, $Q_g$.

(d) Above the attachment position, the horizontal distributions of bubble frequency $f_B$ and gas holdup $\alpha$ are similar. Empirical relations, (3.3) and (3.5), are given for these distributions.

### 3.2.2 *Liquid Flow Characteristics*

In this section, data will be given on the mean velocity and turbulence components of water flow in and near an air–water bubbling jet subjected to the Coanda effect, measured with a two-channel laser Doppler velocimeter. These quantities are closely associated with mixing in metallurgical reactors and the erosion of the side wall of the reactors [22]. Particular attention is paid to whether or not the horizontal distributions of the liquid flow characteristics near the side wall are similar in the vertical region above the attachment position.

#### 3.2.2.1 Experimental Apparatus and Procedure

The experimental apparatus has been described in Sect. 3.2.2. The electroresistivity probe was removed and a two-channel laser Doppler velocimeter was set up to measure the three velocity components of water flow in the bath. The origin of the cylindrical coordinates $(z,\ r,\ \theta)$ was placed at the center of the bath, as shown in Fig. 3.3. The velocity components were designated by $u$, $v$, and $w$, respectively. The components, $u$ and $v$, were measured in the $z-r$ plane including the centerline of the bath and the center of the nozzle exit [21, 23]. Digitized velocity data were decomposed into the mean velocity and turbulence components as follows:

$$u_i = \bar{u} + u_i', \tag{3.7}$$

$$v_i = \bar{v} + v_i', \tag{3.8}$$

$$\bar{u} = \Sigma u_i / N, \tag{3.9}$$

$$\bar{v} = \Sigma v_i / N. \tag{3.10}$$

The over-bar and prime denote the mean velocity and turbulence components, respectively, $N$ is the number of data points, and the subscript $i$ designates the $i$th digitized datum.

Velocity measurements were carried out at four fixed vertical positions ($z = 0.05$, 0.10, 0.15, and 0.19 m) for three gas flow rates $Q_g = 41.4 \times 10^{-6},\ 100 \times 10^{-6}$, and $293 \times 10^{-6}\ \mathrm{m^3/s}$ just like the measurements of bubble characteristics [22]. The root mean square (rms) values of the axial and radial turbulence components, $u'_{\mathrm{rms}}$ and $v'_{\mathrm{rms}}$, and the Reynolds shear stress $\overline{u'v'}$ were calculated from the following equations:

$$u'_{\mathrm{rms}} = \left[\Sigma u_i'^2/N\right]^{1/2}, \tag{3.11}$$

$$v'_{\mathrm{rms}} = \left[\Sigma v_i'^2/N\right]^{1/2}, \tag{3.12}$$

$$\overline{u'v'} = \Sigma u_i' v_i'/N. \tag{3.13}$$

#### 3.2.2.2 Experimental Results

Resultant Mean Velocity Vectors

Figures 3.20 and 3.21 show the resultant vectors of $u$ and $v$ for $Q_g = 41.4 \times 10^{-6}$ and $293 \times 10^{-6}\,m^3/s$ on the $r - z$ plane, respectively. The vectors at the four axial positions are all directed towards the side wall, partly validating the existence of the Coanda effect.

Mean Velocity and Turbulence Components

The experimental results for $Q_g = 41.4 \times 10^{-6}\,m^3/s$ are presented to illustrate the appearance of the Coanda effect. In Fig. 3.22, a peak occurs in every horizontal

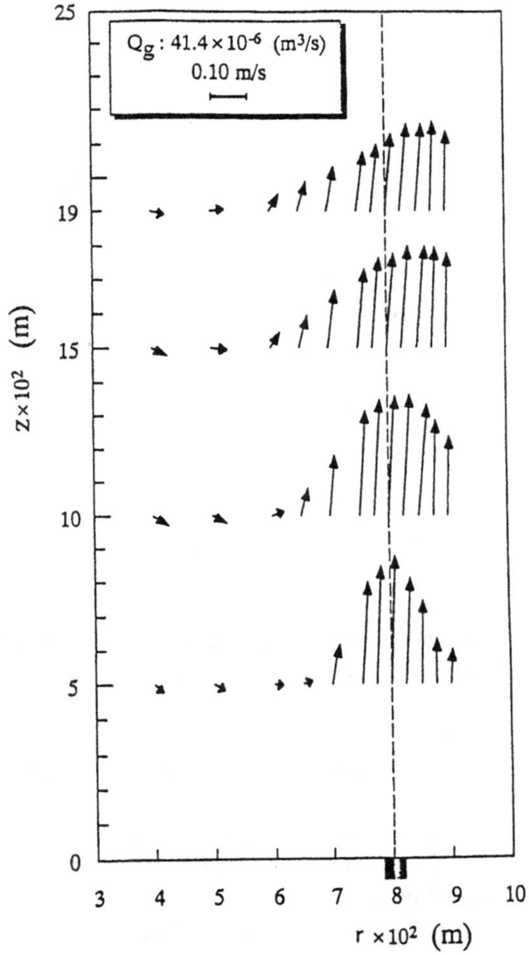


**Fig. 3.20** Resultant velocity vectors for gas flow rate of $Q_g = 41.4 \times 10^{-6}\,m^3/s$

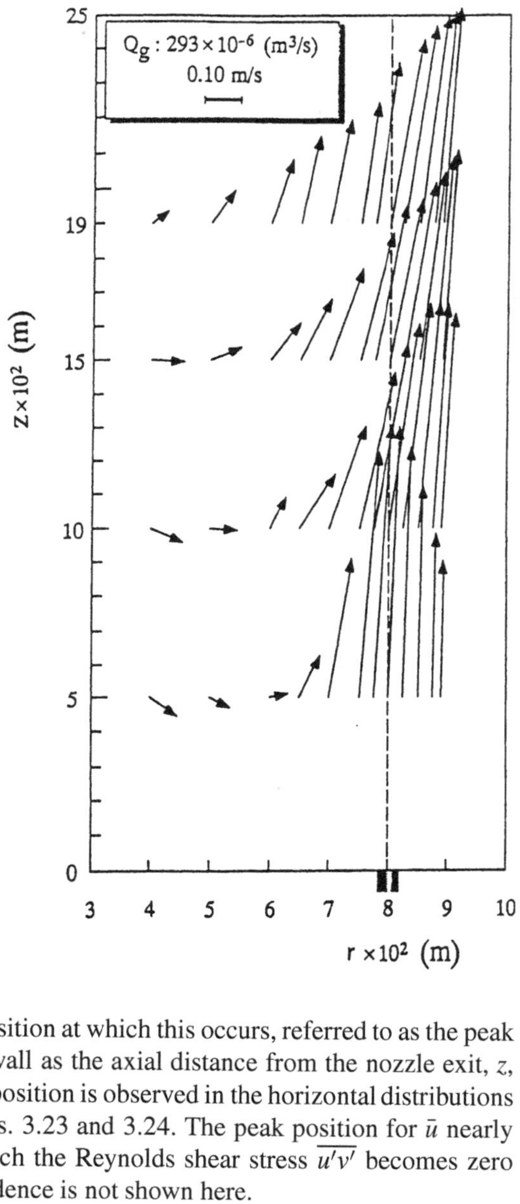


**Fig. 3.21** Resultant velocity vectors for gas flow rate of $Q_g = 293 \times 10^{-6}\,\mathrm{m^3/s}$

distribution of $\bar{u}$. The horizontal position at which this occurs, referred to as the peak position, moves towards the side wall as the axial distance from the nozzle exit, $z$, increases. Also, a shift in the peak position is observed in the horizontal distributions of $u'_{\mathrm{rms}}$ and $v'_{\mathrm{rms}}$, as shown in Figs. 3.23 and 3.24. The peak position for $\bar{u}$ nearly corresponds to the position at which the Reynolds shear stress $\overline{u'v'}$ becomes zero near the side wall although the evidence is not shown here.

### Maximum Axial Mean Velocity and Characteristic Length

The measured values of $\bar{u}_{\mathrm{max}}$ for three gas flow rates are shown plotted against the axial distance $z$ in Fig. 3.25. The solid and dot-dash lines indicate the values of $\bar{u}_{\mathrm{max}}$ measured in the vertical bubbling jets for $Q_g = 41.4 \times 10^{-6}$ and $293 \times 10^{-6}\,\mathrm{m^3/s}$, which are free from the side wall of the vessel [21]. The measured $\bar{u}_{\mathrm{max}}$ values for

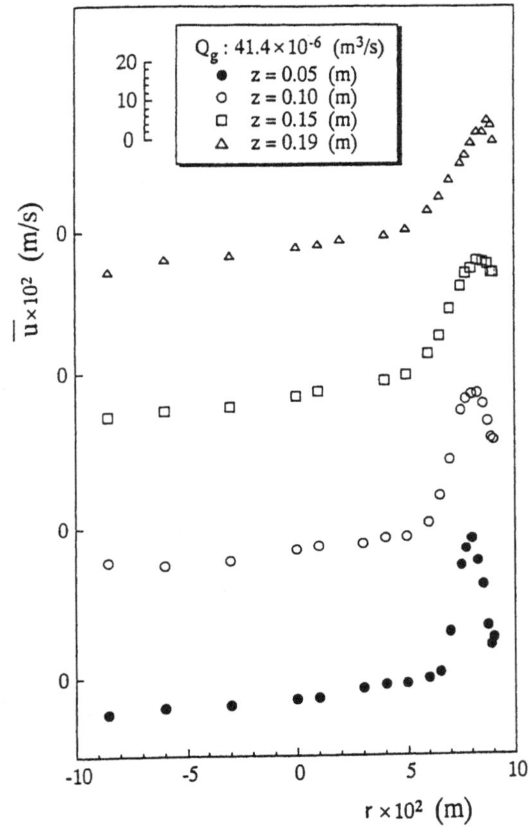


**Fig. 3.22** Axial mean velocity $u$ for gas flow rate of $Q_g = 41.4 \times 10^{-6}\ \mathrm{m^3/s}$

each gas flow rate can be satisfactorily approximated by the $\bar{u}_{max}$ values measured in the vertical bubbling jet.

Concerning the vertical bubbling jets free from the side wall, the half-value radius, $b_u$, of the radial distribution of the axial mean velocity $u$ is approximated by [21]:

$$b_{\underline{u}} = 0.14z. \tag{3.14}$$

If a bubbling jet generated near the side wall of a cylindrical vessel rises straight upward, the half-value width or characteristic length $\eta_{u,\max/2}$, which is defined in Fig. 3.3, could be expressed by

$$\eta_{u,\max/2} = \eta_n + 0.14z. \tag{3.15}$$

This is represented by the solid line in Fig. 3.26. The measured values of $\eta_{u,\max/2}$ are essentially independent of the gas flow rate $Q_g$. In the vicinity of the nozzle exit ($z = 0.05$ m), the measured $\eta_{u,\max/2}$ can be satisfactorily approximated by (3.15), but overestimated at other axial positions. This reconfirms that the bubbling jets do not rise straight upward but instead, rise along the side wall of the vessel.

**Fig. 3.23** The rms value of axial turbulence component, $u'_{\mathrm{rms}}$, for gas flow rate of $Q_g = 41.4 \times 10^{-6}\ \mathrm{m^3/s}$

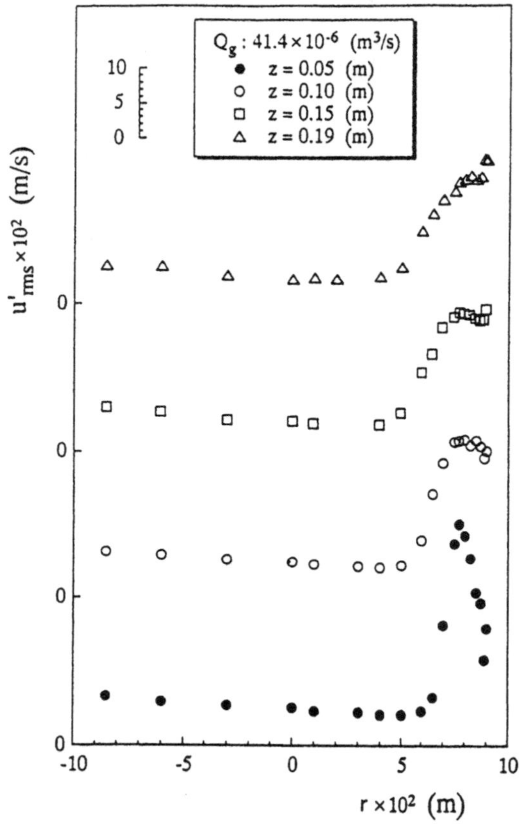


Maximum Turbulence Components and Characteristic Lengths

The measured values of the axial turbulence component $u'_{\mathrm{rms,max}}$ also could be approximated by the data for the vertical bubbling jet, [21] as demonstrated in Fig. 3.27. On the other hand, Fig. 3.28 indicates that $v'_{\mathrm{rms,max}}$ becomes smaller in the bubbling jet subjected to the Coanda effect than in the vertical bubbling jet. This is because the horizontal (radial) turbulence motions are significantly suppressed by the side wall.

The characteristic lengths, $\eta_{u'\mathrm{rms,max}/2}$ and $\eta_{v'\mathrm{rms,max}/2}$, shown in Fig. 3.29 are larger than $\eta_{u,\mathrm{max}/2}$, just like the vertical bubbling jets free from the Coanda effect.

Similarity Distributions of Mean Velocity and Turbulence Components

Figure 3.30 shows nondimensionalized distributions of the axial mean velocity component. The data at three vertical positions above the attachment position, designated by the three open symbols, agree with one another. Consequently, the horizontal distributions of $\bar{u}$ are similar just like the horizontal distributions of

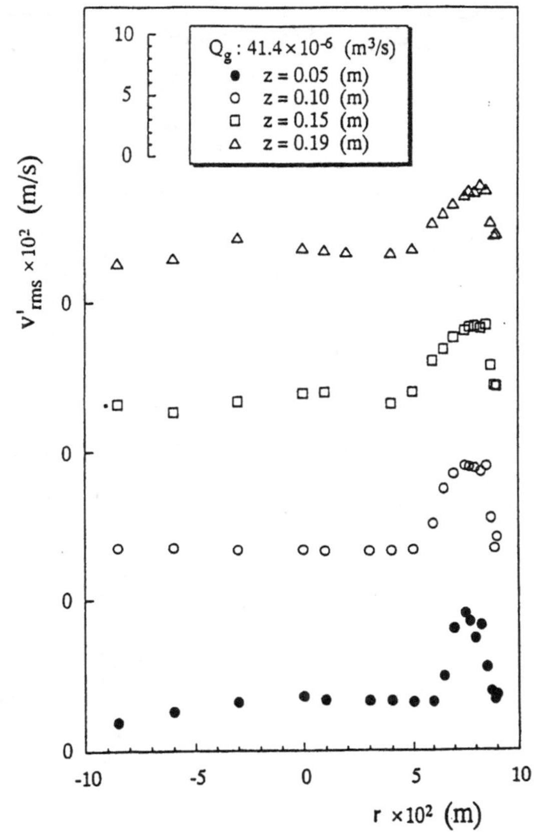


**Fig. 3.24** The rms value of radial turbulence component, $v'_{\mathrm{rms}}$, for gas flow rate of $Q_g = 41.4 \times 10^{-6}\,\mathrm{m^3/s}$

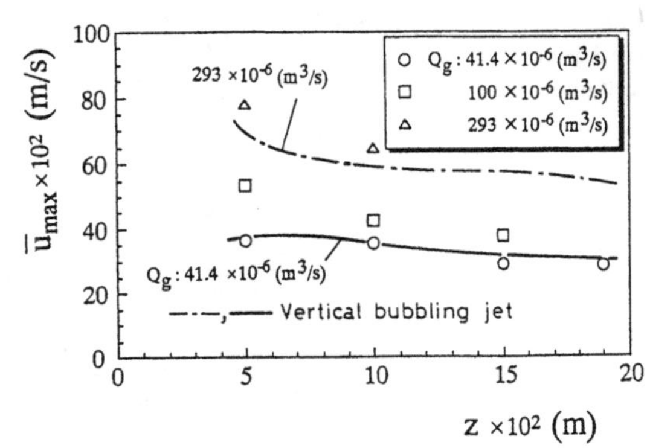


**Fig. 3.25** Maximum value of axial mean velocity

bubble frequency $f_B$ and gas holdup $\alpha$. Thus the similarity distribution of $\bar{u}$ can be expressed by:

$$\bar{u}/\bar{u}_{\max} = 1.52 \exp(-1.1\eta_u'^2) \tan h(2.64\eta_u'), \tag{3.16}$$

$$\eta_u' = \eta/\eta_{u,\max/2}. \tag{3.17}$$

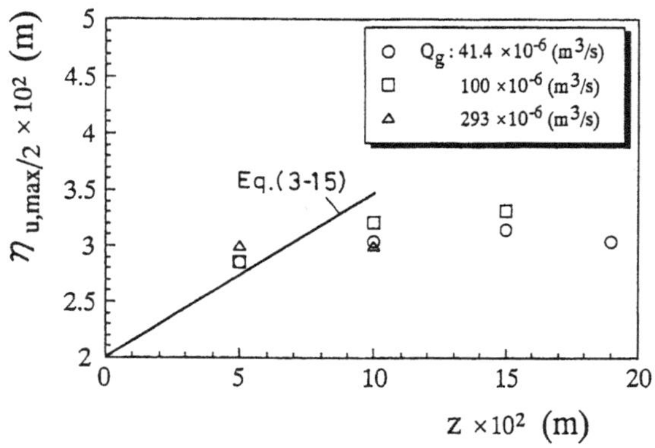


**Fig. 3.26** Characteristic length $\eta_{u,\max/2}$

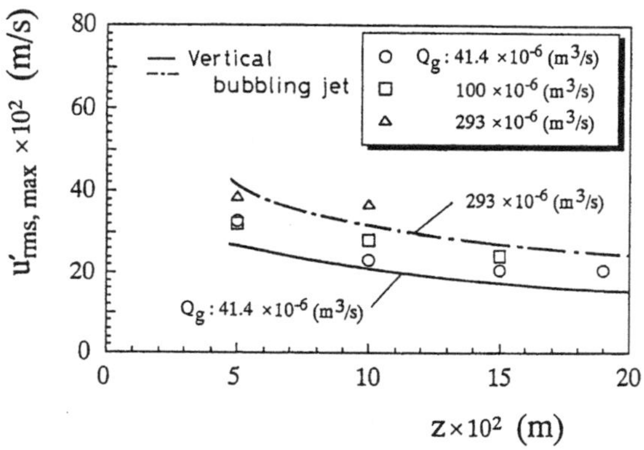


**Fig. 3.27** Maximum value of the rms value of axial turbulence component

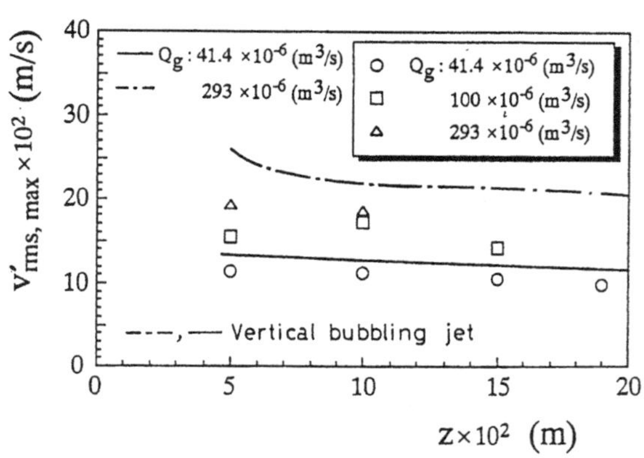


**Fig. 3.28** Maximum value of the rms value of radial turbulence component

Equation (3.16) is represented by the broken line in Fig. 3.30. A slight difference between the measurement and correlation observed at a larger $\eta_u'$ value is attributable to the existence of a recirculating flow outside the bubbling jet.

**Fig. 3.29** Characteristic lengths, $\eta_{u'\text{rms,max}/2}$ and $\eta_{v'\text{rms,max}/2}$

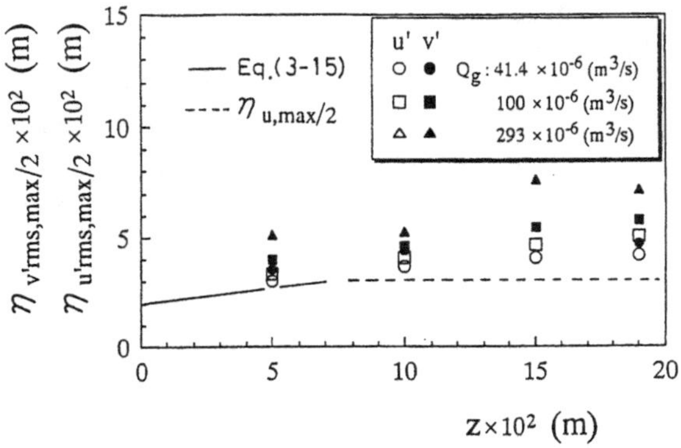


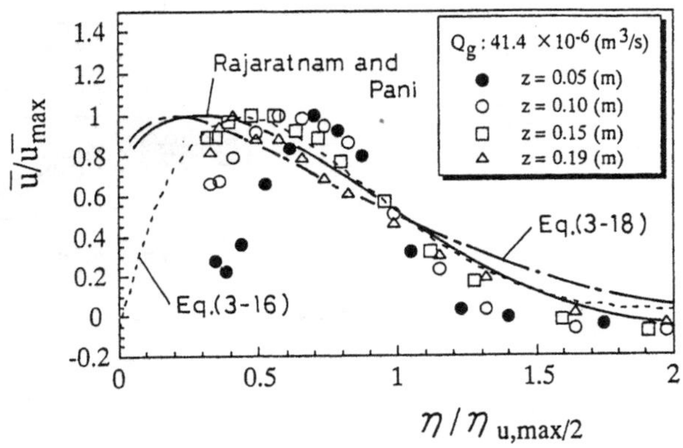


**Fig. 3.30** Horizontal distributions of axial mean velocity for $Q_g = 41.4 \times 10^{-6}\,\text{m}^3/\text{s}$

The dot-dash line in Fig. 3.30 represents the horizontal distribution of $\bar{u}/u_{\max}$ for a two-dimensional, single-phase wall jet [24, 25]. This distribution is expressed by

$$\bar{u}/\bar{u}_{\max} = 1.48\left(\eta_u{}'\right)^{1/7}\left\{1 - \text{erf}\left(0.68\,\eta_u{}'\right)\right\}, \tag{3.18}$$

where erf stands for the error function. Equation (3.18) is significantly different from the similarity distribution of (3.16).

The solid line in Fig. 3.30 represents $\bar{u}/u_{\max}$ values measured by Rajaratnam and Pani [26] for a three-dimensional single-phase wall jet. This curve seems to be closer to the measurement than (3.18), as expected. Nevertheless, there is still a noticeable difference between the two sets of data in Fig. 3.30.

Figure 3.31 shows that the horizontal distributions of $\bar{u}$ measured above the attachment position for $Q_g = 100 \times 10^{-6}$ and $293 \times 10^{-6}\,\text{m}^3/\text{s}$ are similar to that for $Q_g = 41.4 \times 10^{-6}\,\text{m}^3/\text{s}$.

The data for $u'_{\text{rms}}$ and $v'_{\text{rms}}$ also are nondimensionalized by $\bar{u}_{\max}$ and plotted against the nondimensionalized horizontal distance $\eta/\eta_{u,\max/2}$ in Figs. 3.32

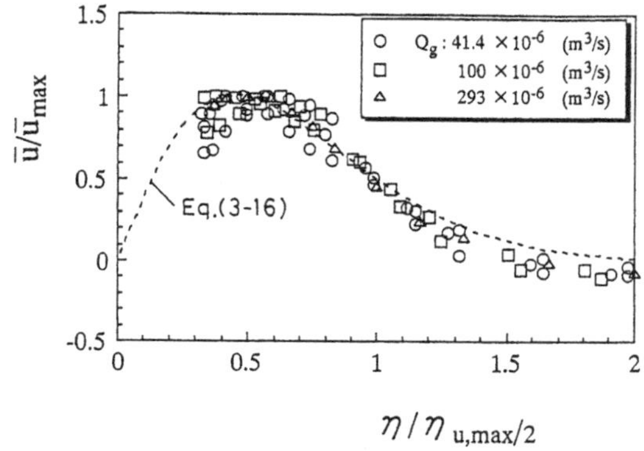


**Fig. 3.31** Similar distribution of $u$

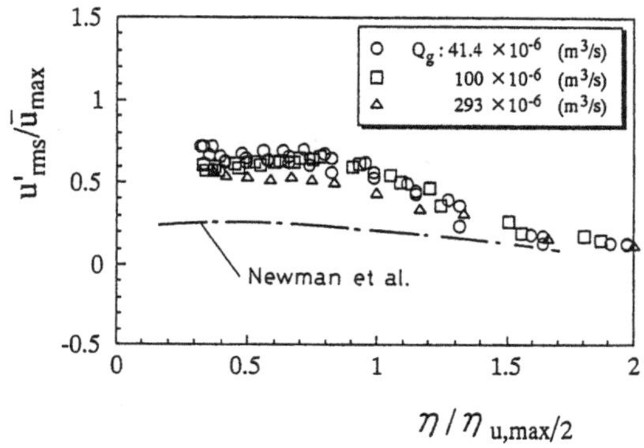


**Fig. 3.32** Similar distribution of $u'$rms

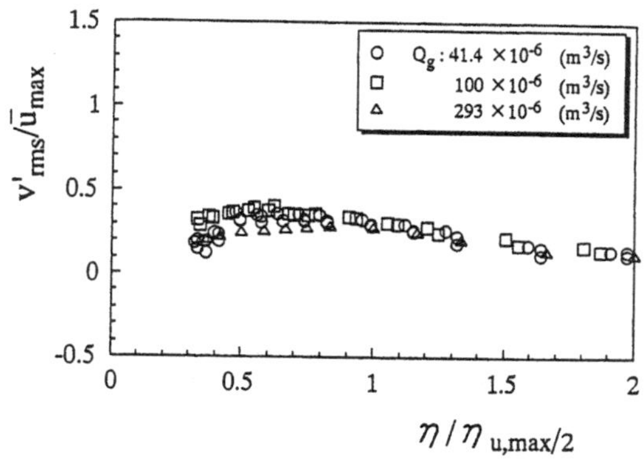


**Fig. 3.33** Similar distribution of $v'$rms

and 3.33, respectively. In each figure the measurement for the three different gas flow rates considered agree with one another. Consequently, the horizontal distributions of these turbulence quantities are similar at a vertical position above the attachment position. It is interesting to note that the values of $u'_{\text{rms}}/\bar{u}_{\max}$ in the bubbling wall jet is approximately twice those in a three-dimensional single-phase wall jet [27]. This difference is attributable to additional turbulence production in the wake of bubbles.

#### 3.2.2.3 Summary of Liquid Flow Characteristics

(a) A description is provided of the measured flow characteristics in an air–water bubbling jet, generated near the side wall of a cylindrical vessel and rising along the side wall through the Coanda effect. The measurement was carried out with a two-channel laser Doppler velocimeter. Above the point of attachment of the bubbling jet to the side wall, the horizontal distributions of the axial mean velocity $\bar{u}$ are similar irrespective of the gas flow rate $Q_g$. Accordingly, an empirical correlation of $\bar{u}$, (3.16), is established. The rms values of the axial and radial turbulence components, $u'_{rms}$ and $v'_{rms}$, also follow distributions.
(b) The maximum values of $\bar{u}$ and $u'_{rms}$ are not influenced by the side wall and are nearly the same as for a vertical bubbling jet. On the other hand, the maximum value of $v'_{rms}$ is lower than that for the vertical bubbling jet due to the confinement effect of the side wall.

## 3.3 Interaction Between Two Bubbling Jets

### *3.3.1 Critical Condition for Merging of Two Bubbling Jets*

The critical condition for a bubbling jet to attach to an adjacent vertical wall was represented by (3.1) as a function of gas flow rate and the distance between the nozzle exit and the side wall. In this section, we will derive a critical condition for two bubbling jets to merge with each other in a water bath.

#### 3.3.1.1 Experimental Apparatus and Procedure

A schematic of the experimental apparatus is shown in Fig. 3.34. The test vessel made of transparent acrylic resin has an inner diameter, $D$, of 0.4 m and a height, $H$, of 1.2 m. The depth of water, $H_L$, is 0.35, 0.50, or 0.65 m. The inner diameter of two top lance nozzles, $d_{ni}$, is $2.0 \times 10^{-3}$ m. The nozzles are made of brass pipe, and the wettability of the nozzle by water is quite good. The horizontal distance between the two nozzles, $L_H$, is varied from 0.02 to 0.10 m at equal intervals of 0.02 m, and the vertical distance between them, $L_v$, is zero, that is, on the same horizontal plane.

The gas flow rates issuing out of the nozzles, $Q_{g1}$ and $Q_{g2}$, are varied from $0.50 \times 10^{-6}$ to $500 \times 10^{-6}$ m$^3$/s. The behavior of the two bubbling jets is then observed by eye inspection. For convenience, interacting patterns of the two bubbling jets are classified into the following four categories.

(a) Jets do not merge and rise straight upward independently.
(b) Jets pull each other, but never merge.
(c) Jets repeat merging (attachment) and detachment.
(d) Jets merge with each other completely and become a larger-scale single bubbling jet.

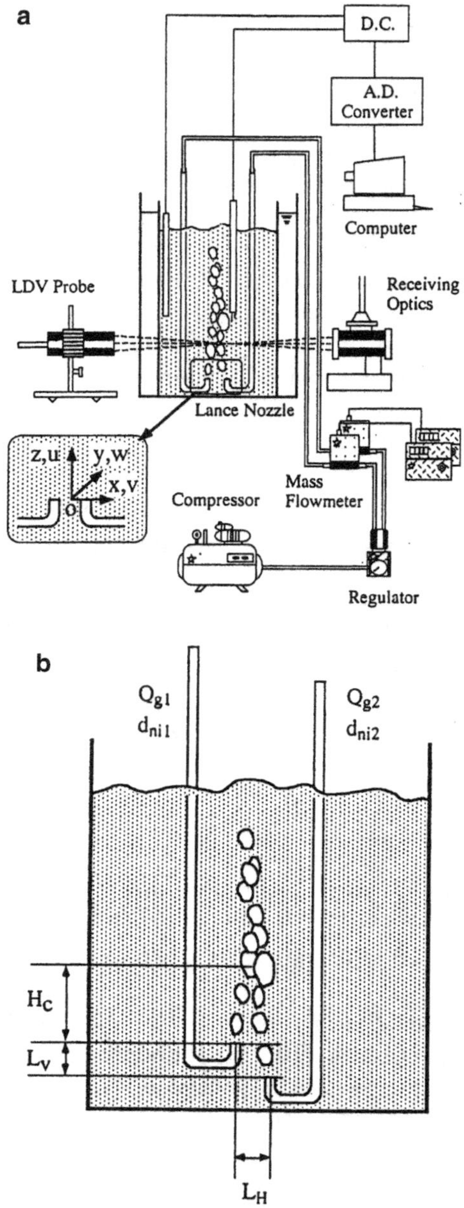


**Fig. 3.34** Experimental apparatus and definition of merging distance: (**a**) experimental apparatus (**b**) definition of merging distance

First, a critical condition describing no merging between the two bubbling jets is determined for $Q_{g1} = Q_{g2}$. That is, the boundary between (b) and (c) is determined. Second, the critical condition is determined for unequal gas flow rates.

#### 3.3.1.2 Experimental Results

Critical Condition for $Q_{g1} = Q_{g2}$

A critical condition for two bubbling jets without merging is determined for equal gas flow rates, $Q_{g1} = Q_{g2}$. The experiments are carried out by changing the bath depth, $H_L$, the horizontal distance between two nozzles, $L_H$, and the gas flow rate, $Q_{g1}(= Q_{g2})$.

It is assumed that the merging between two bubbling jets is related to the extent of bubble dispersion region of each bubbling jet. The half-value radius of gas holdup $\alpha$ at $\alpha = 50\%$, designated by $b_{\alpha,50}$, is approximated by [15, 19]:

$$b_{\alpha,50} = 0.42(Q_g{}^2/g)^{1/5}, \tag{3.19}$$

where $Q_g$ is the gas flow rate. Accordingly, the radius of the bubble dispersion region for $Q_{g1}$ is approximated by $2b_{\alpha,50}$, and the following length scale may be introduced.

$$L_1 = (Q_{g1}{}^2/g)^{1/5}. \tag{3.20}$$

The dispersion radius, $L_1$, is not dependent on the inertia force of injected gas, but on the buoyancy force of the bubbling jet. $L_1$ is used as a characteristic length to correlate the condition without merging of the jets. A critical value of $L_H$ is nondimensionalized by $L_1$ and plotted in Fig. 3.35. The solid line is obtained from the relation:

$$L_1/L_H = 0.098. \tag{3.21}$$

The critical value of $L_H$ is predicted by (3.21) to within a scatter of ±15 pct.

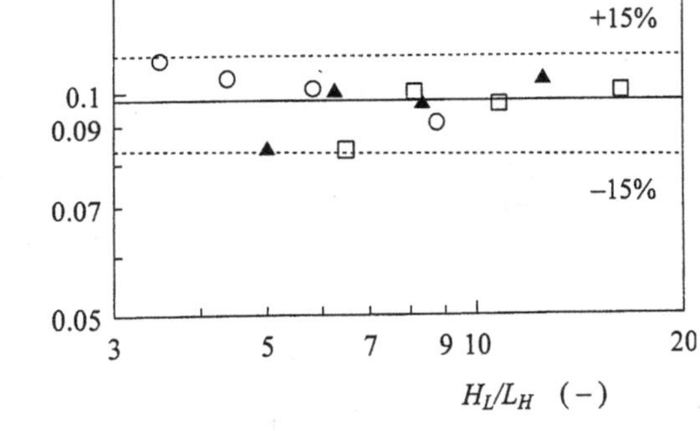


**Fig. 3.35** Nozzle distance without merging of two bubbling jets for the same gas flow rates

This result indicates that the critical condition for merging of two bubbling jets is dependent only on the dispersion radius of the bubbling jet, i.e., only on the gas flow rate.

### Critical Condition for $Q_{g1} \neq Q_{g2}$

The critical condition for $Q_{g1} \neq Q_{g2}$ is determined in the same manner as in Sect. 3.3.1.2. The specific length, $L$, is defined by the following relations:

$$L = (L_1 + L_2)/2, \tag{3.22}$$

$$L_1 = (Q_{g1}{}^2/g)^{1/5}, \tag{3.23}$$

$$L_2 = (Q_{g2}{}^2/g)^{1/5}, \tag{3.24}$$

where $L_1$ and $L_2$ are the specific lengths for the two nozzles. Figure 3.36 shows a nondimensionalized specific length $(L_1 + L_2)/(2L_H)$ against $H_L/L_H$. All the critical $L_H$ values are correlated by the following equation within a scatter of ±30 pct.

$$(L_1 + L_2)/(2L_H) = 0.091. \tag{3.25}$$

The results imply that the critical condition is dependent on the dispersion radius of the two bubbling jets.

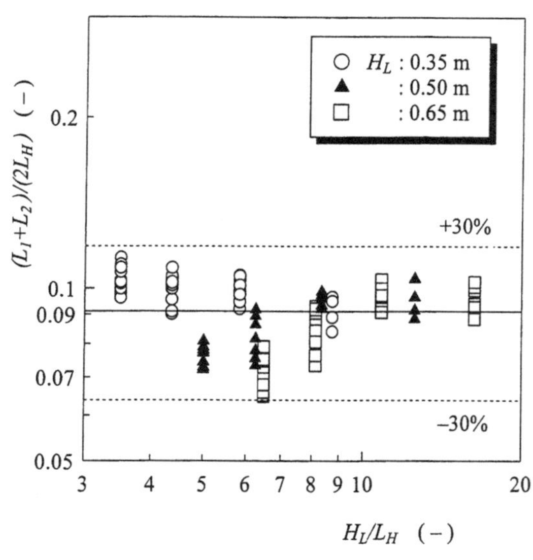


**Fig. 3.36** Nozzle distance without merging of two bubbling jets for different gas flow rates

### 3.3.2 *Merging Length of Two Bubbling Jets*

Multi-tuyere gas injection has been widely used to enhance the intensity of mixing, i.e., to shorten the mixing time in the bath of metallurgical reactors [11]. If bubbling jets generated through multi-tuyeres merge into a single jet just above the tuyeres, the enhancement of the mixing intensity may not be expected. On the other hand, when the bubbling jets merge near the bath surface of the reactors, the interaction between them sometimes causes intense wave motions on the bath surface which may lead to undesirable slopping and spitting [28, 29].

Considering these circumstances, the merging distance $H_c$, is described in this section, based on experimental data obtained from a water model.

#### 3.3.2.1 Experimental Apparatus and Procedure

Two top lances are used to inject air into a water bath contained in a cylindrical vessel (see Fig. 3.34). The vessel has an inner diameter $D$ of $30 \times 10^{-2}$ m and a height $H$ of $40 \times 10^{-2}$ m. Each top lance is bent at an angle of 180°, and its exit is contracted to form a single-hole nozzle. The height from the bottom of the vessel to the nozzle tip is $4.5 \times 10^{-2}$ m. The depth from the nozzle exit to the bath surface, $H_L$, is varied from $10 \times 10^{-2}$ to $30 \times 10^{-2}$ m with equal intervals of $10 \times 10^{-2}$ m. The inner diameter of the nozzle, $d_{ni}$, is $2.0 \times 10^{-3}$ or $4.0 \times 10^{-3}$ m. The gas flow rate $Q_g$ varies from $10 \times 10^{-6}$ to $100 \times 10^{-6}$ m$^3$/s. The horizontal distance between the two nozzles, $L_H$, varies over a wide range. Photographs of the two bubbling jets subjected to the Coanda effect are then taken with a still camera.

#### 3.3.2.2 Experimental Results

Effect of Gas Flow Rate $Q_g$ on Merging Distance

Two typical photographs of interacting bubbling jets are shown in Fig. 3.37. Figure 3.37a shows a photograph for $Q_{g1} = Q_{g2} = 40 \times 10^{-6}$ m$^3$/s. The other experimental conditions are given in the caption. It is evident that the two jets merge into one around $z = 25 \times 10^{-2}$ m where $z$ is the vertical distance from the nozzle exit.

Figure 3.38 shows that measured values of $H_c$ obtained for $Q_{g1} = Q_{g2}$ can be approximated by the following empirical relation within a scatter of ±30% irrespective of the gas flow rate:

$$H_c = 6.2 L_H. \tag{3.26}$$

This relation is valid for the following modified Froude number.

$$Fr_m = \rho_g {Q_g}^2/(\rho_L g {d_{ni}}^5). \tag{3.27}$$

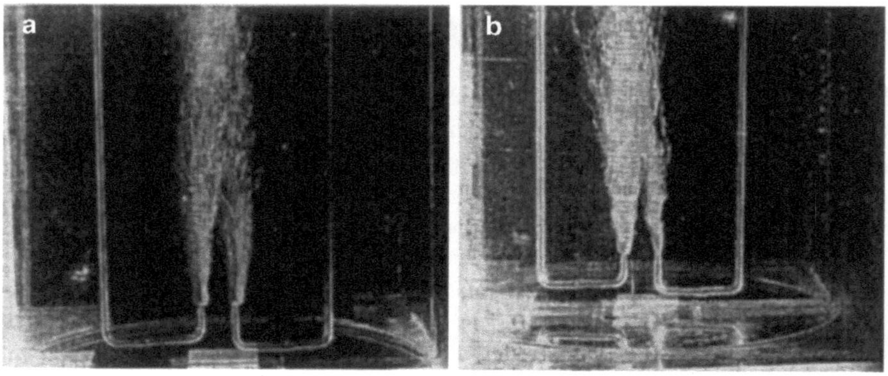


**Fig. 3.37** Photographs of merging bubbling jets. (**a**) $Q_{g1} = Q_{g2} = 40 \times 10^{-6}\,\mathrm{m^3/s}$, $H_L = 30 \times 10^{-2}\,\mathrm{m}$, $d_{ni} = 2.0 \times 10^{-3}\,\mathrm{m}$, $L_H = 3.44 \times 10^{-2}\,\mathrm{m}$, $H_c = 24.6 \times 10^{-2}\,\mathrm{m}$. (**b**) $Q_{g1} = 40 \times 10^{-6}\,\mathrm{m^3/s}$, $Q_{g2} = 10 \times 10^{-6}\,\mathrm{m^3/s}$, $H_L = 30 \times 10^{-2}\,\mathrm{m}$, $d_{ni} = 2.0 \times 10^{-3}\,\mathrm{m}$, $L_H = 2.83 \times 10^{-2}\,\mathrm{m}$, $H_c = 15.9 \times 10^{-2}\,\mathrm{m}$

**Fig. 3.38** Merging distance for $Q_{g1} = Q_{g2}$, $H_L = 30 \times 10^{-2}\,\mathrm{m}$, and $d_{ni} = 2.0 \times 10^{-3}\,\mathrm{m}$

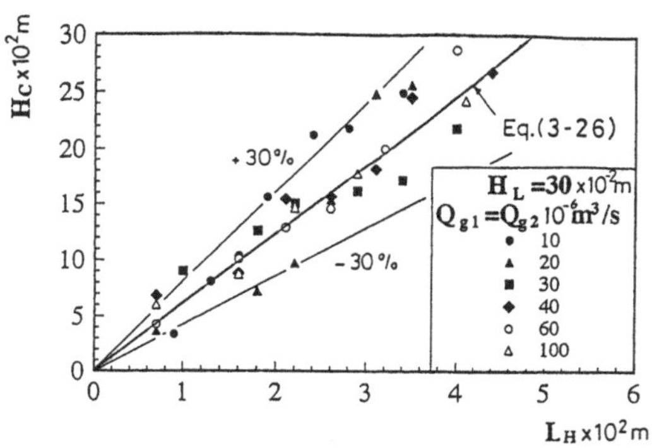


Over the range $0.4 < Fr_m < 40$, where $\rho_L$ and $\rho_g$ are the densities of liquid and gas, respectively, and $g$ is the acceleration due to gravity.

The gas flow rate through the left nozzle, $Q_{g1}$, is kept at $40 \times 10^{-6}\,\mathrm{m^3/s}$, while that through the right nozzle, $Q_{g2}$, is varied from $10 \times 10^{-6}$ to $100 \times 10^{-6}\,\mathrm{m^3/s}$. Figure 3.37b shows a photograph of the merging bubbling jets of different gas flow rates. The jet of the lower gas flow rate is pulled towards that of higher gas flow rate. It is, however, interesting to note that the merging distance, $H_c$, can be approximated by (3.26), as seen in Fig. 3.39.

### Effect of Bath Depth $H_L$ on Merging Distance

The measured values of $H_c$ for different bath depths are also approximated by (3.26), as partly demonstrated in Fig. 3.40. Consequently, the bath depth $H_L$ hardly affects the merging length under the experimental conditions considered. The work done per unit time by the buoyancy force acting on bubbles to liquid around a bubbling jet, $P_B$, is expressed by [30, 31]:

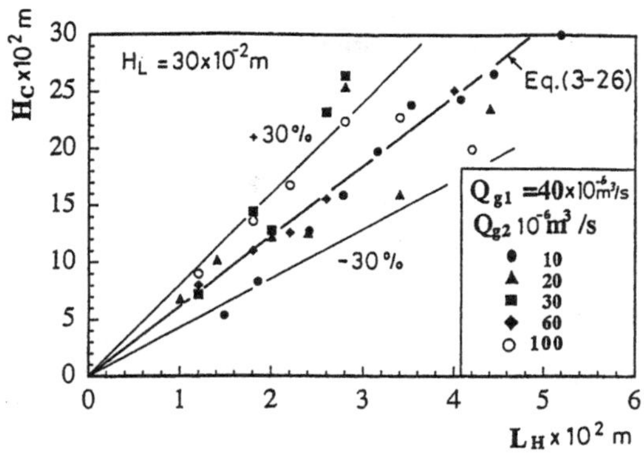


**Fig. 3.39** Merging distance for $Q_{g1} = Q_{g2} = 40 \times 10^{-6}\,\mathrm{m^3/s}$, $H_L = 30 \times 10^{-2}$ m, and $d_{ni} = 2.0 \times 10^{-3}$ m

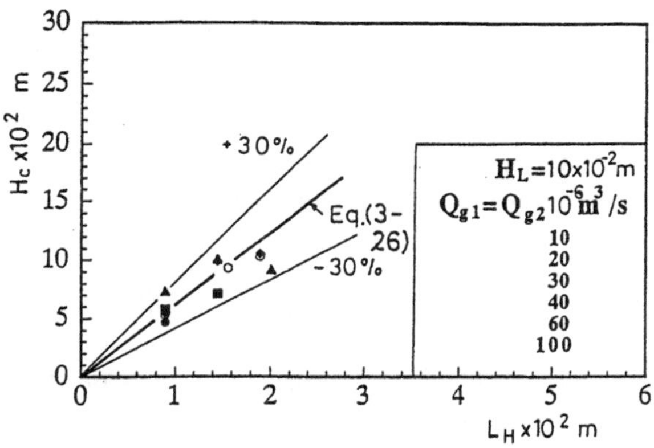


**Fig. 3.40** Merging distance for $Q_{g1} = Q_{g2}$, $H_L = 10 \times 10^{-2}$ m, and $d_{ni} = 2.0 \times 10^{-3}$ m

$$P_B = (\rho_L - \rho_g) g Q_g H_L. \tag{3.28}$$

The experimental results reveal that the buoyancy force does not have much effect on the merging length $H_c$ because $Q_g$ and $H_L$ are nearly independent of $H_c$.

### Effect of Nozzle Inner Diameter $d_{ni}$ on the Merging Distance

Figure 3.41 shows that the measured values of $H_c$ obtained for two nozzles of the same inner diameter $d_{ni} = 4 \times 10^{-3}$ m are also approximated by (3.26). The modified Froude number $Fr_m$ is varied from 0.6 to 1.2.

The work per unit time done by the inertia force of the injected gas, $P_I$, is expressed by [30]:

$$P_I = (1/2)\rho_g u_n{}^2 Q_g, \tag{3.29}$$

$$u_n = Q_g/(\pi d_{ni}{}^2/4), \tag{3.30}$$

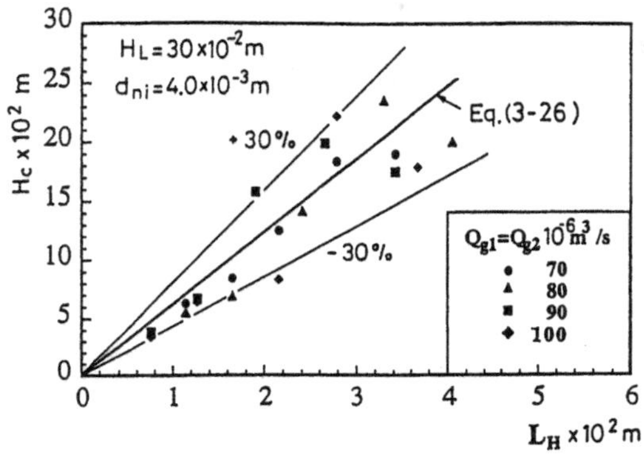


**Fig. 3.41** Merging distance for $Q_{g1} = Q_{g2}$, $H_L = 30 \times 10^{-2}$ m, and $d_{ni} = 4.0 \times 10^{-3}$ m

where $u_n$ is the gas velocity at the nozzle exit. Substitution of (3.30) into (3.29) yields

$$P_I = (1/2)\rho_g Q_g{}^3/(\pi d_{ni}{}^2/4)^2. \tag{3.31}$$

Equation (3.31) and Fig. 3.41 collectively indicate that the inertia force of the injected gas has little influence on the merging distance.

## Mechanism of Interaction of Two Bubbling Jets

An air–water bubbling jet which is not subjected to the Coanda effect is known to rise straight upward while entraining the surrounding water into it [21]. As a result, the horizontal region in which water moves vertically upward spreads as $z$ increases. The extent of this horizontal region can be represented, for example, by the half-value radius, $b_u$, of the horizontal distribution of the axial mean velocity of water flow, $u$ (see Fig. 3.42). Based on existing experimental study, [21] $b_u$ can be approximated by

$$b_u = 0.14z, \tag{3.14}$$

where $z$ is the vertical distance from the nozzle exit. Equation (3.14) states that the half-value radius $b_u$ is independent of the gas flow rate $Q_g$ and the inner diameter of the nozzle, $d_{ni}$.

It can be assumed that the two air–water bubbling jets, shown in Fig. 3.42, begin to interact and pull each other when the above-mentioned upward moving water regions of the jets overlap beyond a certain critical value. This situation can be described by

$$k(2b_u) = L_H \tag{3.32}$$

where $k$ is a constant characterizing the overlapping.

Substituting $z = H_c$ into (3.14) and combining the resulting equation with (3.32) yields,

$$H_c = L_H/(0.28k). \tag{3.33}$$

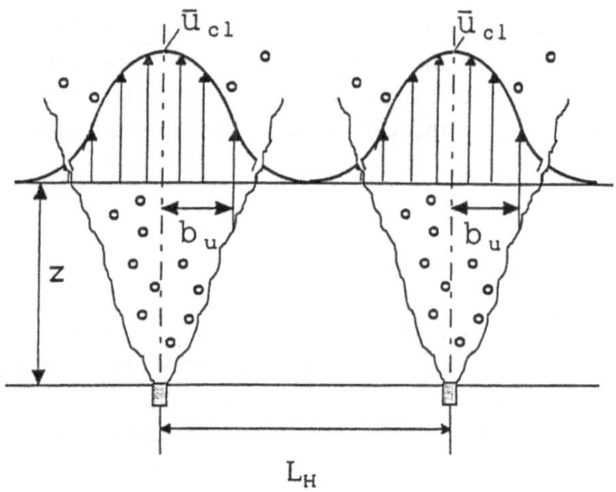


**Fig. 3.42** Schematic illustration of two bubbling jets generated a short distance off

This functional relationship is the same as (3.26). A comparison of (3.33) with (3.26) yields $k = 0.58$. Consequently, the above assumption about the interaction between two bubbling jets seems reasonable.

Interaction of Two Bubbling Jets in Molten Metal Bath

It has been demonstrated that the half-value radius $b_u$ of molten Wood's metal flow induced by He gas injection through a single-hole bottom nozzle can be satisfactorily predicted by (3.14) [32] Accordingly, (3.26) may also be applicable in predicting the interaction between two bubbling jets in a molten metal bath.

#### 3.3.2.3 Summary

The interaction between two air–water bubbling jets generated through two single-hole nozzles in a water bath has been discussed. The two bubbling jets merge into one at a certain distance, $H_c$, from the nozzle exit. This distance is approximated by (3.26) within a scatter of $\pm 30\%$ regardless of the gas flow rate and the inner diameter of the nozzle [33].

### *3.3.3 Bubble Characteristics*

In steel refining processes, gas is often injected into the bath through multi-nozzles or multi-tuyers in order to enhance mixing, chemical reaction rates, and inclusion removal efficiently [34, 35]. If the nozzles are placed adjacent to one another, bubbling jets thus generated would interact and then merge into a larger scale one as discussed in Sect. 3.3.2.3. In this situation, the purpose of multi-nozzle gas injection may not be realized. The bubble characteristics in merging bubbling jets could be described in this section to provide information that may be useful for the design of nozzle configuration [36].

#### 3.3.3.1 Experimental Apparatus and Procedure

Figure 3.34 shows a schematic of the experimental apparatus. The test vessel, made of transparent acrylic resin, has an inner diameter $D$ of $300 \times 10^{-3}$ m and a height $H$ of $400 \times 10^{-3}$ m. Two J-shaped lance nozzles of the same inner diameter of $2 \times 10^{-3}$ m are so placed in the bath that the nozzle exits are on the same horizontal plane. The two lances are made from transparent glass pipe. The midpoint of a line connecting the centers of the two nozzle exits is placed on the vertical axis of the vessel. The origin of the Cartesian coordinates $(x, y, z)$ is located at the midpoint. The distance from the bottom of the bath to the nozzle tips $H_n$ is $50 \times 10^{-3}$ m and the distance from the nozzle exits to the bath surface $H_L$ is $300 \times 10^{-3}$ m. By referring to the results of the merging length in Sect. 3.3.2.2, the distance between the two nozzle exits, $L_H$, is chosen to be $20 \times 10^{-3}$ m.

The air flow rate supplied with a compressor is adjusted with a regulator and a mass flow controller. The gas flow rates at the exits of the left nozzle, $Q_{g1}$, and the right nozzle, $Q_{g2}$, are the same at $40 \times 10^{-6}\,\mathrm{m^3/s}$.

In order to compare the experimental results for dual nozzle gas injection system, the two nozzles are removed, and subsequently gas is injected through a single-hole nozzle of inner diameter $d_{ni}$ of $2.0 \times 10^{-3}$ m. The center of the nozzle exit is placed at the origin of the coordinate system. The air flow rate is set to $80 \times 10^{-6}\,\mathrm{m^3/s}$, being equal to the sum of $Q_{g1}$ and $Q_{g2}$. The center of the nozzle is placed at the vertical axis of the vessel. Bubble characteristics represented by gas holdup $\alpha$, bubble frequency $f_B$, mean bubble rising velocity $u_B$, and mean bubble chord length $L_B$ are measured by making use of a two-needle electroresistivity probe [23, 37].

#### 3.3.3.2 Experimental Results

Merging of Two Bubbling Jets

The two bubbling jets pull each other and then merge into single jet at around $z =$ 15 cm, as shown in Fig. 3.37. The merging length $H_c$ is given by,

$$H_c = 6.2L_H, \tag{3.26}$$

where $L_H$ is the distance between the two nozzle exits. This equation gives $H_c$ of 12.4 cm, which is close to the measured merging length for the two bubbling jets. The resulting larger scale single jet is found to be axisymmetrical for $z > H_c$.

Axial Distribution of Bubble Characteristics

1. Bubble frequency on the centerline, $f_{B,cl}$

Figure 3.43 shows measured values of bubble frequency, $f_{B,cl}$, on the centerline of the bath against the axial distance from the nozzle exit, $z$, where the subscript $cl$ denotes a value on the centerline of the vessel. The bubble frequency is defined as the number of bubbles passing through a point of measurement in 1 s. The frequency $f_{B,cl}$ is much smaller for the dual nozzle gas injection than for the single nozzle gas injection when $z$ is very small because the origin of the Cartesian coordinates is placed at the midpoint of the line connecting the two nozzle exits. Thus no bubble source exists near the origin for the dual nozzle gas injection.

The solid line represents an empirical relation of $f_{B,cl}$ for $Q_g = 80 \times 10^{-6}\,\mathrm{m^3/s}$ proposed by Iguchi et al. [20] for single nozzle gas injection. The broken line is calculated from the same relation by assuming that bubbles do not merge into larger bubbles, although two bubbling jets merge into one. In other words, the bubble frequency in the larger scale bubbling jet is assumed to be two times as large as that calculated from the empirical equation for $Q_g = 40 \times 10^{-6}\,\mathrm{m^3/s}$. The measured values of $f_{B,cl}$ for the single nozzle gas injection, represented by solid symbols, are satisfactorily predicted by the empirical relation, indicating that the experimental accuracy is quite satisfactory.

The chain line in Fig. 3.43 denotes the merging distance from the nozzle exit, calculated from (3.26). For $z > H_c$ the measured values of $f_{B,cl}$ agree well with the broken line, and accordingly, coalescence and breakup of bubbles hardly take place in the merged bubbling jet.

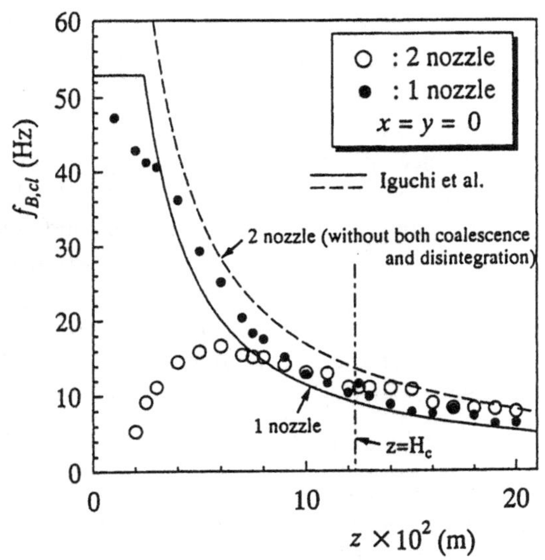


**Fig. 3.43** Axial distribution of bubble frequency

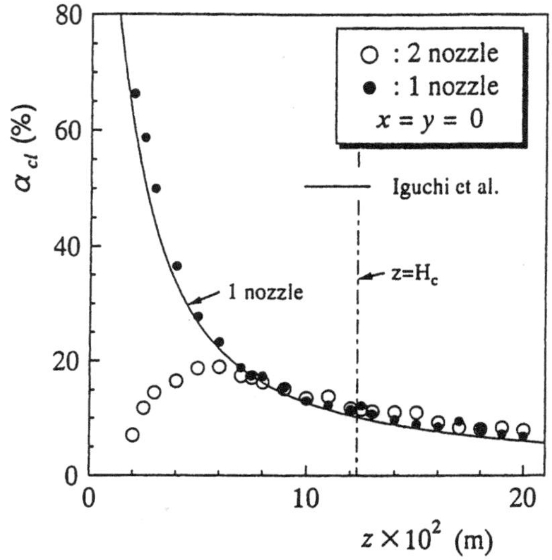


**Fig. 3.44** Axial distribution of gas holdup

2. Gas holdup on the centerline, $\alpha_{\text{cl}}$

The relationship between gas holdup on the centerline, $\alpha_{\text{cl}}$, and the axial distance, $z$, is given in Fig. 3.44. The gas holdup is defined as the time-averaged volumetric fraction of gas in a two phase mixture. In practice, it is determined by dividing the sum of bubble signal periods by total measurement duration. The agreement between the measured and predicted values for the single nozzle gas injection is good. It is interesting to note that the measured values of $\alpha_{\text{cl}}$ for the dual nozzle gas injection are in good agreement with those for the single nozzle gas injection in the axial region with $z > 7 \times 10^{-2}$ m.

3. Mean bubble rising velocity on the centerline, $\bar{u}_{\text{B,cl}}$

Figure 3.45 shows measured values of $\bar{u}_{\text{B,cl}}$ against $z$. Detailed discussion on the distribution of $\bar{u}_{\text{B,cl}}$ near the nozzle cannot be made because of the relatively large scatter in the data points. However, for $z > 5.0 \times 10^{-3}$ m, the values of $\bar{u}_{\text{B,cl}}$ are not influenced by the bubble generating devices and remain unchanged in the axial direction. This result implies that the flow field for $z > 5.0 \times 10^{-2}$ m is governed mainly by the buoyancy force acting on bubbles in the bubbling jet.

An empirical equation for $\bar{u}_{\text{B,cl}}$ proposed by Castello-Branco et al. [38] can predict the measured values except for the data obtained near the nozzle exits. The difference between the measurement and the empirical relation seems to become larger as $z$ increases. This difference is however less than 10%, which is within the applicable scattering range of ±30% inherent in the empirical equation.

4. Mean bubble chord length on the centerline, $\bar{L}_{\text{B,cl}}$

The relationship between $\bar{L}_{\text{B,cl}}$ and $z$ is shown in Fig. 3.46. In the middle gas flow rate regime [30, 39, 40], the mean diameter of bubbles generated with a single-hole nozzle is known to increase with an increase in the gas flow rate $Q_g$ [20]. The middle gas flow rate regime is defined as that in which bubble frequency is dependent on

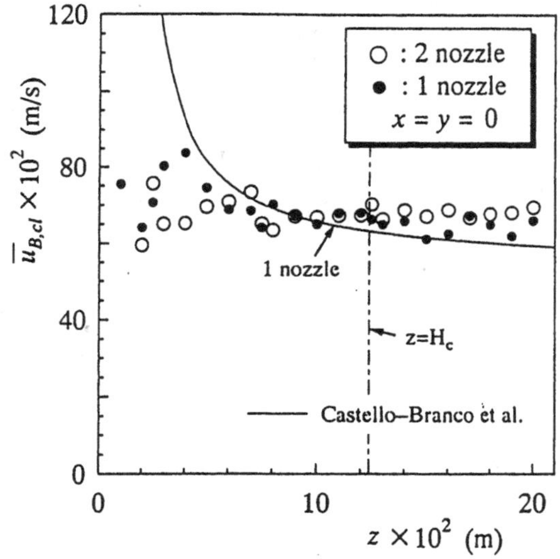


**Fig. 3.45** Axial distribution of mean bubble rising velocity

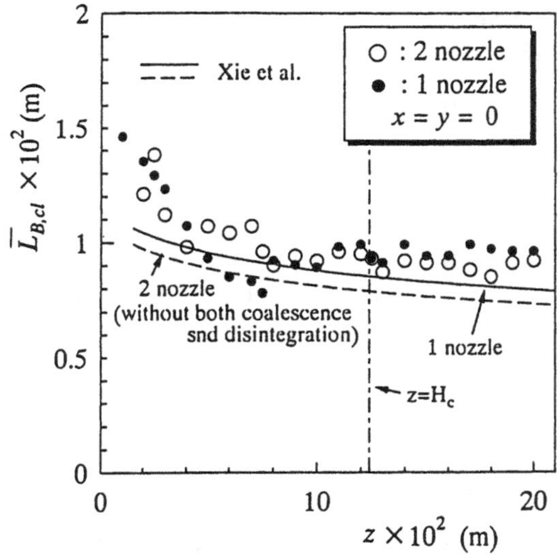


**Fig. 3.46** Axial distribution of mean bubble chord length

the inertia force of injected gas, drag force, buoyancy force, and the surface tension force acting on a generating bubble. As described on the data presented in Fig. 3.43, bubbles generated with the dual nozzles ascend without coalescence and breakup. Accordingly, in the axial region after the two bubbling jets merge ($z > H_c$) the mean bubble chord length $\bar{L}_{B,cl}$ should be greater for the single nozzle than for the dual nozzle. The data shown in Fig. 3.46 confirms this presumption.

Meanwhile, in an axial region of over approximately $z = 5.0 \times 10^{-2}$–$10.0 \times 10^{-2}$ m, $\bar{L}_{B,cl}$ is greater for the single nozzle gas injection than for the dual nozzle gas injection. This seems attributable to the flattening of bubbles generated with the single nozzle.

The solid and broken lines denote predictions based on an empirical equation for $\bar{L}_{B,cl}$ proposed by Xie et al. [40–42]. It is evident that the equation is useful for predicting $\bar{L}_{B,cl}$ under the experimental conditions described here.

Horizontal Distributions of Bubble Characteristics

1. Bubble frequency $f_B$

Figure 3.47 illustrates the distributions of bubble frequency in the $x$ and $y$ directions at $z = 15\times10^{-2}$ m ($> H_c$). The solid and broken lines denote predicted results from the empirical equation [20]. Under the experimental conditions considered, the experimental error inherent in the electroresistivity method is approximately $\pm10\%$ for the bubble frequency. Considering such a level of error, it can be concluded that the measured horizontal distributions of $f_B$ are equal in the two directions, suggesting that the merged bubbling jet is axisymmetrical. In addition, the measured value of $f_B$ near the centerline is larger for the dual nozzle gas injection than for the single nozzle injection, but there is no difference in the outer region of the bubbling jets. The empirical relation can satisfactorily approximate the horizontal distributions of $f_B$ for the two types of gas injection systems.

2. Gas holdup

The distributions of gas holdup $\alpha$ in the $x$ and $y$ directions at $z = 15.0 \times 10^{-2}$ m are shown in Fig. 3.48 The experimental error is also considered to be $\pm10\%$. The horizontal distribution of the gas holdup in the bubbling jet caused by the dual nozzle gas injection is the same as that in the bubbling jet caused by the single nozzle gas injection. In addition, the empirical equation proposed by Iguchi et al. [20] is found to be a good approximation for the gas holdup.

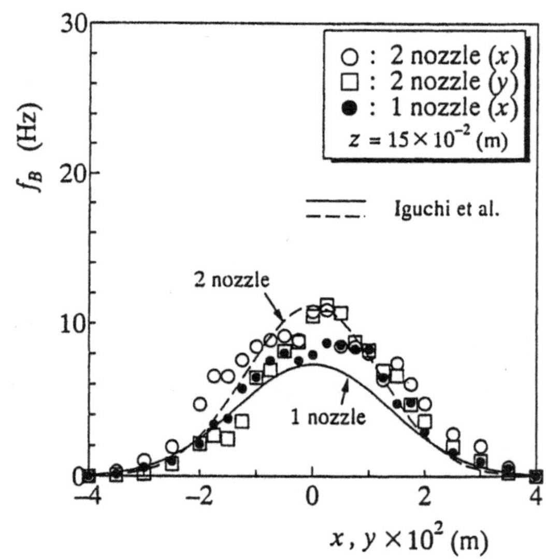


**Fig. 3.47** Horizontal distribution of bubble frequency

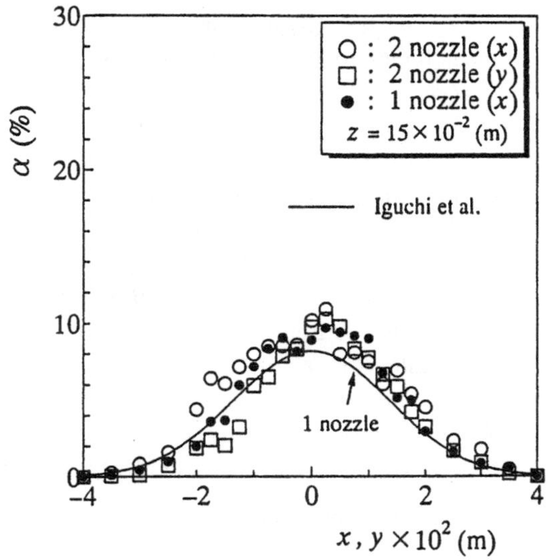


**Fig. 3.48** Horizontal distribution of gas holdup

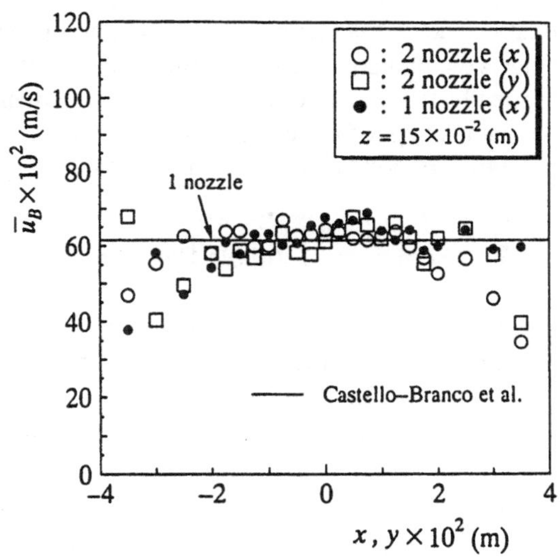


**Fig. 3.49** Horizontal distribution of mean bubble rising velocity

3. Mean bubble rising velocity $\bar{u}_B$

Figure 3.49 shows the horizontal distributions of mean bubble rising velocity $\bar{u}_B$ in the $x$ and $y$ directions for $z = 15.0 \times 10^{-2}$ m. Just like the axial distributions, the horizontal distributions of $\bar{u}_B$ is not influenced by nozzle configuration and is satisfactorily predicted by the empirical equation proposed by Castello-Branco et al. [38] mainly in the central part of each bubbling jet.

As the bath is very shallow, the effect of static pressure change on the expansion of ascending bubbles is negligible. Thus, the total gas flow rate in the axial region above the merging position of two bubbling jets can be expressed by

$Q_g = Q_{g1} + Q_{g2}$. The gas flow rate $Q_g$ is expressed in terms of the gas holdup $\alpha$ and mean bubble rising velocity $u_B$ as follows:

$$Q_g = \int 2\pi r \alpha u_B \, dr/100, \tag{3.34}$$

where, $R$ is the radius of the test vessel. In this experiment, the total gas flow rate for the dual nozzle gas injection is set to be the same as that for the single nozzle gas injection. The distributions of $\bar{u}_B$ for the two types of gas injection systems, therefore, must be equal to each other because the $\alpha$ distributions are the same. This presumption is confirmed by the results in Fig. 3.49.

Meanwhile, the calculated values of $Q_g$ from (3.34) are $78 \times 10^{-6}$ and $62 \times 10^{-6}\,m^3/s$ in the $x$ and $y$ directions, respectively, for the dual nozzle gas injection. The $Q_g$ value is $73 \times 10^{-6}\,m^3/s$ for the single nozzle gas injection. Such a deviation from the $80\,cm^3/s$ is acceptable in this type of measurement.

4. Mean bubble chord length $\bar{L}_B$

The horizontal distributions of $\bar{L}_B$ for $z = 15.0 \times 10^{-2}$ m are shown in Fig. 3.50. In the same manner as in Fig. 3.46, $\bar{L}_B$ is larger for the single nozzle than for the dual nozzles. This fact implies that bubbles generated through dual nozzles hardly coalesce although the two bubbling jets merge into a larger scale single jet. Consequently, the total interfacial area between ascending bubbles and molten metal in the bath is larger for the dual nozzles and thus, chemical reaction rates between the bubbles and molten metal would be enhanced by increasing the number of nozzles.

### 3.3.3.3 Summary of Bubble Characteristics

The bubble characteristics are described for a system in which air is injected into a cylindrical bath through dual nozzles of the same inner diameter at the same gas

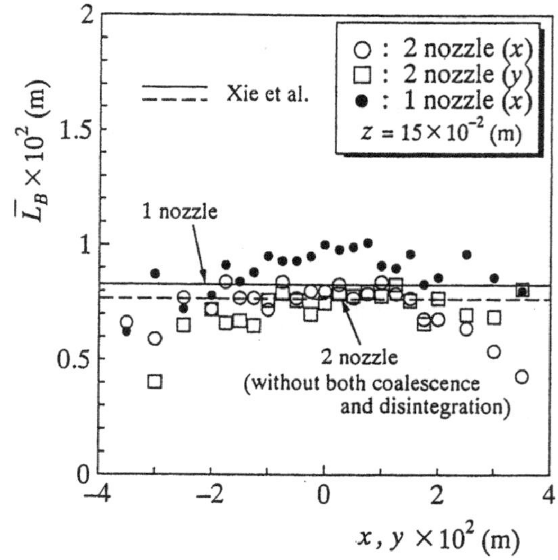


**Fig. 3.50** Horizontal distribution of mean bubble chord length

flow rates. The bubble characteristics in the bubbling jets thus generated are experimentally investigated. The results are then compared with those for a bubbling jet generated through a single nozzle at the same total gas flow rate. The main findings can be summarized as follows:

(a) The measured values of gas holdup $\alpha$ and mean bubble rising velocity $\bar{u}_B$ for the dual nozzle gas injection are in good agreement with their respective values for the single nozzle gas injection. On the other hand, the bubble frequency and mean bubble chord length differ from those for the single nozzle gas injection.
(b) The merged bubbling jet is axisymmetrical and its horizontal extent is nearly the same as that of a bubbling jet generated by single nozzle gas injection. The mean bubble diameter in the merged jet is smaller than that in the bubbling jet generated with the single nozzle, while the number of bubbles is larger in the merged jet. Consequently, a total interfacial area relevant to metallurgical reactions is greater in the merged bubbling jet.

## *3.3.4 Liquid Flow Characteristics*

The liquid flow characteristics specified, for example, by the mean velocity components, the rms values of the turbulence components, and the Reynolds shear stresses are nearly independent of the mean diameter of bubbles provided that the gas flow rate is the same [43]. The same will be shown to be true for the merged bubbling jet. That is, the liquid flow characteristics in the merged jet are not dependent on the bubble diameters. Therefore, if the merging distance, $H_c$, is much smaller than the bath depth, $H_L$, dual nozzle gas injections would not be useful for the enhancement of mixing in the baths.

### 3.3.4.1 Experimental Apparatus and Procedure

The experimental apparatus and conditions are the same as described in Sect. 3.3.2. The velocity components in the axial and horizontal directions are measured with a two channel laser Doppler velocimeter (LDV). Further details of the LDV are given elsewhere [21,23].

### 3.3.4.2 Experimental Results

Axial Distributions

1. Axial mean velocity on the centerline, $\bar{u}_{cl}$

Figure 3.51 shows the axial mean velocity on the centerline of the bath, $\bar{u}_{cl}$, against the axial distance $z$. As in the case of the bubble characteristics, the axial position $z = H_c$ is denoted by a broken line. The measured values of $\bar{u}_{cl}$, for the two gas

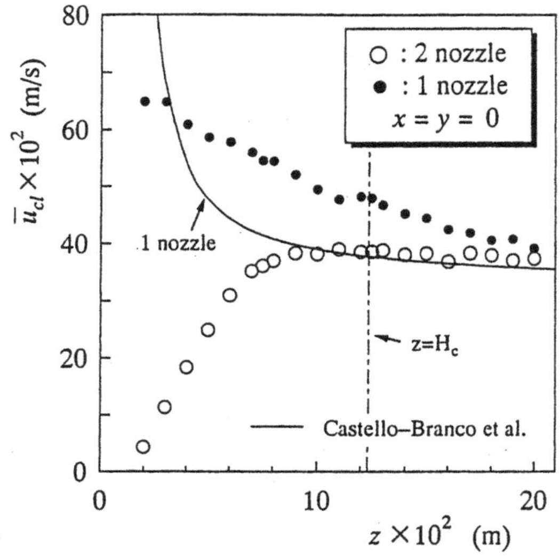


**Fig. 3.51** Axial distributions of axial mean velocity

injection systems approach each other as $z$ increases and approximate the empirical relation proposed by Castello-Branco et al. Because the liquid flow is driven mainly by the buoyancy force acting on bubbles, it is reasonable that the values of $\bar{u}_{\mathrm{cl}}$ for the two gas injection systems begin to agree with each other at a further axial position than those of gas holdup $\alpha_{\mathrm{cl}}$ or mean bubble rising velocity $\bar{u}_{\mathrm{B,cl}}$.

2. The root mean square of centerline turbulence components

Figures 3.52 and 3.53 show the horizontal distributions of the axial and horizontal turbulence components, $u'_{\mathrm{rms,cl}}$ and $v'_{\mathrm{rms,cl}}$, respectively. The $u'_{\mathrm{rms,cl}}$ for the dual nozzle gas injection becomes equal to that for the single nozzle gas injection at around $z = H_{\mathrm{c}}$. A solid line drawn in the figure is expressed by [44–46]

$$u'_{\mathrm{rms,cl}} = 0.5\bar{u}_{\mathrm{cl}}. \tag{3.35}$$

On the other hand, the $v'_{\mathrm{rms,cl}}$ for the two different gas injection systems are the same at a nearer axial position ($z = 7.0 \times 10^{-2}$ m) than that for $u'_{\mathrm{rms,cl}}$. Such a difference in the $u'_{\mathrm{rms,cl}}$ and $v'_{\mathrm{rms,cl}}$ distributions may be attributable to a difference in the diameter of bubbles.

Horizontal Distributions

1. Axial mean velocity

The distributions of $\bar{u}$ in the $x$ and $y$ directions at $z = 15.0 \times 10^{-2}$ m are shown in Fig. 3.54. A solid line denotes a Gaussian error curve expressed by

$$\bar{u} = \bar{u}_{\mathrm{cl}} \exp\left(-\ln 2x^2/b_u^2\right), \tag{3.36}$$

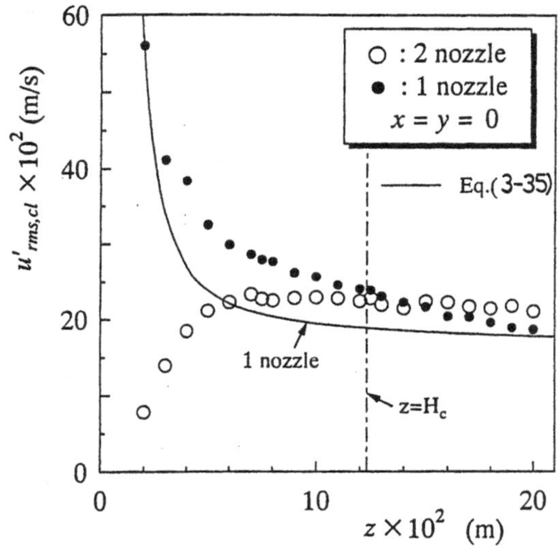


**Fig. 3.52** Axial distributions of the root mean square value of turbulence component in the $z$ direction

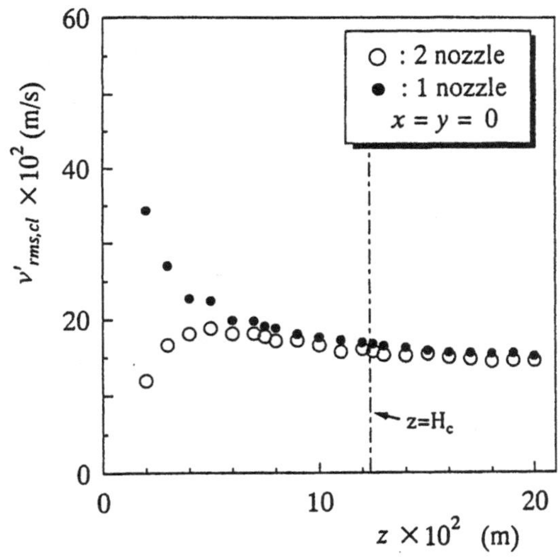


**Fig. 3.53** Axial distributions of the root mean square value of turbulence component in the $x$ direction

where $\bar{u}_{\mathrm{cl}}$ is calculated from the empirical relation by Castello-Branco et al. [38] and the half-value radius of the $u$ distribution, $b_u$, is given by [21],

$$b_u = 0.14z. \tag{3.14}$$

The $\bar{u}$ distribution in the $y$ direction can also be calculated from (3.36) by replacing $x$ by $y$. The measured value of $\bar{u}$ for the single nozzle gas injection is slightly larger than that for the dual nozzle gas injection, but the data for the two gas injection systems become equal at a farther axial position, although the evidence is not shown here.

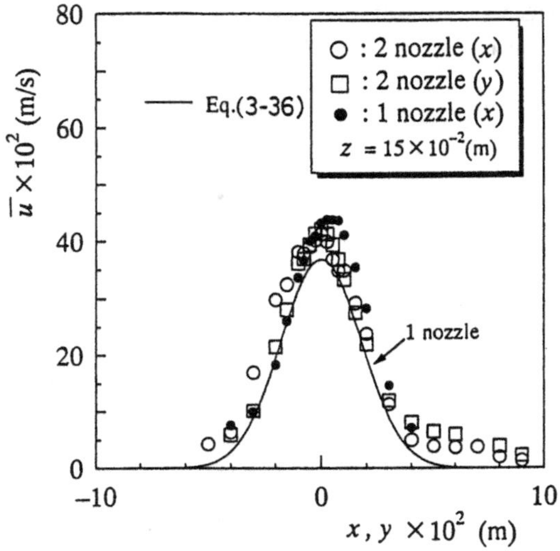


**Fig. 3.54** Horizontal distributions of axial mean velocity

It should be noted that the measured values of $\bar{u}$ do not follow the Gaussian error curve expressed by (3.36) in the outer region of the bubbling jet. This is because the vessel size is finite and a downward flow exists outside the jet. In a strict sense, (3.36) is valid for a bubbling jet in a very large vessel.

2. Root-mean-square turbulence components

Figure 3.55 shows the horizontal distributions of $u'_{\mathrm{rms}}$. The solid line denotes the following Gaussian error curve:

$$u'_{\mathrm{rms}} = u'_{\mathrm{rms,cl}} \exp\left[-\ln 2 \times x^2/{b_{\mathrm{t}}}^2\right], \tag{3.37}$$

$$u'_{\mathrm{rms,cl}} = 0.5\bar{u}_{\mathrm{cl}}, \tag{3.35}$$

$$b_{\mathrm{t}} = 1.8 b_u, \tag{3.38}$$

where the subscript cl represents the value on the centerline of the merged bubbling jet and $b_{\mathrm{t}}$ is the half-value radius.

The measured values for the two gas injection systems agree with each other everywhere in the bubbling jet. They are also predicted satisfactorily by (3.37) in the inner region of the bubbling jets.

Figure 3.56 shows the horizontal distributions of $v'_{\mathrm{rms}}$ together with those of $w'_{\mathrm{rms}}$ for $z = 15.0 \times 10^{-2}$ m. The measured values of $v'_{\mathrm{rms}}$ and $w'_{\mathrm{rms}}$ for the dual nozzle gas injection agree with each other, thus indicating that the merged bubbling jet is axisymmetrical with respect to the centerline of the bubbling jet. In addition, the two rms values are not influenced by the nozzle configuration.

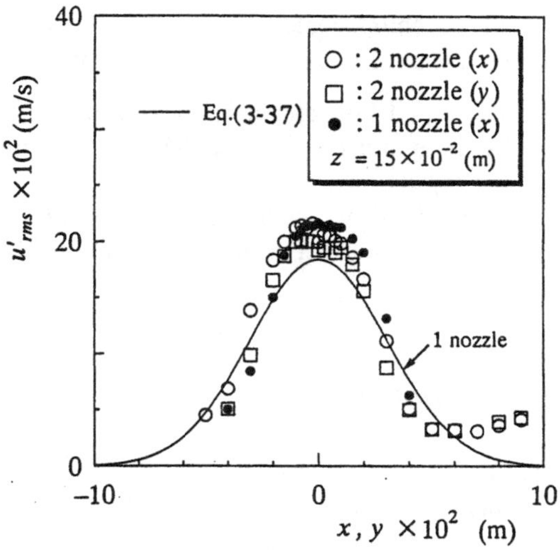


**Fig. 3.55** Horizontal distributions of the root mean square value of turbulence component in the $z$ direction

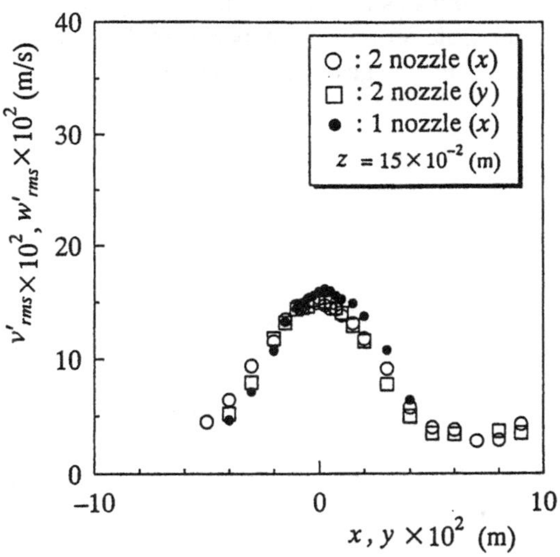


**Fig. 3.56** Horizontal distributions of the root mean square value of turbulence components in the $x$ and $y$ directions

### 3.3.4.3 Summary of Liquid Flow Characteristics

(a) In an axial region of $z > H_c$, the axial mean velocity $\bar{u}$ and the rms value of the axial turbulence component, $u'_{rms}$, are nearly independent of the nozzle configuration as long as two bubbling jets merge into a larger scale one.

On the other hand, the rms values of the horizontal turbulence component, $v'_{rms}$, for the two types of gas injection systems considered are the same at an axial position much nearer the nozzle ($z << H_c$). Such a difference in the axial and horizontal turbulence components cannot be explained from limited experimental results. Further experimental investigation is therefore desirable.

(b) The liquid flow characteristics in the merged bubbling jet are not influenced by the number of nozzles. This result implies that the effect of multi-nozzle gas injection cannot be expected as long as $H_c$ is much smaller than the bath depth $H_L$. However, metallurgical reaction rates would be enhanced because the interfacial area between bubbles and liquid is increased.

### *3.3.5 Mixing Time*

The mixing time in a bath agitated by dual nozzle gas injection is increased under several conditions, and the results are compared with those for a single nozzle gas injection. Representative data on the mixing time are presented in Figs. 3.57 and 3.58. There is little difference between the data for the two types of gas injection systems as long as two bubbling jets generated with dual nozzles merge into a larger scale jet. Further experimental investigations are needed to reveal the effect of multi-nozzle gas injection on the mixing time in real metallurgical processes.

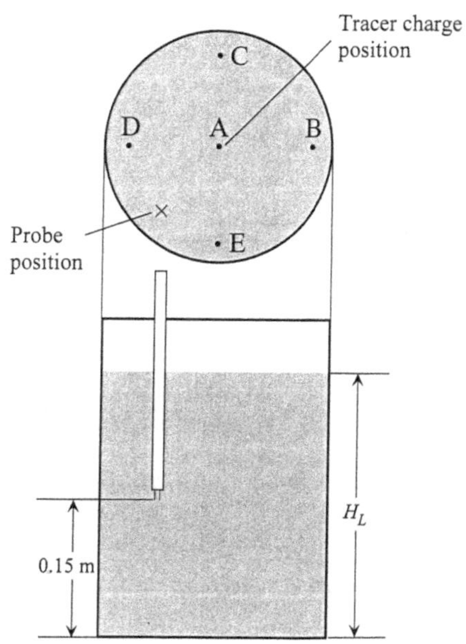


**Fig. 3.57** Tracer charge position and sensor position

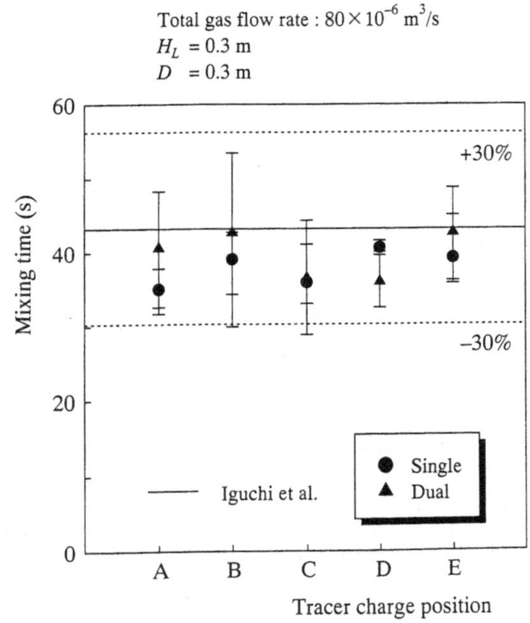


**Fig. 3.58** Mixing time data for single and dual nozzle gas injection systems

# References


1. Kirshner TM (1966) Fluid amplifiers. McGraw-Hill, New York, pp 118–124
2. Kadosh M (1964) Attachment of a jet to a curved wall, In: Proceedings of the fluid amplification symposium, 2nd edn, vol 4. Harry Diamond Laboratories, Washington DC, p 5
3. Japan Hydraulics and Pneumatics Society (1989) Handbook of hydraulics and pneumatics. Ohm Book Co Ltd, Tokyo, p 51,569
4. Bourque C, Newmann BG (1960) Reattachment of a two-dimensional incompressible jet to an adjacent flat plate. Aeronaut Quart 11:201
5. Sawyer RA (1963) Two-dimensional reattaching jet flows including the effects of curvature on entrainment. J Fluid Mech 17:481
6. Shimizu A (1983) Drag and wake modification of axisymmetric bluff bodies using coanda blowing, J Jpn Hydraulics Pneumatics Soc 14:488
7. Freund JB, Mungal MG (1994) Drag and wake modification of axisymmetric bluff bodies using coanda blowing, J Aircraft 31:572–578
8. Wille R, Fernholz (1965) Report on the first European mechanics colloquium, on the Coanda effect. J Fluid Mech 23:801–819
9. Szekeky J, Carlsson G, Helle L (1988) Ladle metallurgy. Springer, New York
10. Szekeky J, Ilegbusi OJ (1989).The physical and mathematical modeling of tundish operations. Springer, New York
11. ISIJ (1984) Recent developments in steelmaking technologies using stirring, 100th and 101th Nishiyama Memorial Lecture. ISIJ, Tokyo
12. Sahai Y, St Pierre GR (1992).Advances in transport processes in metallurgical systems. Elsevier, Amsterdam
13. Mazumdar D, Guthrie RIL (1995) The physical and mathematical modelling of gas stirred ladle systems. ISIJ Int 35:1–20
14. Iguchi M, Kawabata H, Nakajima K, Morita Z (1995) Measurement of bubble characteristics in a molten iron bath at 1600°C using an electroresistivity probe. Metall Mater Trans B 26B: 67–74

15. Tacke TH, Schubert HG, Weber DJ, Schwerdtfeger K (1985) Characteristics of round vertical gas bubble jets. Metal Trans B 16:263–275
16. Sano M, Makino H, Ozawa Y, Mori K (1986) Behavior of gas jet and plume in liquid metal. Trans ISIJ 26:298–304
17. Castillejos AH, Brimacombe JK (1989) Physical characteristics of gas jets injected vertically upward into liquid metal, Metal Trans B 20:595–601
18. Taniguchi S, Kikuchi A, Hatsuzaki H, Bessho N (1988) Dispersion of bubbles and gas-liquid mass transfer in a gas-stirred system. Trans ISIJ 28:262–270
19. Iguchi M, Nozawa K, Morita Z (1991) Bubble characteristics in the momentum region of air-water vertical bubbling jet. ISIJ Int 31:952–959
20. Iguchi M, Nozawa K, Tomida H, Morita Z (1992) Bubble characteristics in the buoyancy region of a vertical bubbling jet. ISIJ Int 32:747–754
21. Iguchi M, Ueda H, Uemura T (1995) Bubble and liquid flow characteristics in a vertical bubbling jet. Int J Multiphase Flow 21:861–873
22. Iguchi M, Sasaki K, Nakajima K, Kawabata H (1998) Coanda effect on bubble characteristics in a bubbling jet rising near the side wall of a cylindrical vessel. ISIJ Int 38(12):1297–1303
23. Iguchi M, Kondoh T, Uemura T (1994) Simultaneous measurement of liquid and bubble velocities in a cylindrical bath subject to centric bottom gas injection. Int J Multiphase Flow 20:753–762
24. Rajaratnam N (1976) Turbulent jet. Elsevier, Amsterdam
25. Verhoff A (1963) The two-dimensional turbulent wall jet without an external free stream, Rep. 626, Princeton University
26. Rajaratnam N, Pani BS (1974) Three dimensional turblent wall jets, Trans ASCE 100(HY1) 69
27. Newman BG, Patel RP, Savage SB, Tjio HK (1972) Three-dimensional wall jet originating from a circular office, Aeronaut Quart 23:189
28. Kato Y, Nakanishi K, Nozaki T, Suzuki K, Emi T (1982) Wave motion of metal bath in bottom blown converter. Tetsu-to-Hagane 68:1604–1612
29. Koch K, Roth C, Peter M (1996) Cold model investigations of fluid flow and mixing characteristics by bottom-blowing in discontinuous metallurgical reactors. ISIJ Int 36(Suppl): 50–53
30. Mori K, Sano M (1981) Process kinetics in injection metallurgy. Tetsu-to-Hagane 67:672–695
31. Mazumdar D, Guthrie RIL (1986) Mixing models for gas stirred metallurgical reactors. Metall Trans B 17B:725–733
32. Iguchi M, Tokunaga H, Tatemichi H (1997) Bubble and liquid flow characteristics in a wood's metal bath stirred by bottom helium gas injection. Metall Mat Trans B 28B:1053–1061
33. Iguchi M, Takahashi K, Kiuchi H (1999) Merging distance of two air-water vertical bubbling jets subjected to coanda effect. ISIJ Int 39:1311–1313
34. Umezawa K (1984) 100th and 101th Nishiyama Memorial Lecture, ISIJ, Tokyo
35. Refining forum for high-temperature process division of ISIJ and reaction group of 19th committee of JSPS (1996) Refining Limit for Impurities. ISIJ, Tokyo, p 151
36. Sasaki K, Iguchi M (1999) Bubble and liquid flow characteristics in interacting two water–air Vertical bubbling jets in a cylindrical vessel. Tetsu-to-Hagane 85:432–438
37. Iguchi M, Sasaki K (1999) Coanda effect on liquid flow characteristics in a bubbling jet rising near the side wall of a cylindrical vessel. ISIJ Int 39:213–218
38. Castello-Branco MASC, Schwerrtdfeger K (1994) Large-scale measurements of the physical characteristics of round vertical bubble plumes in liquids. Metall Mater Trans B 25B:359–371
39. Davidson L, Amick EH (1956) Formation of gas bubbles at horizontal orifices, AIChE J 2: 337–342
40. Xie Y, Orsten S, Oeters F (1992) Behaviour of bubbles at gas blowing into liquid wood's metal. ISIJ Int 32:66–75
41. Iguchi M, Tomida T, Nakajima K, Morita Z (1992) Cold model experiments on mass transfer from a solid body immersed in vertical bubbling jets. Tetsu-to-Hagane 78:1786
42. Iguchi M, Tomida T, Nakajima K, Morita Z (1993) Mass transfer from a solid body immersed in a cylindrical bath with bottom gas injection. ISIJ Int 33:728

43. Iguchi M, Okita K, Nakatani T, Kasai N (1997) Structure of turbulent round bubbling jet generated by premixed gas and liquid injection. Int J Multiphase Flow 23:249–262
44. Iguchi M, Takeuchi H, Morita Z (1990) The flow field in air-water vertical bubbling jets in a cylindrical vessel. Tetsu-to-Hagane 76:699–706
45. Iguchi M, Takeuchi H, Morita Z (1991) The flow field in air-water vertical bubbling jets in a cylindrical vessel. ISIJ Int 31:246–253
46. Sasaki K, Iguchi M (1999) Bubble and liquid flow characteristics in interacting two water–air vertical bubbling jets in a cylindrical vessel. Tetsu-to-Hagane 85:432–438

# Chapter 4
# Interfacial Phenomena

## 4.1 Single Bubble on Flat Plate

### *4.1.1 Overview*

Foaming in steel refining processes and in bath smelting reduction processes adversely impacts the quality and productivity of steel. Suppression of foaming is, therefore, one of the key processes in the steelmaking industry. Foaming is known to be caused through the effect of CO gas generation from decarburization and to be suppressed by adding coke or coal to the upper slag layer, decreasing the viscosity of the slag, or increasing the surface tension of the slag [1–5]. In addition, a low-frequency sonic wave is considered to be effective for suppression of slag foaming [6]. Based on X-ray fluoroscopic observations, Ogawa et al. [7] found that the attachment of CO bubbles to a solid body of poor wettability is extremely effective for suppressing foaming. The CO bubbles attached to the body become larger due to coalescence and are pulled to the bath surface as a result of the increased buoyancy force acting on the bubbles. This result suggests that the attachment of bubbles to solid bodies of poor wettability may be useful for removing fine bubbles contained in molten metals. For example, the method would be applicable to the removal of fine argon bubbles from continuous casting molds and thus aid the production of clean steel by suppressing bubble-induced defects such as pinhole and sliver.

Mizuno et al. [8] undertook fundamental investigations of this problem using a water model. A circular cylinder was coated with liquid paraffin to change its wettability. The advancing contact angle $\theta_a$ was 96°. When this cylinder was horizontally immersed in an air–water bubbling jet generated by bottom gas injection, small bubbles of a mean diameter less than 1 mm were attached to either the front surface or the lower surface of the cylinder. The bubbles moved toward the rear part of the cylinder where some coalesced into a larger single bubble near the rear stagnation point. When the volume of the coalesced bubble reached a certain critical value, it detached from the cylinder.

It is of practical importance to predict the shape and size of such a bubble just before detachment from the cylinder. If the cylinder is immersed in a still water bath, the shape and size of a bubble in contact with it can be accurately predicted

M. Iguchi and O.J. Ilegbusi, *Modeling Multiphase Materials Processes: Gas-Liquid Systems*, DOI 10.1007/978-1-4419-7479-2_4,

by the well-known Laplace method. However, this method is not convenient for consideration of the effect of liquid flow around the bubble. Mizuno et al. [8], therefore, focused on an energy equation for the bubble [9]. This method allows ready evaluation of the effects of flow and bubble coalescence on the shape and size of bubbles attaching to a solid body. The accuracy of the prediction will be discussed in this section by comparing the calculated shape and size of bubbles in a still bath with those calculated using the Laplace method. Hereafter, the energy equation method will be referred to simply as the potential method.

### *4.1.2 Experimental Apparatus and Procedure*

Figure 4.1 shows a schematic of the experimental apparatus used for observing the attachment of a bubble to the upper surface of a flat plate of poor wettability immersed in a still water bath. This system is referred to as a water–air system. The test vessel has a cross section of $250 \times 10^{-3}$ m by $100 \times 10^{-3}$ m and a height of $150 \times 10^{-3}$ m. The vessel was filled with water to a depth of $140 \times 10^{-3}$ m. A flat plate of poor wettability was placed horizontally in the bath. The distance from the upper surface of the plate to the bath surface was $40 \times 10^{-3}$ m. The $\theta_a$ was 96, 104, 116, or 131° for the water–air system. The contact angle of 104° was realized by coating an acrylic flat plate with liquid paraffin while 116 and 131° were obtained by coating a brass plate with two types of fluororesin. The wettability of a plate is regarded as poor when $\theta_a$ falls between 90 and 180° for a liquid.

Air was supplied to the water bath through a small straight hole of inner diameter $1.0 \times 10^{-3}$ m (see part A in Fig. 4.1) by means of a syringe driven by a stepper motor. A bubble thus formed on the upper surface of the plate grew gradually with time. When the volume of the bubble exceeded a certain critical value, it detached from the surface of the plate. The behavior of the growing bubble was observed with a CCD camera from the horizontal direction, and the images were recorded on a personal computer. The shape and size of the bubble were determined by processing these images. The volume of the bubble was obtained by assuming axisymmetry. In a strict sense, the volume was determined by integration using a relationship between the local height of the volume and the radial distance from the center of the bubble. The mean exhausting volumetric flow rate of air was $2.1 \times 10^{-3}$ $m^3/s$, and accordingly, the exhausted volume of air during every stepping of the motor was $0.21 \times 10^{-9}$ $m^3$. Consequently, the effect of the inertia force of the air injected into the bubble was negligible. The temperature of the water was controlled at $25 \pm 2$ °C.

Figure 4.2 shows a schematic of the experimental apparatus used for mercury measurements. Hereafter, this system is referred to as the mercury–air system. The cross section of the vessel made of transparent acrylic resin was $26 \times 10^{-3}$ m by $26 \times 10^{-3}$ m, and the depth was $50 \times 10^{-3}$ m. The vessel was filled with mercury to a depth of $40 \times 10^{-3}$ m. The advancing contact angle of the mercury on the plate, $\theta_a$, was 130°. Air was supplied to the mercury bath through a vertical hole of $1.0 \times 10^{-3}$ m in a manner similar to that in the water–air system in order to form

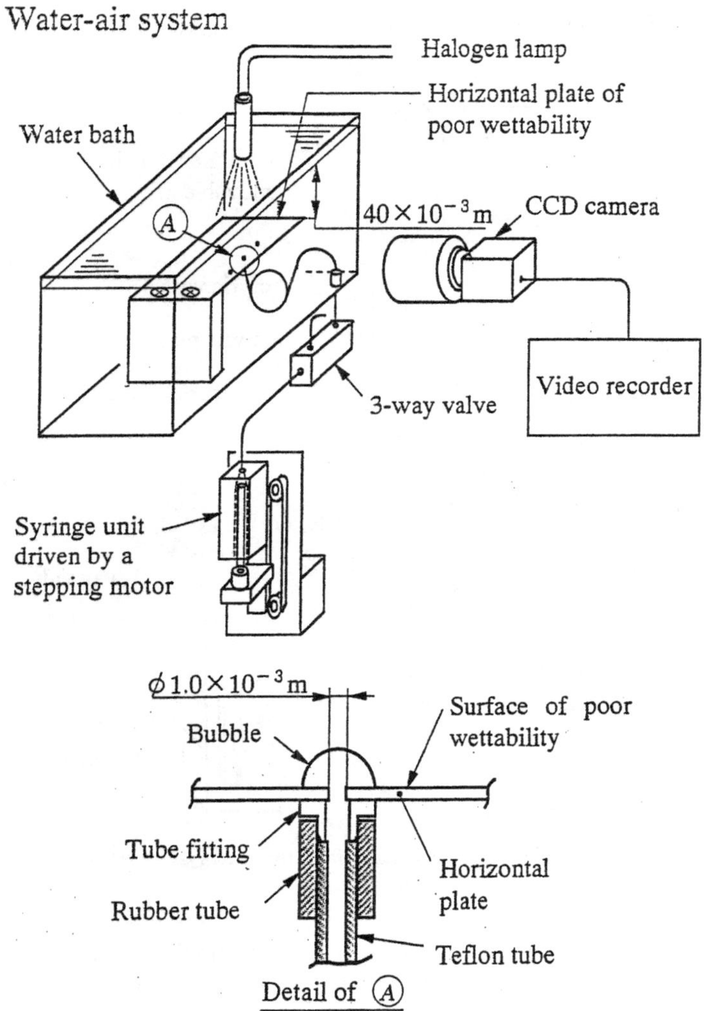


**Fig. 4.1** Schematic diagram of experimental apparatus for water–air system

a bubble on the flat plate. The bottom of the bubble was observed with a CCD camera placed below the plate.

Images of a growing bubble were recorded on a personal computer and then processed to determine its shape and size. The bottom of the bubble thus observed was circular in shape, and the diameter of the circle was designated by $D_B$. There was an inherent error of $\pm 4\%$ in the measurement. The height of the bubble, $H_B$, was measured with a needle equipped with a three-dimensionally traversing bed. The needle was gradually immersed in the mercury bath until its tip came in contact with the top of the bubble. The minimum legible length was $0.01 \times 10^{-3}$ m, and hence, the error inherent in the measurement was $\pm 3\%$.

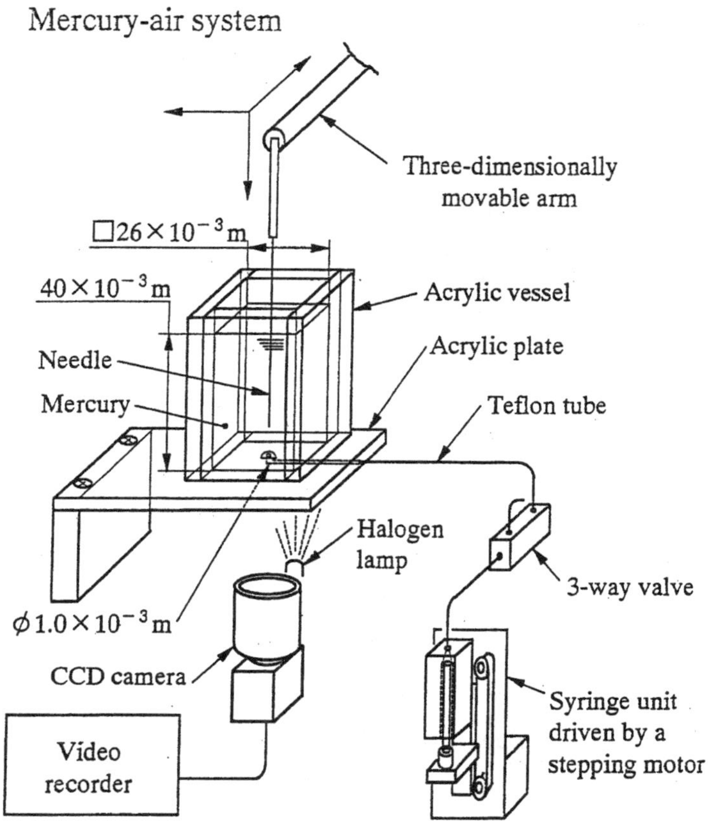


**Fig. 4.2** Schematic diagram of experimental apparatus for mercury–air system

The volume of bubble in the mercury bath was measured with the aforementioned syringe. The volume of air exhausted by the syringe was corrected by taking the static pressure around the bubble into consideration.

## *4.1.3 Experimental Results*

### 4.1.3.1 Bubble Shape Using Potential Method

As mentioned briefly in the introduction, two methods can be used to predict the shape and size of a bubble in contact with a horizontal flat plate of poor wettability. One is the Laplace method which focuses on local pressure balance around a gas–liquid interface [10–13]. The other, referred to as the potential method, calculates the shape of the bubble through energy minimization. Details of the Laplace method is described, for example, in the work of Sano et al. [14] which deals with the formation of bubbles from a single-hole nozzle immersed in a molten iron bath.

Lee et al. [10] reported that the two methods give the same results for the prediction of a droplet on a horizontal flat plate when the droplet is assumed to be ellipsoidal in shape. Accordingly, the potential method would also be useful for predicting the shape of a bubble as long as the shape is adequately modeled.

The Laplace method is not suitable for predicting the shape and size of a bubble attaching to a solid body of poor wettability that is immersed in a liquid flow field. This is due to the difficulty of representing the effects of a change in pressure of flow around a bubble or the coalescence of bubbles. On the contrary, these effects can be readily considered in the potential method, although the treatment is not exact. For example, the effect of hydrodynamic drag acting on a bubble can be approximated by adding the following term to the energy equation.

$$E_{\mathrm{D}} = C_{\mathrm{D}} A_{\mathrm{p}} \rho_{\mathrm{L}} {u_{\mathrm{w}}}^2 z_{\mathrm{G}}/2 \tag{4.1}$$

where $C_{\mathrm{D}}$ is the drag coefficient, $A_{\mathrm{p}}$ is the projected area to a plane perpendicular to the flow direction, $\rho_{\mathrm{L}}$ is the density of water, $u_{\mathrm{w}}$ is the mean velocity of water flow, and $z_{\mathrm{G}}$ is the distance from the wall to the gravitational center of the bubble.

Considering these circumstances, the potential method was chosen to predict the shape of a bubble in contact with a plate of poor wettability, and five simplified models were proposed for the shape of the bubble.

Details of Prediction Method

Figure 4.3 shows a bubble in contact with a horizontal plate of poor wettability. The bubble has three types of energy; potential energy due to gravitational force, surface energy on the gas–liquid interface, and interfacial energy on the gas–solid interface. The following relationship can be obtained [13]. from the energy equation for a droplet on a horizontal flat plate [13]:

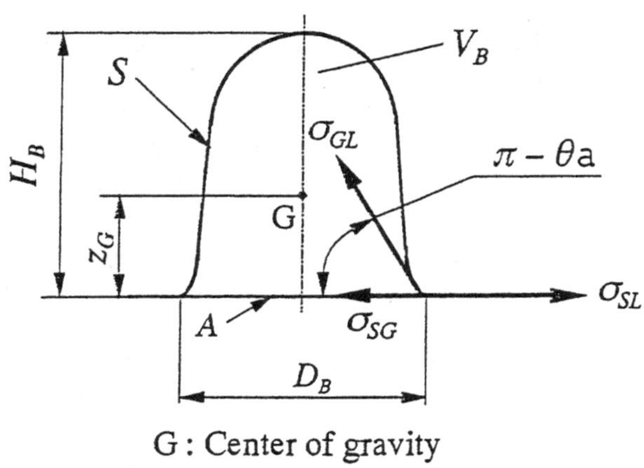


**Fig. 4.3** Explanation of symbols in energy equation for bubble

$$E = -(\rho_L - \rho_g) g V_B z_G + \sigma_{GL} S + \sigma_{GL} A \cos\theta_c \tag{4.2}$$

where $\rho_g$ is the density of gas, $g$ is the acceleration due to gravity, $V_B$ is the volume of the bubble, $\sigma_{GL}$ is the surface tension, $S$ is the interfacial area between the bubble and liquid, and $A$ is the interfacial area between the bubble and liquid.

We propose five different models which involve some parameters specifying the shape of a bubble. These parameters can be determined by minimizing the energy $E$ for a given bubble volume, $V_B$.

Modeling of Bubble Shape

Five different models were proposed for describing the shape of bubble in contact with a flat plate of poor wettability, as shown in Fig. 4.4. The four parameters, $V_B$, $S$, $A$, and $z_G$, in (4.2) are expressed in terms of the bottom diameter $D_B$ and the height of the bubble $H_B$ for Model 1 as follows.

Model 1 (see Fig. 4.4a)

$$V_B = \pi {D_B}^2 H_B/4 - \pi {D_B}^3/24 \tag{4.3}$$

$$S = \pi D_B H_B \tag{4.4}$$

$$A = \pi {D_B}^2/4 \tag{4.5}$$

$$z_G = ({D_B}^2 - 8 D_B H_B + 24 {H_B}^2)/(48 H_B - 8 D_B) \tag{4.6}$$

These parameters can be readily obtained for the remaining four models illustrated in Fig. 4.4b–e. The expressions are very complicated and lengthy, and accordingly, they are not reproduced here for brevity. Interested readers should refer to Mizuno et al. [8].

#### 4.1.3.2 Bubble Volume at Incipient Detachment Using Energy and Force Balance

There are three types of relationships between the aspect ratio $H_B/D_B$ $(= A_B)$ and the bubble volume $V_B$, as shown in Fig. 4.5. Models classified into Type 1 do not have a minimum value of energy $E$ for bubble volume larger than a certain critical value. The Type 2 models have a minimum of $E$ for all $V_B$. In Type 3 models, $E$ does not have a minimum for $V_B$ smaller than a certain critical value.

In Type 1, a bubble is considered to be unstable for the critical value of $V_B$, and assumed to detach from the plate. In fact, the critical values calculated from the models of Types 1 and 3 for a plate of $\theta_a$ of 96° in the air–water system could satisfactorily predict the experimentally measured volume of detached bub-

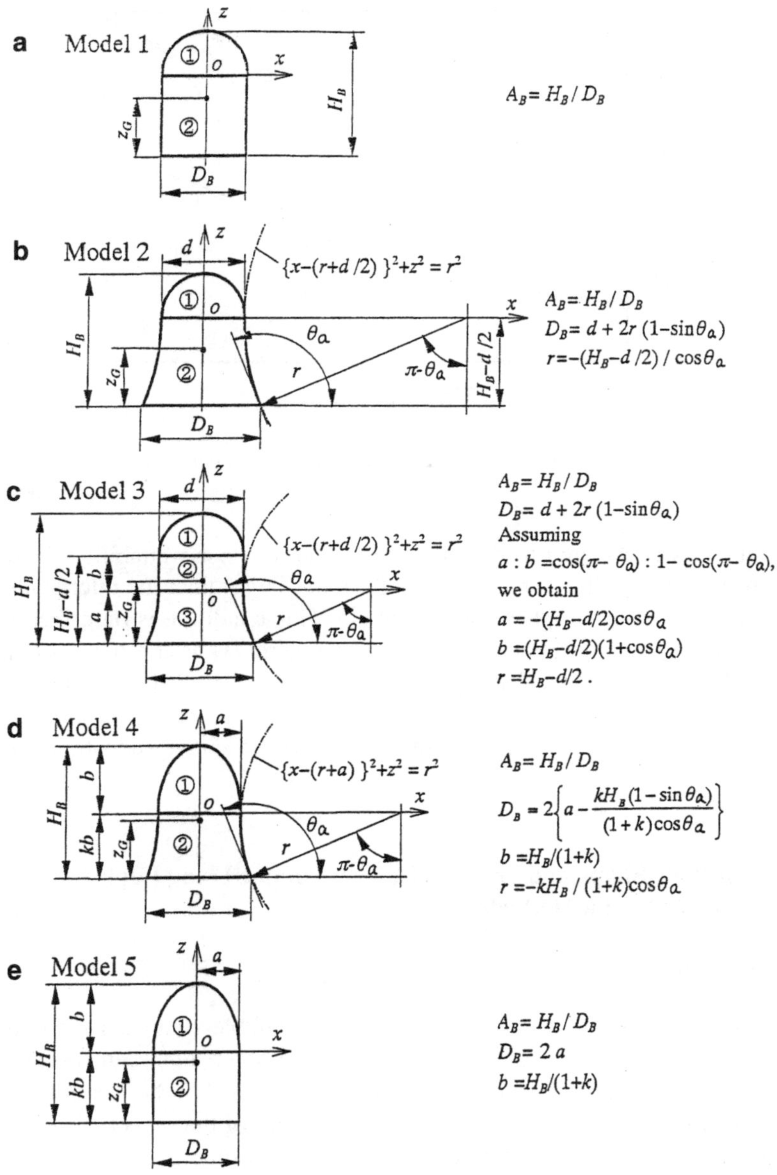


**Fig. 4.4** Models applied to bubble shape

bles. However, both model Types 2 and 4 give a minimum value of $E$ for all ranges of bubble volumes. The detachment of a bubble from the plate, therefore, cannot be determined solely from (4.2). In order to predict the critical volume for detachment of a bubble from the plate, a force balance equation is required in addition.

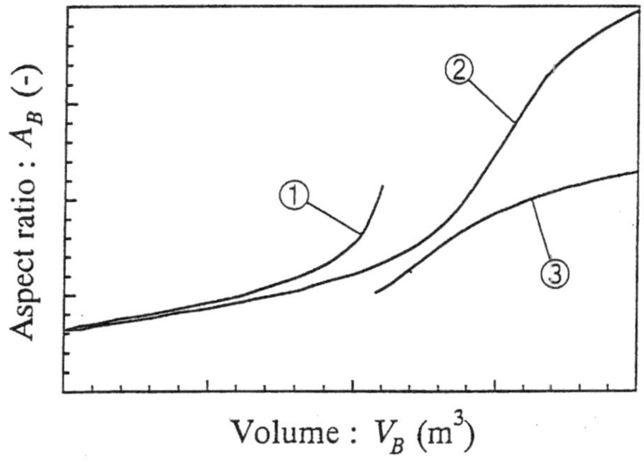


**Fig. 4.5** Relationship between $A_B$ and $V_B$ when bubble energy is minimum

A bubble is considered to detach from a plate of poor wettability as soon as the buoyancy force acting on the bubble becomes larger than the interfacial force acting on the bottom periphery of the bubble. This condition is described by the relationship $(\rho_L - \rho_g)gV_B > \pi D_B \sigma_{GL} \sin\theta_a$, where $D_B$ is the bottom diameter of the bubble. The value of $V_B$ at which the left and right hand sides become equal to each other is defined as the critical value for the detachment of the bubble. $D_B$ should, of course, be evaluated from the energy equation for each model.

#### 4.1.3.3 Bubble Shape and Size and Critical Volume Using Laplace and Potential Methods

1. Bubble shape and size

Figures 4.6–4.9 show the relationship between the bubble volume, $V_B$, and the aspect ratio, $A_B$, for $\theta_a = 96$, 104, 116, and 131° in the water–air system, respectively. A similar relationship is shown in Fig. 4.10 for $\theta_a = 130°$ in the mercury–air system. In each figure, thin lines denote predicted values of the aspect ratio based on the potential method, a thick line denotes predicted results based on the Laplace method, and open and solid circles denote the measured values [8]. The open circles represent bubbles long before detachment and the solid circles denote bubbles just before detachment. The measured data for $\theta_a = 96°$ were obtained by observing bubbles on the upper surface of a rectangular cylinder placed horizontally in a water–air bubbling jet [15]. The remaining data were obtained by using the experimental apparatus shown in Figs. 4.1 and 4.2.

Models of Types 4 and 5 proposed in the preceding section could give the minimum energy for every bubble volume. The difference between Model 4 and the exact value calculated from the Laplace method ranged from −1 to +35%, while

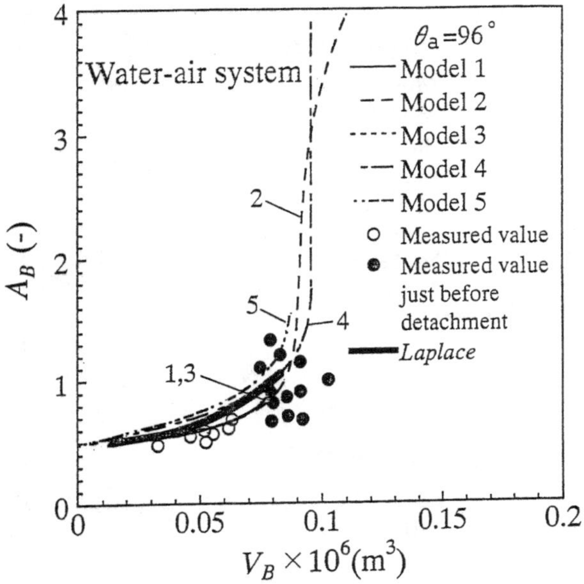


**Fig. 4.6** Comparison of $A_B$ between measured and calculated values at $\theta_a$ of 96° in water–air system

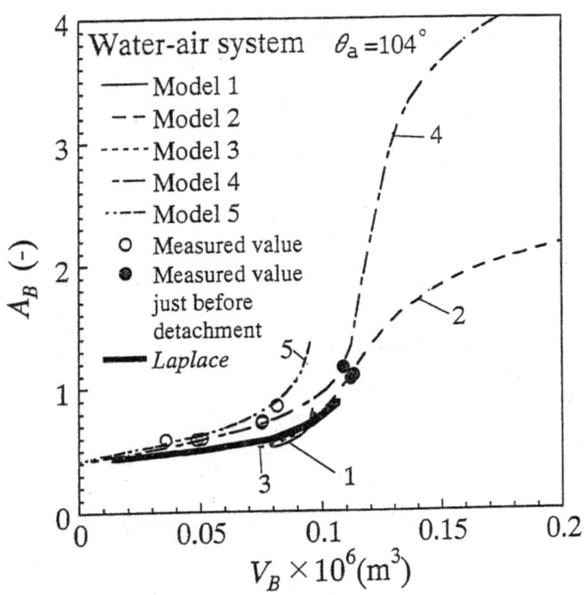


**Fig. 4.7** Comparison of $A_B$ between measured and calculated values at $\theta_a$ of 104° in water–air system

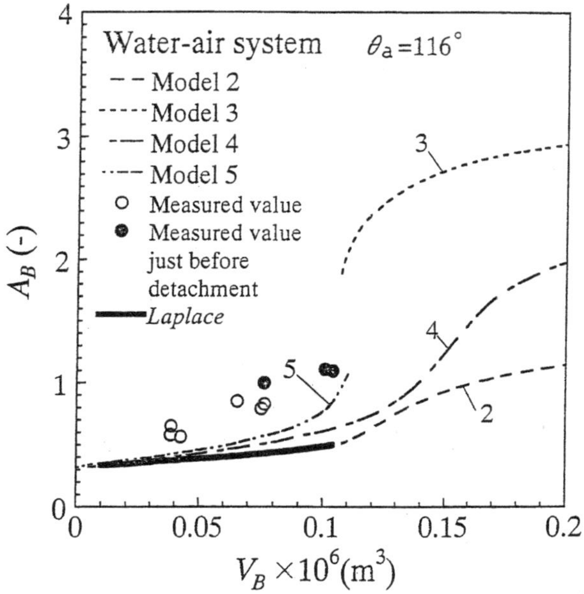


**Fig. 4.8** Comparison of $A_B$ between measured and calculated values at $\theta_a$ of 116° in water–air system

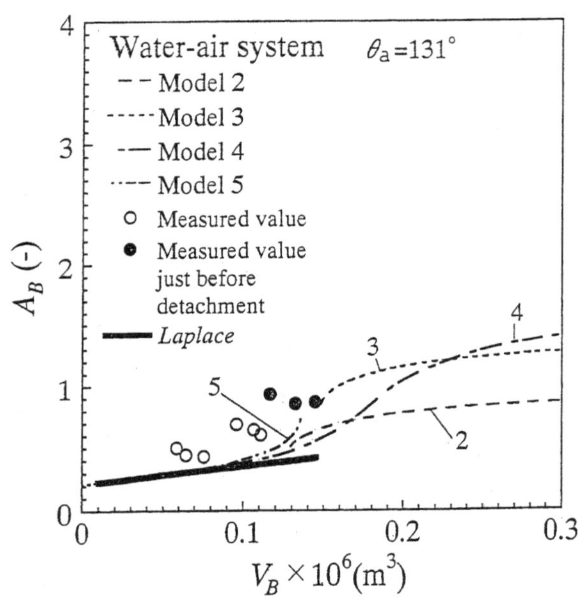


**Fig. 4.9** Comparison of $A_B$ between measured and calculated values at $\theta_a$ of 131° in water–air system

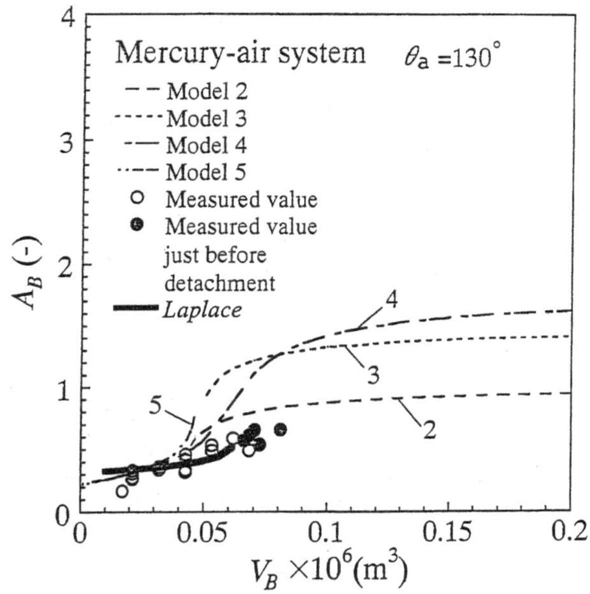


**Fig. 4.10** Comparison of $A_B$ between measured and calculated values at $\theta_a$ of 130° in mercury–air system

the range for Model 5 was −20 to +70%. Such a level of deviation could be reduced by introducing a more appropriate but complex model. The level of scattering for Model 4 could be considered acceptable at the present stage. Model 4 will thus be used in this section for the prediction of the critical bubble volume.

2. Critical bubble volume at incipient detachment

Figure 4.11 compares the critical bubble volumes just before detachment predicted from the potential and Laplace methods. The deviation of Model 4 prediction from that based on the Laplace method was −27 to +18% for an advancing contact angle range from 94 to 130°. This range is typical of many practical applications in the metallurgical industry. Therefore, although such a range of deviation is relatively large, it may be considered acceptable at the present stage.

#### 4.1.3.4 Measured and Predicted Aspect Ratio and Critical Bubble Volume

Figures 4.6–4.10 show that the measured aspect ratio, $A_B$, deviates appreciably (in the range −40 to 100%) from that predicted with the Laplace method. The deviation was −40 to +100%. A slightly smaller deviation of −37 to +5% is observed for the critical bubble volume, $V_B$, just before detachment. The deviation becomes large with an increase in $\theta_a$. Such a trend can be attributed to the following:

First, the measured $V_B$ in the water–air system was numerically determined from the relationship between the radial distance from the bubble center, $r$, and the height of the bubble there, $H(r)$.

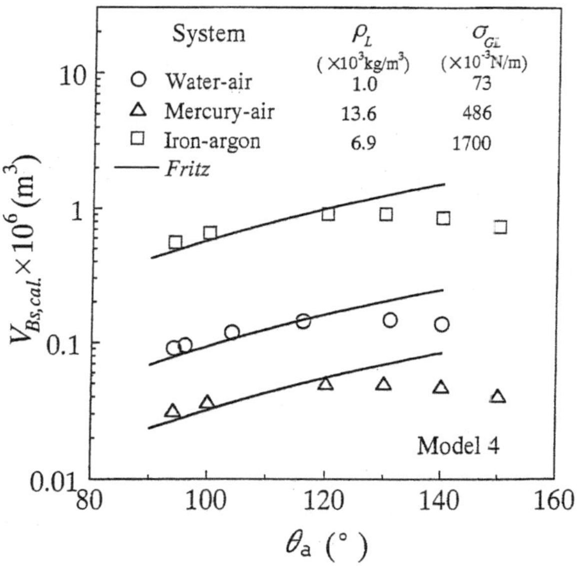


**Fig. 4.11** Comparison of $V_{Bs}$ between present method and previous method by Fritz

$$V_B = \int 2\pi r H(r) dr \tag{4.7}$$

The error inherent with this method is $\pm 21\%$ for $\theta_a = 96°$ [15]. The error inevitably increases with an increase in $\theta_a$ because a small measurement error in $H(r)$ near the edge of the bubble on the plate amplifies the error in $V_B$. This is primarily responsible for the observed deviation of the measured $V_B$ from the value calculated with the Laplace method for an increase in $\theta_a$.

Second, small disturbances originating from the air supply system and the surroundings may affect the shape of a bubble when the surface tension is relatively small, as in the case of the water–air system.

It is interesting to note from Fig. 4.12 that $D_B$ of the bubble in the mercury–air system is $6.5 \times 10^{-3}$ m [8], which agrees with the result of Ozawa et al. [16].

Figures 4.13 and 4.14 illustrate the calculated and observed bubbles in the water–air system and mercury–air system, respectively. As mentioned above, more sophisticated models are desirable for the prediction of the shape of bubbles over a wide range of advancing contact angle.

### 4.1.4 Summary

The shape and size of a bubble in contact with the upper surface of a flat plate of poor wettability placed horizontally in a water bath or a mercury bath were predicted. The error inherent in the potential method employed was evaluated by comparing the results with those calculated from the Laplace method. The results indicate the need for a more sophisticated model for predicting the bubble shape than could be

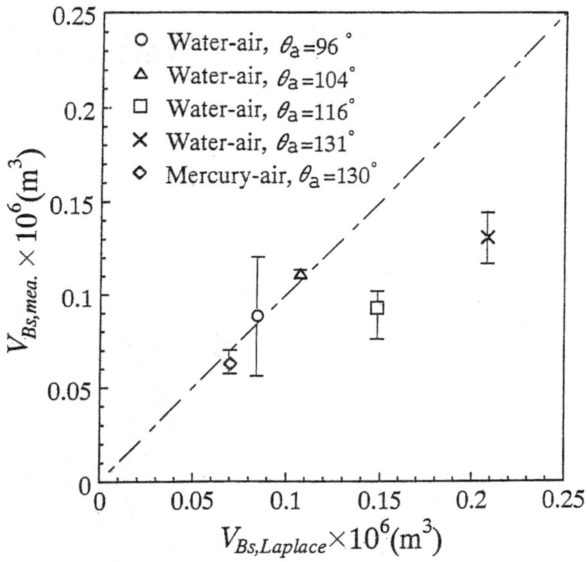


**Fig. 4.12** Comparison of $V_{Bs}$ between measured and calculated values

achieved with the potential method. Nevertheless, from a practical point of view, the model considered (Model 4) provides a useful tool for predicting the critical bubble volume at imminent detachment from a solid body of poor wettability, immersed in fluid flows.

## 4.2 Bubbling Jet Along Vertical Flat Plate

### *4.2.1 Bubble Characteristics*

#### 4.2.1.1 Overview

Dispersion of bubbles in a molten metal bath and the induced flow strongly influence the performance of gas-agitated steelmaking processes. Unfortunately, measurements of the bubble and molten metal flow characteristics are quite difficult in real steelmaking processes. Hence water model experiments have been extensively employed for such investigations, as discussed in detail in Chap. 1 [17, 18]. The vessels used for such model experiments are usually fully wetted by water, whereas in the actual steelmaking processes, the wettability of the reactor wall is generally poor. When the wall material, i.e., refractory, is wetted by the molten metal, some chemical reactions may occur between them, and consequently, the molten metal may be contaminated by the refractory. As mentioned earlier, the wettability is evaluated in terms of $\theta_a$ [19].

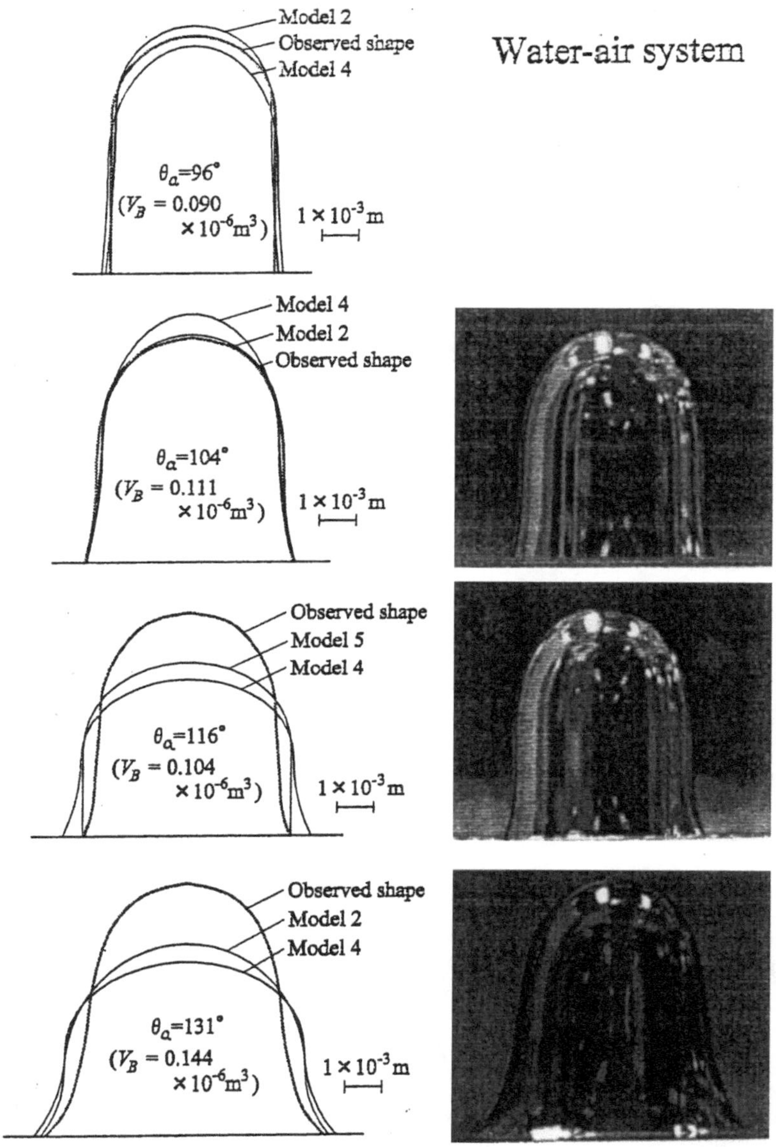


**Fig. 4.13** Observed and calculated bubble shapes in water–air system

Very little is known about the effects of wettability of a plate on bubble and molten metal flow characteristics except for the frequency of bubble formation, $f_B$ [20–22]. These studies have shown that the frequency of bubble formation from a single-hole nozzle or a porous nozzle that is not wetted by the liquid is significantly different from that from a single-hole nozzle of good wettability. This result suggests that the bubble and liquid flow characteristics near a vertical flat plate would

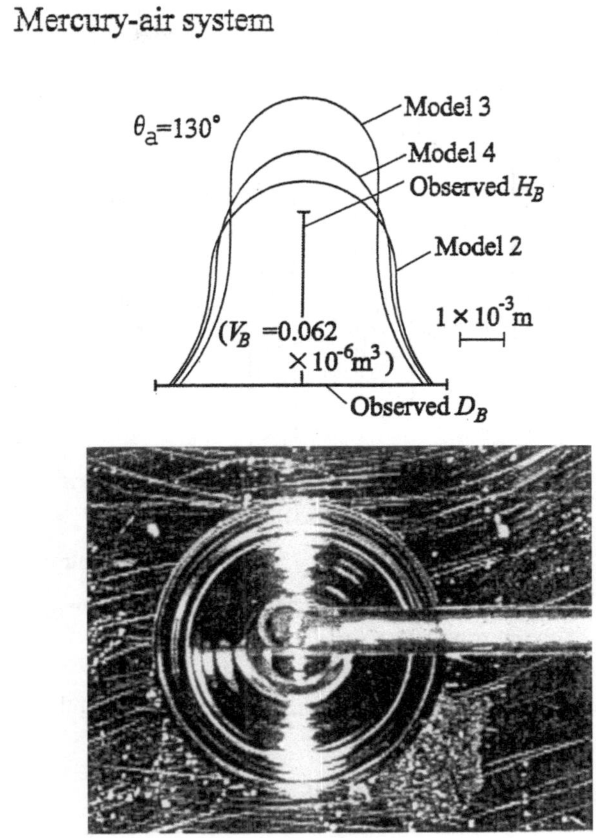


**Fig. 4.14** Observed and calculated bubble shapes in mercury–air system

also be dependent on the wettability of the plate [7, 23]. In this section, particular attention will be given to the bubble characteristics in a bubbling jet rising along a vertical flat plate of varying wettability. Such a jet is defined as a bubbling wall jet. The information obtained on the bubble characteristics may improve fundamental understanding of refining processes such as the RH degassing process [18].

#### 4.2.1.2 Experimental Apparatus and Procedure

Figure 4.15 shows a schematic of the experimental apparatus. The transparent cylindrical vessel had an inner diameter $D$ of $200 \times 10^{-3}$ m and height $H$ of $390 \times 10^{-3}$ m. The vessel was enclosed in a water jacket of square cross section in order to reduce parallax effect as much as possible. The distance from the nozzle tip to the bath surface $H_L$ was $240 \times 10^{-3}$ m. A single-hole nozzle having an inner diameter $d_{ni}$ of $2.0 \times 10^{-3}$ m, an outer diameter $d_{no}$ of $6.0 \times 10^{-3}$ m, and a height of

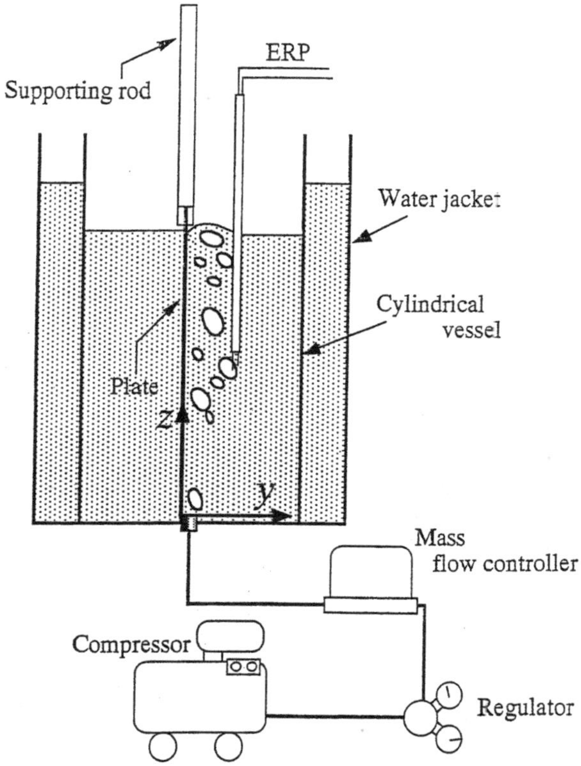


**Fig. 4.15** Experimental apparatus

$6.5 \times 10^{-3}$ m was placed at the center of the bottom of the vessel. A flat plate was vertically situated beside the nozzle. The origin of the Cartesian coordinate system $(x, y, z)$ was placed on the center of the nozzle tip.

Bubble frequency $f_B$, gas holdup $\alpha$, mean bubble rising velocity $\bar{u}_B$, and mean bubble chord length $\bar{L}_B$ were measured with a two-needle electroresistivity probe mainly for gas flow rate $Q_g$ of $41.4 \times 10^{-6}$ m$^3$/s [24]. This gas flow rate was chosen because it was used for previous experimental studies on bubbling jet generated away from a vertical flat plate [25]. The previous results were used, for example, to assess the accuracy of the new measurements. The experiments were repeated at least twice at every measurement position on the $y-z$ plane depending on the scatter of the data. The mean value of the data will be presented in subsequent figures. The duration of measurement was 360 s in the horizontal region ranging from $y = 1 \times 10^{-3}$ to $3 \times 10^{-3}$ m, and 120 s for the remaining measurement positions located at $y > 3 \times 10^{-3}$ m. The outer edge of the region of interest was located where there were more than ten bubbles over 120-s duration. Bubble trapping on the surface of the plate was examined for gas flow rates in the range of $Q_g = 10 \times 10^{-6} - 80 \times 10^{-6}$ m$^3$/s.

The advancing contact angle $\theta_a$ was measured with the sessile drop method [26]. Two flat plates were used: a copper plate with $\theta_a = 62°$ (good wettability) and an acrylic plate coated with paraffin with $\theta_a = 104°$ (poor wettability).

#### 4.2.1.3 Experimental Results

1. Horizontal distribution of bubble frequency and gas holdup

The horizontal distribution of bubble frequency $f_B$ measured at four axial positions $z = 50 \times 10^{-3}$, $100 \times 10^{-3}$, $150 \times 10^{-3}$, and $200 \times 10^{-3}$ m is shown in Fig. 4.16. The abscissa is represented in logarithmic scale in order to amplify the bubble behavior near the wall. In the close vicinity of the plate at $z = 50 \times 10^{-3}$ m, $f_B$ for

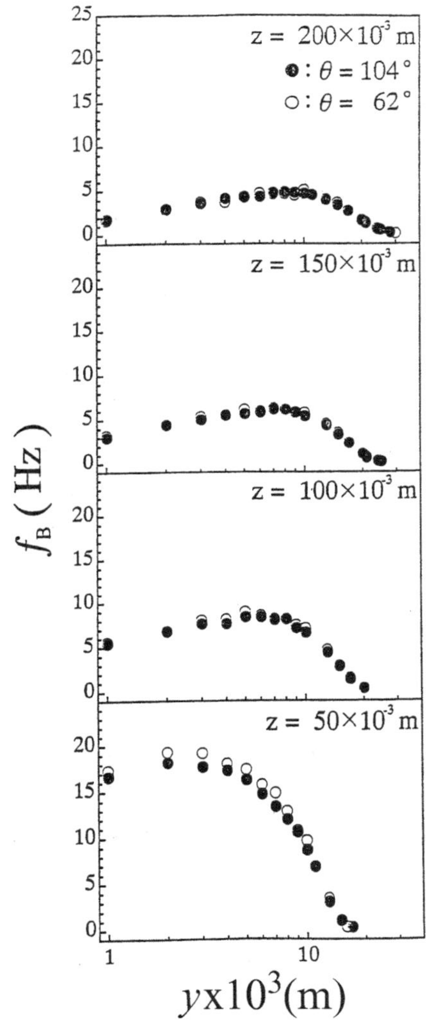


**Fig. 4.16** Horizontal distributions of bubble frequency for $Q_g = 41.4 \times 10^{-6}$ m$^3$/s

the poorly wetted plate is slightly smaller than the results for the plate with good wettability. There is little difference between the horizontal distributions for the two plates at other axial measurement positions $z \geq 100 \times 10^{-3}$ m.

Figure 4.17 shows that at $z = 50 \times 10^{-3}$ m and $y = 1 \times 10^{-3}$ m, the measured gas holdup $\alpha$ is larger for the poorly wetted plate than that of good wettability. However, this trend is reversed in the horizontal region between approximately $y = 2 \times 10^{-3}$ and $10 \times 10^{-3}$ m at the same axial position. The values of $\alpha$ are nearly the same for the two plates for $y > 10 \times 10^{-3}$ m. The difference between the results for the two plates diminishes significantly for $z > 100 \times 10^{-3}$ m. Consequently, the wettability of a vertical flat plate has little effect on the horizontal distribution of $\alpha$ for $z \geq 100 \times 10^{-3}$ m.

The vertical distributions of $f_{\mathrm{B,max}}$ and $\alpha_{\mathrm{max}}$ are given in Fig. 4.18, where the subscript max denotes a maximum value. The wettability of the two plates has no discernible influence on the axial distribution of $f_{\mathrm{B,max}}$ and $\alpha_{\mathrm{max}}$ except at $z = 50 \times 10^{-3}$ m. The results for the plate of good wettability can be correlated as

$$f_{\mathrm{B,max}} = 1.15/z \tag{4.8}$$

$$\alpha_{\mathrm{max}} = 1.69/z \tag{4.9}$$

Two length scales, $y_{f\mathrm{B,max}}$ and $y_{f\mathrm{B,max}/2}$, can be defined to characterize the horizontal distribution of $f_{\mathrm{B}}$ as shown in Fig. 4.19. Those for the horizontal distribution of gas holdup $\alpha$, designated by $y_{\alpha,\mathrm{max}}$ and $y_{\alpha,\mathrm{max}/2}$, are similarly defined as in Fig. 4.19.

The vertical distributions of $y_{f\mathrm{B,max}}$, $y_{f\mathrm{B,max}/2}$, $y_{\alpha,\mathrm{max}}$, and $y_{\alpha,\mathrm{max}/2}$ are shown in Figs. 4.20 and 4.21. These quantities are not significantly influenced by the wettability of the plates at any axial position and can be approximated thus,

$$y_{f\mathrm{B,max}} = 0.03z + 2.53 \times 10^{-3} \tag{4.10}$$

$$y_{f\mathrm{B,max}/2} = 0.05z + 7.59 \times 10^{-3} \tag{4.11}$$

$$y_{\alpha,\mathrm{max}} = 0.02z + 3.74 \times 10^{-3} \tag{4.12}$$

$$y_{\alpha,\mathrm{max}/2} = 0.04z + 8.15 \times 10^{-3} \tag{4.13}$$

It is evident that $y_{f\mathrm{B,max}/2}$ is approximately twice as large as $y_{f\mathrm{B,max}}$. The same relationship holds between $y_{\alpha,\mathrm{max}}$ and $y_{\alpha,\mathrm{max}/2}$. There is no clear explanation for the linear dependence of the four quantities on the axial distance $z$.

2. Similarity laws for horizontal distribution of bubble frequency and gas holdup

The horizontal distribution of bubble frequency in a bubbling free jet which is free from any walls is known to be similar. In other words, the distributions follow a normal (Gaussian) distribution [24, 27, 28]. The horizontal distribution of bubble frequency near a vertical flat plate is also expected to follow another type of similarity distribution. In order to ascertain this anticipation, each $f_{\mathrm{B}}$ value was normalized by the maximum value $f_{\mathrm{B,max}}$ and then plotted against a dimensionless horizontal distance from the plate, $\eta_{f\mathrm{B}} (= y/y_{f\mathrm{B,max}/2})$, in Fig. 4.22. The solid line denotes

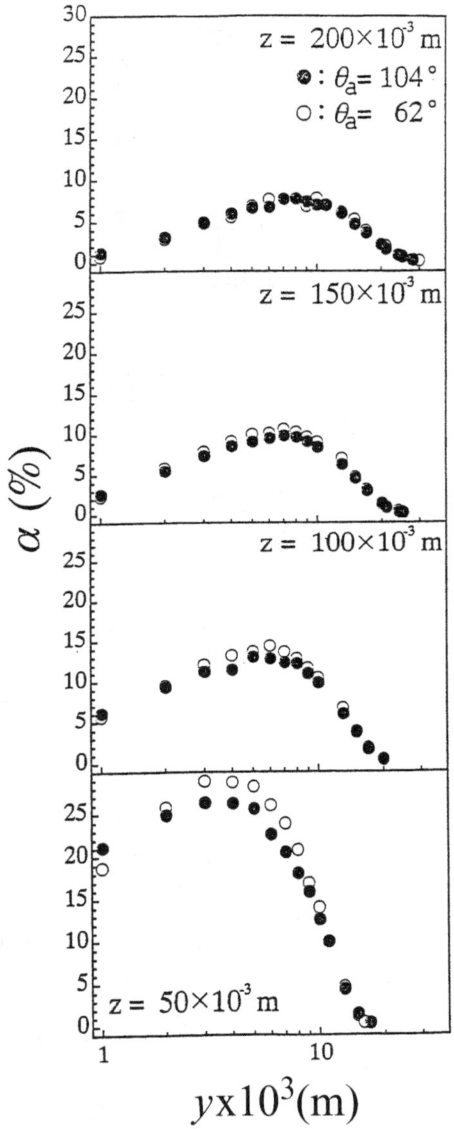


**Fig. 4.17** Horizontal distributions of gas holdup for $Q_g = 41.4 \times 10^{-6}\,\mathrm{m^3/s}$

experimentally derived similarity distribution for $f_B$, being expressed by

$$f_B/f_{B,max} = 0.5\eta_{fB}{}^{0.55}\exp[-1.43\{\eta_{fB}{}^2 - 1\}] \tag{4.14}$$

$$\eta_{fB} = y/y_{fB,max/2} \tag{4.15}$$

The measured values of $f_B$ follow (4.14) for $z \geq 100 \times 10^{-3}$ m, irrespective of the wettability of the plates. On the contrary, the distributions of $f_B$ for the two plates deviate from (4.14) at $z = 50 \times 10^{-3}$ m.

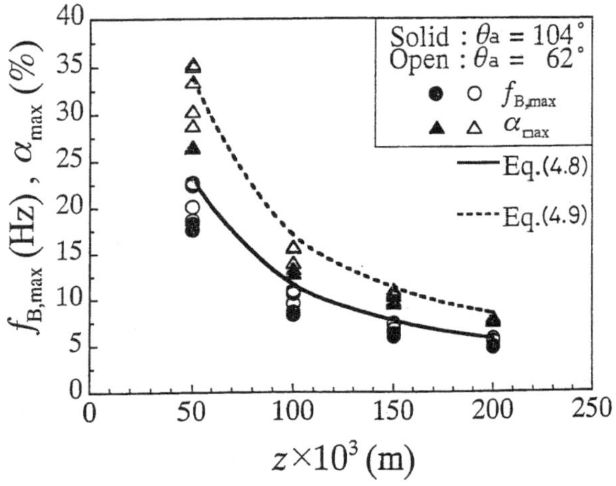


**Fig. 4.18** Definition of length scales

Figure 4.23 indicates that for $z \geq 100 \times 10^{-3}$ m, the gas holdup $\alpha$ obeys a similarity distribution for two plates, irrespective of wettability. This similarity distribution represented by the solid line can be expressed by

$$\alpha/\alpha_{max} = 0.5{\eta_\alpha}^{0.87} \exp[-1.74\{{\eta_\alpha}^2 - 1\}] \tag{4.16}$$

$$\eta_\alpha = y/y_{\alpha,max/2} \tag{4.17}$$

A bubbling free jet that is not influenced by the side wall of the vessel can be classified into four regions with respect to the axial distance from the nozzle tip [29]: the momentum, transition, buoyancy, and bath surface regions. The inertial force of injected gas governs the flow in the momentum region, while the buoyancy force acting on the bubbles is dominant in the buoyancy region. The midpoint of the transition region, $z_t$, is given by [29]

$$z_t = 9.4({Q_g}^2/g)^{1/5} \tag{4.18}$$

where $g$ is the acceleration due to gravity. Equation (4.18) yields $z_t = 105 \times 10^{-3}$ m for the experimental conditions considered. Such flow field classification appears to be valid for bubbling wall jets, because the distributions of $f_B$ and $\alpha$ for $z = 50 \times 10^{-3}$ m do not follow (4.14) and (4.16), respectively. The above $z_t$ position corresponds approximately to the vertical position at which the horizontal distributions of $f_B$ (and $\alpha$) for the two plates of good and poor wettability begin to agree with each other. It can thus be concluded that the wettability of the plates affects the bubble frequency and gas holdup distributions primarily in the momentum region.

3. Horizontal distribution of mean bubble rising velocity $\overline{u}_B$

Figure 4.24 shows the experimental data on mean bubble rising velocity $\overline{u}_B$. The data for the poorly wetted plate decrease monotonically in the $y$ direction at every

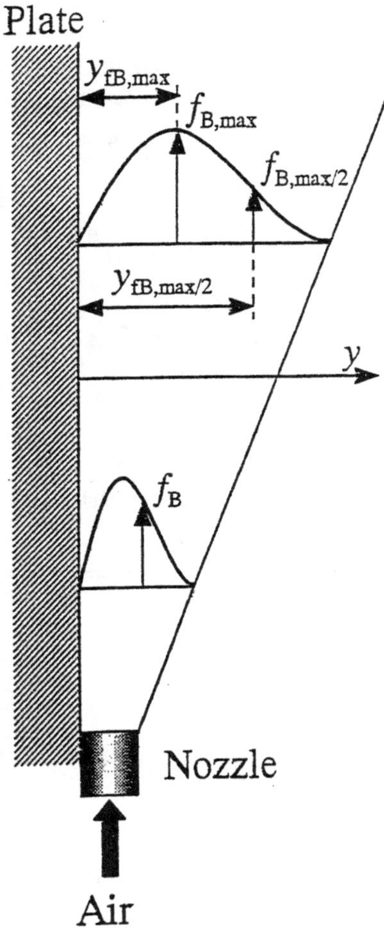


**Fig. 4.19** Axial distributions of the maximum values of bubble frequency and gas holdup for $Q_g = 41.4 \times 10^{-6}\,\text{m}^3/\text{s}$

axial position. On the contrary, a distinct peak appears in the horizontal distribution at $z = 200 \times 10^{-3}$ m for this plate. In the horizontal region $y > 5 \times 10^{-3}$ m, $\overline{u}_B$ is not dependent on the wettability of the plate. However, at locations $y < 5 \times 10^{-3}$ m and $z < 150 \times 10^{-3}$ m, $\overline{u}_B$ is lower for the poorly wetted plate than for the plate with good wettability. This is because the rising velocity of the bubbles decreases due to the attachment of the bubbles to the poorly wetted plate.

The standard deviation of $\overline{u}_B$ is represented by $u_{B'_{rms}}$. The ratio of $u_{B'_{rms}}$ to $\overline{u}_B$ is approximately 50% at $z = 150 \times 10^{-3}$ and $200 \times 10^{-3}$ m, and remains nearly constant in the $y$ direction, as shown in Fig. 4.25. Meanwhile, at $z = 50 \times 10^{-3}$ and $100 \times 10^{-3}$ m, the ratio also is about 50% for $y < 5 \times 10^{-3}$ m. Outside this horizontal position $y > 5 \times 10^{-3}$ m, $u_{B'_{rms}}/\overline{u}_B$ increases abruptly and then decreases with an increase in $y$. Such a drastic change in $u_{B'_{rms}}/\overline{u}_B$ appears to be associated with strong entrainment of the surrounding liquid into the bubbling wall jet. In the close vicinity of the plate at the axial positions $z = 50 \times 10^{-3}$ and $100 \times 10^{-3}$ m, $u_{B'_{rms}}/\overline{u}_B$ is slightly larger for the poorly wetted plate than that of

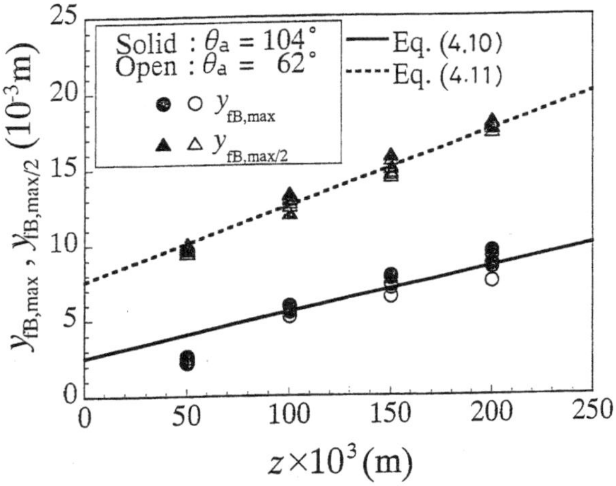


**Fig. 4.20** Axial distributions of length scales, $yf_{B,max}$ and $yf_{B,max/2}$, for $Q_g = 41.4 \times 10^{-6}\,m^3/s$

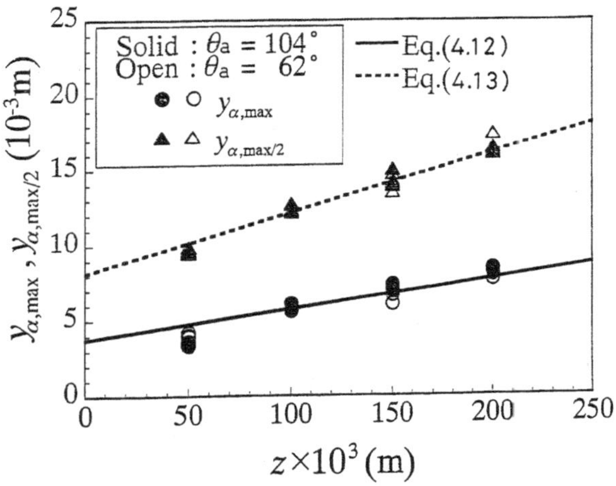


**Fig. 4.21** Axial distributions of length scales, $y_{\alpha,max}$ and $y_{\alpha,max/2}$, for $Q_g = 41.4 \times 10^{-6}\,m^3/s$

the plate with good wettability. This trend can be attributed to the fact that bubbles rising near the poorly wetted plate often attach to it, although they are not always trapped on the surface.

4. Horizontal distribution of mean bubble chord length

The horizontal distributions of mean bubble chord length $\bar{L}_B$ are shown in Fig. 4.26. In the vicinity of the plate, $y < 5 \times 10^{-3}$ m, near the nozzle tip, $z \leq 100 \times 10^{-3}$ m, the measured value of $\bar{L}_B$ is larger for the plate with good wettability than that of the plate with poor wettability. There is a distinct peak in the measured $\bar{L}_B$ distributions for the highly wetted plate at $z = 150 \times 10^{-3}$ and $200 \times 10^{-3}$ m.

The experimental data for $\bar{u}_B$ and $\bar{L}_B$ collectively suggest that bubbles rising near the highly wetted plate for $z \leq 100 \times 10^{-3}$ m are more elongated in the vertical

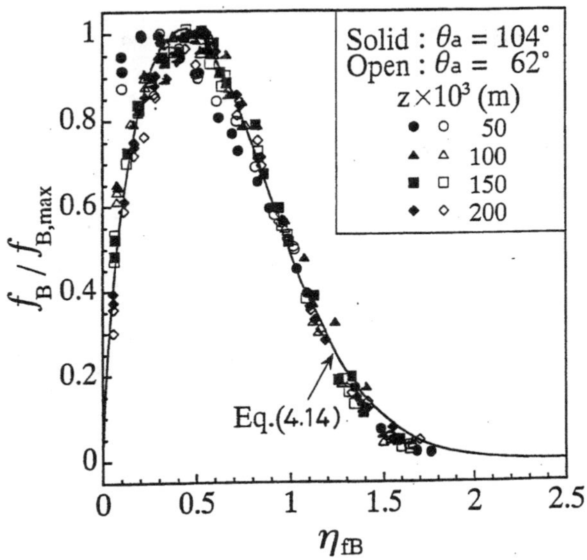


**Fig. 4.22** Similar distributions of bubble frequency for $Q_g = 41.4 \times 10^{-6}\ \mathrm{m^3/s}$

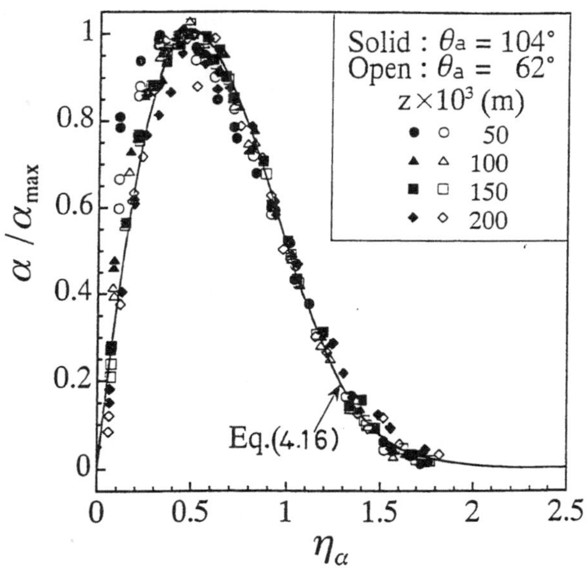


**Fig. 4.23** Similar distributions of gas holdup for $Q_g = 41.4 \times 10^{-6}\ \mathrm{m^3/s}$

direction than those rising near the poorly wetted plate. This result is due to the frequent attachment of bubbles to the latter plate. As $z$ increases, bubbles rising along the highly wetted plate begin to separate from the plate, while those rising along the poorly wetted plate remain near the plate right to the top bath surface.

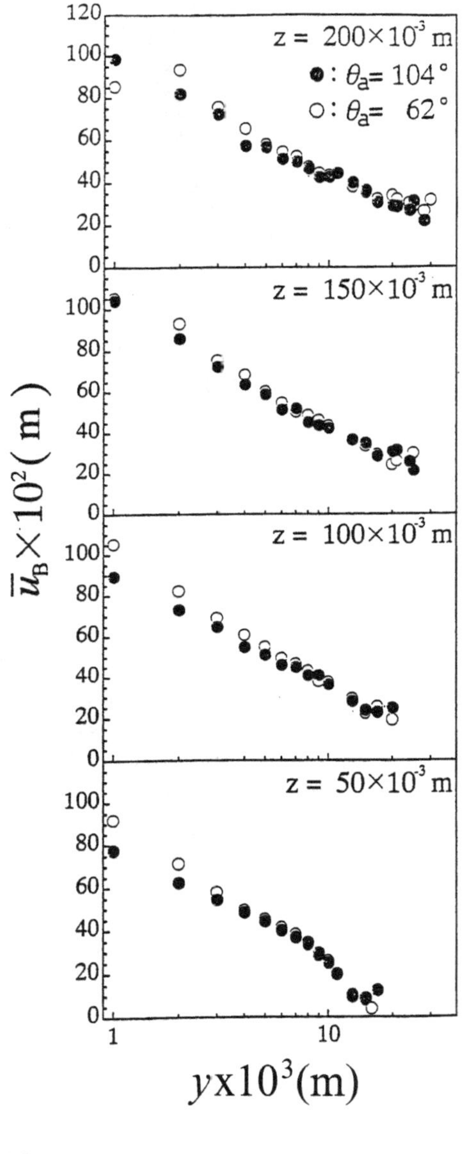


**Fig. 4.24** Horizontal distributions of mean bubble rising velocity for $Q_g = 41.4 \times 10^{-6}\,\mathrm{m^3/s}$

It should be noted that the data presented for only the $y-z$ plane are not sufficient to describe fully the mass balance of the gas phase.

5. Attachment of bubbles to plate of poor wettability

The experimental data indicate that bubbles are not in direct contact with a highly wetted plate. This is evidenced by the existence of a corresponding image on the right hand side for each bubble in Fig. 4.27a. This is because a water film exists between the bubble and the plate. On the contrary, Fig. 4.27b shows that bubbles sometimes are in direct contact with the poorly wetted plate, as indicated by an arrow. The image of each bubble cannot be seen, as the plate coated with paraffin does not work like a mirror.

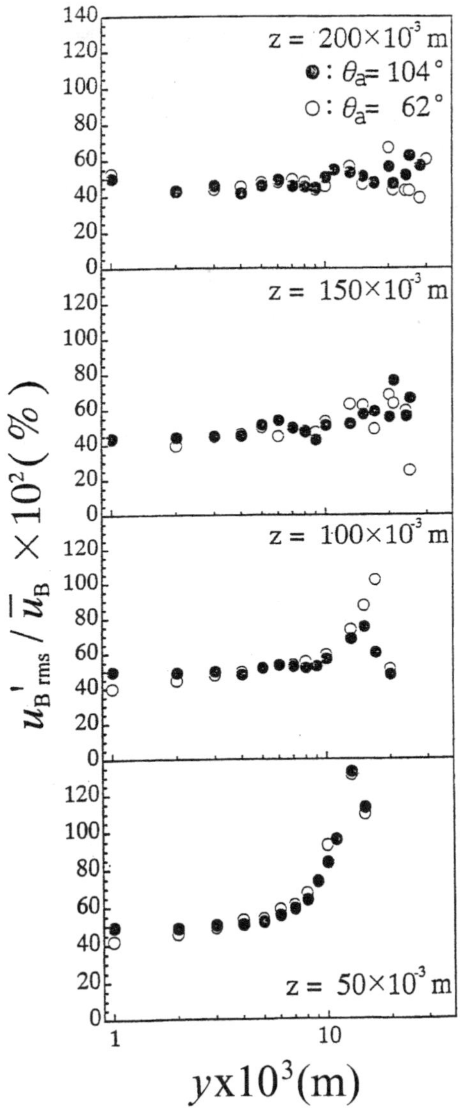


**Fig. 4.25** Horizontal distributions of the standard deviation of bubble rising velocity divided by mean value for $Q_g = 41.4 \times 10^{-6}\,\mathrm{m^3/s}$

As discussed earlier, the differences in the bubble characteristics between the two types of plates are most noticeable in the vicinity of the plates. These differences are closely associated with the attachment of bubbles to the poorly wetted plate and otherwise in the highly wetted plate. Small bubbles were trapped on the former plate after the termination of air injection, as shown in Fig. 4.28. Such small bubbles are produced due to disintegration of larger bubbles being detached from the plate. Such trends are not evident for the plate with good wettability.

The attachment of small bubbles to a poorly wetted plate would significantly influence heat transfer between the plate and molten metal. Therefore, the total contact area of trapped bubbles $A_{tB}$ was determined by using a still camera. The measured

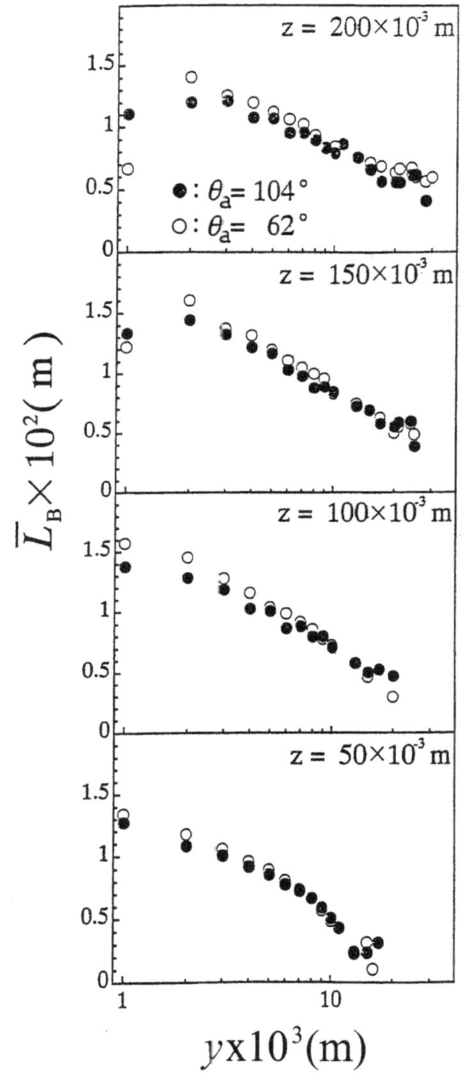


**Fig. 4.26** Horizontal distributions of mean bubble chord length for $Q_g = 41.4 \times 10^{-6}\,\mathrm{m^3/s}$

value of $A_{tB}$ was normalized by the area of the bubble dispersion region projected onto the plate, $A_{pB}$, giving the trapped bubble area ratio $r_{TB}$. The projected area $A_{pB}$ was determined by processing the contour of the bubble dispersion region that was recorded on a video camera using a personal computer.

Figure 4.29 shows the variation of the measured values of $r_{TB}$ with gas flow rate $Q_g$. For $Q_g = 10 \times 10^{-6}\,\mathrm{m^3/s}$, the turbulent motion of water induced by the bubbles is not sufficiently strong to cause instability of the water film between the bubble and the poorly wetted plate. As a result, the total number of trapped bubbles is small and $r_{TB}$ remains at a low level. With an increase in $Q_g$, the measured value of $r_{TB}$ increases significantly due to enhancement of the horizontal turbulence component of water flow. For $Q_g$ larger than $41.4 \times 10^{-6}\,\mathrm{m^3/s}$, $r_{TB}$ is nearly invariant

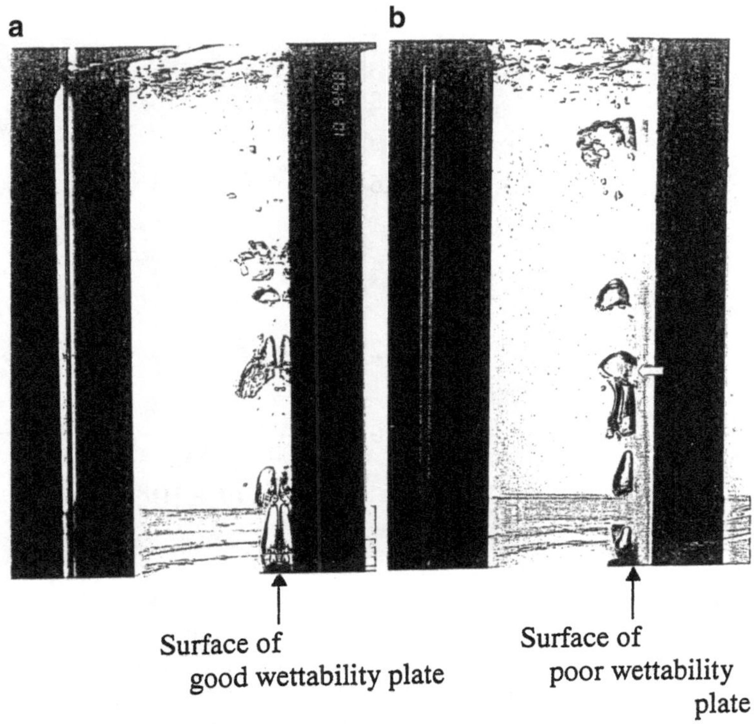


**Fig. 4.27** Photographs showing bubbles rising along a flat plate for $Q_g = 41.4 \times 10^{-6}\ m^3/s$. (**a**) Good wettability plate; (**b**) poor wettability plate

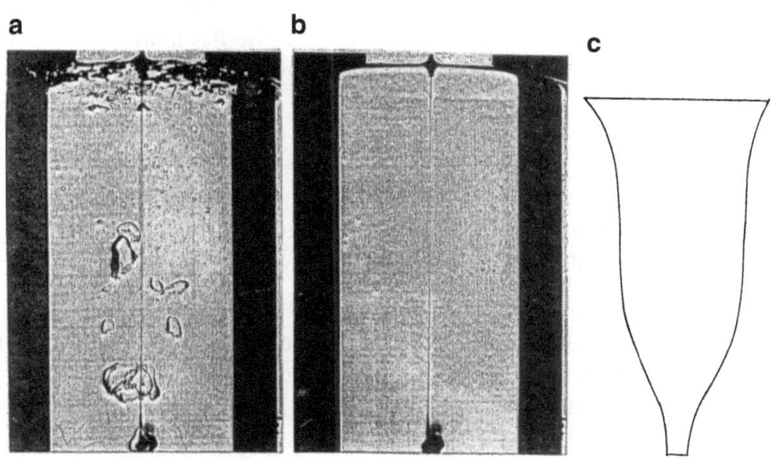


**Fig. 4.28** Photographs showing trapped bubbles on the flat plate surface for $Q_g = 41.4 \times 10^{-6}\ m^3/s$. (**a**) during gas injection; (**b**) after termination of gas injection; (**c**) outer edge of bubble dispersion region

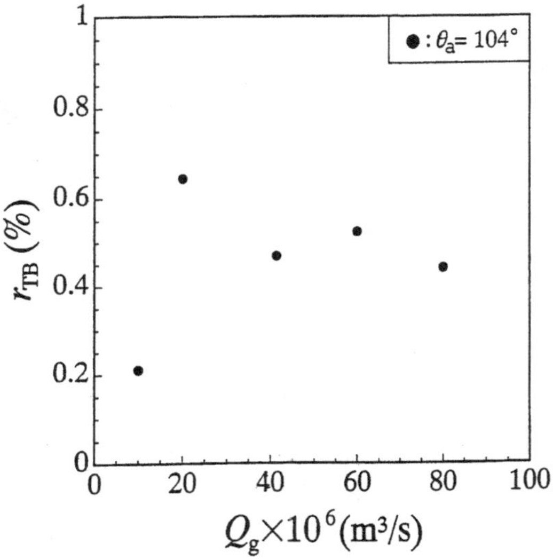


**Fig. 4.29** Trapped bubble area ratio

with $Q_g$. This equilibrium state may be explained as follows. The number of bubbles attaching to the plate increases with $Q_g$, but at the same time, the drag force acting on each bubble increases. Such an elevated drag force enhances bubble detachment from the plate, and thus results in the aforementioned equilibrium state.

The roughness of the plate may influence bubble entrapment to the plate. The advancing contact angle is known to be dependent on the roughness of the plate. Therefore, roughness effect can be partly evaluated in terms of the advancing contact angle when the roughness height is much smaller than the liquid film thickness. If the roughness height is equal to or larger than the film thickness, another type of interaction may occur between the bubbles and the flat plate. This subject is, however, beyond the scope of the present consideration.

#### 4.2.1.4 Summary

The effects of the wettability of a flat plate on a bubbling wall jet rising along the plate can be summarized as follows:

1. Bubbles attach to the poorly wetted plate ($\theta_a = 104°$), some of which are trapped on the surface. No bubble entrapment occurs on the highly wetted plate ($\theta_a = 62°$). The bubbles in the latter case are more elongated in the vertical direction and rise faster than those near the poorly wetted plate in the momentum region ($z < 100 \times 10^{-3}$ m). With a further increase in axial position $z$ in the buoyancy region ($z > 100 \times 10^{-3}$ m), bubbles rising along the highly wetted plate begin to migrate away from the plate, while those rising along the poorly wetted plate remain near the plate right through to the bath surface.
2. As a consequence of bubble attachment to the poorly wetted plate, the measured bubble frequency $f_B$, gas holdup $\alpha$, mean bubble rising velocity $\bar{u}_B$, and mean

bubble chord length $\bar{L}_B$ measured in the vicinity of the poorly wetted plate differ from their respective values for the highly wetted plate. In particular, the bubble frequency and gas holdup are different in the momentum region located near the nozzle tip. Differences in the mean bubble rising velocity and mean bubble chord length are observed between the two flat plates over the entire axial region ranging from the nozzle tip to the bath surface.

3. Both the bubble frequency and gas holdup exhibit similarity distributions in the buoyancy region, $z \geq 100 \times 10^{-3}$ m. The similarity distributions are independent of the wettability of the plates.
4. The total contact area of trapped bubbles to the poor wettability plate is evaluated in terms of the trapped bubble area ratio $r_{TB}$.

### *4.2.2 Liquid Flow Characteristics*

The precise information on molten metal flow induced by bubbles is essential for improving current refining processes [30]. Such molten metal flow characteristics appear to be significantly affected by boundary conditions on the inner walls of the reactor, in addition to the shape and size of the reactor and the injection method. The refractories used in these processes are usually not wetted by the molten metal [31]. Yet most previous model studies on molten metal flow characteristics were carried out using vessel materials that are wetted by the liquid [30]. Experiments on the effects of wettability on molten metal flow characteristics are quite limited [20–22, 32].

Iguchi et al. [32] focused on the relationship between the wettability of a vertical flat plate and the behavior of bubbles in a bubbling wall jet rising along the plate. It was found that bubbles rising along a poorly wetted plate repeatedly attach and detach from the plate. $\bar{u}_B$ and $\bar{L}_B$ in the close vicinity of the plate were dependent on the wettability of the plate in the whole vertical region extending from the nozzle tip to the bath surface. Meanwhile, the bubble frequency and gas holdup were influenced by the wettability solely in the vertical region near the nozzle tip where the inertial force of the injected gas was dominant. This section describes the behavior of water flow induced by bubbles rising along a vertical flat plate of different wettability.

#### 4.2.2.1 Experimental Apparatus and Procedure

Figure 4.30 shows a schematic diagram of the experimental apparatus. The dimensions of the vessel and nozzle as well as the experimental conditions mirror those described in the preceding section [32]. The horizontal and vertical coordinates are denoted by $y$ and $z$, respectively. The corresponding velocity components of water flow are designated by $v$ and $u$. The gas flow rate $Q_g$ was set at $41.4 \times 10^{-6}$ m$^3$/s, the same value used for the bubble characteristic measurements [32].

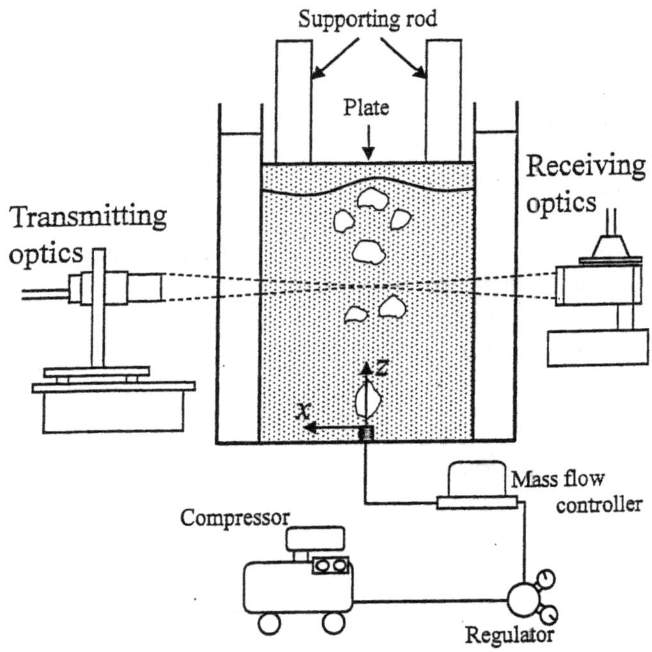


**Fig. 4.30** Experimental apparatus

The bubble characteristics, represented by the bubble frequency $f_B$, gas holdup $\alpha$, mean bubble rising velocity $\bar{u}_B$, and mean bubble chord length $\bar{L}_B$, were measured mainly on the $y-z$ plane with a two-needle electroresistivity probe. Water velocity was measured with a two-channel laser Doppler velocimeter [25]. The locations at which measurements were made were chosen by reference to the results for the bubble characteristics. The characteristics of water flow were represented by the vertical mean velocity $\bar{u}$, the root-mean-square value of the vertical turbulence component $u'_{\mathrm{rms}}$, and the skewness and flatness factors of the vertical turbulence component, $S_u$ and $F_u$. Data aquisition time was 600 s for each position.

#### 4.2.2.2 Results

1. Classification of flow field

The flow field above the nozzle can be classified broadly into two regions: the momentum region near the nozzle and the buoyancy region far from the nozzle [32]. In the former region, the inertial force of the injected gas governs the flow field, while the buoyancy force acting on the bubbles determines the latter. The two regions share a boundary at approximately $z = 100 \times 10^{-3}$ m.

2. Horizontal distribution of vertical mean velocity of water flow

The horizontal distribution of $\bar{u}$ was obtained at four vertical positions $z = 50 \times 10^{-3}$, $100 \times 10^{-3}$, $150 \times 10^{-3}$, and $200 \times 10^{-3}$ m, as shown in Fig. 4.31.

**Fig. 4.31** Horizontal distributions of vertical mean velocity

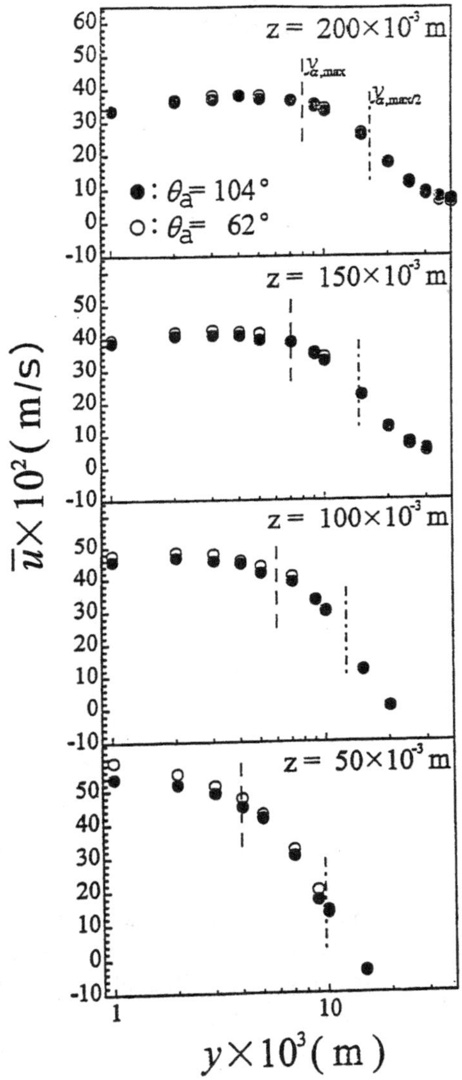


At $z = 50 \times 10^{-3}$ m within the momentum region, the measured mean values of $\overline{u}$ are slightly larger for the plate with good wettability than for the plate with poor wettability in the horizontal region $y < y_{\alpha,\max/2}$, where $y_{\alpha,\max/2}$ is the half-value distance of the horizontal gas holdup distribution [32]. For $y > y_{\alpha,\max/2}$, the wettability effect is insignificant. In the buoyancy region $z \geq 100 \times 10^{-3}$ m, the difference between the measured values of $\overline{u}$ for the two types of plates diminishes. In effect, the influence of wettability on $\overline{u}$ is important only in the momentum region.

There appears a peak in the horizontal distribution of $\overline{u}$ in the buoyancy region. The peak value is denoted by $\overline{u}_{\max}$. In order to describe quantitatively the horizontal

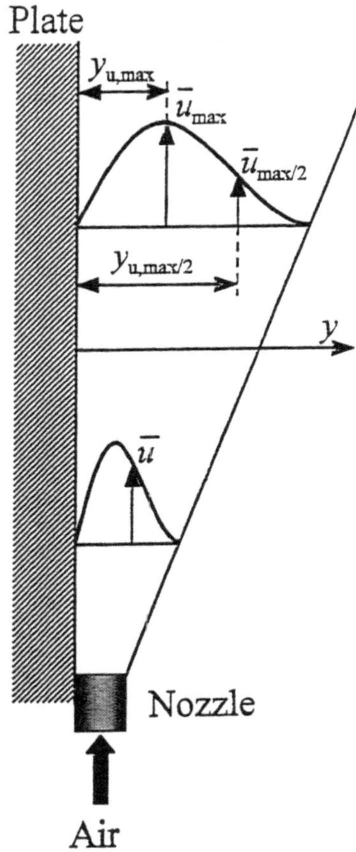


**Fig. 4.32** Definition of velocity and length scales

distribution of $\overline{u}$, two length scales $y_{u,\max/2}$ and $y_{u,\max}$ are defined, as shown in Fig. 4.32. For the highly wetted plate, the following empirical equation was derived for $\overline{u}_{\max}$ (see Fig. 4.33), although the maximum velocity of single-phase wall jet is proportional to $z^{-1}$ [33],

$$\overline{u}_{\max} = 0.160z^{-1/2} \tag{4.19}$$

The measured values of $\overline{u}_{\max}$ for the poorly wetted plate are also scattered around (4.19). However, the decreasing rate of $\overline{u}_{\max}$ in the vertical direction is slightly smaller than that for the wetted plate. This is probably because of the attachment of bubbles to the plate [32].

Figure 4.34 shows the variation of the two length scales, $y_{u,\max/2}$ and $y_{u,\max}$, with the vertical distance $z$. The dotted and solid lines are represented by the following equations, respectively:

$$y_{u,\max/2} = 0.075z + 4 \times 10^{-3} \tag{4.20}$$

$$y_{u,\max} = 0.02z \tag{4.21}$$

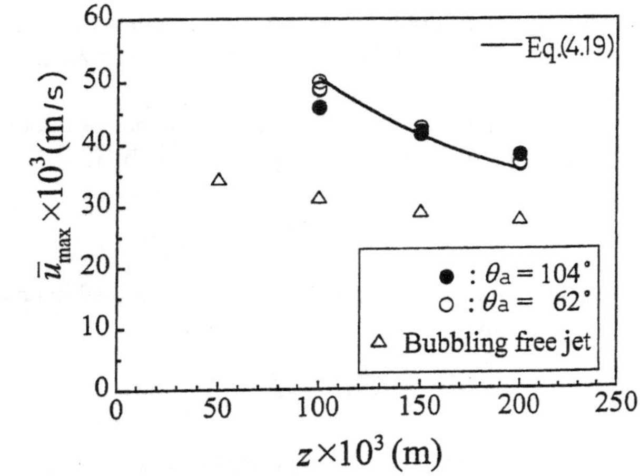


**Fig. 4.33** Relationship between maximum value of vertical mean velocity and vertical distance

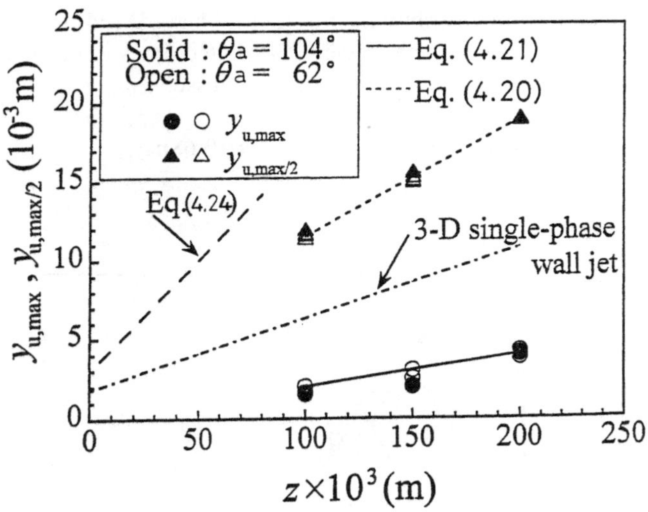


**Fig. 4.34** Relationship between length scales and vertical distance

There is little difference between the measured length scales for the two types of plates considered. The half-value distance for three-dimensional (3D) single-phase wall jets is given by [34]

$$y_{u,\max/2}/d_{\mathrm{ni}} = 0.045z/d_{\mathrm{ni}} + 0.90 \tag{4.22}$$

The half-value distance for bubbling wall jets was found to be much larger than that of single-phase jet. This was attributed to the effect of horizontal dispersion of bubbles.

It is interesting to examine whether the half-value distance of bubbling free jets, $b_u$, is applicable to bubbling wall jets for which $b_u$ is given by [25]

$$b_u = 0.14z \tag{4.23}$$

In applying this equation to bubbling wall jets along a vertical flat plate, $y_{u,\max/2}$ should be expressed as follows:

$$y_{u,\max/2} = 0.14z + d_{no}/2 \tag{4.24}$$

Unfortunately, (4.24) overestimates the measured values of $y_{u,\max/2}$ for the bubbling wall jet.

3. Similarity law for horizontal distribution of vertical mean velocity

Figure 4.35 shows the nondimensional distributions of $\overline{u}$. A dotted line and a chain line are used to denote the distributions for 2D [35] and 3D [34] single-phase wall jets along highly wetted plate, respectively. The dotted line was calculated from an empirical relationship derived by Verhoff [35], expressed as

$$\overline{u}/\overline{u}_{\max} = 1.48{\eta_u}^{1/7}\{1 - \text{erf}(0.68\eta_u)\} \tag{4.25}$$

$$\eta_u = y/y_{u,\max/2} \tag{4.26}$$

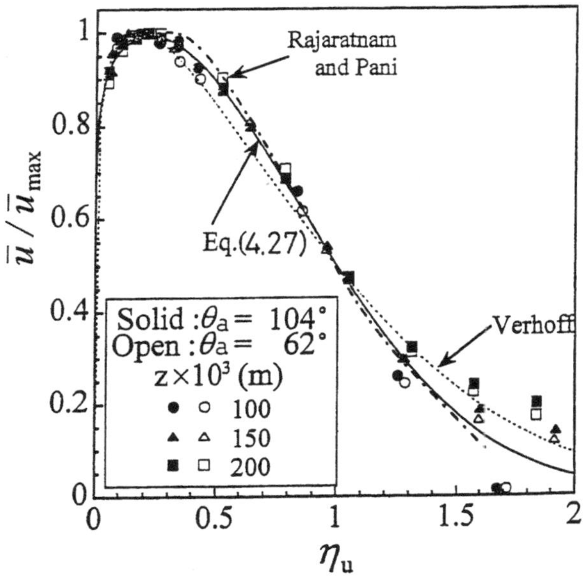


**Fig. 4.35** Similar distribution of vertical mean velocity

The chain line was drawn so as to pass through the mean of the values measured by Rajaratnum and Pani [34].

In the buoyancy region, the nondimensional distribution of $\overline{u}/\overline{u}_{max}$ in the bubbling wall jets along the vertical flat plate with good wettability is approximated by

$$\overline{u}/\overline{u}_{max} = 0.5\eta_u{}^{0.081} \exp\{-0.85(\eta_u{}^2 - 1)\} \quad (4.27)$$

Equation (4.27) is also applicable to the $\overline{u}$ distribution for the poorly wetted plate. It can thus be concluded that the distributions of $\overline{u}$ for the bubbling wall jets in the buoyancy region are self-similar. They also agree favorably with the similarity distribution for single-phase wall jets along a highly wetted plate.

4. Horizontal distribution of relative velocity

The horizontal distributions of relative velocity $\overline{u}_r(= \overline{u}_B - \overline{u})$ are shown in Fig. 4.36. At every measurement position in the horizontal region $y > y_{\alpha,max/2}$, the relative velocity is approximately 20 cm/s, which is the same as in a stagnant liquid [36] regardless of the wettability of the plate. On the contrary, in the region $y < y_{\alpha,max/2}$, the relative velocities at $z = 50 \times 10^{-3}$, $100 \times 10^{-3}$, and $150 \times 10^{-3}$ m are higher for the wetted plate than those for the poorly wetted plate. Further up ($z = 200 \times 10^{-3}$ m), there is a peak in the distribution for the wetted plate in the same manner as the $\bar{u}_B$ shown in the previous section [32]. This result is due to the fact that larger bubbles with higher velocity tend to migrate away from wetted a plate, while they rise near poorly wetted wall with repeated attachment and detachment from the plate.

5. Horizontal distribution of root-mean-square turbulence component

Figure 4.37 shows the horizontal distributions of $u'_{rms}$. At every vertical position, the measured values of $u'_{rms}$ are nearly the same for both types of plates. In the buoyancy region, $u'_{rms}$ remains unchanged in the horizontal region $y < y_{\alpha,max}$. The horizontal distributions of the vertical turbulence intensity $u'_{rms}/\overline{u}_{max}$ are shown in Fig. 4.38. The data on $u'_{rms}/\overline{u}_{max}$ for the plate with good wettability are approximated by

$$u'_{rms}/\overline{u}_{max} = 0.47 \exp(-0.23\eta_u{}^3) \quad (4.28)$$

The turbulence intensity of the bubbling wall jet is about two times as large as that of single-phase wall jet along a wetted plate (4.34, 4.35). Accordingly, the production in the wake of bubbles is largely attributable to the turbulence production in the bubbling wall jet.

Hetsroni [37] reported that turbulence is generated in the wake of bubbles when the bubble Reynolds number $Re_B$ is higher than 400. According to Kawakami et al. [24], the mean bubble diameter $\bar{d}_B$ is 1.5 times as large as $\bar{L}_B$, provided that the bubbles are spherical in shape, and thus the bubble Reynolds number $Re_B$ is defined as

$$Re_B = 1.5\bar{L}_B\overline{u}_r/\nu \quad (4.29)$$

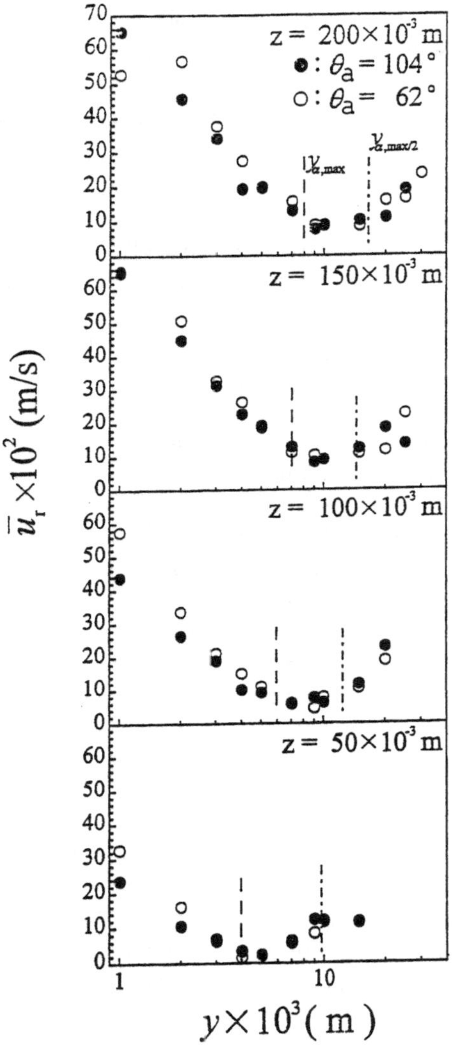


**Fig. 4.36** Horizontal distributions of relative velocity

In this study, $Re_B$ ranged from 1,900 to 14,000, and hence, turbulence can be generated in the wake of bubbles.

6. Horizontal distribution of skewness and flatness factors

High-order turbulence correlations such as skewness and flatness factors are commonly introduced to understand the detailed structure of turbulence. The skewness factor $S_u$ and the flatness factor $F_u$ for the vertical turbulence component, $u'$, are defined as

$$S_u = \{\Sigma(u_i'/u'_{\text{rms}})^3\}/N \tag{4.30}$$

$$F_u = \{\Sigma(u_i'/u'_{\text{rms}})^4\}/N \tag{4.31}$$

$$N > 15,000 \tag{4.32}$$

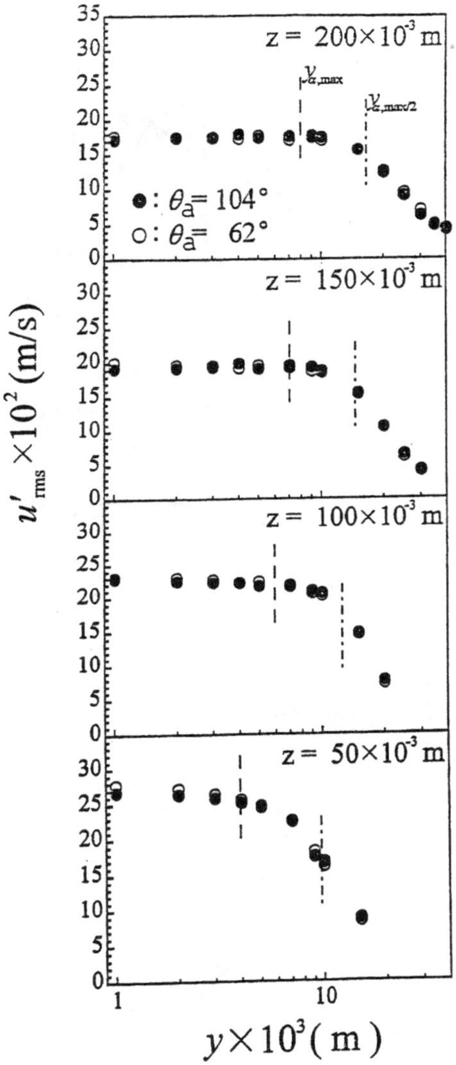


**Fig. 4.37** Horizontal distributions of the root-mean-square value of vertical turbulence component

Here, $i$ denotes the $i$th datum on $u'$ and $N$ is the number of velocity data. The measured values of $S_u$ and $F_u$ are shown in Figs. 4.39 and 4.40, respectively. At $z = 50 \times 10^{-3}$ m in the momentum region, $S_u$ is slightly larger for the poorly wetted plate than that for the plate with good wettability in the close vicinity of the plates, $y < y_{\alpha,\max}$. There is no significant difference between the values of $S_u$ for the two plates at the three vertical measurement positions located in the buoyancy region ($z > 100 \times 10^{-3}$ m).

The flatness factor $F_u$ is nearly independent of the wettability of the plate. These measurements for $S_u$ and $F_u$ differ from their respective values for a Gaussian distribution ($S_u = 0$ and $F_u = 3$). Such discrepancy is associated with the turbulence production in the wake of bubbles.

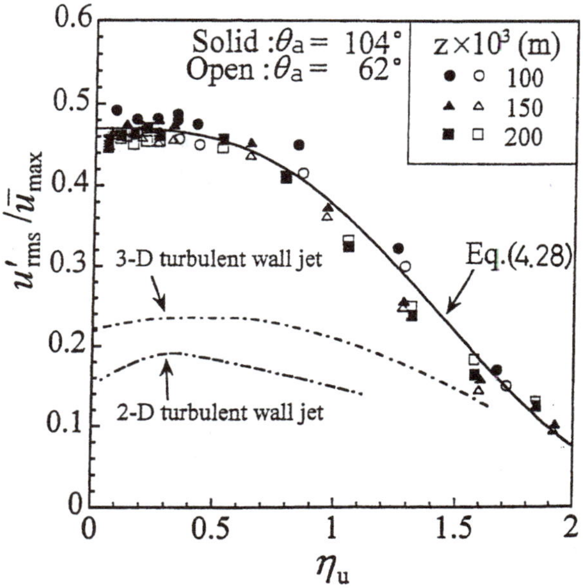


**Fig. 4.38** Similar distribution of the root-mean-square value of vertical turbulence component

#### 4.2.2.3 Summary

The effect of wettability is noticeable primarily on $\bar{u}$ and $S_u$ of the vertical turbulence component for water flow in the momentum region located near the nozzle tip. This is the region where the inertial force of the injected gas is dominant [38]. In the buoyancy region located above the momentum region, the water flow characteristics are nearly independent of the wettability of the plate.

The vertical mean velocity $\bar{u}$ and the root-mean-square value of the vertical turbulence component $u'_{rms}$ exhibit similarity distributions in the buoyancy region. Empirical relations (4.27) and (4.28) are proposed for the horizontal distributions of $\bar{u}$ and $u'_{rms}$.

## 4.3 Bubble Shape and Size

The refractory used for metal refining processes and continuous casting molds is commonly poorly wetted by the molten metal [39]. Water model experiments have shown that bubbles ascending near a wall of poor wettability sometimes attach to the wall [32] which differs markedly from bubbles ascending near a wall of good wettability. Although the interaction between a wall and bubbles has been widely studied, focus has been on wetted surfaces [40–44]. In contrast, there is little information on the attachment and detachment of bubbles from a wall of poor wettability [32].

**Fig. 4.39** Horizontal distributions of skewness factor

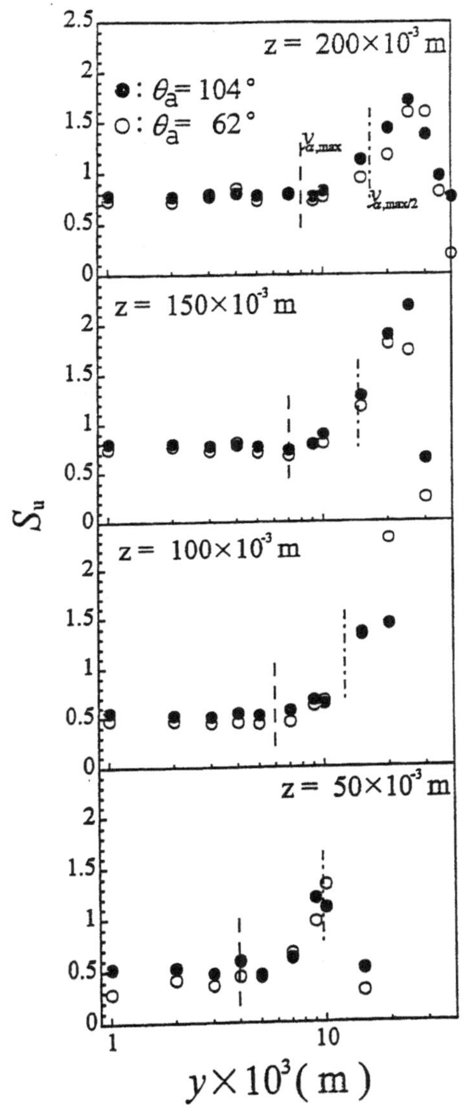


The interaction between refractory and bubbles in metals refining processes significantly affects the life of the refractory. The local erosion of the refractory is known to be closely associated with the generation sites and the generation rates of bubbles [45]. In addition, entrapment of small argon bubbles in solidifying shells in continuous casting causes bubble-induced defects such as sliver and pin-hole. These defects reduce the quality of the products [46, 47]. Investigation of the shape and size of bubbles in contact with a poorly wetted wall as well as the behavior of bubbles ascending in the vicinity of a liquid/solid interface is, therefore, important to elucidate the mechanism of generation of bubble-induced defects.

**Fig. 4.40** Horizontal distributions of flatness factor

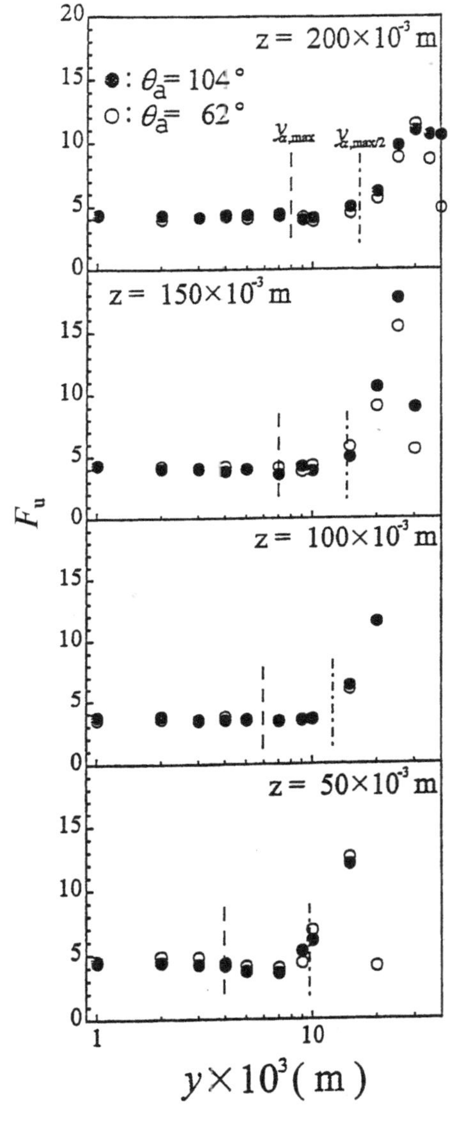


Due to the typically high melting points of molten metals (often > 1,000°C) and their opacity, direct observation of bubbles in the molten metals is quite difficult. An X-ray fluoroscope is a powerful tool [21] that can be used, but the spatial and temporal resolutions of bubble images are not sufficiently high. Thus a number of cold model experiments have been carried out by using water and silicone oils to simulate molten metals [48, 49]. Such model liquids are easy to handle, cheap and transparent. In most of the previous model experiments, the wall of the vessel is usually wetted by the liquid, which is a major departure from reality.

In this section, the behavior of a bubble colliding with an acrylic flat plate coated with paraffin is considered. The shape and size of a bubble in contact with the plate are also investigated in detail.

### 4.3.1 Experimental Apparatus and Procedure

Figure 4.41 shows a schematic of the experimental apparatus. The vessel had a square cross section of 0.260 m by 0.260 m and a height of 0.600 m. It was filled with water to a prescribed depth of 0.500 m. An acrylic flat plate was immersed in the bath such that its center was 0.100 m below the bath surface. The angle of inclination of the plate to the horizontal plane, $\theta_p$, was varied from 0 to 80°.

A predetermined volume of air was fed with a syringe into a cap-like container of diameter $4.1 \times 10^{-3}$ m from the bottom of the vessel. The container was then turned over to release the air into the bath. This allowed precise continuous control of the volume of bubble thus generated.

The bubble behavior was observed with a high-speed video camera at 200 frames per second. The images were recorded on a personal computer and then processed to determine the shape and size of the bubble at every instant. The horizontal and vertical sizes of the bubble before collision with the plate were denoted by $d_{By}$ and

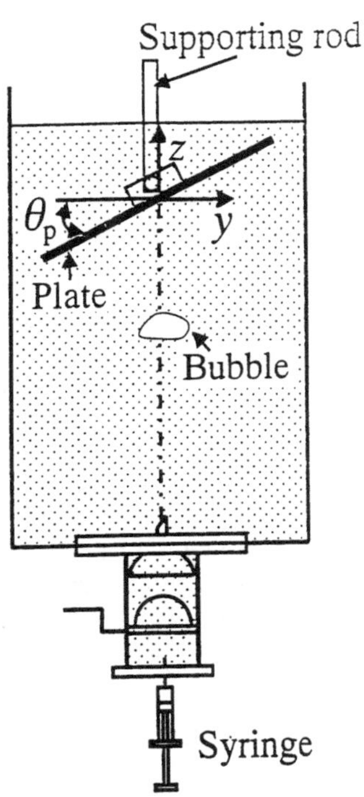


**Fig. 4.41** Schematic of experimental apparatus

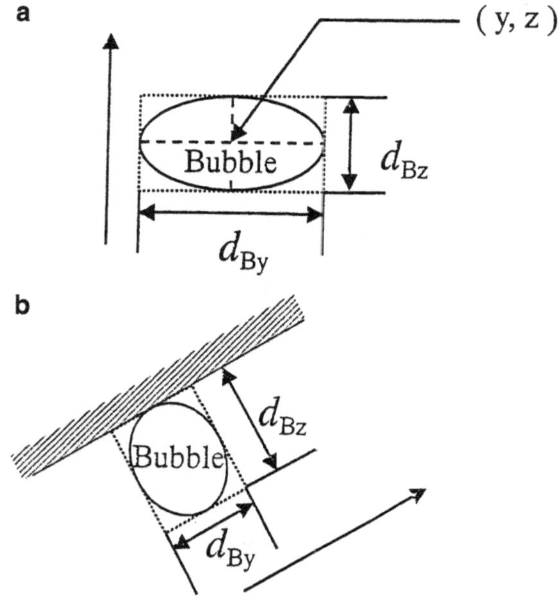


**Fig. 4.42** Definitions of bubble diameter and position. (**a**) Rising bubble toward a flat plate; (**b**) rising bubble along a flat plate or a sessile bubble beneath it

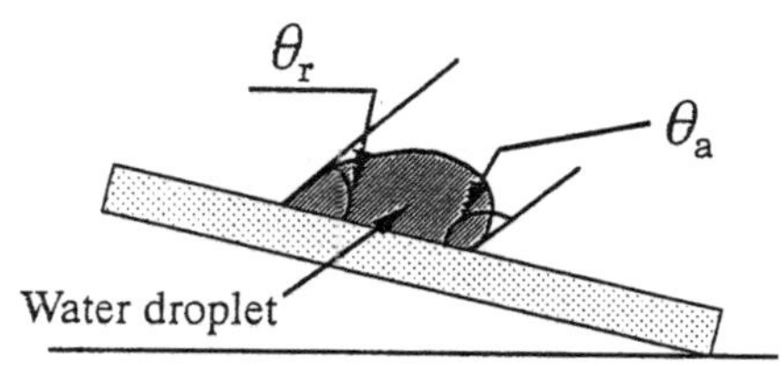


**Fig. 4.43** Schematic of the inclined plate setup for contact angle measurement

$d_{Bz}$, respectively, as shown in Fig. 4.42a. A representative diameter of the bubble, $d_B$, was defined as

$$d_B = (d_{By} d_{Bz})^{0.5} \tag{4.33}$$

The difference between $d_B$ and the volume-equivalent diameter was found to be less than 20%. This result will be discussed in detail in a subsequent section.

Experiments were carried out for representative diameter $d_B$ ranging from $1.0 \times 10^{-3}$ to $15.0 \times 10^{-3}$ m. The rising velocity of bubble was calculated by dividing the vertical displacement of the center of the bubble $(y, z)$ by a predetermined time interval. This velocity became constant (i.e., reached the terminal velocity $u_{B,\infty}$) at a location far below the flat plate. Therefore, the bubble collided with the plate at this terminal velocity. For convenience, the size of a stationary bubble or a sessile bubble just beneath the plate was modeled as shown in Fig. 4.42b.

The wettability of a flat plate was quantitatively evaluated in terms of $\theta_a$. This angle and the receding contact angle $\theta_r$ were determined by the inclined plate method, as shown in Fig. 4.43. A water droplet of known volume was placed on a flat plate, and the plate was gradually inclined. The shape of the bubble was observed with the high-speed video camera. The two contact angles, $\theta_a$ and $\theta_r$, were determined from the image of the droplet just before it started to slide [50–52]. The two angles are influenced by the surface roughness of the plate in addition to the interfacial energies.

Acrylic flat plate coated with paraffin was used to represent a plate of poor wettability. A flat brass plate was also chosen as a plate of good wettability. The observed advancing and receding contact angles of the paraffin-coated plate were 104 and 50°, respectively, while those of the brass plate were 80 and 17°, respectively.

## *4.3.2 Experimental Results*

### 4.3.2.1 Bubble Attachment to Flat Plate

Figure 4.44 shows the shape of a sessile bubble just below a flat plate for an inclination angle of $\theta_p = 0°$. It should be noted here that bubbles never attach directly to the flat plate of good wettability ($\theta_a = 80°$). There exists a thin water layer between the plate and the upper part of the bubble (see Fig. 4.44a–c). When the volume of the bubble is small, it is almost spherical in shape. As the volume increases, the effect of the static pressure around the bubble becomes larger than that of the

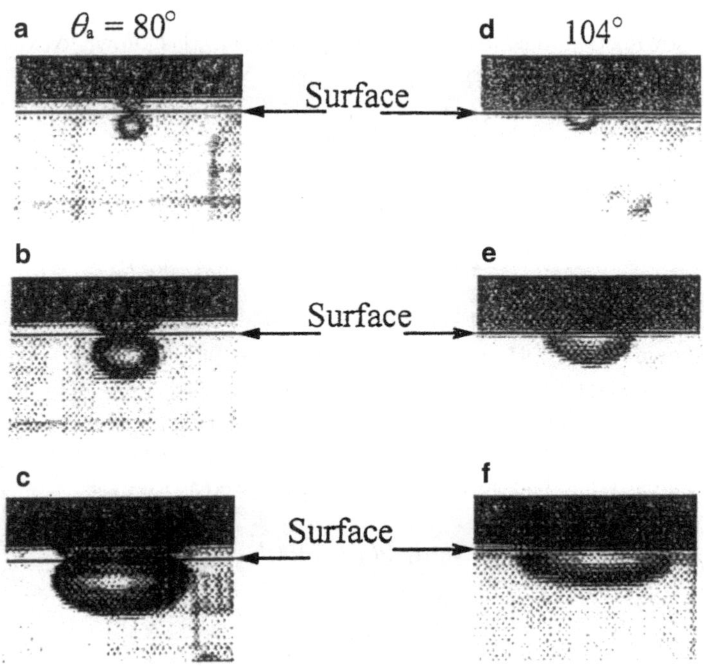


**Fig. 4.44** Sessile bubble beneath horizontal plates of different wettability

surface tension force acting on the bubble, and consequently, the bubble is flattened as shown in Fig. 4.44b, c. On the contrary, bubbles always attach directly to the plate of poor wettability (see Fig. 4.44d–f). In the same manner as for the plate of good wettability, bubbles become flattened as their volume increase.

A sessile bubble is considered to be stable when its energy is minimum. The energy consists of the potential energy due to gravitational force and surface energy. For a sessile bubble just beneath a highly wetted plate, the energy can be expressed by the following equation since there is no direct contact between the plate and the bubble:

$$E = (\rho_L - \rho_g)V_B g h_G + S\gamma_{Lg} \tag{4.34}$$

Here, $\rho_L$ is the density of liquid, $\rho_g$ is the density of gas, $V_B$ is the volume of the bubble, $g$ is the acceleration due to gravity, $h_G$ is the distance from the gravitational center of the bubble to the plate, $S$ is the gas/liquid interfacial area, and $\gamma_{Lg}$ is the gas/liquid interfacial energy.

In a similar manner, the energy of a sessile bubble just below a plate of poor wettability is given by

$$E = (\rho_L - \rho_g)V_B g h_G + S\gamma_{Lg} + rA(\gamma_{Sg} - \gamma_{SL}) \tag{4.35}$$

where $A$ is the interfacial area between the bubble and the plate, the surface of which is regarded as being completely smooth; $r$ is the roughness parameter; $rA$ is the actual gas/solid interfacial area; $\gamma_{Sg}$ is the gas/solid interfacial energy; and $\gamma_{SL}$ is the liquid/solid interfacial energy.

The following relationship first derived by Wensel [53] is known to exist between $\theta_a$ and the two interfacial energies, $\gamma_{Sg}$ and $\gamma_{SL}$,

$$r(\gamma_{Sg} - \gamma_{SL}) = \gamma_{Lg}\cos\theta_a \tag{4.36}$$

Combination of (4.35) and (4.36) yields

$$E = (\rho_L - \rho_g)V_B g h_G + S\gamma_{Lg} + A\gamma_{Lg}\cos\theta_a \tag{4.37}$$

The shape of a sessile bubble below each of the two types of plates is assumed to be as shown in Fig. 4.45. Specifically, disk-type bubble having an edge radius of $0.5d_{By}$ was assumed for the plate of good wettability [54]. A model of the shape of a sessile bubble beneath the plate of poor wettability is somewhat complicated, as shown in Fig. 4.45b [55]. In each model, the gas/liquid interfacial area $S$, gas/solid interfacial area $A$, and bubble volume $V_B$ are expressed as follows:

$$\begin{aligned} S &= \pi(d_{By} - d_{Bz})^2/2 + \pi^2 d_{Bz}[d_{Bz}/\pi + (d_{By} - d_{Bz})/2] \\ &\quad (\theta_a < \pi/2) \\ &= 2\pi r_c[(d_{By} - 2r_c\sin\theta_a)(\pi - \theta_a)/2 + r_c(1 + \cos\theta_a)] \\ &\quad + \theta(d_{By} - 2r_c\sin\theta_a)^2\pi/4(\theta_a \geq \pi/2) \end{aligned} \tag{4.38}$$

$$A = 0(\theta_a < \pi/2) = \pi d_{By}{}^2/4(\theta_a \geq \pi/2) \tag{4.39}$$

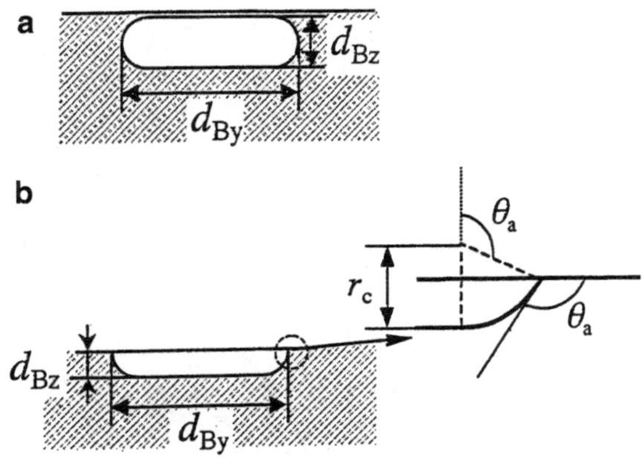


**Fig. 4.45** Bubble profile beneath a horizontal plate (**a**) $0° \le \theta_a < 90°$; (**b**) $90° \le \theta_a \le 180°$

$$
\begin{aligned}
V_B &= \pi(d_{By} - d_{Bz})^2 d_{Bz}/4 + \pi^2 d_{Bz}{}^2[2d_{Bz}/(3\pi) + (d_{By} - d_{Bz})/2] \\
&\quad (\theta_a < \pi/2) \\
&= (\pi/4)[\{(d_{By} - 2r_c \sin\theta_a)^2 + 4r_c{}^2\} r_c(1 + \cos\theta_a) \\
&\quad -(4r_c{}^3/3)(1 + \cos^3\theta_a) + \{2(\pi - \theta_a) + \sin 2\theta_a\} r_c{}^2 d_{By} - 2\{2(\pi - \theta_a) \\
&\quad + \sin 2\theta_a\} r_c{}^3 \sin\theta_a)](\theta_a \ge \pi/2)
\end{aligned}
\tag{4.40}
$$

where $r_c$ is indicated in Fig. 4.45b.

Here, the following aspect ratio $R$ is introduced:

$$
R = d_{By}/d_{Bz} \tag{4.41}
$$

Equations (4.38)–(4.40) are expressed as functions of $R$ as follows:

$$
\begin{aligned}
S &= (\pi/2)d_{Bz}{}^2\{R^2 + (\pi - 2)R + 3 - \pi\}(\theta_a < \pi/2) = (\pi d_{Bz}{}^2/4)[R^2 \\
&\quad + 4(\pi - \theta_a - \sin\theta_a)R/(1 + \cos\theta_a) - 4\sin\theta_a\{2(\pi - \theta_a) \\
&\quad - \sin\theta_a\}/(1 + \cos\theta_a)^2 + 8/(1 + \cos\theta_a)](\theta_a \ge \pi/2)
\end{aligned}
\tag{4.42}
$$

$$
A = 0(\theta_a < \pi/2) = \pi d_{Bz}{}^2 R^2/4(\theta_a \ge \pi/2) \tag{4.43}
$$

$$
\begin{aligned}
V_B &= \pi d_{Bz}{}^3/4\{R^2 + (\pi/2 - 2)R + 5/3 - \pi/2\}(\theta_a < \pi/2) = (\pi/4) \\
&\quad d_{Bz}{}^3[R^2 + \{2(\pi - \theta_a) + \sin 2\theta_a)/(1 + \cos\theta_a)^2 - 4\sin\theta_a)/(1 + \cos\theta_a)\} \\
&\quad R + 4(1 + \sin^2\theta_a)/(1 + \cos\theta_a)^2 - 4(1 - \cos\theta_a + \cos^2\theta_a)/\{3(1 + \cos\theta_a)^2\} \\
&\quad -2\sin\theta_a\{2(\pi - \theta_a) + \sin 2\theta_a\}/(1 + \cos\theta_a)^3](\theta_a \ge \pi/2)
\end{aligned}
\tag{4.44}
$$

A value of $R$ minimizing $E$ for a given bubble volume $V_B$ will be determined in the following manner. It is evident from (4.44) that $d_{Bz}$ is a function of $V_B$ and $R$. By substituting $d_{Bz}$ obtained from (4.44) into (4.42) and (4.43), both $S$ and $A$ can be expressed as functions of $R$. Accordingly, the bubble energy $E$ becomes a sole function of $R$ if $V_B$ is given.

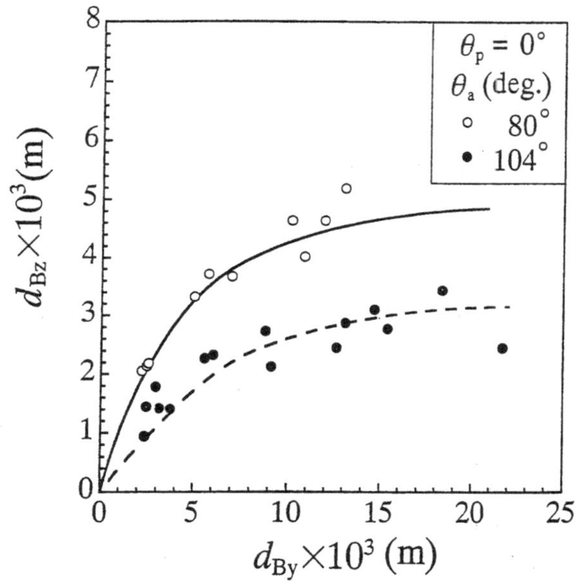


**Fig. 4.46** Horizontal and vertical dimensions of a bubble beneath horizontal plates of different wettability

The minimum $E$ value was determined iteratively by changing $R$ at 0.001 intervals. In these calculations, the surface tension of water, $\gamma_{Lg}$; the density of water, $\rho_L$; and the density of air, $\rho_g$, are 72.75 mN/m, 998.2 kg/m$^3$, and 1.20 kg/m$^3$, respectively. The calculated values are compared with the measured data in Fig. 4.46. Both sets of results agree well regardless of the wettability of the plate. Therefore, the bubble shape models presented above are useful for the prediction of the shape of a sessile bubble below a horizontal flat plate.

#### 4.3.2.2 Bubble Collision with Flat Plate

Figure 4.47 shows representative behavior of a bubble just after collision with each flat plate for an inclination angle of $\theta_p = 40°$. For the wetted plate, a small bubble ascends the plate, maintaining a spherical shape (Fig. 4.47a). As the volume of the bubble increases, the bubble begins to flatten, as shown in Fig. 4.47b, c. Figure 4.47d, e show the attachment of a bubble to the plate of poor wettability for $\theta_p = 40°$. In Fig. 4.47e, the bubble ascends along the plate due to buoyancy. When the bubble is large (Fig 4.47f) or when the inclination angle $\theta_p$ is large, no attachment of the bubble to the poorly wetted plate is observed. The bubble ascends near the wall in the same manner as that for the plate of good wettability.

The behavior of bubbles colliding with the plate of poor wettability ($\theta_a = 104°$) can be classified into the following five types with respect to the bubble diameter $d_B$ and the inclination angle $\theta_p$, as shown in Fig. 4.48.

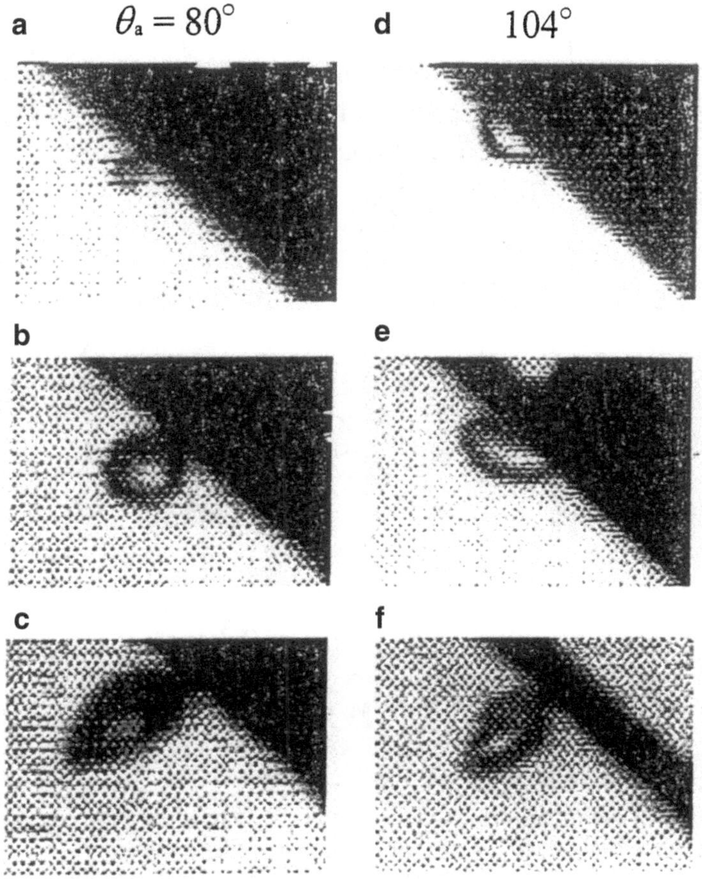


**Fig. 4.47** Shape of bubbles rising along an inclined plate of $\theta_p = 40°$. (**a**) $d_B = 1.6 \times 10^{-3}$ m; (**b**) $d_B = 3.96 \times 10^{-3}$ m; (**c**) $d_B = 7.76 \times 10^{-3}$ m; (**d**) $d_B = 2.16 \times 10^{-3}$ m; (**e**) $d_B = 4.46 \times 10^{-3}$ m; (**f**) $d_B = 8.16 \times 10^{-3}$ m

1. Bubble attaches to the plate after bouncing off it.
2. Bubble attaches to the plate without bouncing off.
3. Bubble ascends, sliding below the plate after attaching to the plate.
4. Bubble ascends, sliding along the plate without attaching to the plate.
5. Bubble ascends with repeated bouncing but never attaches to the plate.

Attachment of a bubble to the plate of poor wettability was observed even at $\theta_p = 60°$.

It is known that a bubble bounces off a flat plate of good wettability for a Weber number, *We*, larger than 0.3 [43]. Under the experimental conditions considered here, a bubble of approximately $2 \times 10^{-3}$ m in diameter was observed to bounce off a poorly wetted plate. For example, the Weber number for the bubble shown

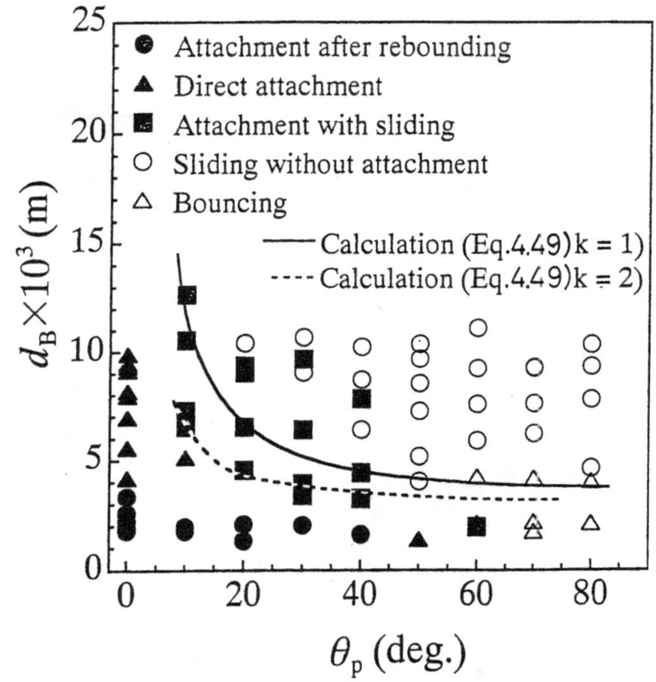


**Fig. 4.48** Behavior of an impacting bubble to a poor wettability plate

in Fig. 4.49a was 2.0. However, there was no observed bouncing for a bubble of diameter much larger than $2 \times 10^{-3}$ m, as shown in Fig. 4.49b.

The mechanism of the bouncing of a bubble from a plate can be explained as follows. Before the bubble comes in contact with the plate, the water just above the bubble first collides with the plate. Consequently, the pressure in the thin water layer between the plate and the bubble is thereby elevated. When the diameter of the bubble is small, such as $2 \times 10^{-3}$ m, the force acting downwards on the bubble due to the pressure increase inside the water layer is higher than the buoyancy force, and the surface tension force is sufficiently large to prevent a breakup of the bubble. This results in the bouncing of the bubble off the plate. This phenomenon occurs irrespective of wettability of the plate. The implication of this result is that the bouncing of a bubble off a plate is always accompanied by the existence of a thin liquid layer between the bubble and the plate even if the plate is not wetted by the liquid.

With increase in both $\theta_p$ and $d_B$, a bubble attached to the poorly wetted plate starts to slide along the plate due to the buoyancy force acting on the bubble. The point of incipient slide is represented in Fig. 4.48 by the solid and broken lines. The method used to determine these lines are described below.

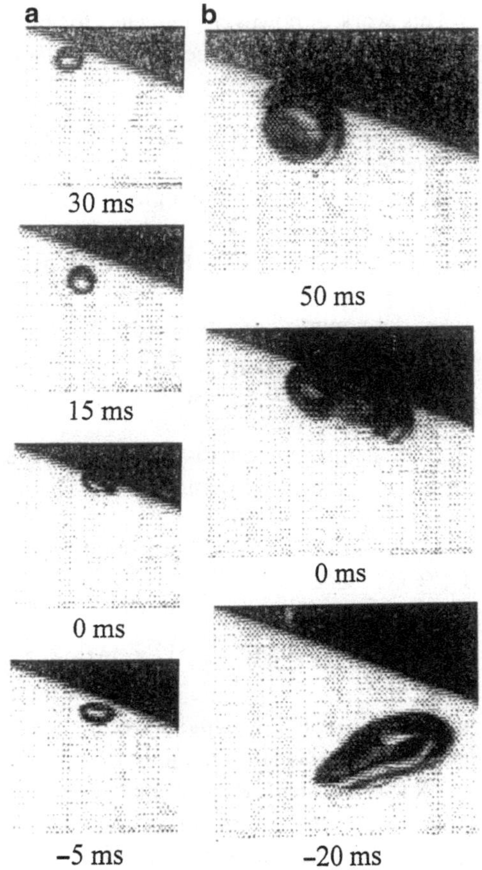


**Fig. 4.49** Impact sequence of a bubble to a flat plate coated with paraffin wax. (**a**) $d_B = 2.66 \times 10^{-3}$ m, $u_B = 0.238$ m/s; (**b**) $d_B = 5.76 \times 10^{-3}$ m, $u_B = 0.22$ m/s

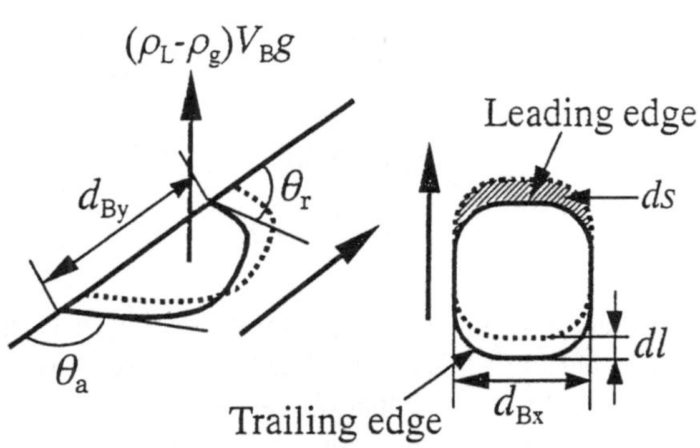


**Fig. 4.50** Profile of a bubble sliding on a poor wettability plate

The shape of a bubble which starts to ascend and slide along the flat plate is assumed, as shown in Fig. 4.50. The work required to replace a solid/liquid interface of area d$s$ by a solid/gas interface just before the leading edge of the bubble is given by

$$W_A = -\gamma_{Lg}\{1 - (\cos\theta_r)/r\}r\,ds \tag{4.45}$$

This work is however neglected for $\theta_r \geq 90°$. Similarly, the work required to replace a solid/gas interface of areas d$s$ located near the trailing edge of the bubble is expressed by

$$W_T = -\gamma_{Lg}\{1 - (\cos\theta_a)/r\}r\,ds \tag{4.46}$$

Combination of (4.45) and (4.46) yields the total work required for the bubble to ascend along the inclined flat plate as follows:

$$W = W_A + W_T = \gamma_{Lg}(\cos\theta_r - \cos\theta_a)ds \tag{4.47}$$

On the contrary, the work done by the buoyancy force acting on the bubble is given by

$$W = k(\rho_L - \rho_g)V_B g(\sin\theta_p)dl \tag{4.48}$$

where d$l$ is the apparent displacement of the bubble along the plate and coefficient $k$ represents the effect of surface roughness of the plate. For a completely smooth plate, $k = 1$.

Assuming $ds \cong d_{Bx}\,dl$, we obtain the following force balance equation from (4.47) and (4.48):

$$k(\rho_L - \rho_g)V_B g\sin\theta_p = \gamma_{Lg}(\cos\theta_r - \cos\theta_a)d_{Bx} \tag{4.49}$$

When the plate is completely smooth, i.e., $k = 1$, (4.49) reduces to that describing the force balance for a liquid droplet on a flat plate [51, 56–58].

If $V_B$ is constant, there remain three unknown parameters, $\theta_p$, $k$ and $d_{Bx}$, in (4.49). Furthermore, assuming $d_{Bx} = d_{By}$ for convenience, $d_{By}$ can be determined from (4.37) as the bubble volume $V_B$ is given. This reduces the number of unknown parameters to two. The experimental conditions considered, however, indicate that a bubble of relatively large volume is no longer spherical. Thus the relationship between $d_B$ and $V_B$ must be obtained a priori. One empirical relation is expressed by

$$d_B = (6V_B/\pi)^{1/3} \quad (V_B <= 1.5 \times 10^{-8}\,\text{m}^3) \tag{4.50}$$

$$d_B = 115{V_B}^{2/3} + 2.32 \times 10^{-3} \quad (V_B > 1.5 \times 10^{-8}\,\text{m}^3) \tag{4.51}$$

Figure 4.51 shows the relationship between $V_B$ and $d_B$, as well as that between (4.50) and (4.51). Equation (4.51) compares favorably with the measured values. These results indicate that $d_B$ can indeed be determined provided that the volume of a bubble is given. In other words, if the roughness parameter $k$ is known a priori, the relationship between $d_B$ and $\theta_p$, which gives the critical condition for a bubble to start to slide along the plate, can be determined from (4.37), (4.49), and (4.51).

The solid line in Fig. 4.48 denotes the calculated value for $k = 1$. Clearly, this line is not satisfactory for predicting the boundary between the sessile bubble regime and the sliding bubble regime. The value of $k$ giving the best fit to the boundary was found to be 2, as demonstrated by the broken line in Fig. 4.48.

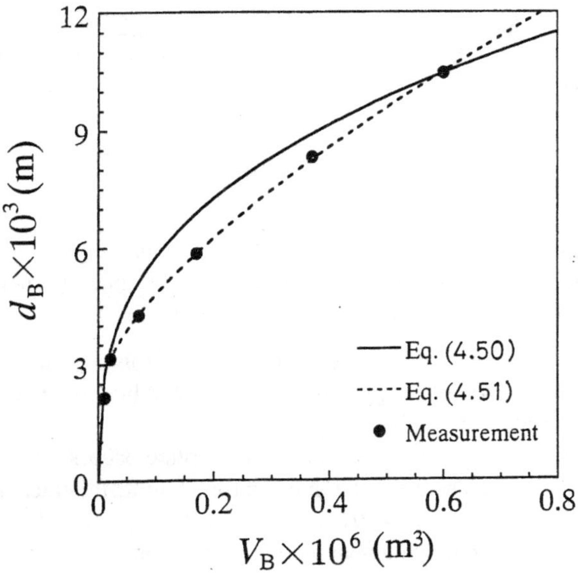


**Fig. 4.51** Relationship between bubble volume $V_B$ and bubble diameter

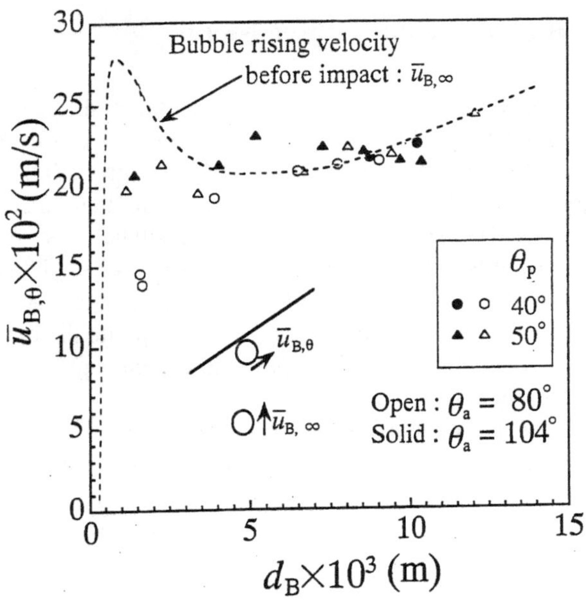


**Fig. 4.52** Bubble rising velocity along an inclined plate

The mean rising velocity, $\overline{u_{B,\theta}}$, of bubbles ascending without contact with an inclined flat plate is shown in Fig. 4.52. There is no difference between the measured values for the plates of good and poor wettability. For a $d_B$ smaller than

$4 \times 10^{-3}$ m, $\overline{u_{B,\theta}}$ is lower than the terminal velocity $\overline{u_{B,\infty}}$. Both velocities become close as $d_B$ increases. This trend is consistent with the experimental results of Maxworthy [42].

#### 4.3.2.3 Summary

Results have been presented of water model experiments on bubble collision and subsequent attachment or detachment on an inclined flat plate immersed in a water bath. The main findings can be summarized as follows:

1. The shape of a sessile bubble placed beneath a horizontal plate can be predicted satisfactorily by using an energy equation for the bubble, irrespective of the wettability of the plate.
2. Bubble attachment to a poorly wetted inclined plate occurs until the inclination angle $\theta_p$ reaches 60°. The critical condition for a bubble to attach and stick to the plate can be represented by (4.49) as a function of $\theta_a$, $\theta_r$, and $k$.
3. The velocity of a bubble rising near an inclined flat plate without attaching to the plate is essentially independent of wettability.

## 4.4 Bubble Removal from Molten Metal

Removal of small bubbles from molten metal is essential for the production of clean steel as well as suppression of foaming in metal refining processes [59–62], as described in Sect. 4.1. In continuous casting molds, Ar gas is introduced into the submerged entry nozzle (SEN) to prevent the attachment of alumina ($Al_2O_3$) to the inner part of the SEN. The Ar gas disintegrates into small bubbles over a wide range of diameters due to highly turbulent motion caused by molten steel flow issuing from the SEN. Large bubbles are lifted up due to buoyancy force and are removed from the mold surface, while smaller bubbles are carried deep into the mold and trapped in the solidified steel as pin-holes. Such pin-hole defects significantly reduce the quality of the steel product.

In this section, experimental results will be presented on the attachment of bubbles to a circular cylinder placed horizontally in a water–air vertical bubbling jet [15].

### *4.4.1 Experimental Apparatus and Procedure*

Figure 4.53 shows a schematic of the experimental apparatus. The acrylic cylindrical vessel had an inner diameter $D$ of $200 \times 10^{-3}$ m and a height $H$ of $400 \times 10^{-3}$ m. The vessel was filled with water to a depth $H_L$ of $300 \times 10^{-3}$ m. Air was injected into the bath through a perforated nozzle. The exit of the nozzle has a square cross

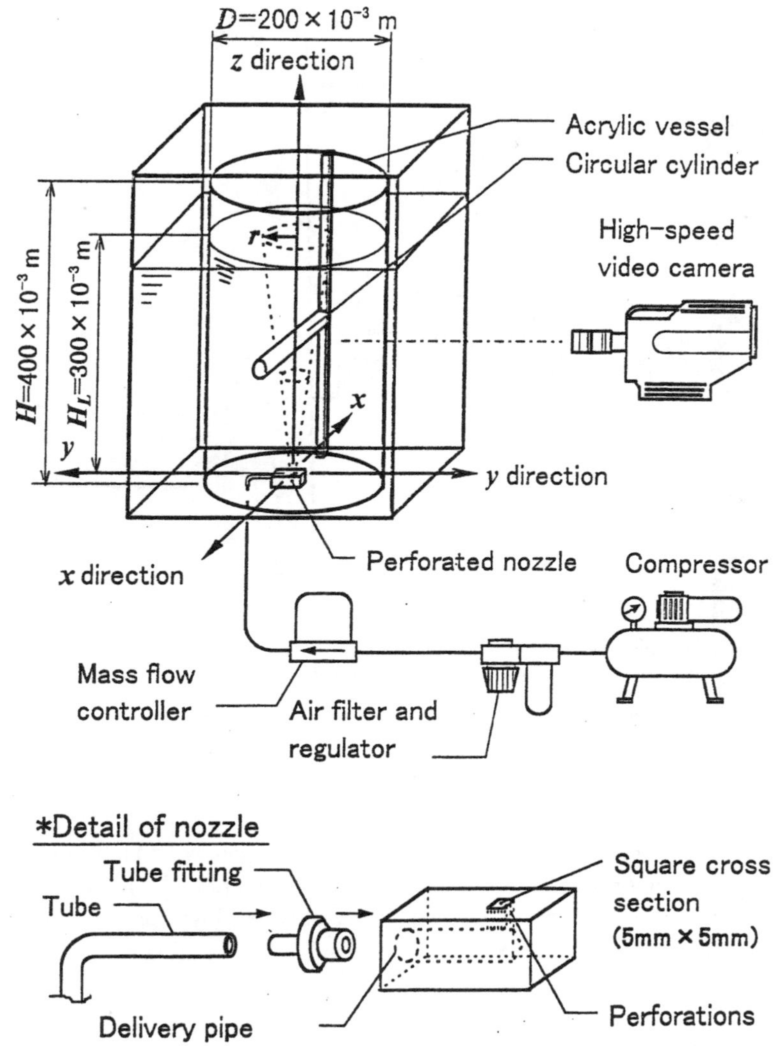


**Fig. 4.53** Schematic diagram of the experimental apparatus and coordinate system

section of $5 \times 10^{-3}$ m by $5 \times 10^{-3}$ m. The air flow rate was adjusted with a mass flow controller over the range $0.5 \times 10^{-6}$–$100 \times 10^{-6}$ m$^3$/s. A circular cylinder was placed horizontally above the nozzle, as shown in Fig. 4.54. The distance from the nozzle tip to the leading edge, i.e., the lower stagnation point, of the cylinder was set at $150 \times 10^{-3}$ m. The vessel diameters were $10 \times 10^{-3}$, $15 \times 10^{-3}$ or $20 \times 10^{-3}$ m. The effective diameters were $11 \times 10^{-3}$, $16 \times 10^{-3}$, or $21 \times 10^{-3}$ m after each cylinder was coated with paraffin. The $\theta_a$ of the original aluminum cylinders was 63° and that of the paraffin-coated cylinders was 96°. Accordingly, the cylinders of $\theta_a = 63°$ were wetted by water, while those of $\theta_a = 96°$ were not

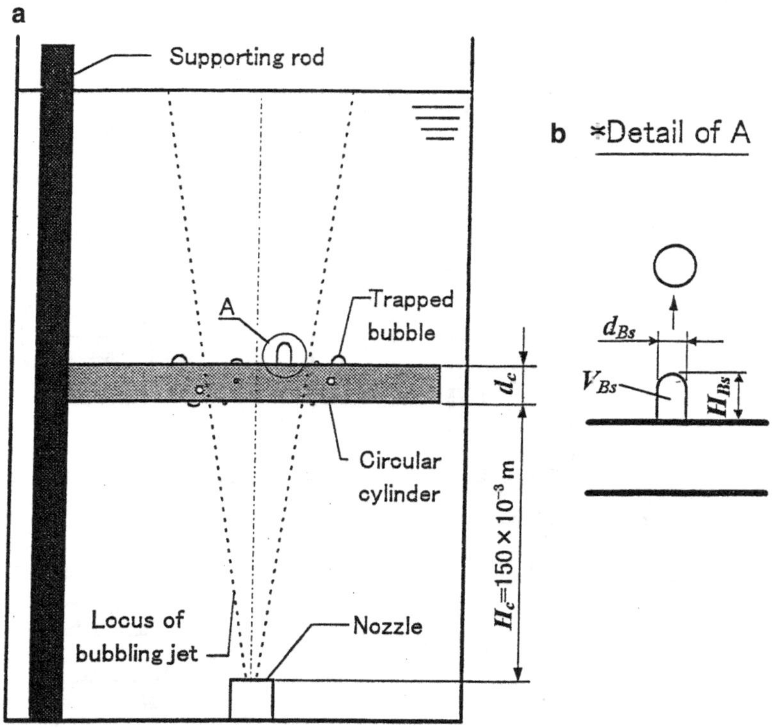


**Fig. 4.54** Position of circular cylinder and explanation of symbols

wetted by water. The mean bubble diameter, bubble frequency, and the width of bubble dispersion region were determined with a still camera and a high-speed video camera at 200 frames per second. The velocity of water flow was measured with a two-channel laser Doppler velocimeter.

## *4.4.2 Experimental Results*

### 4.4.2.1 Behavior of Bubbling Jet Approaching Horizontal Cylinder

1. Classification of bubbling jets generated from perforated nozzle

Before discussing the attachment and detachment behavior of bubbles to a horizontal cylinder, the characteristics of bubbling jets approaching the cylinder were first investigated [15]. From previous studies on bubbling jets generated from a porous nozzle [63, 64], bubble formation patterns can be classified into three types with respect to the gas flow rate $Q_g$: low, medium, and high gas flow rate regimes, as schematically shown in Fig. 4.55. The gas flow rates indicating the boundaries

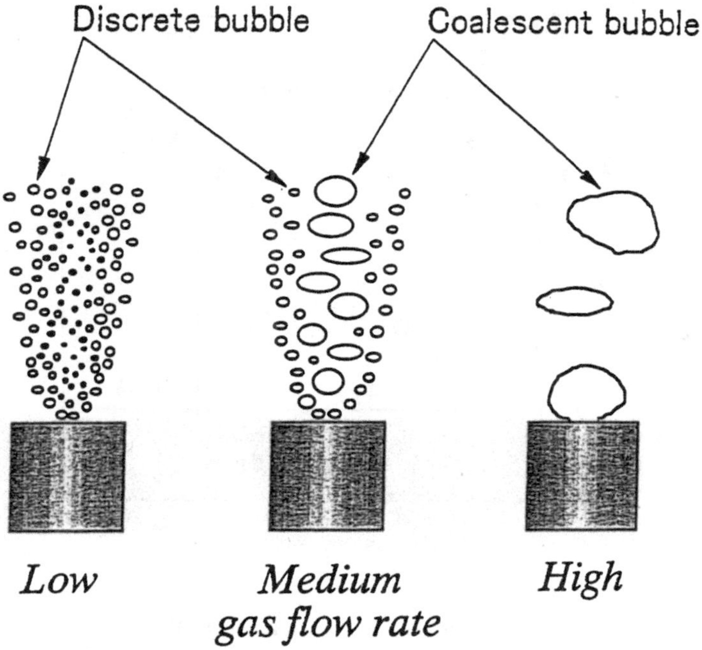


**Fig. 4.55** Classification of bubble formation patterns

between the three flow regimes were $Q_g = 6 \times 10^{-6}$ and $Q_g = 30 \times 10^{-6}\,\mathrm{m^3/s}$, respectively. In the low gas flow rate regime, small discrete bubbles are generated that rise vertically upward without coalescence and disintegration. In the medium gas flow rate regime, small bubbles are also generated, but some of them coalesce into lager bubbles near the centerline of the bubble dispersion region. In the high gas flow rate regime, a large bubble covering the entire nozzle surface is repeatedly generated.

2. Mean bubble diameter in bubbling jets approaching cylinder

The mean bubble diameter, $\overline{u_B}$, measured in a bubbling jet approaching a cylinder is given in Fig. 4.56. The boundaries between the aforementioned three gas flow rate regimes are also indicated in the figure. Open symbols in the low and medium gas flow rate regimes denote discrete bubbles generated from the perforated nozzle, while shaded symbols in the medium and high gas flow rate regimes denote coalescent bubbles. In the regime ranging from approximately $6 \times 10^{-6}$–$30 \times 10^{-6}\,\mathrm{m^3/s}$, there are two distinct bubble mean diameters typical of the medium gas flow rate regime. The diameter of the discrete bubbles increases with an increase in the gas flow rate in the low gas flow rate regime. On the contrary, in the medium gas flow rate regime, the diameter decreases slightly with $Q_g$. The mean diameter of the coalesced bubbles increases monotonically with an increase in $Q_g$. In the high gas

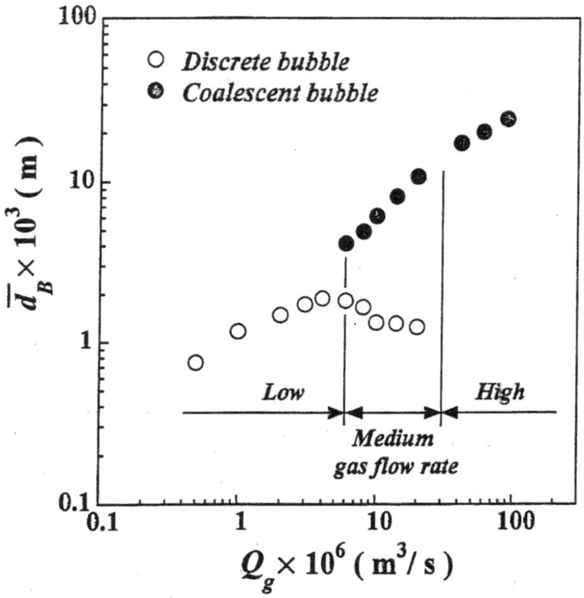


**Fig. 4.56** Mean bubble diameter in a bubbling jet approaching a cylinder

flow rate regime, the diameter can be satisfactorily approximated by the following empirical relation for the diameter of bubbles generated from a single-hole nozzle [26, 65]; thus,

$$d_B = 0.54[Q_g(d_{ni})^{1/2}]^{0.289} \tag{4.52}$$

where $d_{ni}$ is the inner diameter of the nozzle, with a value of $5 \times 10^{-3}$ m in the experimental study.

Figure 4.57 shows that the mean rising velocity of small discrete bubbles is nearly independent of the gas flow rate, while that of large coalesced bubbles increases monotonically with $Q_g$. The mean rising velocity of the large coalescent bubbles can be approximated by the following empirical equation proposed for bubbles generated from a single-hole nozzle [66]:

$$\overline{u_B}/(g^{0.4}{Q_g}^{0.2}) = 1.60(z/z_0)^{-2.04} + 1.82(z/z_0)^{-0.08} \tag{4.53}$$

$$z_0 = 6.8d_{ni}[\rho_g {Q_g}^2/(\rho_L g {d_{ni}}^5)]^{0.272} \tag{4.54}$$

where $z$ is the vertical distance from the nozzle exit and $z_0$ is the $z$ value at which gas holdup on the centerline of the bubbling jet is 50%.

(a) Axial mean velocity of water flow approaching cylinder

Figure 4.58 shows the variation with $Q_g$, of the axial mean velocity $\overline{u_{cl}}$ and the root-mean-square value of axial turbulence component $u'_{rms,cl}$. Both variables increase with $Q_g$. The turbulence intensity increases from 30% in the low gas flow

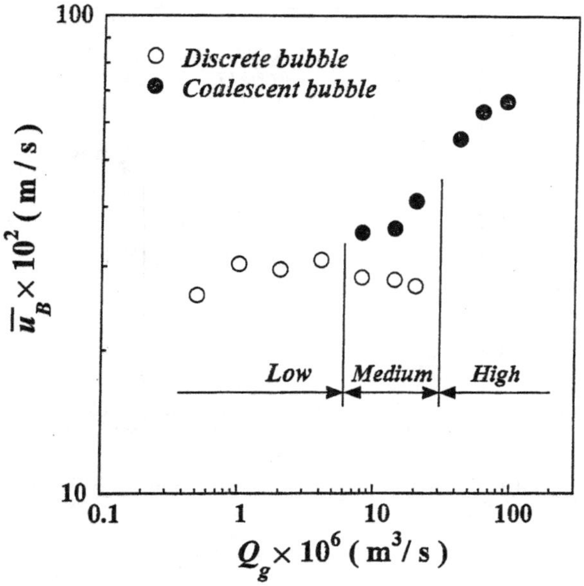


**Fig. 4.57** Mean bubble rising velocity in a bubbling jet approaching a cylinder

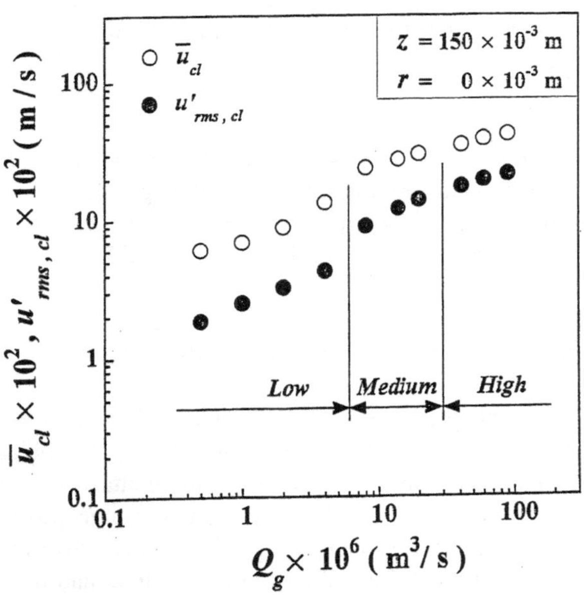


**Fig. 4.58** Axial mean velocity and the root-mean-square value of axial turbulence component of water flow approaching a cylinder

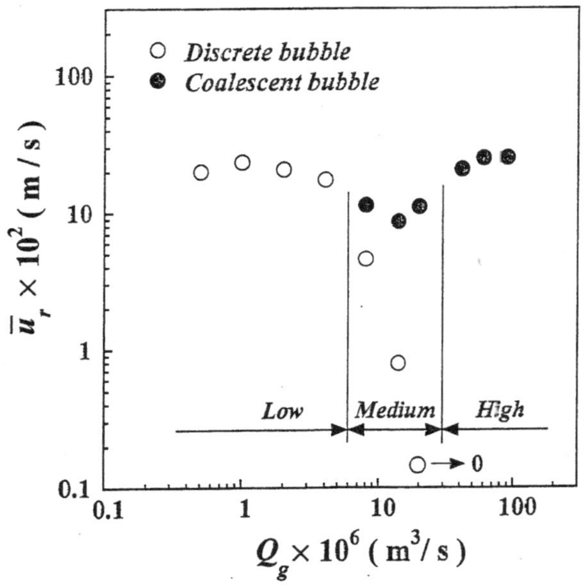


**Fig. 4.59** Relative velocity

rate regime to 50% in the medium and high gas flow rate regimes. The turbulence intensity is known to be approximately 50% in a bubbling jet produced using a single-hole nozzle.

(b) Relative velocity

The relative velocity, $\overline{u_r}$, defined by $\overline{u_B} - \overline{u_{cl}}$, is approximately $20 \times 10^{-2}$ m/s in the low and high gas flow rate regimes. In the medium gas flow rate regime, $\overline{u_{yr}}$ for large coalescent bubbles exhibits a minimum of $10 \times 10^{-2}$ m/s and that for small discrete bubbles becomes very small, as shown in Fig. 4.59.

#### 4.4.2.2 Behavior of Bubbles on Cylinder Surface

Figure 4.60a–c show photographs of three circular cylinders of good wettability placed in a bubbling jet. There is no observed attachment of bubbles to any of the cylinders. On the contrary, bubbles are observed to be attached to the cylinder of poor wettability (see Fig. 4.61a) in the low gas flow rate regime. Most of the attached bubbles are subsequently expelled from the cylinder surface due to strong turbulence induced by the wakes of the rising bubbles in the medium and high gas flow rate regimes, as can be seen in Fig. 4.61b, c.

The mechanism of bubble attachment to a solid body of poor wettability can be explained as follows. When a bubble approaches the solid body for which the advancing contact angle falls between 90 and 180°, the bubble pushes water away from the solid body and attaches to it (see Fig. 4.62). On the contrary, the bubble

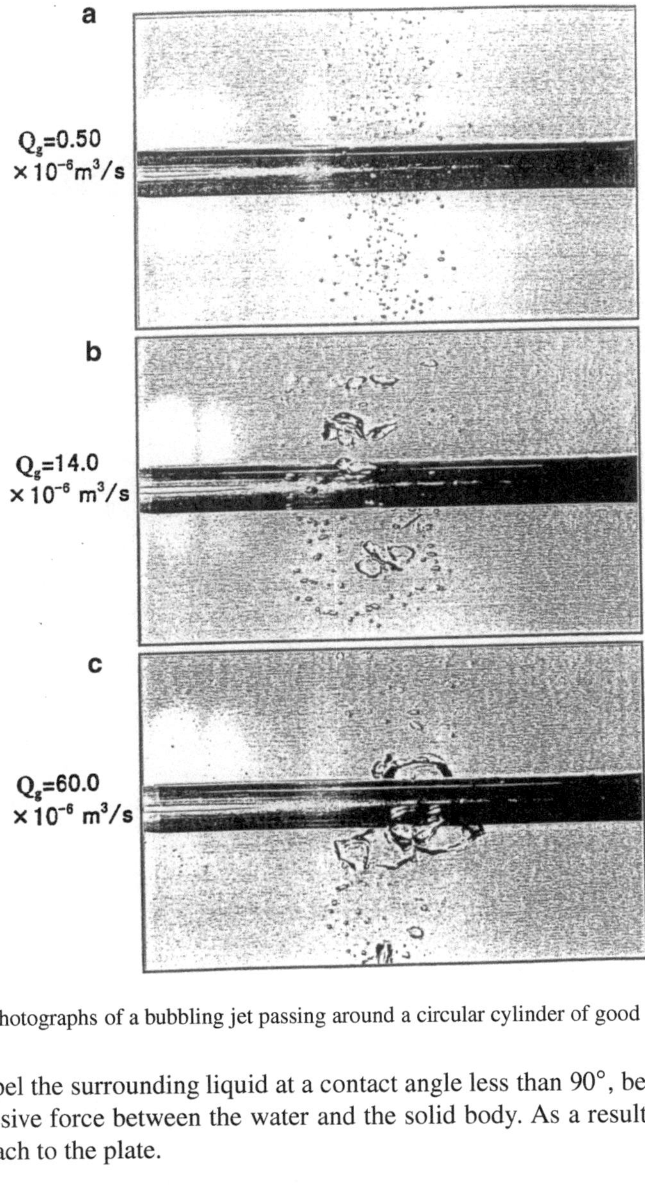


**Fig. 4.60** Photographs of a bubbling jet passing around a circular cylinder of good wettability

cannot expel the surrounding liquid at a contact angle less than 90°, because of the large adhesive force between the water and the solid body. As a result, the bubble cannot attach to the plate.

#### 4.4.2.3 Stem Diameter and Stem Height of Trapped Bubble

The measured stem diameter $\overline{d_{BS}}$ and stem height $\overline{H_{BS}}$ of a trapped bubble at incipient detachment from the rear stagnation point of a cylinder are given in Figs. 4.63 and 4.64, respectively. The values of both $\overline{d_{BS}}$ and $\overline{H_{BS}}$ remain unchanged with respect to gas flow rate $Q_g$ and cylinder diameter, $d_c$, in the low gas flow rate regime. $d_c$ has a negligible effect on both $\overline{d_{BS}}$ and $\overline{H_{BS}}$. In the medium gas flow rate regime,

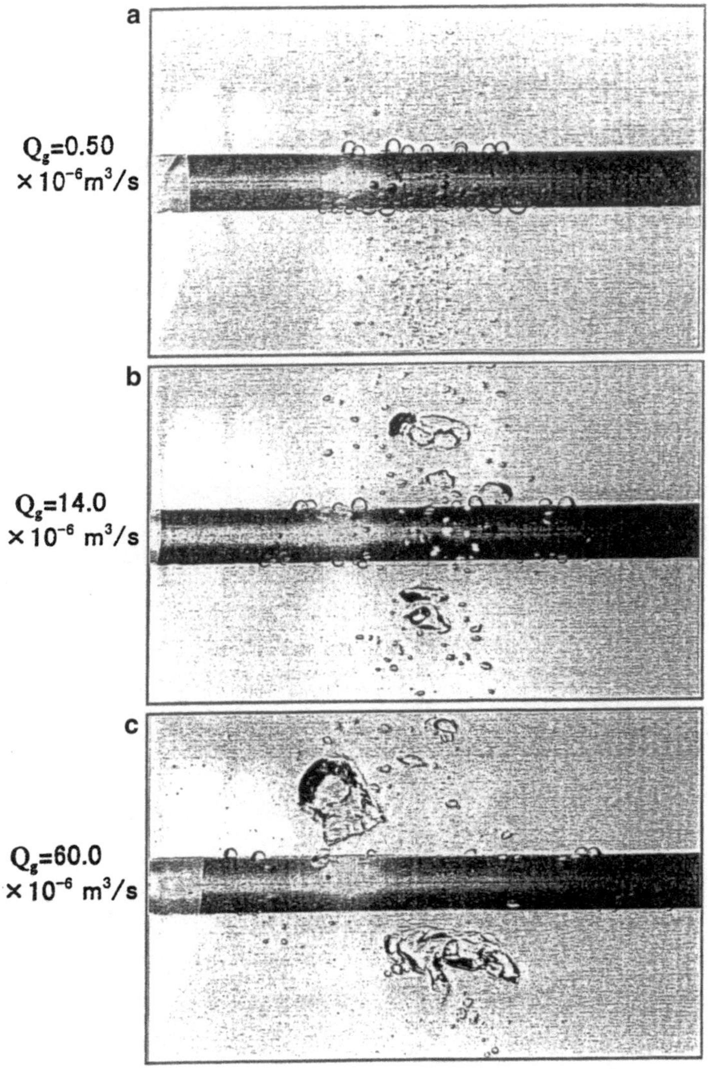


**Fig. 4.61** Photographs of a bubbling jet passing around a circular cylinder of poor wettability

$\overline{d_{BS}}$ and $\overline{H_{BS}}$ decrease abruptly due to the highly turbulent motions induced by large coalescent bubbles. The solid line in each figure represents potential flow calculations based on Model 1 (see Fig. 4.4). The calculated and measured values agree quite well in the low gas flow rate regime. Model 1 that was already described fully in Sect. 4.1 is chosen here because although it is the simplest of the five models, it adequately predicts the shape of bubbles for $\theta_a$ of about 96°.

Figure 4.65 presents the volume of a bubble $V_{BS}$ just before detachment from the circular cylinder. The solid line indicates the value predicted based on the potential flow method. The measurement error for $V_{BS}$ was ±20%. In the low gas flow rate

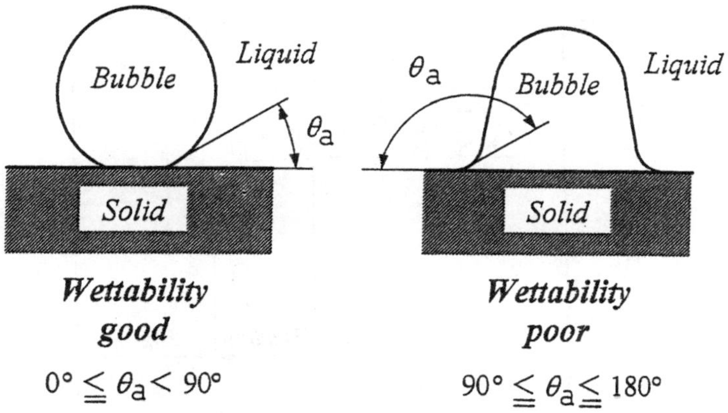


**Fig. 4.62** Mechanism of bubble attachment

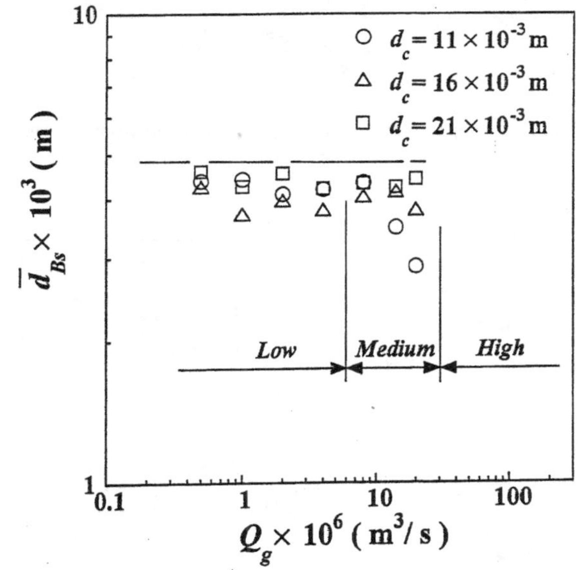


**Fig. 4.63** Stem diameter of a trapped bubble just before detachment from the rear stagnation point of a cylinder

regime, $V_{BS}$ is essentially independent of the gas flow rate, but slightly increases with an increase in $d_c$. This volume decreases drastically in the medium gas flow rate regime and nearly diminishes in the high gas flow rate regime.

Figure 4.66 shows a relationship between the aspect ratio $A_{Bs}$ and $V_B$. The solid line again denotes the value calculated based on Model 1 of Fig. 4.4. The measured values are satisfactorily approximated by the potential method.

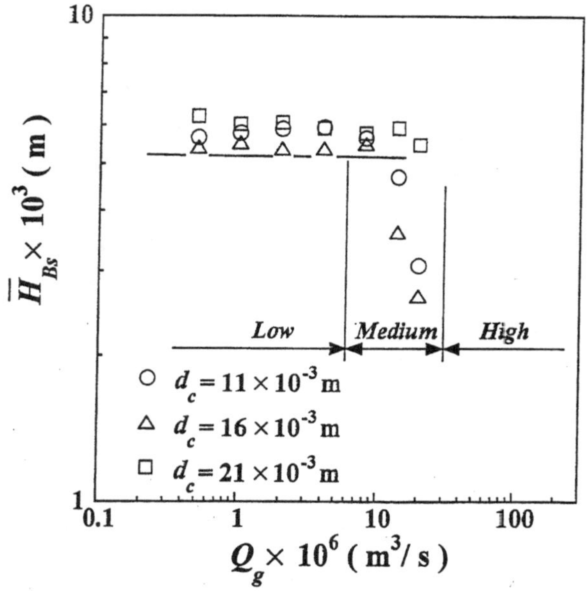


**Fig. 4.64** Stem height of a trapped bubble just before detachment from the rear stagnation point of a cylinder

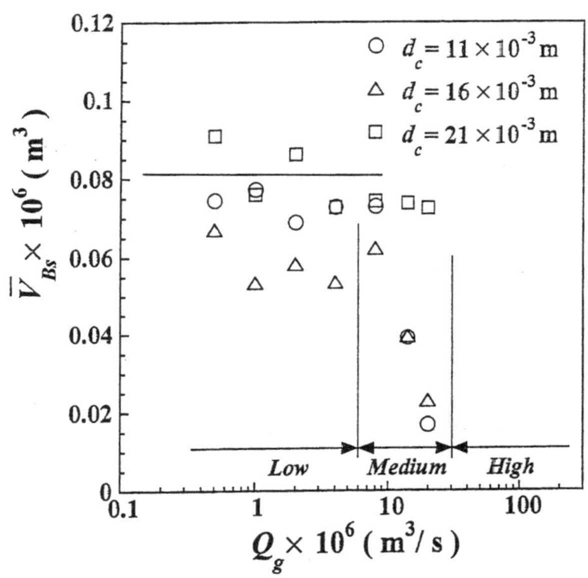


**Fig. 4.65** Volume of a trapped bubble just before detachment from the rear stagnation point of a cylinder

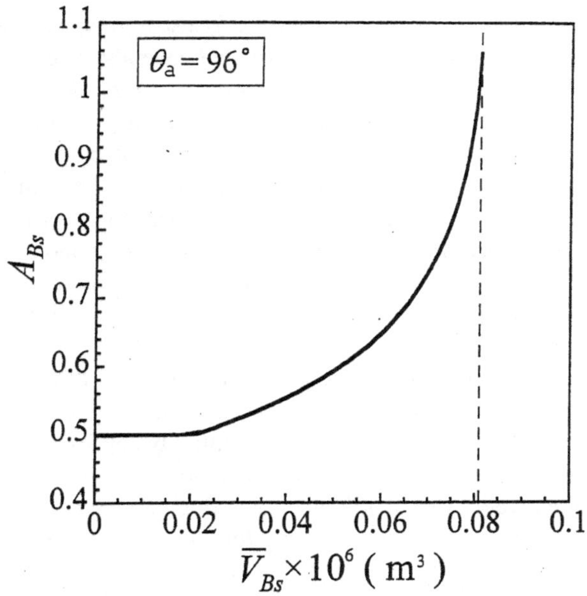


**Fig. 4.66** Relationship between the aspect ratio and the volume of a trapped bubble when the bubble energy is minimum

#### 4.4.2.4 Summary

The behavior of bubbles in a bubbling jet approaching a horizontal circular cylinder, and that of bubbles attaching to and detaching from the cylinder were observed with a still camera and a high-speed video camera [15]. The bubbling jets approaching the cylinder could be classified into three types with respect to the gas flow rate; low, medium, and high gas flow rate regimes. The characteristics of bubbles attaching to and detaching from the cylinder are strongly dependent on the behavior of bubbles in the jet approaching the cylinder.

Only bubbles with volumes less than the critical bubble volume, $V_{Bs}$, can be trapped on the upper surface of a circular cylinder. In a water–air system, the volume equivalent diameter of the trapped bubble was approximately 5 mm. As the critical bubble volume increases with $\theta_a$ of a solid body, larger bubbles can be trapped on the surface of the body by increasing $\theta_a$. The potential method satisfactorily predicts the shape and size of a bubble trapped by a circular cylinder of poor wettability placed in a water–air bubbling jet.

## 4.5 Flow Distribution in Vertical Pipes

Considerable effort has been devoted to understanding the characteristics of gas–liquid two-phase flows in vertical and horizontal pipes. Such systems are used in a variety of engineering fields, including mechanical, chemical, and atomic energy engineering [67–70]. Most studies have focused on pipes of good wettability, i.e., pipes wetted by the liquid. Extensive data have also been accumulated on the flow pattern, velocities of liquid and gas, pressure losses, and heat transfer in the pipes. However, investigations on the effects of wettability on flow characteristics are quite limited [71] in spite of the wide application of poorly wetted pipes in materials engineering [31] and atomic energy engineering. The limited research is partly because the liquids (molten metal) used in these engineering fields are not transparent and partly because the melting temperature of the molten metal is typically very high, so that measuring the flow characteristics is difficult and dangerous. Consequently, there is no adequate sensor for measuring the characteristics under such severe conditions.

Model experiments, therefore, are expected to be used, but it is difficult to control and maintain the wettability of a pipe wall over a sufficiently long duration even in the model experiments. Surface treatment is commonly applied to change the wettability of the pipe wall. The advancing contact angle, which is used to represent the wettability quantitatively [26], rapidly decreases due to contamination and in most cases, the pipe becomes wet with liquid only for a relatively short time. Therefore, long-range experiments are difficult except in very rare cases.

Terauchi et al. [72] carried out model experiments on the flow pattern in a vertical circular pipe of poor wettability. $\theta_a$ of an acrylic pipe was changed by coating a hydrophilic substance or liquid paraffin on the inner wall of the pipe. Three different values of $\theta_a$ were realized; $\theta_a = 36, 77$, and 104°. The flow pattern was observed with a still camera and a high-speed video camera to understand the effects of pipe wettability. The results of this study will be presented in the following.

### 4.5.1 Experimental Apparatus and Procedure

Figure 4.67 shows a schematic of the experimental apparatus. Three transparent acrylic pipes with an inner diameter $D$ of 5.0, 10.0, or 15.0 mm were used. The original pipe had a $\theta_a$ of 77°, and, accordingly, it was wetted by water. The wettability of the pipe was varied by coating with a hydrophilic substance or liquid paraffin. The $\theta_a$ was 36° for the hydrophilic substance coating and 104° for the liquid paraffin coating. The former and the latter pipes are classified into a pipe of good wettability and a pipe of poor wettability, respectively. The lifetime of each coating was long enough to carry out systematic experiments. Although purified water was initially used, it became fully contaminated before the measurements were carried out. The water was circulated with a pump, and air was supplied through a porous nozzle settled flush on the inner wall of the lower part of each pipe. The diameter of the nozzle, $d_p$, was 4.0 mm, and it had a pore diameter of 270 μm and porosity of 25%. Pulsation arising from the compressor was satisfactorily suppressed with a filter. The shape and size of bubbles and slugs rising in the fully developed region $L/D \geq 53$ for the three pipes were recorded with a still camera and a high-speed video camera at 200 frames per second.

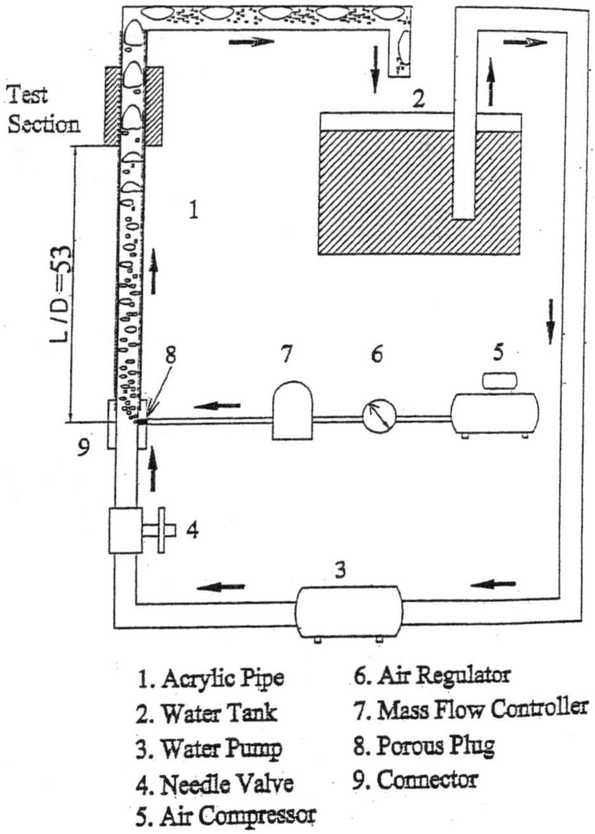


**Fig. 4.67** Experimental apparatus

## 4.5.2 *Experimental Results*

### 4.5.2.1 Flow Distribution

The change in the flow patterns of air–water two-phase flows in the three pipes of different contact angles is shown in Fig. 4.68, where $j_G$ and $j_L$ denote the superficial velocities of air and water, respectively [72]. Each shaded portion appearing in the upper row of the figure indicates the part of a bubble or a slug in contact with the pipe wall. The flow patterns of air–water two-phase flows in the pipe of $\theta_a = 36°$ were nearly identical to those in the pipe of $\theta_a = 77°$. This result implies that the flow pattern of air–water two-phase flows in a vertical pipe of good wettability is not significantly affected by the contact angle at least over the range 36–77°. For the sake of simplicity and brevity, only the experimental results obtained for the pipes of $\theta_a = 77$ and 104° are presented to examine the effects of pipe wettability on the flow pattern.

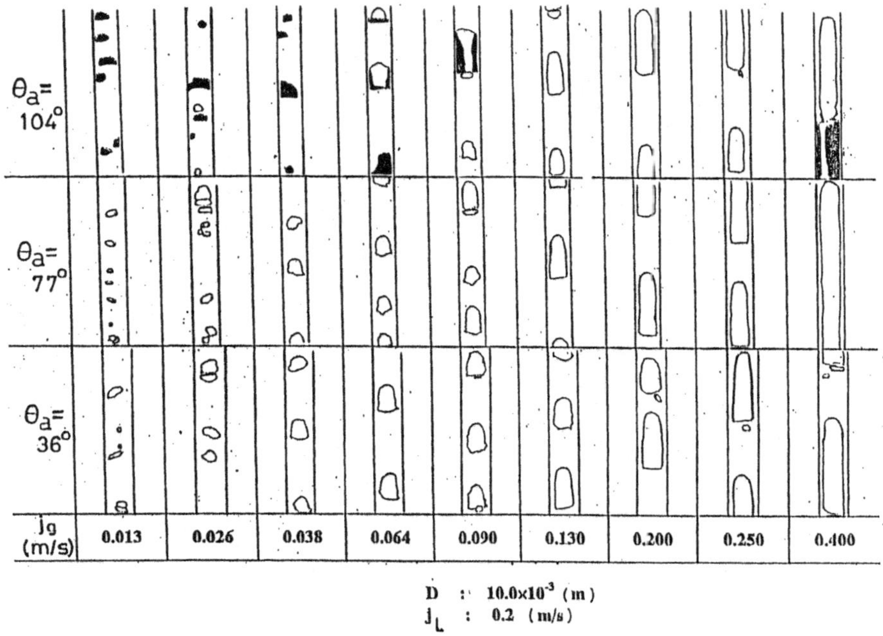


**Fig. 4.68** Flow patterns in pipes of different contact angles

When the pipe was poorly wetted by the liquid and $j_L$ was lower than $j_{L,cr}$, bubbles and slugs frequently attached to the pipe wall. These bubbles do not remain stationary but rather ascend in the pipe and repeatedly attach to and detach from the pipe wall. Such a behavior has never been observed in a wetted pipe. Bubbly flows in the poorly wetted pipe for $j_L < j_{L,cr}$ can, therefore, be further classified into two categories, while slug flows in the same pipe for $j_L < j_{L,cr}$ can be classified into three types [72]. There is yet no detailed information on the boundaries of the bubbly and slug flow regimes. For a better understanding of the flow patterns in a vertical pipe of poor wettability, the definitions of the bubbly and slug flows will be described in this section as well as the critical superficial velocity of water $j_{L,cr}$.

1. Classification of bubbly flow in a pipe of poor wettability for $j_L < j_{L,cr}$
   Two types of bubbly flows were observed in the pipe of $\theta_a = 104°$.

   (a) *Bubbly flow a.* Small bubbles approximately ellipsoidal in shape ascend the pipe and repeatedly attach to and detach from the pipe wall.
   (b) *Bubbly flow b.* When relatively large bubbles attach to the pipe wall, they spread in the circumferential (horizontal) direction and sometimes become piston-like or donut-like bubbles. However, these bubbles never remain on the pipe wall. Their vertical length is less than the pipe diameter.

2. Classification of slug flow in a pipe of poor wettability for $j_L < j_{L,cr}$

   (a) *Slug flow a.* Bullet-like slugs spread on the pipe wall as they attach to the pipe wall and become bullet-like in shape again as they detach from the wall. Such slugs resemble bubbles classified into the aforementioned bubbly flow B when in contact with the pipe wall, but their sizes are much larger than those of the bubbles.

(b) *Slug flow b*. Bullet-like slugs rise without attaching to the pipe wall, but small bubbles behind them repeat attachment to and detachment from the pipe wall.

(c) *Slug flow c*. Bullet-like slugs ascend as their rear parts repeatedly attach to and detach from the pipe wall.

#### 4.5.2.2 Bubbly Flow–Slug Flow Regime Boundary

Wetted Pipe

Many empirical relations have been proposed to describe the boundary between the bubbly flow and slug flow regimes in a vertical pipe of good wettability [66,73–75]. The boundary, determined by Iguchi et al. [76], is represented by the solid line in Fig. 4.69. This boundary can be approximated by the following empirical relation proposed by Taitel et al. [73]:

$$j_L = 3.0 j_G - 1.15(\sigma g \Delta\rho/\rho_L{}^2)^{1/4} \qquad (4.55)$$

where $\sigma$ is the surface tension of liquid, $g$ is the acceleration due to gravity, $\Delta\rho$ is the density difference, and $\rho_L$ is the density of the liquid.

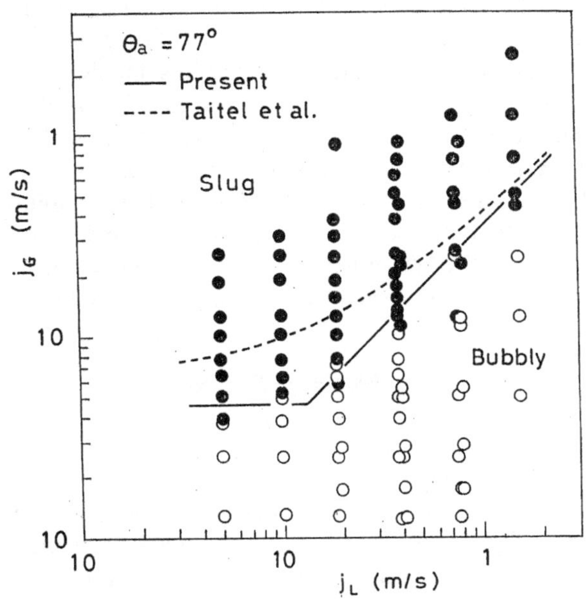


**Fig. 4.69** Boundary between bubbly flow and slug flow regimes in a pipe of good wettability

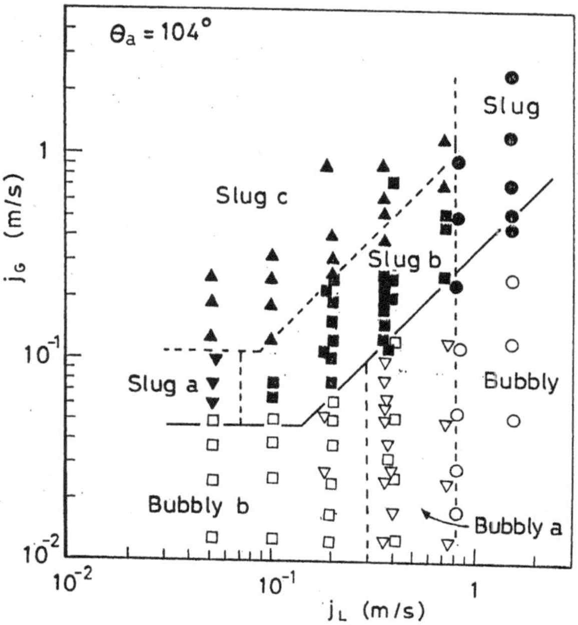


**Fig. 4.70** Boundaries among bubbly and slug flow regimes in a pipe of poor wettability

Poorly Wetted Pipe

Figure 4.70 shows the boundaries between the three types of bubbly flows and four types of slug flows. The boundary between the bubbly flow regimes denoted by open symbols and the slug flow regimes denoted by solid symbols is represented by the solid line. This boundary was found to be almost identical to that for the wetted pipe of part (a) mentioned earlier and shown in Fig. 4.69. Thus the boundary is not significantly affected by the wettability of the pipe. For $j_L > 0.8\,\text{m/s}$, the behavior of bubbles and slugs is also essentially independent of the wettability.

As the pipe diameter decreases, capillary forces become dominant, especially in the poorly wetted pipe. The Weber number similitude can, therefore, be introduced to correlate the boundaries between the two regimes. The two kinds of Weber numbers defined are

$$We_{SG} = \rho_G D {j_G}^2/\sigma \tag{4.56}$$

$$We_{SL} = \rho_L D {j_L}^2/\sigma \tag{4.57}$$

where $\rho_G$ is the density of gas.

The boundaries among the three types of bubbly flows and four types of slug flows are shown in Fig. 4.71. The effect of the wettability of the pipe on the behavior of bubbles and slugs is insignificant when the Weber number, $We_{SL}$, exceeds approximately 100, which thus represents the critical Weber number,

$$We_{SL,cr} = \rho_L D {j_{L,cr}}^2/\sigma = 100 \tag{4.58}$$

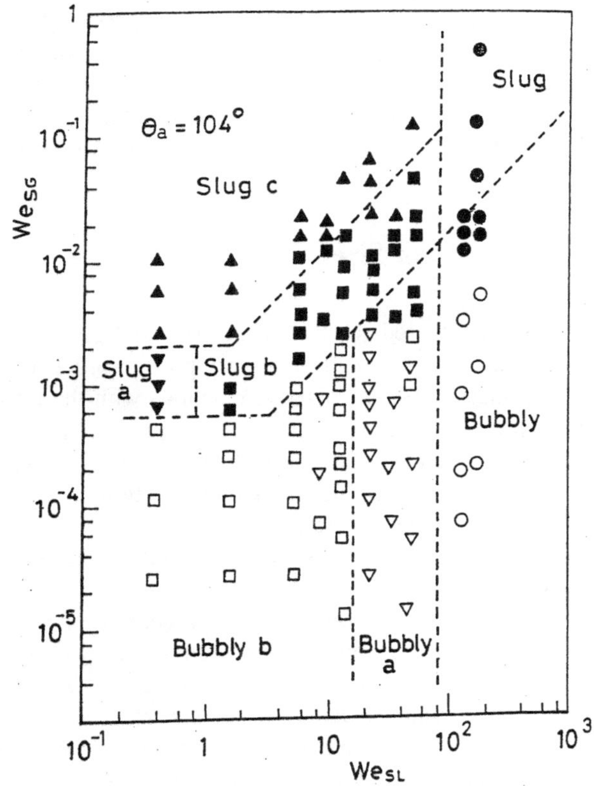


**Fig. 4.71** Correlation of boundaries among bubbly and slug flow regimes in terms of Weber number similitude

Therefore, the critical superficial velocity of water, $j_{\mathrm{L,cr}}$, is approximated by

$$j_{\mathrm{L,cr}} = 10[\sigma/(\rho_{\mathrm{L}} D)]^{1/2} \tag{4.59}$$

It should be remarked that the validity of this relation beyond the specific experimental conditions considered requires further investigation.

#### 4.5.2.3 Summary

The effects of the wettability of a vertical pipe on the flow pattern of air–water two-phase flows in the pipe have been described [76]. When the pipe is poorly wetted with water ($\theta_{\mathrm{a}} = 104°$) and the superficial velocity of water $j_{\mathrm{L}}$ is lower than a critical value, $j_{\mathrm{L,cr.}}$, bubbles and slugs ascend the pipe and repeatedly attach to and detach from the pipe wall. Such a phenomenon does not occur in the highly wetted pipe for $j_{\mathrm{L}} < j_{\mathrm{L,cr}}$. The findings were used to classify bubbly flows in the poorly wetted pipe into two types and slug flows into three. When the superficial velocity of water is higher than the critical value, bubbly and slug flows typical of two-phase flows are observed in a wetted pipe. The boundary between the bubbly flow and slug flow regimes is essentially independent of the pipe wettability. The boundaries among three types of bubbly flows and four types of slug flows were determined as shown in Figs. 4.70 and 4.71. An empirical relation, (4.59), was proposed for $j_{\mathrm{L,cr}}$.

## 4.5.3 *Bubble Velocity and Size*

The wettability of materials used for the refractories and pipes in current metal processing systems is usually designed to be very poor in order to avoid contamination of the molten metal through chemical reactions with the materials. Model studies on the behavior of bubbles rising along a flat plate [32] or rising in a vertical pipe [72] have demonstrated that the bubbles frequently attach to the walls. Such a phenomenon would cause changes in the momentum, heat, and mass transfer from the walls. Information on the interaction between bubbles and poorly wetted walls, therefore, is of practical significance for the redesign of materials processing systems.

In the previous section, the flow patterns of air–water two-phase flows in a vertical poorly wetted pipe were presented. This section discusses the mean rising velocity $\overline{u_{\mathrm{G}}}$ and the mean vertical length $\overline{L_{\mathrm{G}}}$ of bubbles and slugs in such pipes measured with a high-speed video camera. The results are compared with those obtained in a wetted pipe ($\theta_{\mathrm{c}} = 77°$). An empirical equation is proposed for $\overline{L_{\mathrm{G}}}$ in the slug flow regime. It should be noted that the symbol $\overline{u_{\mathrm{G}}}$ is used here to denote the mean rising velocity of dispersed gas phase, although $\overline{u_{\mathrm{B}}}$ was used in the preceding chapters to denote the mean rising velocity of bubbles.

### 4.5.3.1 Experimental Apparatus and Procedure

The details of the experimental apparatus have been given in the description of Fig. 4.67. A still camera and a high-speed video camera were used to observe the behavior of bubbles and slugs in the fully developed region in the pipe. The rising velocity of a bubble or a slug was calculated by dividing its vertical displacement by a prescribed time interval. The mean rising velocity was determined by averaging more than 50 rising velocity data. The vertical length of a bubble or a slug was determined from its image recorded on the high-speed video camera.

### 4.5.3.2 Experimental Results

Mean Rising Velocity of Bubbles and Slugs

The measured values of $\overline{u_{\mathrm{G}}}$ on the pipe axis are shown in Figs. 4.72–4.74. Three arrows in each figure denote the boundaries between the bubbly and slug flow

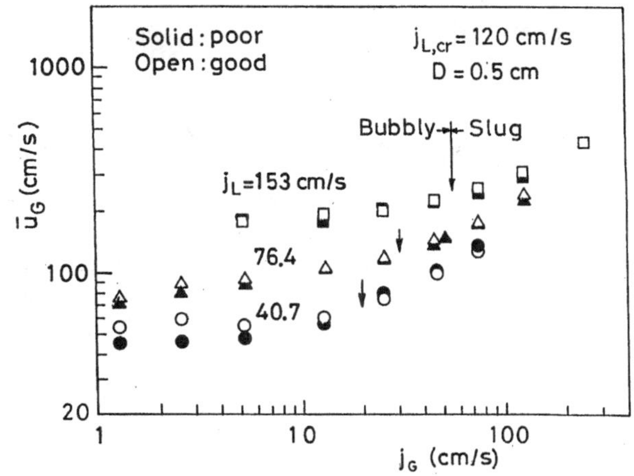


**Fig. 4.72** Relationship between mean rising velocity, $u_G$, and superficial velocity of gas, $j_G$, for a pipe diameter of $D = 0.5\,\text{cm}$

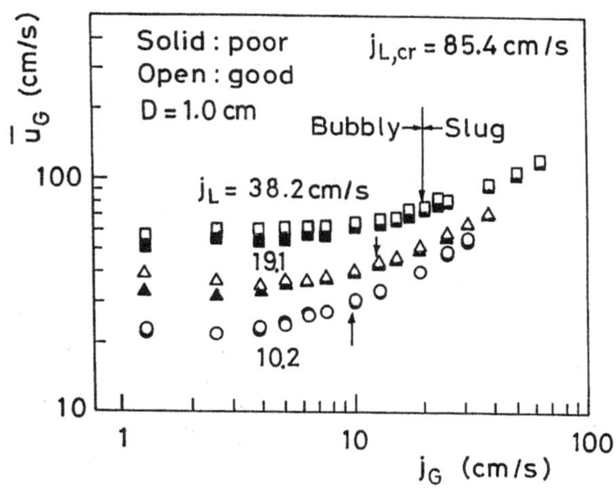


**Fig. 4.73** Relationship between mean rising velocity, $u_G$, and superficial velocity of gas, $j_G$, for a pipe diameter of $D = 1.0\,\text{cm}$

regimes for three different $j_L$ values calculated from (4.55). In the slug flow regime, $\overline{u_G}$ is independent of the pipe wettability. However, in the bubbly flow regime for $j_L \leq j_{L,cr}$, $\overline{u_G}$ is slightly smaller in the poorly wetted than that in the highly wetted pipe. This is because bubbles in the former pipe are likely to attach to the wall and, as a result, their vertical motions are suppressed.

Figures 4.75 and 4.76 show the measured $\bar{u}_G$ in the wetted and poorly wetted pipes, respectively. Allowing a scatter of $\pm 30\%$, the measured values of $\overline{u_G}$ both

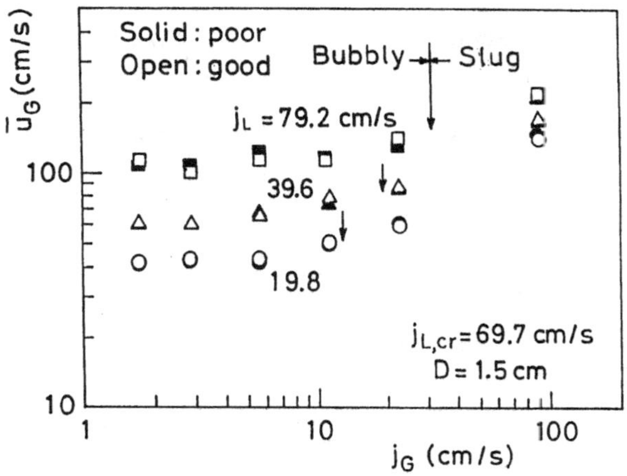


**Fig. 4.74** Relationship between mean rising velocity, $u_G$, and superficial velocity of gas, $j_G$, for a pipe diameter of $D = 1.5\,\text{cm}$

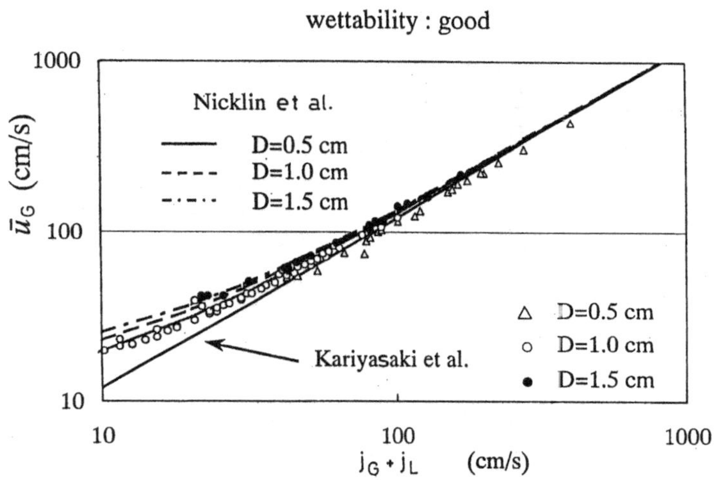


**Fig. 4.75** Comparison of measured with predicted values of mean rising velocity of bubbles and slugs in a good wettability pipe

in the bubbly flow and slug flow regimes can be approximated by the following equation [77], irrespective of the wettability of the pipe:

$$\overline{u_G} = 1.2(j_G + j_L) + 0.35(gD)^{1/2} \tag{4.60}$$

In the slug flow regime, the following empirical equation proposed by Kariyasaki et al. [78] is also valid:

$$\overline{u_G} = 1.2(j_G + j_L) \tag{4.61}$$

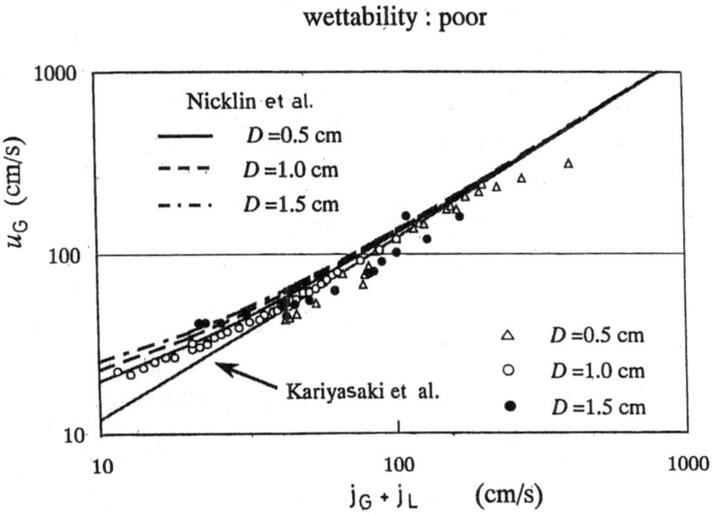


**Fig. 4.76** Comparison of measured with predicted values of mean rising velocity of bubbles and slugs in a poor wettability pipe

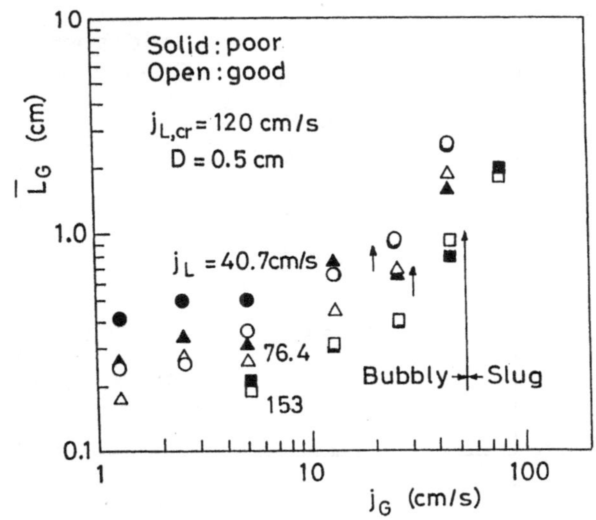


**Fig. 4.77** Relationship between mean vertical length on the pipe center line, $L_G$, and superficial velocity of gas, $j_G$, for a pipe diameter of $D = 0.5$ cm

## Mean Vertical Length of Bubbles and Slugs

Figures 4.77–4.79 show the measured values of the mean vertical lengths of bubbles and slugs in the wetted and poorly wetted pipes. The values of $\overline{L_G}$ in the bubbly flow regime are larger in the latter than those in the former pipe. This result may be

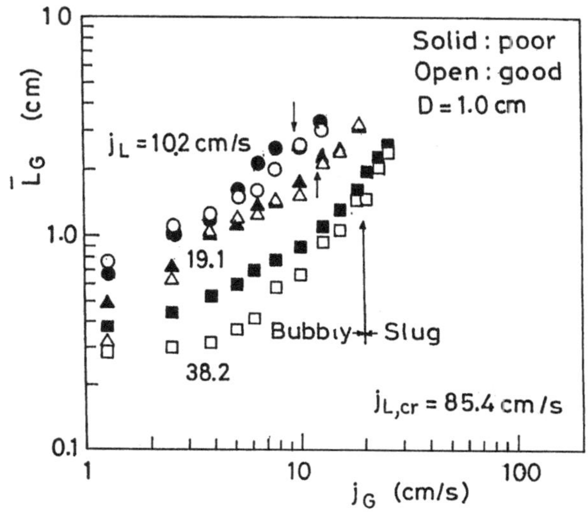


**Fig. 4.78** Relationship between mean vertical length on the pipe center line, $L_G$, and superficial velocity of gas, $j_G$, for a pipe diameter of $D = 1.0\,\mathrm{cm}$

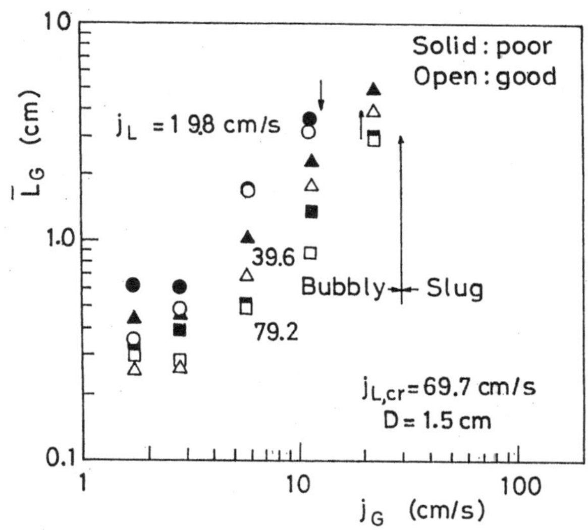


**Fig. 4.79** Relationship between mean vertical length on the pipe center line, $L_G$, and superficial velocity of gas, $j_G$, for a pipe diameter of $D = 1.5\,\mathrm{cm}$

attributed to the fact that bubbles generated in the poorly wetted pipe are likely to coalesce on the wall while ascending the pipe. The values of $\overline{L_G}$ of slugs require a more detailed examination as given below.

*Wetted Pipe*

Akagawa and Sakaguchi [70, 79] proposed the following empirical relation:

$$\overline{L_G} = j_G{}^{1.1} \times 10^{2.16-0.8\, j_L} \quad (4.62)$$

Street and Tek [80] also independently derived the following equation [70]:

$$\overline{L_G} = 0.29 j_G/(j_L + 0.12) \quad (4.63)$$

The units of $\overline{L_G}$, $j_G$, and $j_L$ in (4.62) and (4.63) are m, m/s, and m/s, respectively. However, (4.62) and (4.63) cannot adequately represent the data obtained on $\overline{L_G}$ in the experiments presented here. For example, the measured value of $\overline{L_G}$ is approximately 1.5 cm when $D = 10.0$ mm, $j_G = 20.0$ cm/s, and $j_L = 38.2$ cm/s, while (4.62) and (4.63) predict 12.1 and 6.3 cm, respectively. This large discrepancy between predictions and measurements indicates that a new empirical relation is required for $\overline{L_G}$ in the slug flow regime.

In this regime specified by (4.55), $\overline{L_G}$ is proportional to $j_G$ (see Fig. 4.80) but is inversely proportional to $j_L{}^{3/4}$ (see Fig. 4.81). The relationship between $\overline{L_G}$ and $D$ is shown in Fig. 4.82. It is evident that $\overline{L_G}$ is proportional to $D^{7/8}$. These results collectively suggest that $\overline{L_G}$ is proportional to $DFr_G/Fr_L{}^{3/4}$ in the slug flow regime because $DFr_G/Fr_L{}^{3/4}$ is expressed by

$$DFr_G/Fr_L{}^{3/4} = j_G g^{-1/8} D^{7/8} / j_L{}^{3/4} \quad (4.64)$$

$$Fr_G = j_G/(gD)^{1/2} \quad (4.65)$$

$$Fr_L = j_L/(gD)^{1/2} \quad (4.66)$$

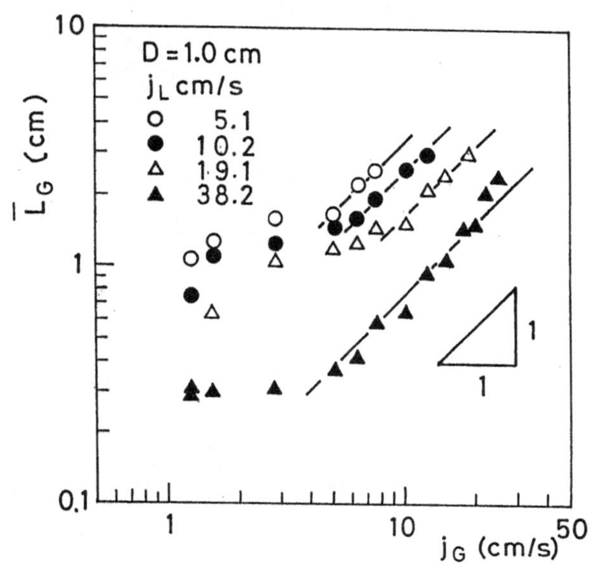


**Fig. 4.80** Mean vertical length data re-plotted from Fig. 4.77

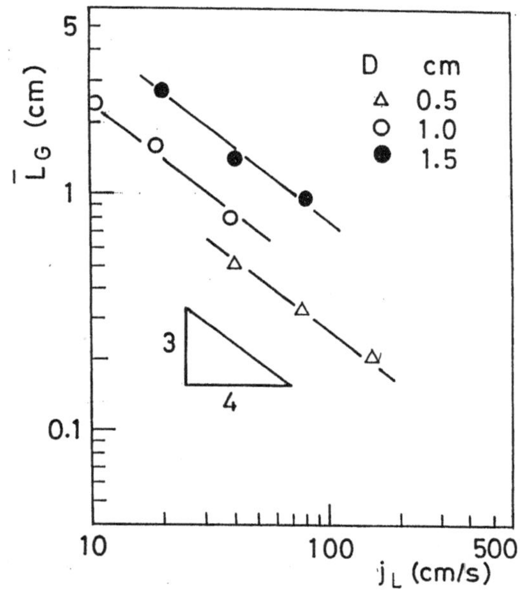


**Fig. 4.81** Relationship between mean vertical length on the pipe center line, $L_G$, and superficial velocity of liquid, $j_L$, for a pipe diameter of $D = 1.0\,\mathrm{cm}$

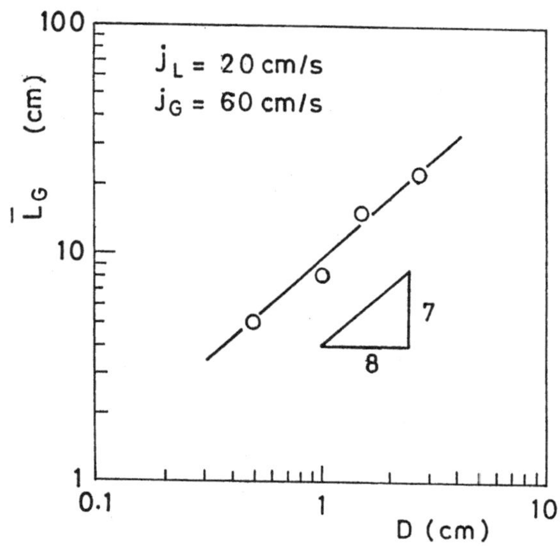


**Fig. 4.82** Relationship between mean vertical length on the pipe center line, $L_G$, and pipe diameter $D$

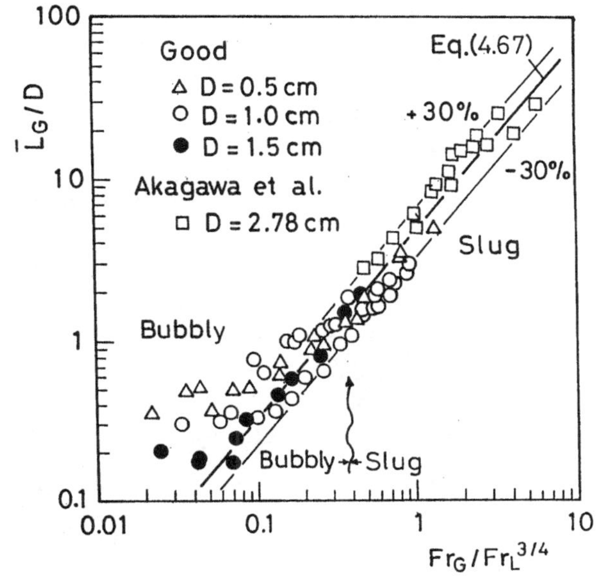


**Fig. 4.83** Correlation of mean vertical length data in a good wettability pipe

The measured values of $\overline{L_G}/D$ are plotted against $Fr_G/Fr_L{}^{3/4}$ in Fig. 4.83 together with those obtained by Akagawa and Sakaguchi [79]. All the measured values in this slug flow regime are adequately correlated by the proposed methodology, resulting in the following empirical relation:

$$\overline{L_G}/D = 4.7[Fr_G/Fr_L{}^{3/4}]^{1.17} \tag{4.67}$$

This equation approximates the measured values within a scatter of $\pm 30\%$. It should be noted that the dependence of $\overline{L_G}$ on $j_G$, $j_L$, and $D$ described by (4.67) is slightly different from that described by (4.64). Additional data are needed on $\overline{L_G}$ in the bubbly flow regime to derive a similar empirical relation.

*Poorly Wetted Pipe*

In the slug flow regime, $\overline{L_G}/D$ is not significantly influenced by the wettability of the pipe. It can thus be approximated by (4.67), as demonstrated in Fig. 4.84.

#### 4.5.3.3 Summary

The effects of the wettability of a vertical pipe on the characteristics of air–water two-phase flows in the pipe can be summarized as follows.

The mean rising velocity of slugs, $\overline{u}_G$, follows the empirical relation proposed by Nicklin et al. [77] irrespective of the wettability of the pipe wall. The mean rising

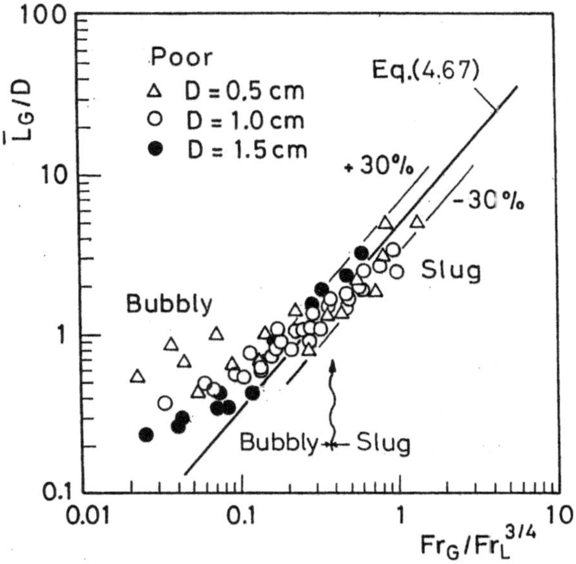


**Fig. 4.84** Correlation of mean vertical length data in a poor wettability pipe

velocity of bubbles is slightly smaller in a poorly wetted than in a highly wetted pipe. Allowing for a scatter of ±30%, the mean rising velocity of bubbles can be predicted by the empirical relation of Nicklin et al. [77].

The mean vertical length of slugs, $\overline{L_G}$, in a wetted pipe is adequately represented by the empirical correlation of Iguchi et al. [81]. The measured values of $\overline{L_G}$ are essentially independent of the pipe wettability. On the contrary, the mean vertical length of bubbles in the poorly wetted pipe is slightly larger than the values for the highly wetted pipe. This result is attributable to the possible coalescence of bubble in the poorly wetted pipe.

In this section, experimental results on the mean rising velocity and mean vertical length of bubbles and slugs were measured only on the pipe center line. Further experimental investigations on the distributions of these quantities in the whole cross section are required for a full understanding of the effect of pipe wettability on the characteristics of bubbles and slugs.

## References

1. Ito K, Fruehan RJ (1989) Study on the foaming of CaO-SiO2-FeO Slags. Part II. Dimensional analysis and foaming in iron and steelmaking processes. Metall Trans B 20B:515
2. Ghag SS, Hayes PC, Lee H-G (1998) Model development of slag foaming. ISIJ Int 38:1208
3. Hing L, Hirasawa M, Sano M (1998) Behavior slags by of slag foaming with graphite. ISIJ Int 38:1339

4. Matsuo M, Hirata H, Katayama H, Ishikawa H, Kajioka H, Tokumitsu N (1986) Tetsu-to-Hagane 72:S970
5. Ito K, Fruehan RJ (1989) Study on the foaming of CaO-SiO2-FeO Slags: Part I. Foaming parameters and experimental results. Metall Trans B 20B:509
6. Komarov SV, Kuwabara M, Sano M (1990) Mechanism of acoustic defoaming by applying low frequency sound. CAMP ISIJ 12:721
7. Ogawa Y, Tokumitsu N (1990) Observation of slag foaming by X-ray fluoroscopy, In: Proceedings of the 6th international iron and steel congress. ISIJ, Tokyo, p 147
8. Mizuno Y, Shimizu T, Sonoyama N, Iguchi M (2000) Attachment of bubbles to a horizontal circular cylinder of poor wettability. Jpn J Multiphase Flow 14:166
9. Sonoyama N, Iguchi M, Sasaki Y, Ishii K (2000) Behavior of a single bubble colliding with a flat plate of poor wettability. Tetsu-to-Hagane 86:203
10. Lee C, Takahashi I, Miyata H (1995) Proposal on evaluation method of adhesion tension based on analytical estimation of configuration of sessile drops. Trans Jpn Soc Mech Eng (B) 61:4386
11. Fritz W (1935) Maximum volume of vapor bubbles, Physik Zeitschr 36:379
12. Bashforth F, Adams JC (1883) An attempt to test the theories of capillary action. Cambridge University Press, London
13. Murakami K, Sunada T, Ueki N (1998) Behavior of residual liquid on disk for mist flow injected between horizontal parallel disks. Trans Jpn Soc Mech Eng (B) 64:2071
14. Sano M, Mori K, Sato T (1977) Bubble formation at single nozzles immersed in molten iron. Tetsu-to-Hagane 63:2308
15. Mizuno Y, Inoue H, Sonoyama N, Iguchi M (2000) Size and shape of a bubble on the upper surface of a horizontal plate of poor wettability. Tetsu-to-Hagane 86:709–716
16. Ozawa Y, Mori K, Sano M (1981) Behavior of injected gas observed at the exit of a submerged orifice in liquid metal. Tetsu-to-Hagane 16:2655
17. Asai S (1984) 100th and 101st Nishiyama memorial lecture. ISIJ, Tokyo, pp 67
18. Sahai Y, St. Pierre GR (1992) Advances in transport phenomena in metallurgical systems. Elsevier, Amsterdam
19. The 140th Committee of Japan Society for Promotion of Science (1988) Handbook of physicochemical properties at high temperatures. ISIJ, Tokyo
20. Mori K, Sano M (1981) Process kinetics in injection metallurgy. Tetsu-to-Hagane 67:672
21. Iguchi M, Chihara T, Takanashi N, Ogawa Y, Tokumitsu N, Morita Z (1995) X-ray fluoroscopic observation ot bubble characteristics in a molten iron bath. ISIJ Int 35:1354
22. Irons GA, Guthrie RIL (1978) Bubble formation at nozzles in pig iron. Metall Trans 9B:101
23. Zhe W, Mukai K, Yamaguchi K, Lee J (1998) Influence of interfacial properties on the formation of bubble-related defects in slab. CAMP ISIJ 11:24
24. Kawakami M, Tomimoto, Itoh K (1982) Statistical analysis of gas bubbles dispersion in liquid phase. Tetsu-to-Hagane 68:774
25. Iguchi M, Ueda H, Uemura T (1995) Bubble and liquid flow characteristics in a vertical bubbling jet. Int J Multiphase Flow 21:861
26. Iida T, Guthrie RIL (1988) The physical properties of liquid metals. Oxford University Press, Oxford
27. Castillejos AH, Brimacombe JK (1987) Measurement of physical characteristics of bubbles in gas-liquid plumes. Part I1. Local properties of turbulent air-water plumes in vertically injected jets. Metall B 18B:659
28. Sano M, Makino H, Ozawa Y, Mori K (1986) Behavior of gas jet and plume in liquid metal. Trans Iron Steel Inst Jpn 26:298
29. Iguchi M, Nozawa K, Tomida H, Morita Z (1992) Bubble characteristics in the buoyancy region of a vertical bubbling jet. ISIJ Int 32:747
30. Mazumdar D, Guthrie RIL (1995) The physical and mathematical modelling of gas stirred ladle systems. ISIJ Int 32:747
31. Sugita K (1998) Various phenomena observed at refractory-iron melt interfaces: A condensed history of studies on their interactions. Bul ISIJ (Ferrum) 3:891
32. Sonoyama N, Iguchi M (1999) Wettability effects on bubble characteristics in a bubbling wall jet along a vertical flat plate. ISIJ Int 39:673–679

33. Rajaratnam N (1981) Turbulent jet (trans: Nomura Y). Morikita Publishing Co. Ltd., Tokyo
34. Rajaratnam N, Pani BS (1994) Three-dimensional turbulent wall jets, Trans ASCE 100(HY1):69
35. Verhoff A (1963) The two-dimensional turbulent wall jet with and without an external stream. Rep. 626, Princeton University, Princeton, NJ
36. Clift R, Grace JR, Weber ME (1978) Bubbles, drops, and particles. Academic, New York
37. Hetsroni G (1989) Particles-turbulence interaction. Int J Multiphase Flow 15:735
38. Iguchi M, Sonoyama N (2000) Wettability effects on liquid flow characteristics in a bubbling wall jet along a vertical flat plate. ISIJ Int 40:1–6
39. Shinozaki N, Echida N, Mukai K, Takahashi Y, Tanaka Y (1994) Wettability of Al2O3-MgO, ZrO2-CaO, Al2O3-CaO substrates with molten iron. Tetsu-to-Hagane 80:748–753
40. Takahira H, Fujikawa S, Akamatsu T (1989) Collapse motion of a single gas bubble near a plane or curved rigid wall. Trans Jpn Soc Mech Eng B 55:2720–2728
41. Shopov PJ, Minev PD, Bazhlekov IB, Zapryanov ZD (1990) Interaction of a deformable bubble with a rigid wall at moderate Reynolds numbers. J Fluid Mech 219:241–271
42. Maxworthy T (1991) Bubble rise under an inclined plate. J Fluid Mech 229:659–674
43. Tsao H, Koch DL (1997) Observations of high Reynolds number bubbles interacting with a rigid wall. Phys Fluids 9:44–56
44. Wang QX, Yeo KS, Khoo BC, Lam KY (1998) Toroidal bubbles near a rigid boundary. Theoret Comput Fluid Dyn 12:29–51
45. Tao Z, Mukai K, Takahashi T (1999) Mechanism of local corrosion of MgO-C refractory at slag-metal interface. CAMP ISIJ 12:208
46. Abbel G, Damen W, de Gendt G, Tiekink W (1996) Argon bubbles in slabs. ISIJ Int 36:S219–S222
47. Kasai N, Watanabe Y, Kajiwara T, Toyoda M (1997) Mechanism of fine bubbles entrapment beneath the surface of continuously cast slabs. Tetsu-to-Hagane 83:24–29
48. Iguchi M, Nakatani T, Ueda H (1997) Model study of turbulence structure in a bottom blown bath with top slag using conditional sampling. Metall Trans B 28:87–94
49. Zhe W, Mukai K, Matsuoka K (1997) Water model experiment for the behaviors of bubbles and liquid flow on the inside of the nozzle and mold of continuous casting process. CAMP ISIJ 10:68–71
50. MacDougall G, Ockrent C (1942) Surface energy relations in liquid/solid systems. I. The adhesion of liquids to solids and a new method of determining the surface tension of liquids. Proc R Soc Lond Ser A 180:151–173
51. Shoji M, Zhang XY (1992) Study of contact angle hysteresis: In relation to boiling surface wettability. Trans Jpn Soc Mech Eng B 58:1853–1859
52. Fukai J, Shibata Y, Yamamoto T, Miyatake O, Poulikakos D, Megaridis CM, Zhao Z (1995) Wetting effects on the spreading of a liquid droplet colliding with a flat surface: Experiment and modeling. Phys Fluids 7:236–247
53. Wenzel RN (1936) Resistance of solid surfaces to wetting by water. Ind Eng Chem 28:988–994
54. Fujita M, Madarame H (1983) Bubble dynamics in a horizontal rectangular pipe with rough wall. Trans Jpn Soc Mech Eng B 49:1676–1684
55. Murakami K, Sunada T, Ueki N (1998) Behavior of residual liquid on disk for mist flow injected between horizontal parallel disks. Trans Jpn Soc Mech Eng B 64:2071–2078
56. Kawasaki K (1960) Study of wettability of polymers by sliding of water drop. J Colloid Sci 15:402–407
57. Furmidge CGL (1962) Studies at phase interfaces. I. The sliding of liquid drops on solid surfaces and a theory for spray retention. J Colloid Sci 17:309–324
58. Larkin BK (1967) Numerical solution of the equation of capillarity. J Colloid Interface Sci 23:305–312
59. Ogawa Y, Huin D, Gaye H, Tokumitsu N (1993) Physical model of slag foaming. ISIJ Int 33(1):224–232
60. Sano N, Fruehan RJ, Cramb AW (1991) Computer aided interfacial measurements, US-Japan joint seminar, fundamental of bath smelting and clean steel production. Myrtle Beach, 7–9 Oct

61. Fukui K (1996) A study on the new iron ore smelting reduction processes. Tetsu-to-Hagane 82(1):1–7
62. Iguchi M, Kondoh T, Uemura T (1994) Simultaneous measurement of liquid and bubble velocities in a cylindrical bath subject to centric bottom gas injection. Int J Multiphase Flow 20:753–762
63. Anagbo PE, Brimacombe JK (1990) Plume characteristics and liquid circulation in gas injection through a porous plug. Metall Mater Trans B 21B:637–648
64. Iguchi M, Kaji M, Morita Z (1998) Effects of pore diameter, bath surface pressure, and nozzle diameter on the bubble formation from a porous nozzle. Metall Mater Trans B 29B:1209–1218
65. Davidson L, Amick EH Jr (1956) Formation of gas bubbles at horizontal orifices. AIChE J 2:337–342
66. Castello-Branco MASC, Schwerdtfeger K (1994) Large-scale measurements of the physical characteristics of round vertical bubble plumes in liquids. Metall Mater Trans B 25B:359–371
67. Akagawa K (1980) Gas–liquid two-phase flow. Corona Publishing Co. Ltd, Tokyo
68. Hetsroni G (1982) Handbook of multiphase systems. Hemisphere Publishing Corporation, Washington
69. Ueda T (1989) Gas–liquid two-phase flow (Fluid flow and heat transfer). Yokendo Publishing Co. Ltd, Tokyo
70. The Japan Society of Mechanical Engineers (1989) Handbook of gas–liquid two-phase flow technology. Corona Publishing Co. Ltd., Tokyo
71. Barajas AM, Panton RL (1993) The effects of contact angle on two-phase flow in capillary tubes. Int J Multiphase Flow 19:337–346
72. Terauchi Y, Iguchi M, Kosaka H, Yokoya S, Hara S (1999) Wettability effect on the flow pattern of air–water two-phase flows in a vertical circular pipe. Tetsu-to-Hagane 85(9):645–651 (in Japanese)
73. Taitel Y, Barnea D, Dukler AE (1980) Modelling flow pattern transitions for steady upward gas-liquid flow in vertical tubes. AIChE J 26(3):345–354
74. Weisman J, Kang SY (1981) Flow pattern transitions in vertical and upwardly inclined lines. Int J Multiphase Flow 7:271–291
75. Mishima K, Ishii M (1984) Flow regime transition criteria for upward two-phase flow in vertical tubes. Int J Heat Mass Transfer 27:723–737
76. Iguchi M, Terauchi Y (2001) Boundaries among bubbly and slug flow regimes in air–water two-phase flows in vertical pipe of poor wettability. Int J Multiphase Flow 27:729–735
77. Nicklin DJ, Wilke JO, Davidson JF (1962) Two-phase flow in vertical tubes. Trans Inst Chem Eng 40(1):61
78. Kariyasaki A, Fukano T, Outsuka A, Kagawa M (1992) Isothermal air-water two-phase up-and downward flows in a vertical capillary tube (1st report, flow pattern and void fraction). Trans Jpn Soc Mech Eng 58(553):2684–2690
79. Akagawa K, Sakaguchi T (1965) Fluctuation of void ratio in a two-phase flow : 3rd report, absolute velocities of slugs and small bubbles. Trans Jpn Soc Mech Eng 31(224):601–607
80. Street JR, Tek MR (1965) Dynamics of bullet shaped bubbles encountered in vertical gas liquid slug flow. AIChE J 11(4):644–650
81. Iguchi M, Terauchi Y (2000) Rising behavior of air–water two-phase flows in vertical pipe of poor wettability. ISIJ Int 40:567–571

# Chapter 5
# Swirling Flow and Mixing

## 5.1 Rotary Sloshing of Liquid in Cylindrical Vessel

A liquid in a rigid vessel subjected to external forced excitation becomes unstable under certain conditions due to strong nonlinear effects [1–12]. A variety of modes of liquid surface oscillations occur such as rotary sloshing or swirl. Such oscillations exert significant influence on the safety of cylindrical tanks preserving petroleum, tankers carrying petroleum, rocket boosters, and liquid containers in many chemical and mechanical plants. When the amplitudes of the oscillations are sufficiently large, the vessels are sometimes destroyed or ruptured, resulting in huge economic losses. A number of theoretical and experimental investigations have been done to understand the oscillation modes of liquids in circular cylindrical tanks [6, 10], spherical tanks [7, 10], sector-compartmented circular cylindrical tanks [3, 8], and long rectangular tanks [12]. Some aspects of the rotary sloshing or swirl are briefly reviewed in this section.

### *5.1.1 Linear Theory*

At the earlier stage, linear theory was developed to estimate unsteady dynamic pressure distributions on the walls of tanks in the presence of liquid surface oscillations of relatively small amplitudes. A dynamic force acting on the wall can be calculated using this pressure distribution.

According to this theory [13], the angular frequency of the $i$th tangential oscillation mode of liquid, $\omega_i$, contained in a circular cylindrical vessel shown in Fig. 5.1 can be given by the relation,

$$\omega_i = [2g\varepsilon_i \tan h(2\varepsilon_i H_\mathrm{L}/D)/D]^{1/2}, \tag{5.1}$$

where, $g$ is the acceleration due to gravity, $\varepsilon_i$ is the $i$th positive zero of the Bessel function $J_1{}'(\varepsilon)$ [14], and $D$ is the vessel diameter. For $i = 1$, (5.1) reduces to [15]:

$$\omega_1 = [3.68g \tan h(3.68H_\mathrm{L}/D)/D]^{1/2}. \tag{5.2}$$

M. Iguchi and O.J. Ilegbusi, *Modeling Multiphase Materials Processes: Gas-Liquid Systems*, DOI 10.1007/978-1-4419-7479-2_5,

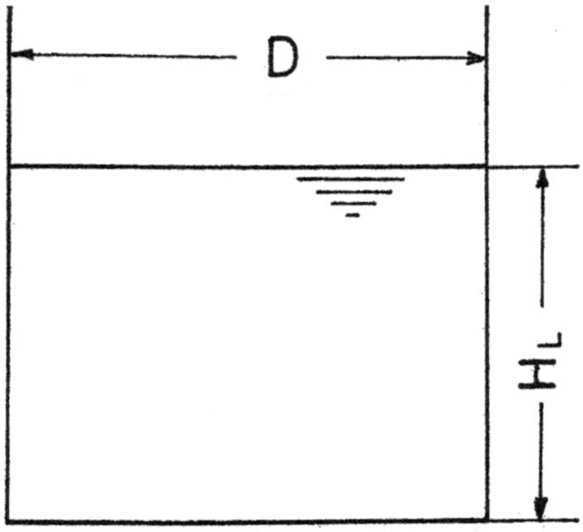


**Fig. 5.1** Rotary sloshing in circular cylindrical vessel

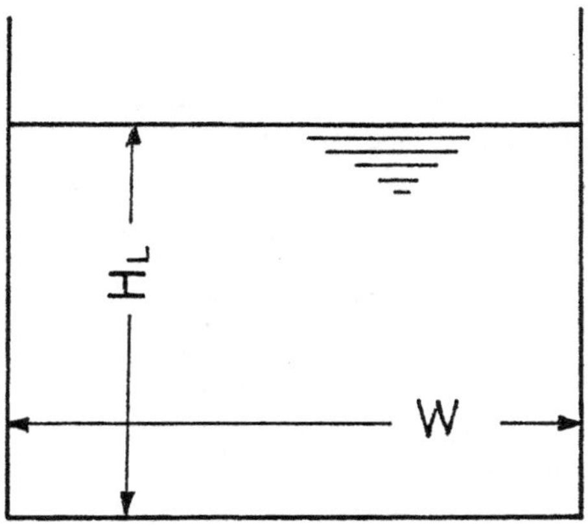


**Fig. 5.2** Liquid surface oscillation in rectangular tank

For a rectangular tank (see Fig. 5.2) the angular frequency of the first oscillation mode which is, of course, not rotary sloshing is expressed as:

$$\omega_1 = [3.16g \tan h(3.16H_\mathrm{L}/W)/W]^{1/2} \tag{5.3}$$

where, $W$ is the width of the tank [13].

### 5.1.2 Nonlinear Theory

The linear theory has been found to be applicable to limited conditions. Nonlinear theories therefore have been developed, and this has allowed prediction of nearly all liquid surface oscillations encountered in practical applications. Majority of previous investigations were concerned with liquids in circular cylindrical tanks [1,6,10]. Abramson et al. [10] focused on nonlinear oscillations of liquid contained

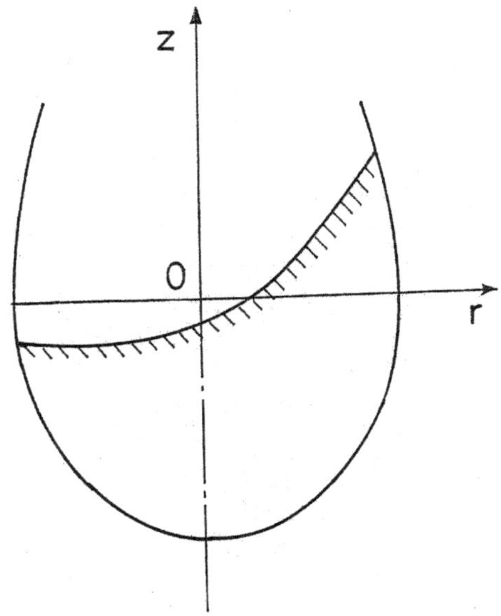


**Fig. 5.3** Liquid surface oscillation in axisymmetrical tank of arbitrary geometry

in circular, cylindrical, and spherical tanks and compared the calculations with measurement. Faltinsen [12] carried out experimental and theoretical investigations on liquid surface oscillations in rectangular tanks.

Kimura and Ohashi [16] succeeded in developing a nonlinear theory for liquids contained in axisymmetrical tanks of arbitrary geometries (see Fig. 5.3) using the calculus of variation and derived the equations governing the motions of liquid surfaces. This theory was further verified by comparing the results using the finite element method with the experimental data for a circular cylindrical tank and a spherical tank undergoing horizontal excitation [17].

The natural frequency, $f(=1/T_s)$, of liquid surface oscillations in a circular cylindrical tank and a spherical tank are presented in Figs. 5.4 and 5.5, respectively. In these figures, $H_L$ is the bath depth, $D$ is the vessel diameter, $H_L/D$ is the aspect ratio, $m$ in the parentheses denotes the $m$th mode of surface oscillation in the tangential direction and $n$ is the $n$th mode in the radial direction. The tangential mode, i.e., rotary sloshing appears first in both tanks. Further information about this study is available elsewhere [13].

### *5.1.3 Summary*

Liquids in tanks of various geometries would be strongly mixed in the presence of rotary sloshing, but at the same time, sloping and splashing would become serious. This is due to the violent wave motions accompanying rotary sloshing. Accordingly, rotary sloshing may be efficiently used to enhance mixing in metallurgical reactors, provided that the sloping and splashing are suppressed. Similar sloshing is

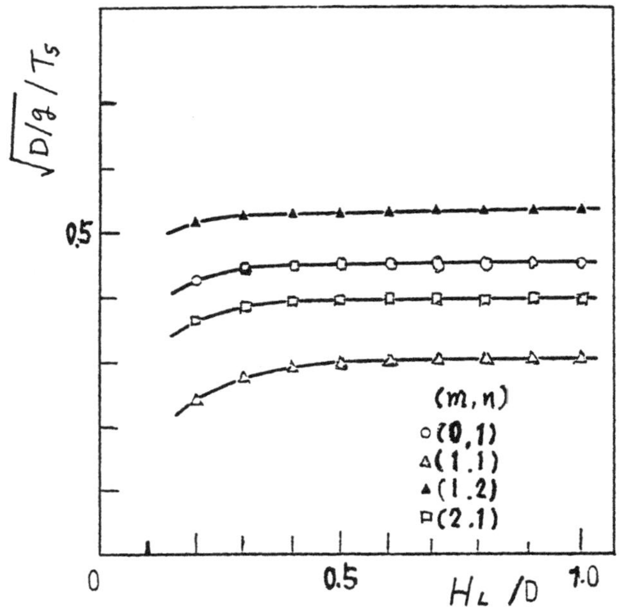


**Fig. 5.4** Surface oscillation modes in circular cylindrical vessel

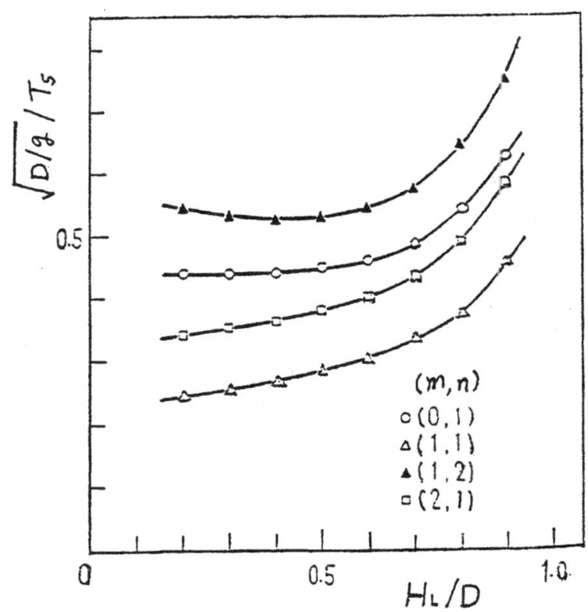


**Fig. 5.5** Surface oscillation modes in spherical vessel

known to appear when a bath is agitated by gas injection. In the following sections, a description will be provided for rotary sloshing or swirl motion in a circular cylindrical vessel induced by gas injection.

## 5.2 Swirl Motion of Bubbling Jet

### *5.2.1 General Features*

A bubbling jet formed in a bottom blown bath does not always rise directly upward. For example, molten metal baths in bottom blown converters are highly excited by gas injection and result in steady oscillations of varying modes under certain blowing conditions [18]. Such oscillations significantly impact refining efficiency and erosion of refractories [18, 19]. On the basis of water model experiments, Kato et al. found that two types of oscillations occur in a bath with bottom gas injection through multi-tuyeres arranged along two parallel lines on the bottom plate [18]. First, small-scale bubbling jets form above the tuyeres and unite into a large-scale jet. The large-scale jet oscillates in the radial direction like the motion of a liquid column in a U-shaped tube. This motion is termed "A-type oscillation". Second, each small-scale bubbling jet rises independently without coalescing. The jets generated from the tuyeres along one line oscillate in the same direction and in phase with one another. Similar results are observed for jets along the other line. However, there is a phase difference of 180° between the former and the latter oscillations (referred to as a B-type oscillation).

Similar bath oscillations have also been observed when liquid in a circular cylindrical vessel is drained through a centric single-hole bottom nozzle [20, 21] or when a circular cylindrical bath is subjected to external forced oscillation, as described in detail in Sect. 5.1.3 [6, 10, 13, 16, 17]. The bath oscillation due to external forced oscillation is called sloshing. In particular, a purely tangential mode of sloshing is referred to as rotary sloshing [6].

Although the above observations show that bath oscillations are not unusual events [22–24], the characteristics of these oscillations induced by gas injection are not fully understood.

In this section, swirl motions, i.e., tangential oscillation modes of a bubbling jet in a cylindrical bath with centric bottom gas injection are described. The oscillations are classified into two categories and empirical correlations are presented for the critical bath depths for the initiation and cessation of swirl motion and the swirl period.

#### 5.2.1.1 Classification of Swirl Motion

Swirl motion of a bubbling jet can be classified into two, as shown in Fig. 5.6. One occurs when the bath depth $H_L$ is smaller than the bath diameter $D$. The radial

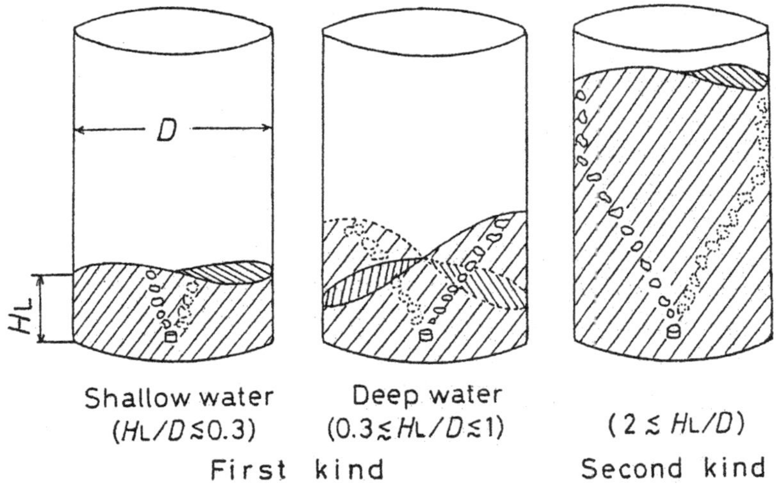


**Fig. 5.6** Classification of the swirl motions of bubbling jet in a cylindrical vessel

displacement of the jet on the bath surface is relatively small and the swirl period of the jet is nearly equal to the period of the rotary sloshing. The wave motion of the liquid is also quite similar to that of the rotary sloshing. This type of swirl motion is caused by internal excitation of liquid due to the quasi-periodic generation of bubbles at the nozzle exit. In a strict sense, rotary sloshing is an oscillation of liquid caused by external forced oscillation, whereas the swirl motion observed here is related to the hydrodynamic instability of two-phase flow of gas and liquid contained in the circular cylindrical vessel. Therefore, the swirl motion of a bubbling jet under consideration is distinguished from rotary sloshing and referred to, for convenience, the first kind of swirl motion.

The first kind of swirl motion is further divided into two categories with respect to the bath depth, $H_L$: shallow water type and deep water type. The boundary between the two is represented by $H_L/D \cong 0.3$.

The other swirl motion of the second kind occurs for a bath depth larger than about twice the bath diameter, $D$. The radial displacement of the jet becomes large and the jet approaches the side wall of the vessel. The swirl period therefore becomes much longer than the period of the first kind of swirl motion. The liquid at a radial location opposite to the jet falls as it swirls. The bath would be highly mixed by such a large-scale swirl motion, and hence, the mixing time would become shorter. In fact, Murthy et al. [25] reported that the mixing time has a minimum at $H_L/D \cong 2$. This motion is caused by the Coanda effect [26–28] and called the second kind of swirl motion. The Coanda effect has been fully discussed in Chap. 3.

#### 5.2.1.2 The First Kind of Swirl Motion

Critical Bath Depth for Initiation of Motion

The critical bath depth $H_{L,1,s}$ for the initiation of the first kind of swirl motion in an air–water system was measured five times under a range of experimental conditions. The arithmetic mean of the data is plotted in Figs. 5.7 and 5.8. The scatter in the experimental data was within ±25%. The results indicate that $H_{L,1,s}$ is about 25 mm irrespective of the bath diameter $D$, the inner diameter of nozzle, $d_{ni}$, and air flow rate $Q_g$. This trend suggests that the inertial force of the injected gas has no effect on the initiation of the first kind of swirl motion.

Swirl Period

Figure 5.9 shows the relationship between the period of the first kind of swirl motion, $T_s$, and the bath depth, $H_L$, for an air–water system. The solid and broken lines denote the periods of the fundamental and second harmonics of the rotary sloshing, respectively. The angular velocity of the $i$th harmonics of the rotary sloshing, $\omega_i$, can be derived using the inviscid linear theory mentioned previously [13] thus,

$$\omega_i = [(2g\varepsilon_i/D)\tan h(2\varepsilon_i H_L/D)]^{1/2}, \qquad (5.1)$$

where, $\varepsilon_i$ is the $i$th positive zero of the Bessel function $J_1{}'(\varepsilon)$ and $g$ is the acceleration due to gravity.

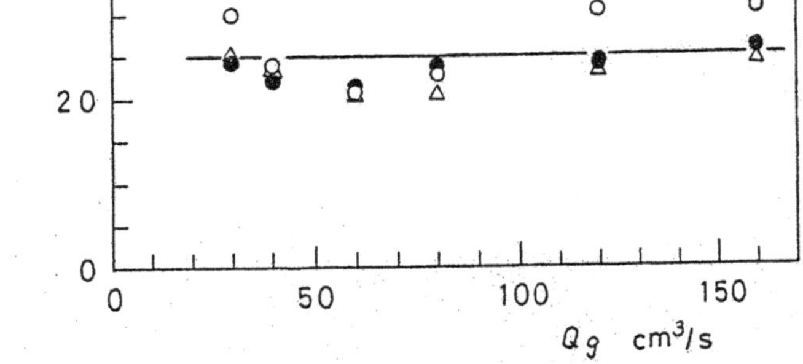


**Fig. 5.7** Critical water bath depth at which the first kind of swirl motions occur for $D = 200$ mm

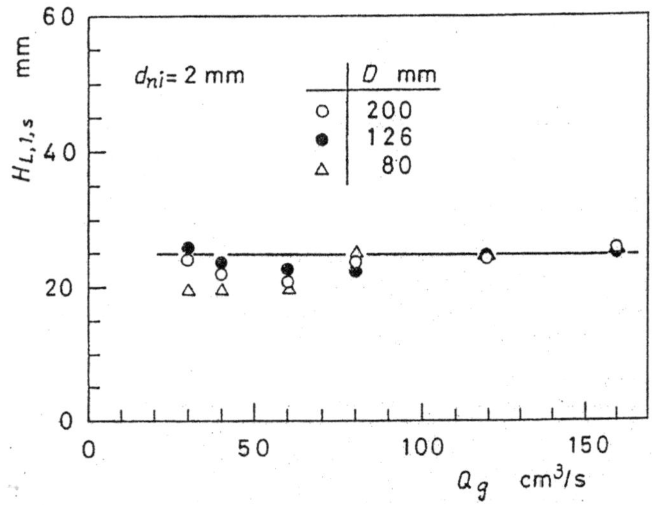


**Fig. 5.8** Critical water bath depth at which the first kind of swirl motions occur for $d_{ni} = 2$ mm

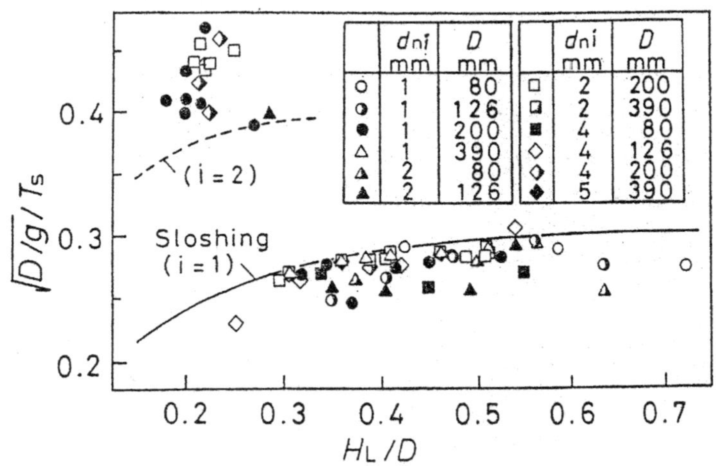


**Fig. 5.9** Comparison of the period of the first kind of swirl motion with the period of rotary sloshing in water bath

The measured swirl period could be adequately approximated by (5.1) and divided into two categories, the limit being at $H_L/D \cong 0.3$. The swirl period agrees with the period of fundamental harmonics of the rotary sloshing ($i = 1$) for $H_L/D > 0.3$ within a scatter of $-20$ to 0%, whereas agreement with the period of the second harmonics ($i = 2$) for $H_L/D < 0.3$ is within a scatter of 0 to +25%. This limit seems to be associated with the limit between the shallow water wave and the deep water wave in a cylindrical vessel [16, 17] given by $H_L/D = 0.3$. In this case, a wave affected significantly by the bottom wall of the vessel is termed shallow

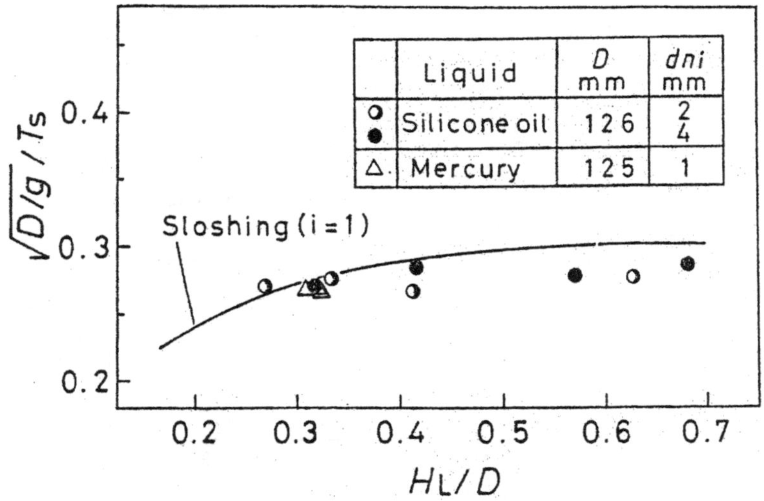


**Fig. 5.10** Comparison of the period of the first of kind swirl motion with the period of rotary sloshing in silicone oil and mercury baths

water wave, while the deep water wave is essentially unaffected by the bottom wall. In the deep water regime, only the fundamental oscillation mode was observed, and the swirl period of the bubbling jet was identical to the period of the liquid wave motion. On the other hand, the liquid wave motion exhibits higher radial oscillation modes in the shallow water regime.

Experimental results for silicone oil and mercury baths are shown in Fig. 5.10. For $H_L/D > 0.3$, the correlation of the oscillation period is similar to that for the water baths.

## Critical Bath Depth for Cessation of Motion

Figures 5.11 and 5.12 show the measured critical bath depth, $H_{L,1,f}$, for cessation of the first kind of swirl motion for the air–water system as a function gas flow rate $Q_g$. Unlike the bath diameter, $D$, the inner diameter of the nozzle, $d_{ni}$, has a negligible effect on $H_{L,1,f}$. Experimental results for the air–water system are shown in Fig. 5.13. The measured data are well correlated within a scatter of ±30%. Hence, the following empirical relation has been proposed:

$$\log(H_{L,1,f}/D) = 0.05 - 1.35(X+6)/[\exp(X+5)](-4 < X < 0), \tag{5.4}$$

$$We = \rho_L Q_g^2/\sigma_L D^3, \tag{5.5}$$

where, $X = \log We$ and $We$ is the Weber number.

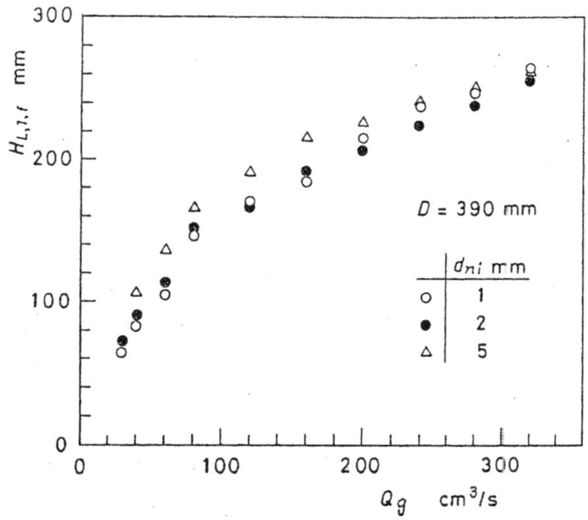


**Fig. 5.11** Critical water bath depth at which the first kind of swirl motion ceases for $D = 390$ mm

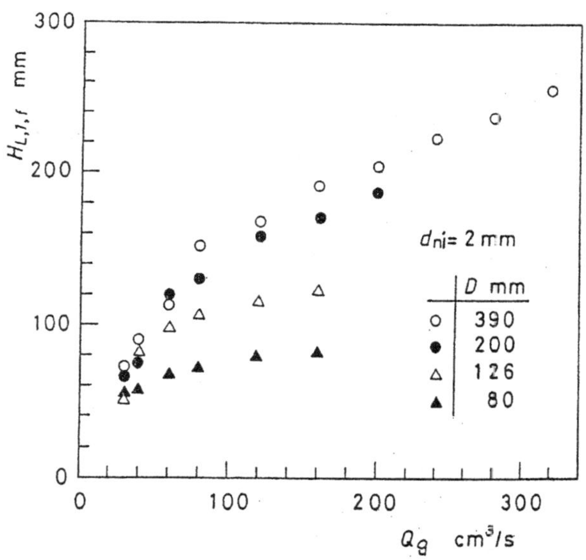


**Fig. 5.12** Critical water bath depth at which the first kind of swirl motion ceases for $d_{ni} = 2$ mm

Experimental results for mercury, *n*-pentane, and silicone oil are shown in Fig. 5.14. The physical properties of the liquids are listed in Table 5.1. These data are similarly approximated by (5.4) within a scatter of ±30%.

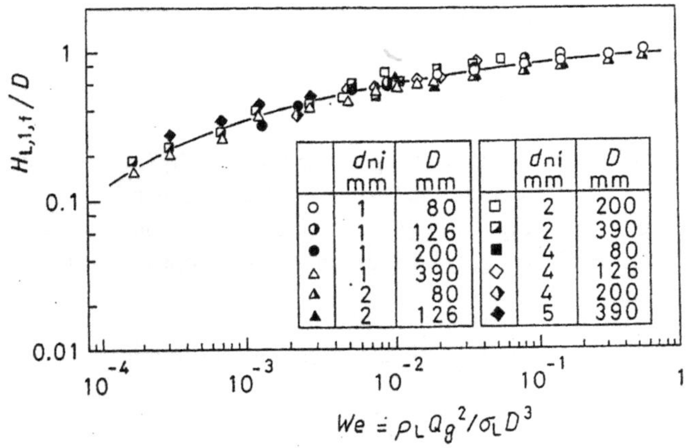


**Fig. 5.13** Correlation of the critical water bath depth at which the first kind of swirl motion ceases

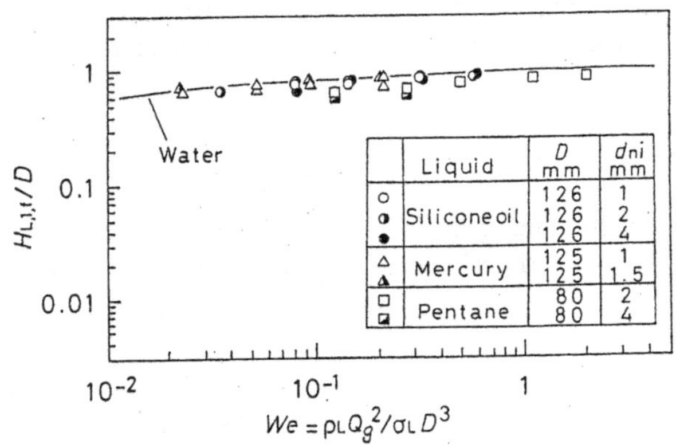


**Fig. 5.14** Correlation of the critical depth at which the first kind swirl of motion ceases in silicone oil, mercury, and normal pentane baths

**Table 5.1** Physical properties of liquids at 298 K

| Liquid | Density $\rho_L(kg/m^3)$ | Kinematic viscosity $\nu_L(mm^2/s)$ | Surface tension $\sigma(mN/m)$ |
|---|---|---|---|
| Water | 997 | 0.891 | 72.7 |
| $n$-pentane | 620 | 0.37 | 15.5 |
| Silicone oil | 935 | 12 | 21 |
| Mercury | 13,600 | 0.11 | 482 |
| Aqueous glycerol solution 1 | 1,170 | 9.4 | –– |
| Aqueous glycerol solution 2 | 1,110 | 4.5 | –– |

#### 5.2.1.3 The Second Kind of Swirl Motion

Critical Bath Depth for Initiation of Motion

Figures 5.15 and 5.16 present measured critical bath depth, $H_{L,2,s}$, normalized by the bath diameter, $D$, as a function of gas flow rate, $Q_g$. In general, $H_{L,2,s}/D$ decreases with an increase in $Q_g$ in Fig. 5.15, while it varies in a complex manner in Fig. 5.16. As shown in Fig. 5.17, $H_{L,2,s}/D$ is reasonably well correlated by the following empirical relation:

$$H_{L,2,s}/D = 0.758 Fr_H^{-0.313}, \tag{5.6}$$

$$Fr_H = (Q_g^2/gH_{L,2,s}^5)^{1/5}, \quad 0.01 < Fr_H < 0.08. \tag{5.7}$$

where, $Fr_H$ is a modified Froude number.

Figure 5.18 shows that the experimental results for other liquids are also adequately approximated by (5.6) proposed originally for the air–water system.

Swirl Period

Figures 5.19 and 5.20 indicate that the swirl period $T_s$ is a decreasing function of gas flow rate $Q_g$. The swirl period is non-dimensionalized in terms of the Strouhal

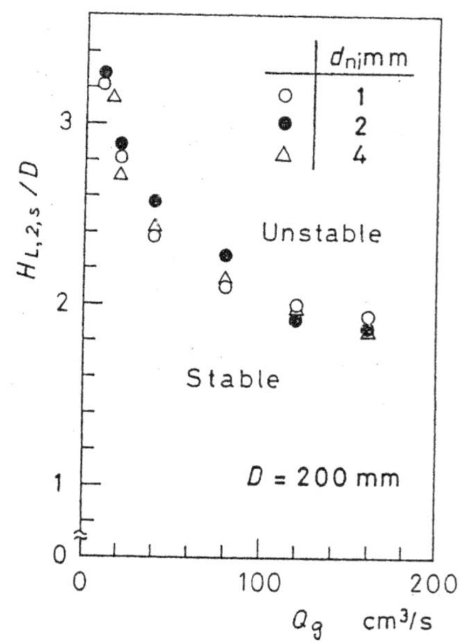


**Fig. 5.15** Relation between air flow rate and the critical bath depth at which the second kind of swirl motion occurs for $D = 200$ mm

**Fig. 5.16** Relation between air flow rate and the critical bath depth at which the second kind of swirl motion occurs for $D = 80\,\mathrm{mm}$

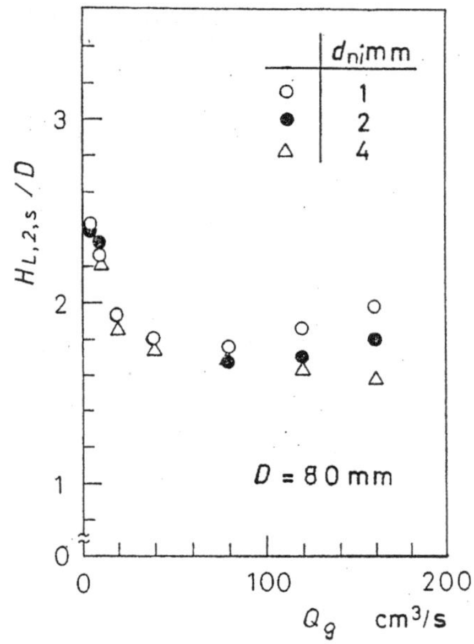


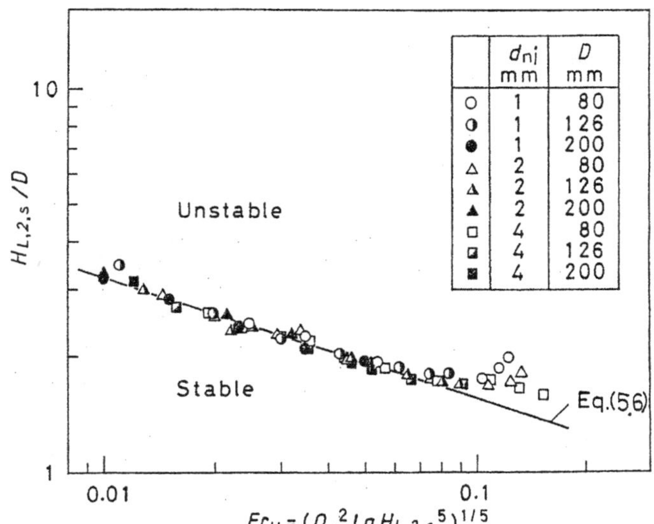


**Fig. 5.17** Correlation of the critical depth at which the second kind of swirl motion occurs in water bath

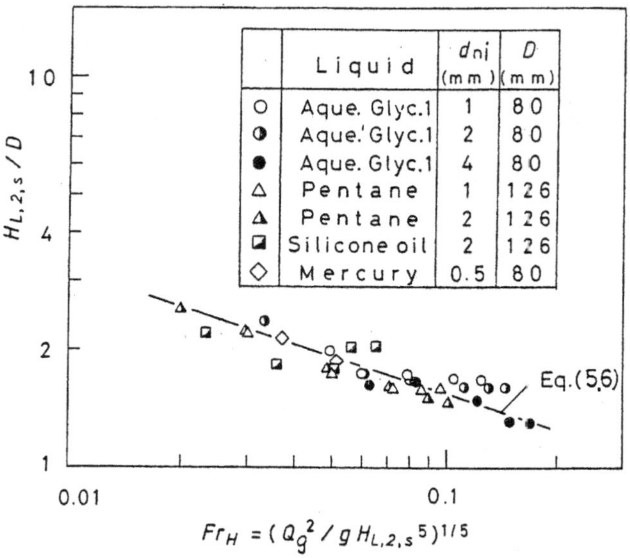


**Fig. 5.18** Relation between the critical bath depth at which the second kind of swirl motion occurs and the dimensionless parameter $Fr_H$

**Fig. 5.19** Relation between the period of the second kind of swirl motion and air flow rate for $D = 200$ mm

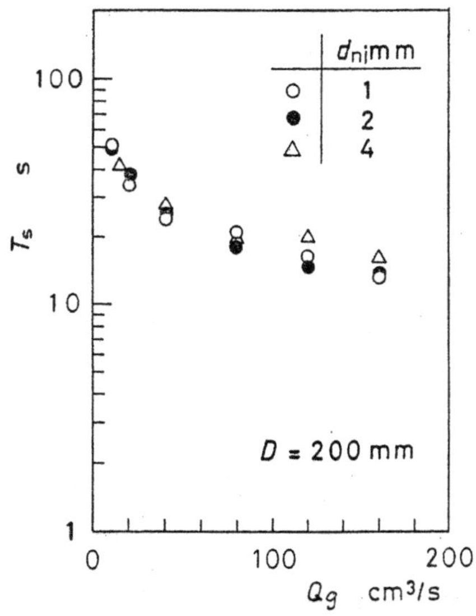

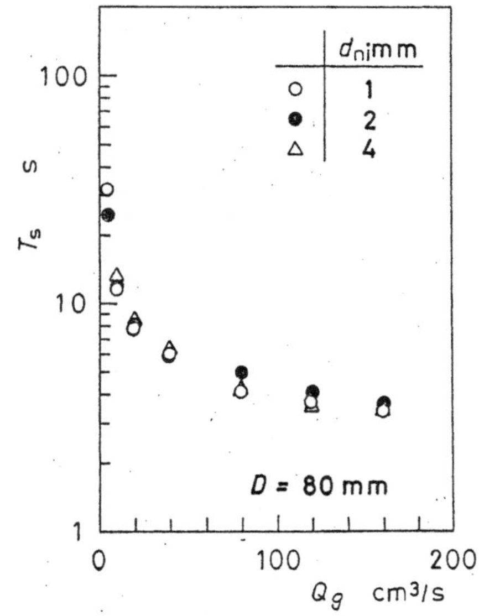


**Fig. 5.20** Relation between the period of the second kind of swirl motion and air flow rate for $D = 80\,\mathrm{mm}$

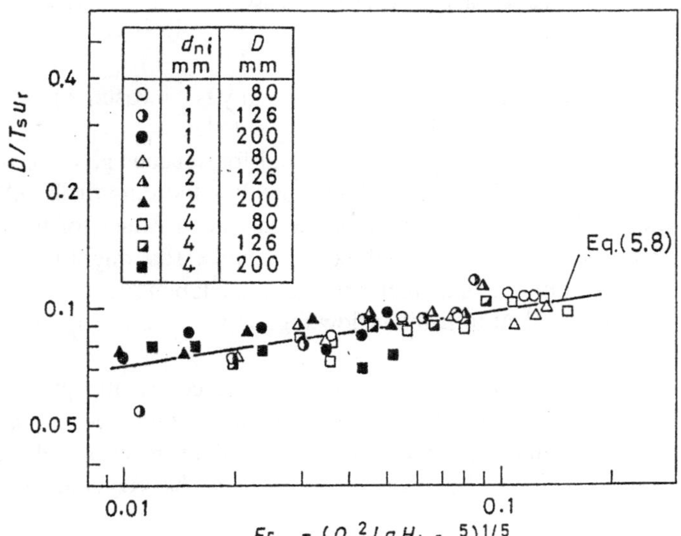


**Fig. 5.21** Correlation of the period of the second kind of swirl motion as a function of $Fr_H$ in water bath

number and plotted against the Froude number $Fr_H$ in Fig. 5.21. The data on $T_s$ are correlated by the following empirical relation within a scatter of $\pm 20\%$.

$$D/(T_s u_r) = 0.137 Fr_H^{0.140}, \tag{5.8}$$

$$u_r = (gQ_g/H_{L,2,s})^{1/3}, \tag{5.9}$$

$$0.01 < Fr_H < 0.15,$$

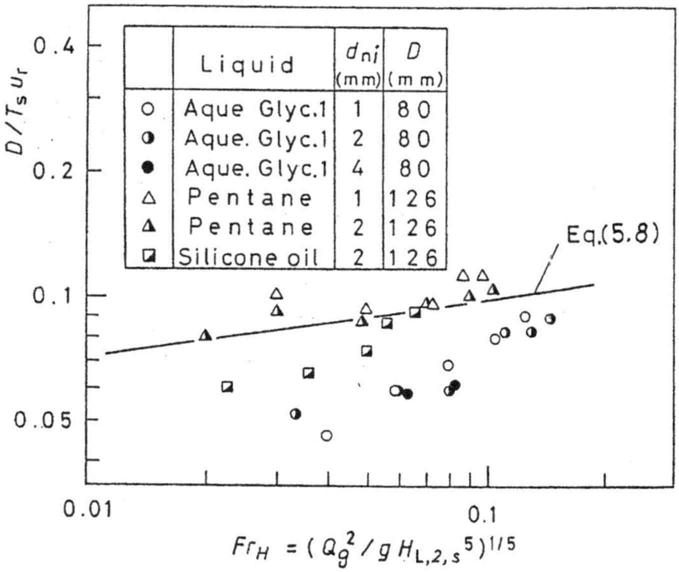


**Fig. 5.22** Correlation of the period of the second kind of swirl motion as a function of $Fr_H$

where $u_r$ has the dimension of velocity. This quantity is a measure of the centerline velocity of upward rising liquid flow induced by bubbles.

Figure 5.22 shows the swirl period measured for aqueous glycerol solution 1, $n$-pentane, and silicone oil. The solid line denotes the empirical (5.8) for the air–water system. The agreement between the measured values for the $n$-pentane and (5.8) is quite good, but poor for the other liquids. This might be because the flow in the recirculation region, which encloses the bubbling jet region, undergoes re-transition from turbulent to laminar flow regime with increasing kinematic viscosity, $\nu_L$.

The measured values of $D/(T_s u_r)$ are normalized by the predicted value based on (5.8) and replotted against the Reynolds number $Re(= u_r D/\nu_L)$ in Fig. 5.23. The solid line represents the best fit of the measured data. The ratio $[D/(T_s u_r)]_{mea}/[D/(T_s u_r)]_{est}$ clearly depends on the Reynolds number for $Re < 2{,}000$.

#### 5.2.1.4 Summary

Swirl motion of circular cylindrical baths agitated by gas injection is not unusual events. It occurs over a variety of modes depending on the blowing condition [29, 30]. The swirl motion is not welcome for safety and stable operations of metallurgical processes, but from a different point of view it is beneficial to the mixing of metallurgical processes. Some features of the first kind of swirl motion will be present in the following sections.

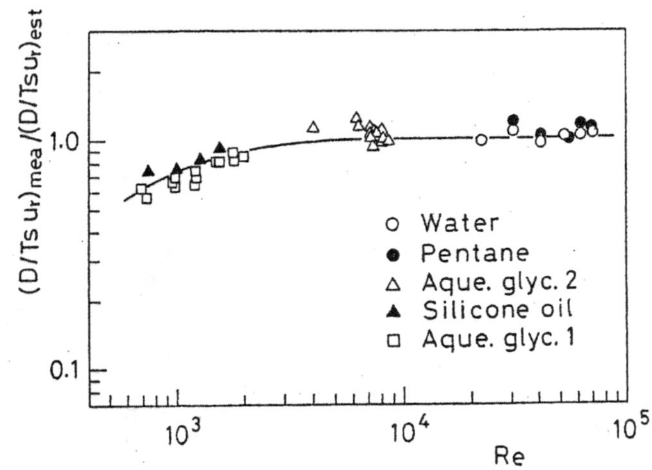


**Fig. 5.23** The ratio of measured to estimated values of $D/(T_s u_r)$ against Reynolds number

### *5.2.2 Operation Under Reduced Surface Pressure*

A variety of swirl motions are known to occur in a bath agitated by gas injection when the bath surface is exposed to the atmosphere, as described in Sect. 5.2.1.4 [18, 23, 29–37]. In particular, two types of swirl motions typically occur in a circular cylindrical bath agitated by single-nozzle bottom gas injection, as schematically illustrated in Fig. 5.6 [29, 30] One is observed over an aspect ratio, $H_L/D$, from approximately 0.2–1.0. The other appears for $H_L/D > 2$. No swirl motion occurs when the aspect ratio falls in the range of 1.0–2.0. The former swirl motion is caused by bath surface oscillations due to quasi-periodic generation and subsequent arrival of bubbles at the bath surface. It resembles the rotary sloshing of a water bath contained in a circular cylindrical vessel [16, 17, 38]. The latter is caused by the Coanda effect [26], which appears when a bubbling jet approaches the side wall of the vessel [29, 30, 39].

The mixing time in a cylindrical water bath with aspect ratio between $H_L/D \cong$ 0.2–1.0 is significantly shortened when the bath is accompanied by the first kind of swirl motion [35, 37]. This motion may thus be beneficial for shortening the mixing time in real refining processes agitated by bottom gas injection. Most of these processes are operated under reduced pressure on the bath surface. However, measurements of the characteristics of the swirl motion and the relationship between the mixing time and the swirl motion in model experiments have been carried out solely under atmospheric pressure on the bath surface.

It is quite difficult and dangerous to carry out systematic experiments to study swirl motion using real processes under reduced pressures on the bath surface. Thus model experiments are conducted using water models under two different reduced pressures on the bath surface in order to accumulate basic data for future applications of the swirl motion to real refining processes. The results are presented in the following.

#### 5.2.2.1 Experimental Apparatus and Procedure

Figure 5.24 shows a schematic of the experimental apparatus. The test vessel had an inner diameter $D$ of $200 \times 10^{-3}$ m and a height $H$ of $400 \times 10^{-3}$ m. The upper part of the vessel was covered with a flat plate, and the pressure on the bath surface, $P_s$, was reduced with a vacuum pump. The details of the procedure can be found in Iguchi et al. [40]. The aspect ratio $H_L/D$ employed ranged from approximately 0.1–1.2. Experiments were conducted with bath surface pressures of 20, 51 and 101 kPa and bath temperature $Te_B$ in the range 293–298 K. The lowest pressure of 20 kPa, was chosen to be larger than the vapor pressure, to suppress cavitation. The volumetric gas flow rate $Q_g$ was varied from $10 \times 10^{-6}$ to $300 \times 10^{-6}$ m$^3$/s. Starting time $T_{s,s}$ is defined as a period from the start of gas injection until the initiation of swirl motion, and damping time $T_{s,d}$ is defined as a period from the stoppage of gas injection until the swirl motion is damped out. The two time scales were determined by eye inspection and with a high-speed video camera (200 frames/s). In addition, the period and amplitude of the swirl motion were similarly obtained.

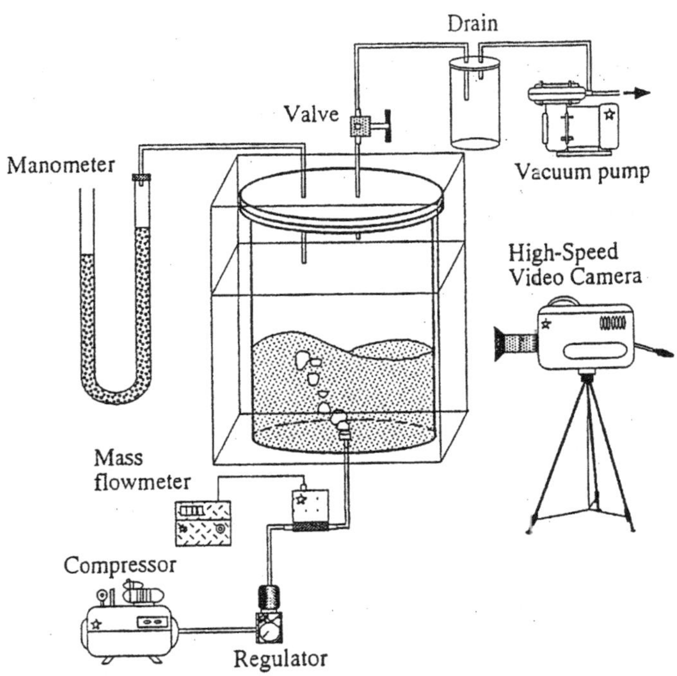


**Fig. 5.24** Experimental apparatus

### 5.2.2.2 Experimental Results and Discussion

Starting Time of Swirl Motion

Figure 5.25 shows the measured values of the starting time of swirl motion, $T_{s,s}$, for three different pressures on the bath surface as a function of the normal gas flow rate $Q_{gN}$. The latter is evaluated at a temperature $Te_N$ of 298 K and a pressure $P_N$ of 101 kPa. The starting time obtained under atmospheric pressure on the bath surface decreases with an increase in the normal gas flow rate. This is because the energy supplied by gas into the water bath increases with $Q_{gN}$. A similar trend is observed for the two reduced pressures on the bath surface. In addition, $T_{s,s}$ becomes shorter as the pressure on the bath surface, $P_s$, decreases.

The relationship between the volumetric gas flow rate at the nozzle tip $Q_g$ and $P_s$ can be expressed by,

$$Q_g = P_N Q_{gN} Te_B / [(P_s + \rho_L g H_L) Te_N], \tag{5.10}$$

where, $\rho_L$ is the density of liquid and $g$ is the acceleration due to gravity. When the normal gas flow rate $Q_{gN}$ is the same for different values of $P_s$, the volumetric gas flow rate $Q_g$ becomes high as $P_s$ decreases due to gas expansion. This is primarily responsible for the observed decrease in $T_{s,s}$ with decreasing $P_s$.

The data shown in Fig. 5.25 are replotted against $Q_g$ in Fig. 5.26. All the measured values of $T_{s,s}$ can be correlated in terms of $Q_g$ irrespective of the bath surface pressure $P_s$.

The following empirical relation has been reported for the starting time $T_{s,s}$ in water and molten metal baths under atmospheric pressure at the bath surface [36],

$$T_{s,s}(g/D)^{1/2} = 1.5\left[Re^{1/2}\left(H_L Q_g / D^{7/2} g^{1/2}\right)\right]^{-1}, \tag{5.11}$$

$$Re = \left(Q_g^2/g\right)^{2/5} (g/D)^{1/2} / \nu_L, \tag{5.12}$$

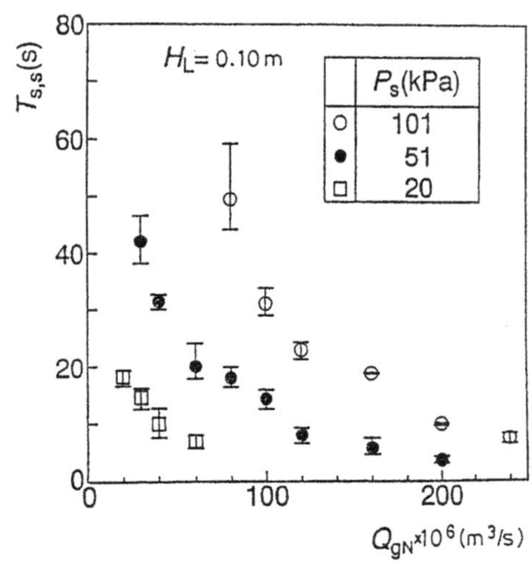


Fig. 5.25 Measured values of the starting time against normal gas flow rate

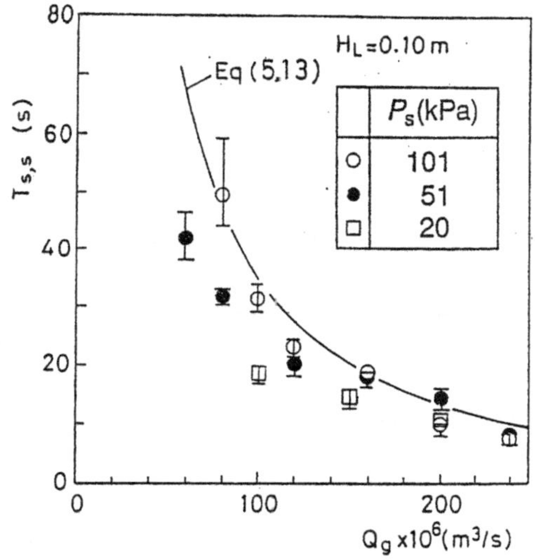


**Fig. 5.26** Measured values of the starting time against volumetric gas flow rate

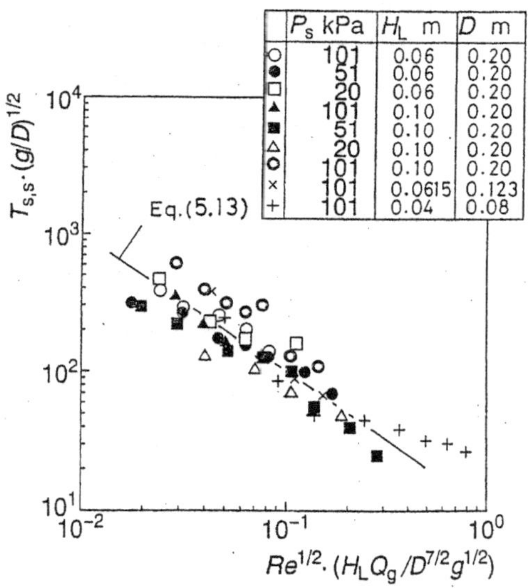


**Fig. 5.27** Comparison of (5.13) with measured values of the starting time

where, *Re* is the Reynolds number and $\nu_L$ is the kinematic viscosity of liquid. The Reynolds number *Re* is defined in terms of a representative length $\left(Q_g{}^2/g\right)^{1/5}$ and a representative velocity $\left(Q_g{}^2/g\right)^{1/5}(g/D)^{1/2}$. The length scale $\left(Q_g{}^2/g\right)^{1/5}$ is a measure of the horizontal extent of a bubbling jet and $(D/g)^{1/2}$ is a measure of the period of the swirl motion.

This functional relationship was also satisfactorily applicable to measured values of $T_{s,s}$ under reduced pressures, as can be seen in Fig. 5.27. However, Fig. 5.27

indicates that the coefficient of 1.5 in (5.11) should be replaced by 1.0. Wave motions of water induced by rupture of bubbles at the bath surface are superimposed on the swirl motion of water. These wave motions become more intense as the pressure on the bath surface decreases, and accordingly, the accuracy of the measurement of $T_{s,s}$ decreases. The following empirical relation has therefore been proposed for $T_{s,s}$ thus,

$$T_{s,s}(g/D)^{1/2} = 1.0\left[Re^{1/2}\left(H_L Q_g/D^{7/2}g^{1/2}\right)\right]^{-1}, \qquad (5.13)$$
$$0.02 < Re < 0.3.$$

Equation (5.13) can be used to correlate all the measured values within a scatter of −40% to +60%, as shown in Fig. 5.27. Such a large scatter is inevitable in this kind of measurement. Equation (5.13) has also been suggested to be applicable to molten metal baths [36].

### Period of Swirl Motion

Figure 5.28 shows the relationship between the period of the swirl motion, $T_s$, and the normal gas flow rate $Q_{gN}$ with the pressure on the bath surface, $P_s$, as a parameter. The measured values of $T_s$ are independent both of $Q_{gN}$ and $P_s$.

Figure 5.29 illustrates the relationship between $(D/g)^{1/2}/T_s$ and the aspect ratio $H_L/D$. The solid line indicates the period of rotary sloshing occurring in a cylindrical water bath subjected to externally forced oscillation. The angular frequency of the fundamental wave of the rotary sloshing, $\omega_1 (= 2\pi/T_s)$, is given by [18, 38]:

$$\omega_1 = [(2g\varepsilon_1/D)\tan h(2\varepsilon_1 H_L/D)]^{1/2}, \qquad (5.14)$$

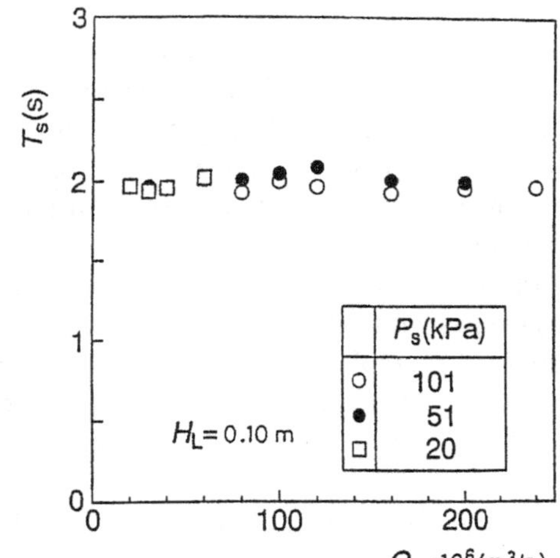


**Fig. 5.28** Measured values of the period of swirl motion against normal gas flow rate

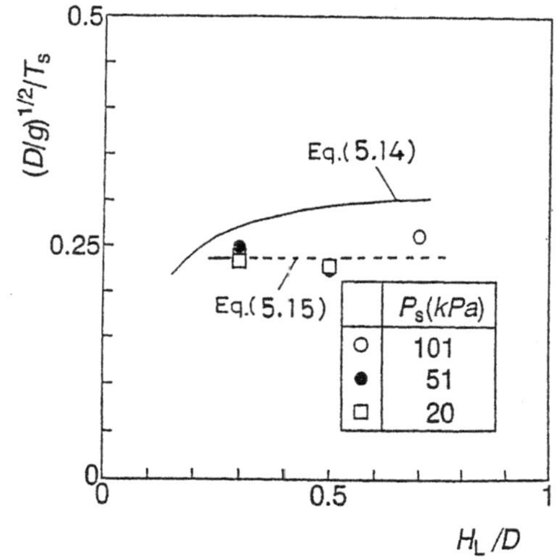


**Fig. 5.29** Correlation of the period of swirl motion

where, $\varepsilon_1$ is the first zero of the Bessel function $J_1{}'(\varepsilon)$. The measured values are slightly smaller than the solid line and essentially independent of the aspect ratio. The following simplified expression therefore has been derived:

$$(D/g)^{1/2}/T_s = 0.23. \tag{5.15}$$

This equation could also correlate the measured $T_s$ obtained in other studies [29,30]. Equation (5.15) is valid for molten metal baths agitated by gas injection under the atmospheric pressure condition on the bath surface [29, 30]. Accordingly, it would also be applicable to molten metal baths under reduced pressures on the bath surface.

## Amplitude of Swirl Motion

The amplitude of the swirl motion was defined and determined in the following manner. The swirl motion was recorded on the high-speed video camera. The vertical distance between the highest and the lowest water levels on the side wall of the vessel was divided by 2, being the amplitude of the swirl motion.

The measured amplitudes of the swirl motion for $H_L = 0.10$ m are shown against the normal gas flow rate $Q_{gN}$ in Fig. 5.30 for three different pressures on the bath surface. The amplitude $A$ increases with an increase in the normal gas flow rate $Q_{gN}$ as well as with a decrease in the pressure on the bath surface, $P_s$. The measured values of $A$ are replotted against the volumetric gas flow rate in Fig. 5.31. These values can be approximately correlated irrespective of the pressure on the bath surface. The following equation was derived with reference to a previous study [2]:

$$A = 1.65\{(Q_g - Q_{gc})^2/g\}^{1/5}\left[1 - \{2/(0.2 - a_{su})\}^2\{H_L/D - (a_{su} + 0.2)/2\}^2\right], \tag{5.16}$$

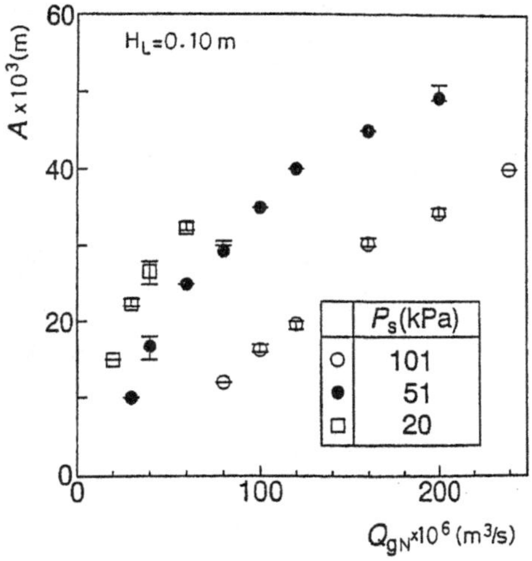


**Fig. 5.30** Measured values of the amplitude of swirl motion against normal gas flow rate

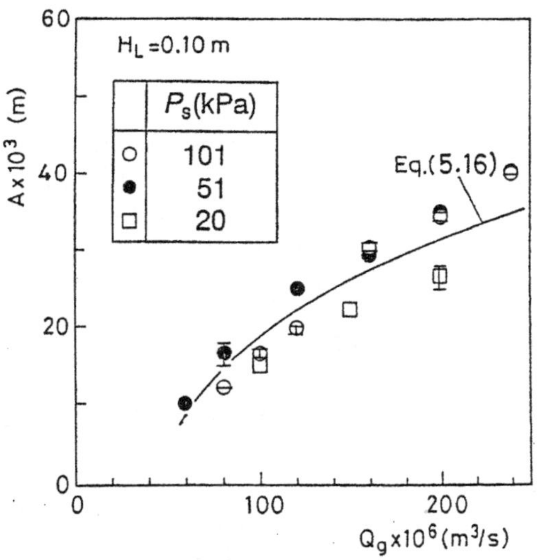


**Fig. 5.31** Measured values of the amplitude of swirl motion against volumetric gas flow rate

$$Q_{gc} = 0.0555(\sigma D^3/\rho_L)^{1/2}, \tag{5.17}$$

$$a_{su} = 10[0.05 - 1.35(X + 6)/\exp(X + 5)], \tag{5.18}$$

$$X = \log We, \tag{5.19}$$

$$We = \rho_L Q_g^2/(\sigma D^3), \tag{5.5}$$

where, $\sigma$ is the surface tension, $a_{su}$ is the upper limit at which the first kind of swirl motion ceases, and *We* is the Weber number. This equation was devised so that the amplitude $A$ becomes 0 at $H_L/D = 0.2$ and $H_L/D = a_{su}$, because the swirl motion ceased for $H_L/D \leq 0.2$ and for $H_L/D \geq a_{su}$.

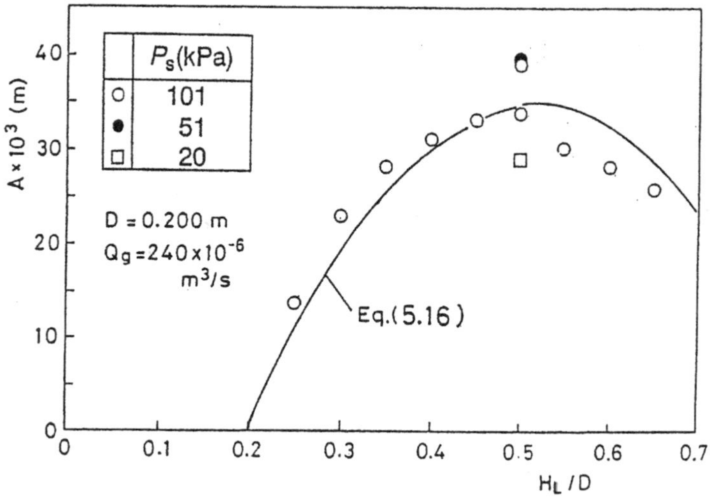


**Fig. 5.32** Comparison of (5.16) with measured values of the amplitude of swirl motion for different aspect ratios

As a representative example, the measured values of $A$ for a gas flow rate $Q_g$ of $240 \times 10^{-6}$ m$^3$/s are shown in Fig. 5.32. Equation (5.16) is represented by the solid line. The agreement between the measured values and (5.16) is quite satisfactory for every aspect ratio considered.

## Damping Time of Swirl Motion

Figure 5.33 shows the variation of the measured damping time of the swirl motion, $T_{s,d}$, with the normal gas flow rate $Q_{gN}$. The damping time increased either by increasing $Q_{gN}$ or by decreasing the pressure on the bath surface, $P_s$. The damping time is associated with the dissipation energy of water flow in the bath. Under steady state condition the dissipation energy is equal to the energy supplied by the injected gas per unit time and mass, $\varepsilon_d$, and approximated by the following expression [29, 30],

$$\varepsilon_d = g(\rho_L - \rho g) Q_g H_L / (\pi D^2 H_L \rho_L / 4). \tag{5.20}$$

Equation (5.20) states that $\varepsilon_d$ is proportional to the gas flow rate $Q_g$. Accordingly, it is reasonable that $T_{s,d}$ is an increasing function of $Q_g$.

The measured values of $T_{s,d}$ are independent of the pressure on the bath surface when plotted against $Q_g$, as demonstrated in Fig. 5.34. The solid line indicates the following empirical relation [36],

$$T_{s,d} Q_g / D_3 = 1.45 \times 10^{-6} Re^{1.74}. \tag{5.21}$$

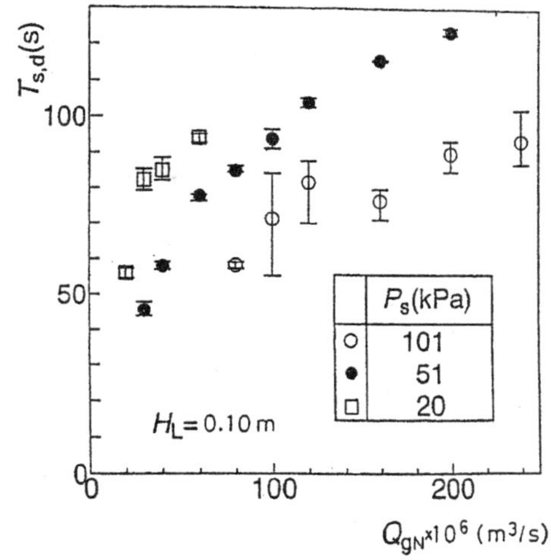


**Fig. 5.33** Measured values of the damping time against normal gas flow rate

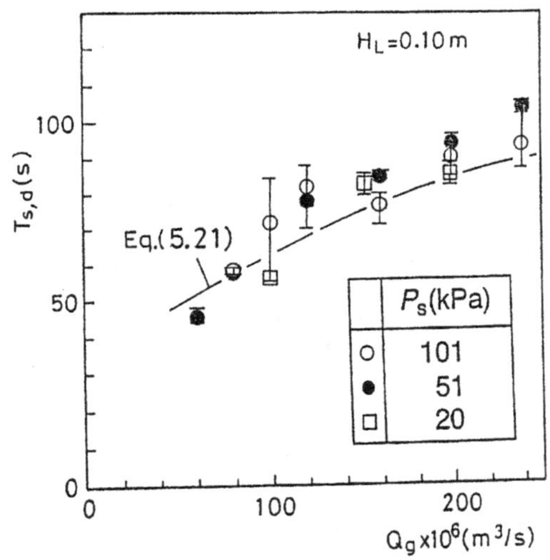


**Fig. 5.34** Measured values of the damping time against volumetric gas flow rate

A variety of data can be approximated by (5.21) within a scatter of ±60%, as shown in Fig. 5.35. The damping time, $T_{s,d}$, for reduced pressures on the bath surface can be approximated by (5.21) taking the expansion of the injected gas at the nozzle tip into consideration.

The following relation has also been proposed for $T_{s,d}$.

$$T_{s,d}Q_g/D^3 = 11Re^{1/2}\left(H_L Q_g/D^{7/2}g^{1/2}\right). \tag{5.22}$$

This equation, however, is less accurate than (5.21) in correlating the experimental data of $nT_{s,d}$.

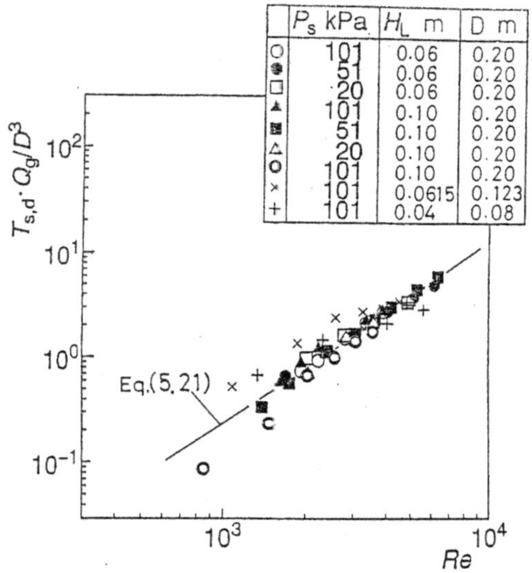


**Fig. 5.35** Comparison of (5.21) with measured values of damping time

#### 5.2.2.3 Summary

Swirl motions occurring in a water bath of aspect ratio ranging approximately from 0.2 to 1.0 were characterized by the starting time $T_{s,s}$, period $T_s$, amplitude $A$, and damping time $T_{d,s}$. The effects of reduced pressure on the bath surface, $P_s$, on these quantities were experimentally investigated [41]. The measured values of these quantities for gas injection under reduced pressure on the bath surface agreed favorably with their respective values for gas injection under atmospheric pressure condition on the bath surface provided that the volumetric gas flow rates at the nozzle tip, $Q_g$, were the same. Empirical relations were proposed for $T_{s,s}$, $T_{s,d}$, $T_s$, and $A$.

It should be noted that the work described here is a first step towards understanding the behavior of swirl motion under reduced pressure on the bath surface. The flow behavior for $P_s$ lower than 20 kPa may not be adequately represented by the empirical relations proposed due to intense rupture of bubbles at the bath surface. Further work is desirable under such conditions ($P_s < 20$ kPa). Silicone oils would be a suitable liquid for such investigation because of their very low vapor pressure.

### 5.2.3 *Mixing Time*

When gas is injected into a bath through a single-hole nozzle placed on the bottom center of a cylindrical vessel, a vertical bubbling jet is generated in the bath. Under certain blowing conditions, the jet rises, swirling around the vessel axis [29, 30, 35–42]. A swirling motion is also induced in the liquid, being in phase with the swirling motion of the jet. This type of swirl motion of the liquid, called the first kind of swirl motion, occurs for aspect ratio, $H_L/D$, ranging from approximately 0.2–1.0. It is very similar to the well-known rotary sloshing observed for a bath contained in a cylindrical vessel oscillating in the horizontal or vertical direction. The period of the liquid swirl motion is approximately equal to that of the rotary sloshing, while the amplitude becomes large with an increase in the gas flow rate.

The liquid contained in a cylindrical vessel is violently agitated in the presence of the swirl motion. This also results in spitting, splashing, and slopping. Due to these disadvantageous effects, the swirl motion has not been used practically in the steelmaking industries. From a different point of view, the swirl motion would also produce excellent mixing of the bath. In fact, it has been used to provide very high melting rate in a snow-melting process [43].

If the spitting, splashing, and slopping can be avoided, for example, by shielding the upper part of a bath, a novel refining process with high mixing efficiency could be developed. However, there exists no information on the mixing time in the presence of the swirl motion [44–55]. Water model experiments have been performed and an empirical relation proposed for the mixing time in a water bath accompanied by the first kind of swirl motion. The details of the model experiments will be presented below.

#### 5.2.3.1 Experimental Apparatus and Procedure

Figure 5.36 shows a schematic of the experimental apparatus. Three cylindrical vessels of different sizes were used. The inner diameter $D$ and the height $H$ of the vessels were $0.125 \times 0.400$ m, $0.200 \times 0.400$ m, and $0.400 \times 0.800$ m. Each vessel was filled with deionized water to a prescribed depth, $H_L$. Air was supplied with a compressor, and the air flow rate, $Q_g$, was adjusted with a regulator and a mass flow controller from $40 \times 10^{-6}$ to $800 \times 10^{-6}$ m$^3$/s. The air was injected into the bath

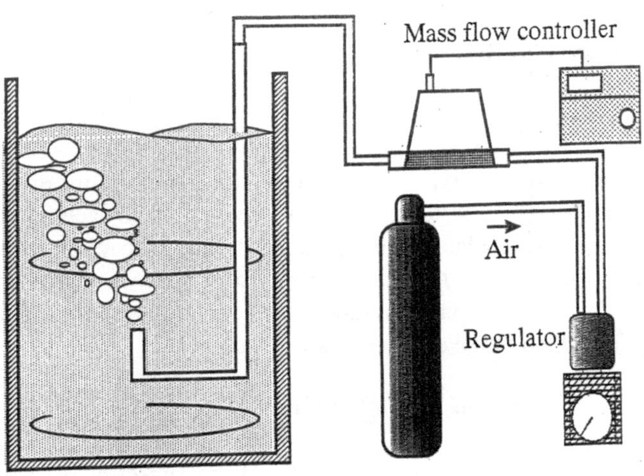


**Fig. 5.36** Experimental apparatus

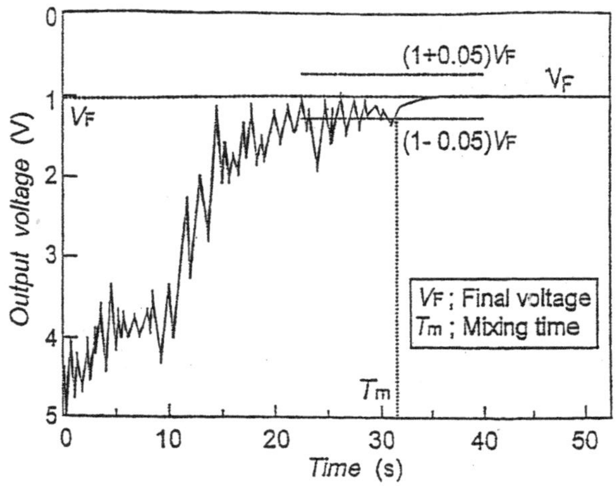


**Fig. 5.37** Definition of mixing time

through either a bottom nozzle of inner diameter $d_{ni}$ of 2 mm or a J-shaped top lance of the same inner diameter. The inner diameter of a bottom nozzle has a negligible effect on the mixing time [44, 45]. The critical condition for the existence of the swirl motion has been established [29, 30, 42].

Mixing time was determined on the basis of the history of the electrical conductivity of liquid in the bath. A dilute aqueous KCl solution was used as tracer and charged from the top onto the bath surface. The electrical conductivity was measured with an electrical conductivity sensor having a time constant of 0.25 s. The output voltage approaches a constant value, $V_F$, with time. The mixing time is defined as the period from the moment of tracer charge to the moment at which the electrical conductivity finally crosses 0.95 $V_F$ or 1.05 $V_F$, as shown in Fig. 5.37. Further details of the experimental method for the mixing time are given elsewhere [52].

### 5.2.3.2 Experimental Results

Bottom Gas Injection

The variation of measured mixing time with gas flow rate for three different vessels are presented in Figs. 5.38, 5.39 and 5.40. The straight line in each figure represents the best fit of the measured data for the same aspect ratio. The mixing time decreases with an increase in the gas flow rate, but increases with an increase in the bath diameter. The following empirical relation has been proposed for the mixing time in a bottom blown bath in the absence of swirl motion [52],

$$T_{mB} = 1{,}200 Q_g^{-0.47} D^{1.97} H_L^{-1} \nu_L^{0.47}, \tag{5.23}$$

where, $\nu_L$ is the kinetic viscosity of liquid and the subscript B denotes bottom blowing.

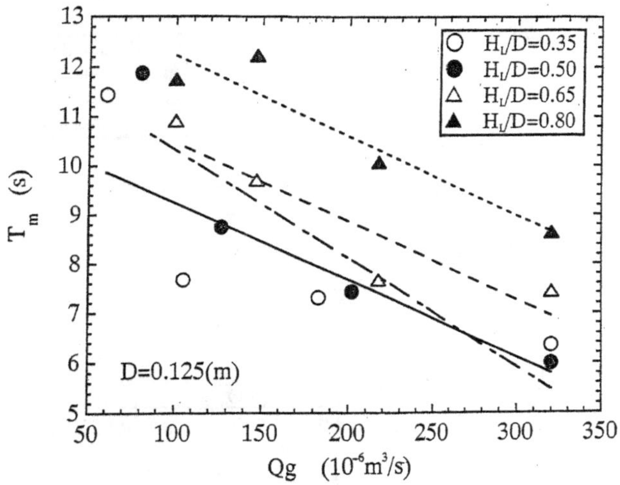


**Fig. 5.38** Relationship between mixing time $T_{\mathrm{m}}$ and gas flow rate for $D = 0.125\,\mathrm{m}$

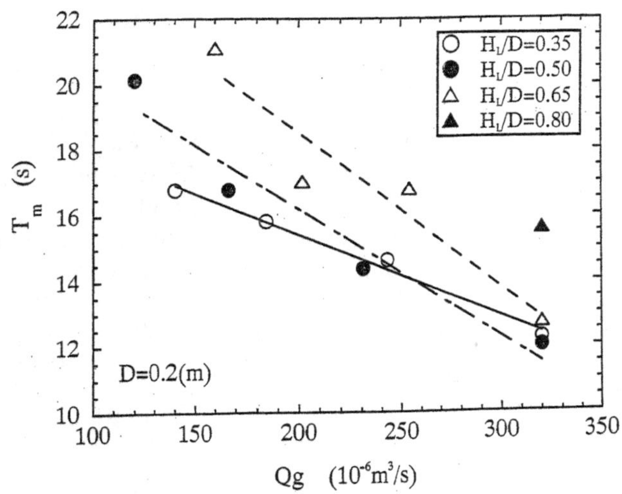


**Fig. 5.39** Relationship between mixing time $T_{\mathrm{m}}$ and gas flow rate for $D = 0.200\,\mathrm{m}$

Equation (5.23) can be rewritten in a dimensionless form as follows:

$$T_{\mathrm{mB}}(H_{\mathrm{L}}/D)(g/D)^{1/2} = 4.21 \times 10^3 Re^{-0.47}, \tag{5.24}$$

$$Re = Dv_{\mathrm{sp}}/\nu_{\mathrm{L}}, \tag{5.25}$$

$$V_{\mathrm{sp}} = 4Q_{\mathrm{g}}/(\pi D^2). \tag{5.26}$$

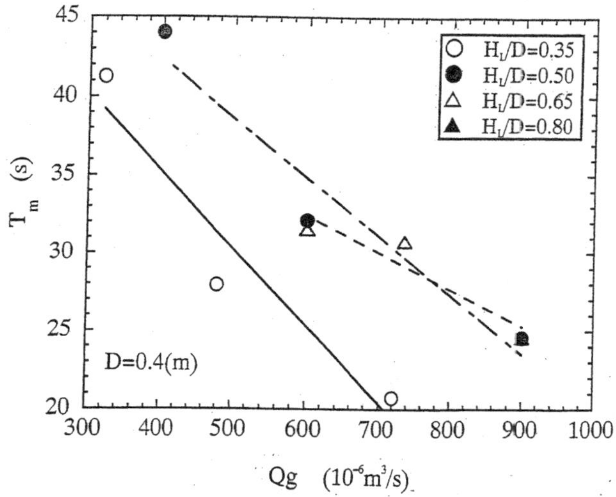


**Fig. 5.40** Relationship between mixing time $T_m$ and gas flow rate for $D = 0.400$ m

Equation (5.24) is valid for

$$30 < Re < 3{,}000. \tag{5.27}$$

In the above equations, $g$ is the acceleration due to gravity and $Re$ is the Reynolds number based on the superficial velocity of gas, $V_{sp}$. The conditions describing the onset of swirl motion have been given Sect. 5.2.3.1 [29, 30].

Equation (5.23) is used to correlate the measured mixing time in the presence of swirl motion. All the mixing time values are divided by (5.23) and replotted against the aspect ratio $H_L/D$ in Fig. 5.41. For each vessel, a straight line is drawn through the mean of the data as close as possible. The gradient of each straight line is 4/3, implying that the mixing time is proportional to $(H_L/D)^{4/3}$. Therefore the following relationship has been assumed:

$$T_m = a_k(H_L/D)^{4/3}T_{mB}. \tag{5.28}$$

Inspection of Fig. 5.41 suggests that the coefficient $a_k$ is a function of the vessel diameter $D$.

Referring to Stokes's first problem [56], we introduce the following representative length to non-dimensionalize the vessel diameter as follows

$$L = (\nu_L T_s)^{1/2}, \tag{5.29}$$

$$T_s = k(D/g)^{1/2}, \tag{5.30}$$

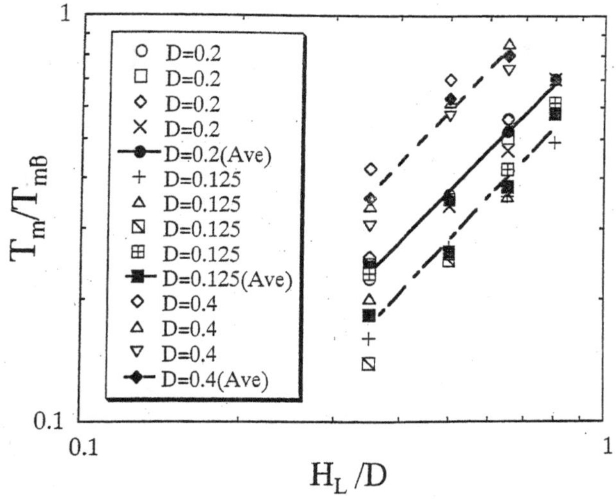


**Fig. 5.41** Relationship between non-dimensionalized mixing time and aspect ratio

where, $T_s$ is the period of the swirl motion and $k$ is a constant. It is known that $k$ is approximately 0.27 [29, 30]. Equation (5.29) characterizes a distance for the momentum transfer from the vessel wall during one period of the swirl motion.

For convenience, $k$ is assumed to be unity and (5.30) is substituted into (5.29) to give

$$L = (\nu_L^2 D/g)^{1/4}. \quad (5.31)$$

The values of $a_k$ can be obtained directly from Fig. 5.41. These values are plotted against $D/L$ in Fig. 5.42, where $D/L$ is expressed by

$$D/L = \left(D^3 g/\nu_L^2\right)^{1/4}. \quad (5.32)$$

The measured values of $a_k$ are approximated by

$$a_k = 4.1 \times 10^{-3} (D^3 g/\nu_L^2)^{7/32}. \quad (5.33)$$

Combination of (5.23), (5.28) and (5.33) yields,

$$T_m = 4.92 (D^3 g/\nu_L^2)^{7/32} (H_L/D)^{4/3} Q_g^{-0.47} D^{1.97} H_L^{-1} \nu_L^{0.47}. \quad (5.34)$$

All the mixing time values obtained in the presence of swirl motion are compared with (5.34) in Fig. 5.43 It is evident that (5.34) can approximate the measured data within a scatter of $-13$ to $+20\%$.

Equation (5.34) can be rewritten as,

$$T_m = 4.92 g^{0.22} Q_g^{-0.47} D^{1.29} H_L^{0.33} \nu_L^{0.03}. \quad (5.35)$$

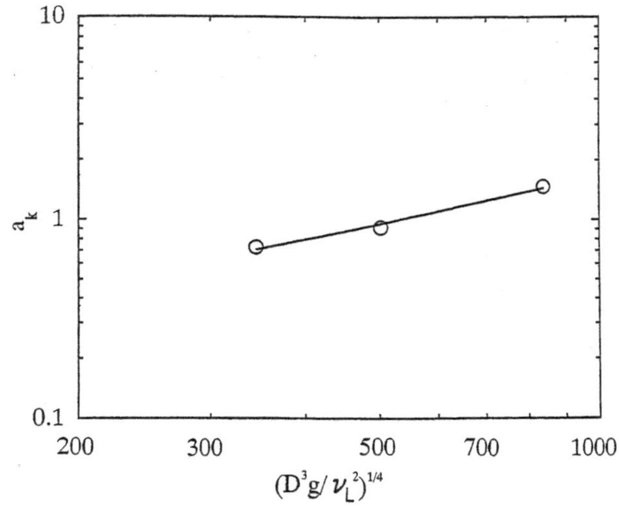


**Fig. 5.42** Coefficient $a_k$ as a function of $\left(D^3g/\nu_L{}^2\right)^{1/4}$

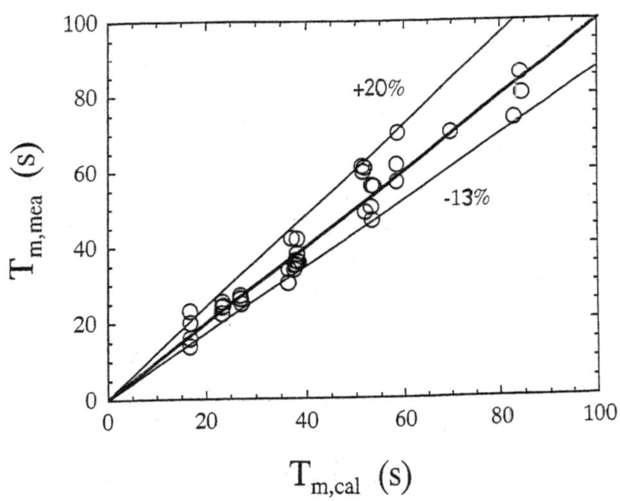


**Fig. 5.43** Comparison of measured mixing time values with (5.34)

The contribution of the kinematic viscosity of liquid, $\nu_L$, was found to be very small and negligible. This result implies that a bath subjected to swirl motion of a bubbling jet is primarily mixed by the strong wave motions induced by the swirl because the wave motions are not affected by the kinematic viscosity of the liquid.

J-Shaped Top Lance Gas Injection

The mixing time in a bath agitated by gas injection through a J-shaped top lance in the absence of the swirl motion can be represented by the following empirical relation [57]:

$$T_{\mathrm{mJ}} = 1,200 Q_{\mathrm{g}}^{-0.47} D^{1.97} H_{\mathrm{L}}^{-1} \nu^{0.47} (H_{\mathrm{in}}/H_{\mathrm{L}})^{0.7-1.34HL/D}, \tag{5.36}$$

where, $H_{\mathrm{in}}$ is the distance from the nozzle tip to the bath surface and the subscript J denotes the J-shaped top lance. Referring to (5.35) and (5.36), we assume that the mixing time in a bath in the presence of the swirl motion can be approximated by

$$T_{\mathrm{m}} = 4.92 \left(D^3 g/\nu_{\mathrm{L}}^2\right)^{7/32} (H_{\mathrm{L}}/D)^{4/3} Q_{\mathrm{g}}^{-0.47} D^{1.97} H_{\mathrm{L}}^{-1} \nu_{\mathrm{L}}^{0.47} (H_{\mathrm{in}}/H_{\mathrm{L}})^{0.7-1.34HL/D} \tag{5.37}$$

In order to examine the applicability of (5.37), additional experiments were carried out in a bath agitated by a J-shaped top lance in the presence of the swirl motion. Representative results for $D = 0.200$ m are shown in Fig. 5.44. Each line is drawn so as to best fit the measured data obtained under the same experimental condition. Mixing time was also measured for the remaining two vessels of $D = 0.125$ and 0.400 m. The experimental results are compared with (5.37) in Fig. 5.45. The measured values for the three vessels of different diameters can be approximated by this equation to within a scatter of −30% to +40%. Such a scattering level is considered acceptable for this type of measurement [46, 53].

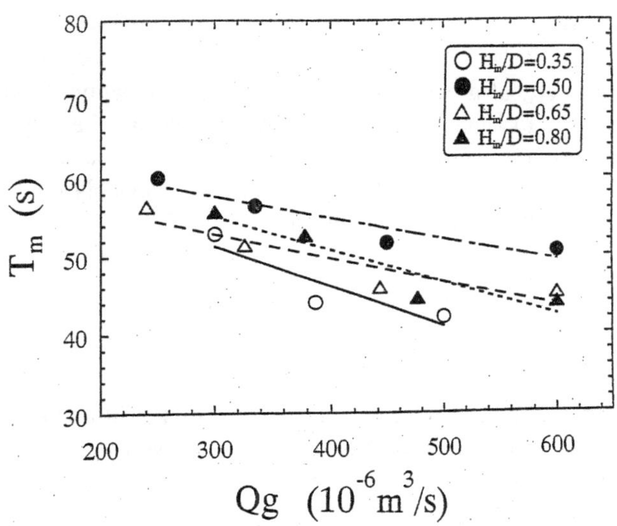


**Fig. 5.44** Relationship between mixing time $T_{\mathrm{m}}$ and gas flow rate for $D = 0.200$ m

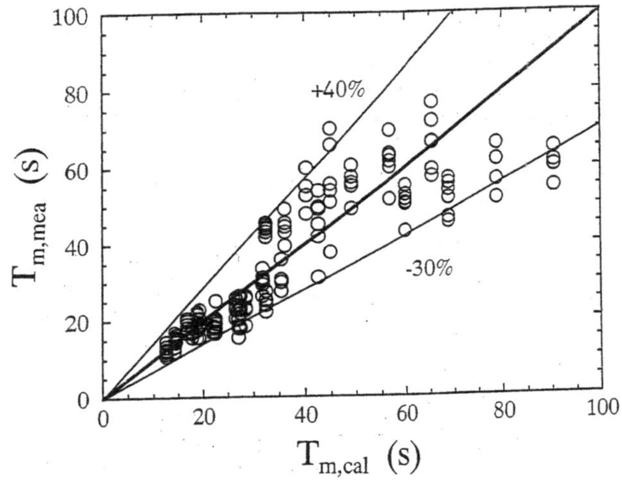


**Fig. 5.45** Comparison of measured mixing time values with (5.36)

Applicability of (5.37) to Molten Metal Bath

Empirical relations for mixing time derived for water baths agitated by bottom gas injection are known to be approximately applicable to molten steel baths [46]. This is because the kinematic viscosity of water is approximately equal to that of molten steel. Accordingly, (5.37) would be applicable to molten steel baths in the presence of swirl motion of liquid.

#### 5.2.3.3 Summary

The mixing time, $T_m$, of a water bath agitated by gas injection through a bottom nozzle or a J-shaped top lance was measured with an electrical conductivity sensor. An empirical relation (5.37) was proposed for $T_m$ in the presence of the first kind of swirl motion. All the existing $T_m$ data could be reasonably correlated by this empirical relation within a scatter of $-30$ to $+40\%$.

### *5.2.4 Effect of Top Slag*

Swirl motion of a bubbling jet occurs even in the presence of a thin slag layer on a molten metal layer. The bath surface oscillations, however, are strongly modulated compared with those in the absence of the slag layer. This section presents the effects of the top slag layer on the first kind of swirl motion of a bubbling jet and the behavior of the slag layer.

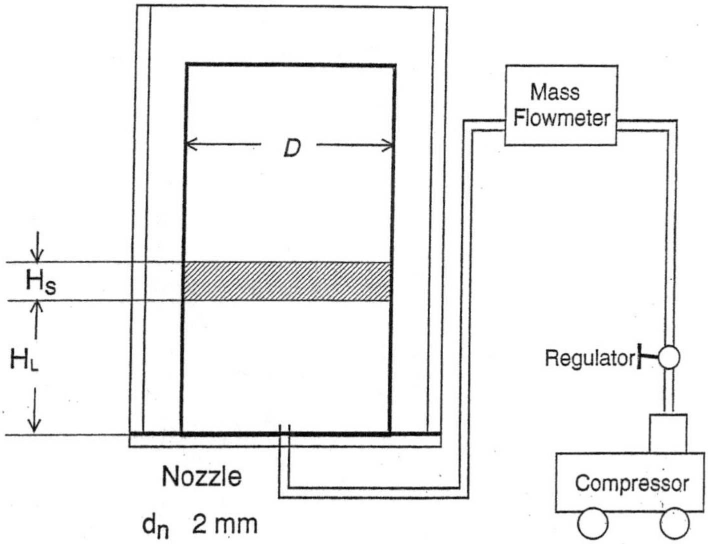


**Fig. 5.46** Experimental apparatus with top slag layer

#### 5.2.4.1 Experimental Apparatus and Procedure

Figure 5.46 shows a schematic of the experimental apparatus. Two vessel diameters $D$ of 125 or 200 mm were employed. The inner diameter of the nozzle $d_{ni}$ was 2 mm. Water and silicone oil were used as models for molten metal and slag, respectively. The physical properties of water are listed in Table 5.1. The density, kinematic viscosity, and surface tension of the silicone oil are 968 kg/m$^3$, 100 mm$^2$/s, and 53 mN/m, respectively. The thickness of the upper slag layer is denoted by $H_s$ and that of the lower molten metal layer by $H_L$. The slag layer thickness was varied up to 2.5 cm, while $H_L$ was chosen to fall in the range $0.3\ H_L/D < 1.0$ in which the first kind of swirl motion occurred. The bath surface oscillations were observed with a commercial video camera. The starting time, $T_{s,s}$, period, $T_s$, amplitude, $A$, and damping time, $T_{s,d}$, of the swirl motion were determined.

#### 5.2.4.2 Experimental Results

Classification of Slag Motion

The motions of the slag layer were classified into three categories, as shown in Figs. 5.47a–c. When the gas flow rate $Q_g$ was lower than a certain critical value, the first kind of swirl motion was not observed (Fig. 5.47a). When $Q_g$ exceeded the critical value, the slag layer disintegrated into small droplets (reverse emulsification). On average, the droplets did not rotate around the vessel axis, i.e., they remained at their initial position, but the bubbling jet rotated in the plume eye (Fig. 5.47b).

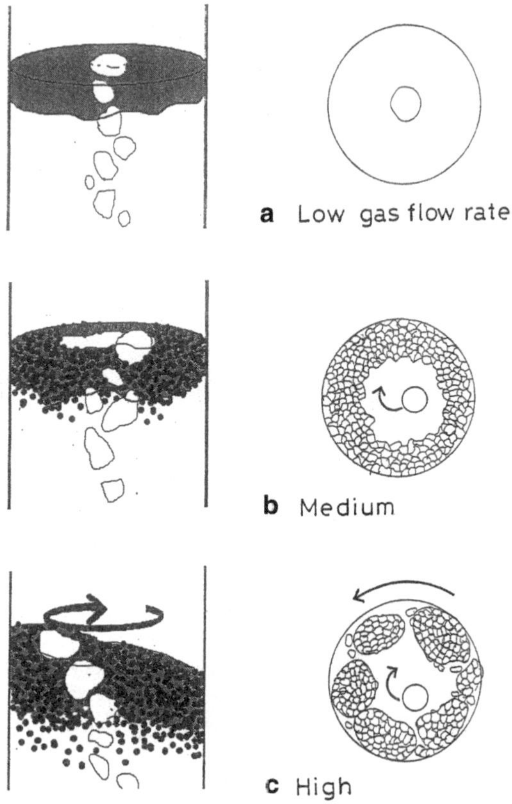


**Fig. 5.47** Classification of motions of top slag layer

As $Q_g$ increased further, the disintegrated slag layer clustered and rotated in the direction opposite to the direction of the bubbling jet (Fig. 5.47c). This result may be explained as follows.

As gas is injected vertically upward into a still water bath, the initial angular momentum of liquid in the bath is zero. If a bubbling jet rotates, for example, counter clockwise, the water and silicone oil outside the bubbling jet rotate clockwise in order to satisfy the conservation law of angular momentum [35, 37].

The three types of motions of the silicone oil layer can be mapped as a function of gas flow rate as shown in Fig. 5.48. It should be stressed that reverse emulsification, i.e., the formation of silicone oil droplets in the lower water layer occurs in the motions of Types 2 and 3.

## Critical Gas Flow Rate

The critical gas flow rate, $Q_{gc}$, for the initiation of the first kind of swirl motion increased with an increase in the thickness of the upper silicone oil layer. No correlation has been proposed yet for the critical gas flow rate.

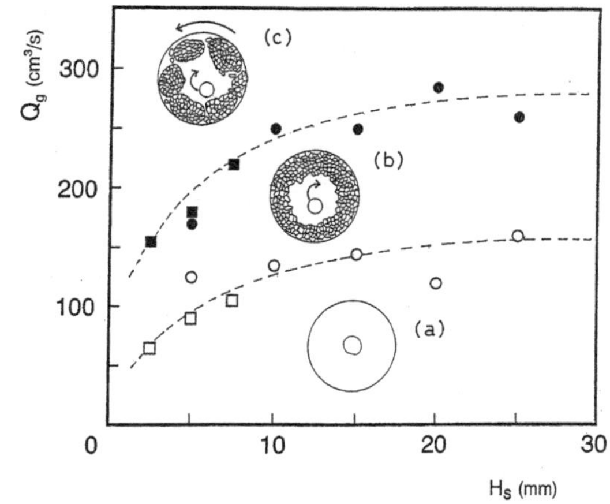


**Fig. 5.48** Three types of motions of top slag layer

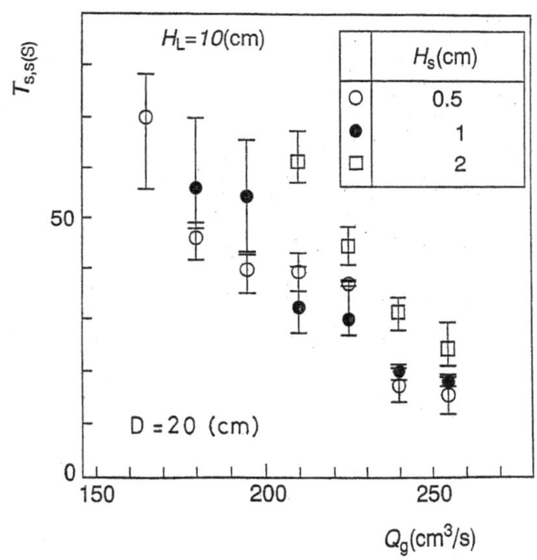


**Fig. 5.49** Relationship between starting time and gas flow rate

## Starting Time of Swirl Motion

The starting time, $T_{s,s}$, is defined as the time duration from the start of gas injection to the moment at which the swirl motion of a bubbling jet reaches steady state. It is one of the important parameters used for the design of metallurgical processes. The thickness of the upper silicone oil layer lengthens the starting time, as shown in Figs. 5.49 and 5.50. Note that $T_{s,s}$ has a minimum at the aspect ratio of approximately 0.5 ($H_L = 10\,\text{cm}$).

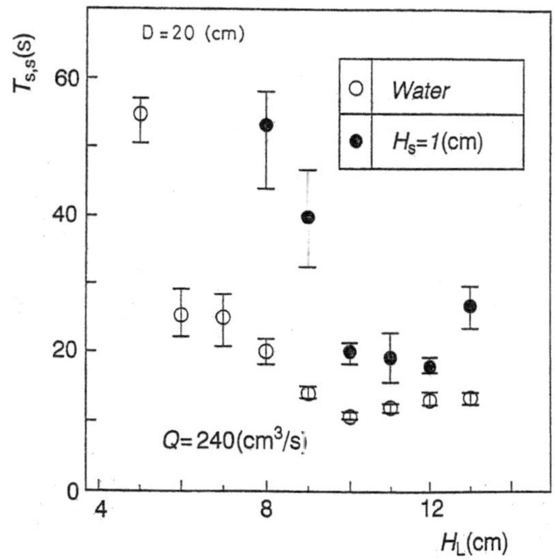


**Fig. 5.50** Relationship between starting time and thickness of lower molten metal layer

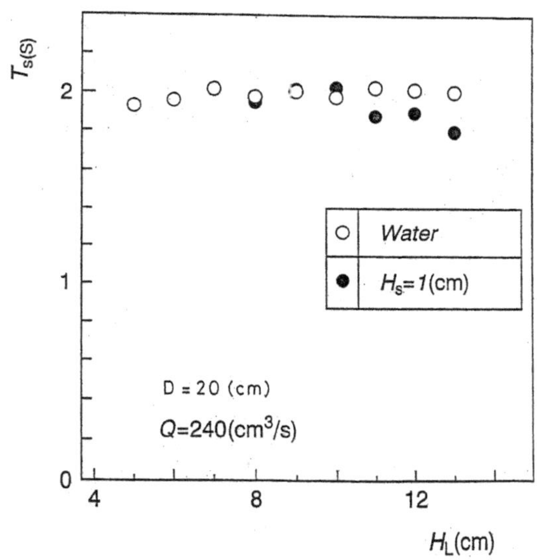


**Fig. 5.51** Relationship between swirl period and gas flow rate

## Period of Swirl Motion

Figures 5.51 and 5.52 show that the swirl period of a bubbling jet is not sensitive to the existence of the upper silicone oil layer. This trend is irrespective of whether the swirl motion of the layer is classified into Types 2 and 3.

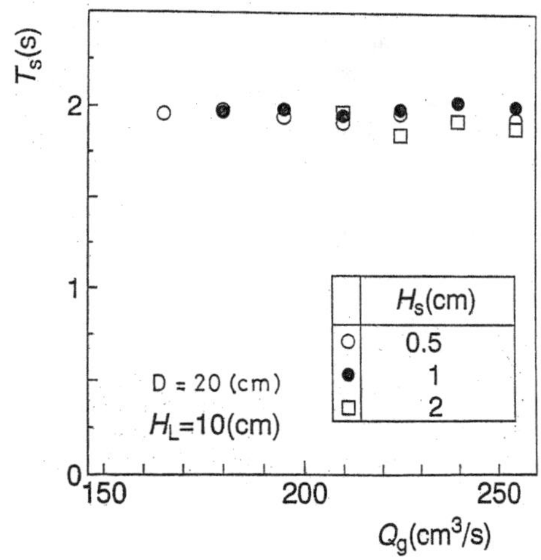


**Fig. 5.52** Relationship between swirl period and thickness of lower molten metal layer

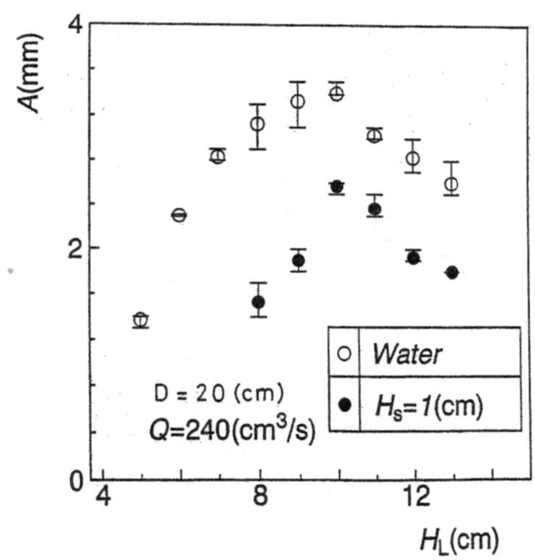


**Fig. 5.53** Relationship between swirl amplitude and gas flow rate

Amplitude of Swirl Motion of Liquid on Side Wall

As the surface of the bath is covered with silicone oil droplets, the amplitude of the swirl motion is highly attenuated in Types 2 and 3 (see Figs. 5.53 and 5.54).

Damping Time of Swirl Motion

The damping time $T_{s,d}$ is defined as the time duration from the stoppage of gas injection to the moment at which the swirl motion of a bubbling jet ceases completely. According to the experimental results shown in Figs. 5.55 and 5.56, $T_{s,d}$ is a weak

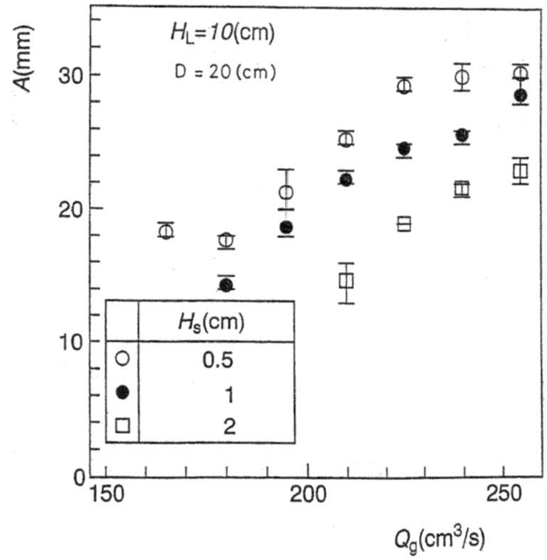


**Fig. 5.54** Relationship between swirl amplitude and thickness of lower molten metal layer

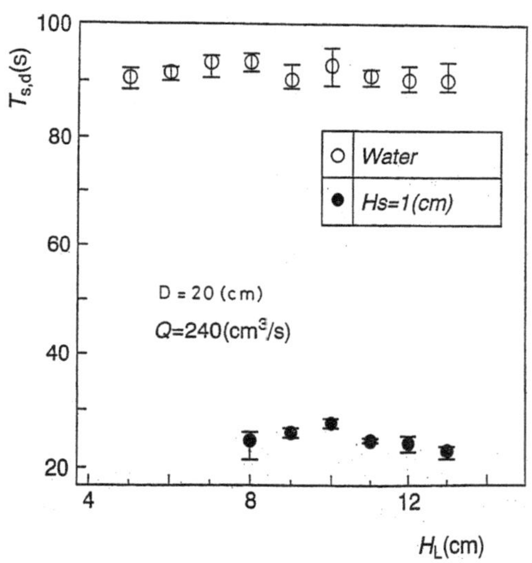


**Fig. 5.55** Relationship between damping time and gas flow rate

function of gas flow rate $Q_g$ and nearly independent of the thickness of the lower water layer, $H_L$. However, $T_{s,d}$ is significantly shortened in the presence of the upper silicone oil layer.

#### 5.2.4.3 Summary

A thin top silicone oil layer was found to significantly affect the first kind of swirl motion in a bottom blown bath with the exception of the swirl period. Although the kinematic viscosity of the silicone oil layer was fixed in this section, the behavior of

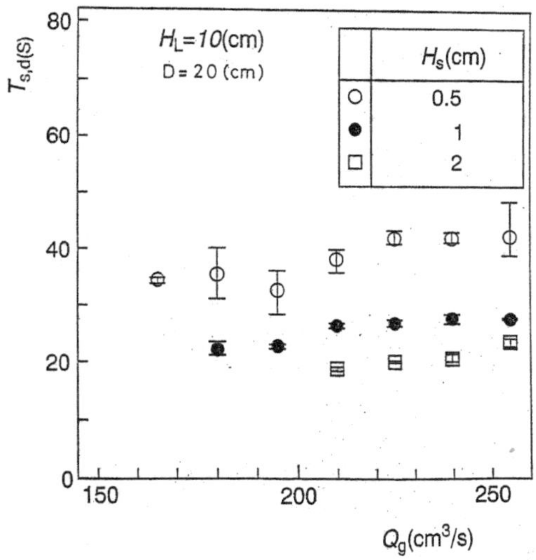


**Fig. 5.56** Relationship between damping time and thickness of lower molten

the swirl motion would be affected to a greater extent by the top silicone oil layer as the kinematic viscosity of the silicone oil is further increased.

Additional experimental investigations are needed for deriving correlations for the starting time, $T_{s,s}$; damping time, $T_{s,d}$; period, $T_s$; and amplitude, $A$, of the first kind of swirl motion in the presence of the top slag layer.

### *5.2.5 Effect of Offset Gas Injection*

The locus of a swirling bubbling jet generated through a centric bottom nozzle is circular in shape on the bath surface. As the nozzle moves toward the side wall, the locus becomes modulated as follows: circular, ellipsoidal, crest-like, as shown in Fig. 5.57, where the subscript $e$ denotes an offset position. For example, $e = 1/4$ means that the nozzle is placed at an offset position $R/4$ away from the center of the vessel, where $R$ is the radius of the vessel. It can be concluded that the behavior of a swirling bubbling jet for $e < 1/3$ is ultimately the same as that of a swirling bubbling jet generated through a centric bottom nozzle ($e = 0$). For $e = 1/3$ the period of swirl motion is approximately the same as that for $e = 0$ although the locus of the bubbling jet is ellipsoidal in shape.

Unfortunately, comprehensive information on the period, amplitude, starting time, and damping time of the swirl motion is not available. In addition, mixing time has not been measured. Experiments in a bath agitated by bottom gas injection through an offset single hole nozzle shows that the mixing time in the absence of the swirl motion has a minimum for $e = 1/3$ [55]. Similar shortening of the mixing time would be expected for an offset gas injection accompanied by the first kind of swirl motion.

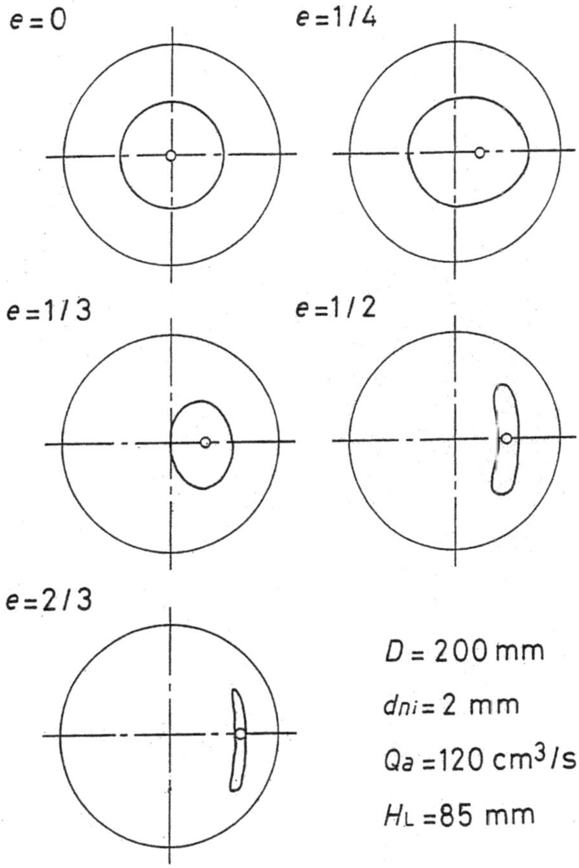


**Fig. 5.57** Locus of bubbling jet on the bath surface

## *5.2.6 Effect of Dual Jet Sources*

Swirl motion of bubbling jets occurs even when gas is injected through dual nozzles, as shown in Figs. 5.58, 5.59 and 5.60, where $e(1)$ and $e(2)$ denote the dimensionless offset positions of the nozzles. When the nozzles are sufficiently separated, each jet rises without merging and swirls independently. In this case, the locus of each jet at the bath surface is approximately circular in shape if the nozzles are placed far away from the side wall ($e < 1/3$). Dual nozzle gas injection accompanied by swirl motion would be useful for controlling as well as enhancing the mixing time in the metallurgical processes. There is yet no available information on the characteristics of such a system and additional experimentation is required.

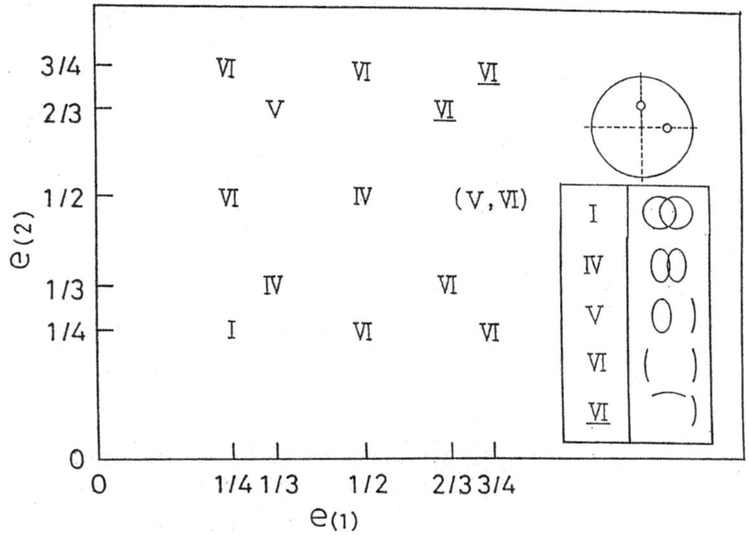


**Fig. 5.58** Loci of bubbling jets generated with dual nozzles (1)

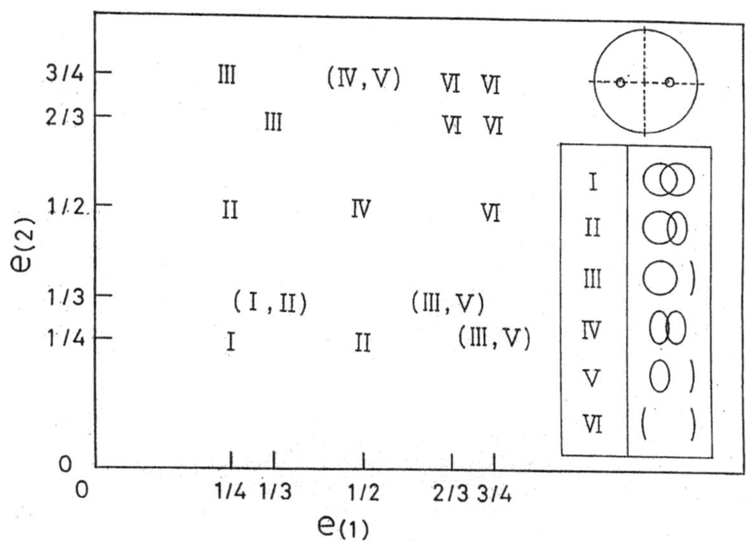


**Fig. 5.59** Loci of bubbling jets generated with dual nozzles (2)

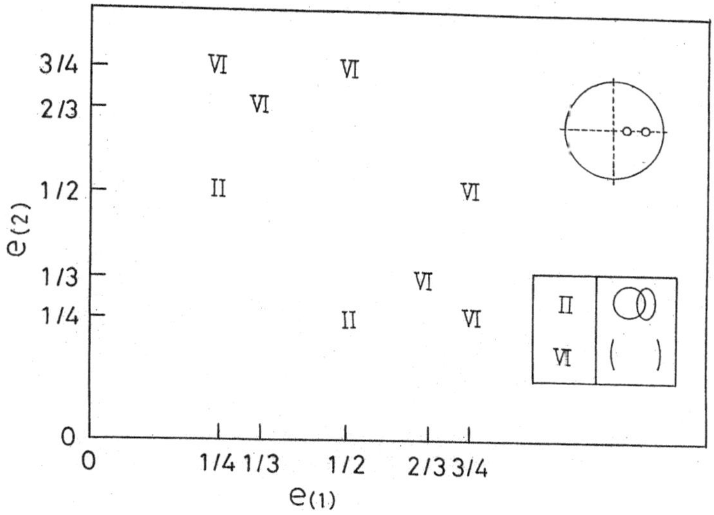


**Fig. 5.60** Loci of bubbling jets generated with dual nozzles (3)

## References

1. Berlot RR (1959) Production of rotation in a confined liquid through translational motions of the boundaries. J Appl Mech Trans ASME 81:513–516
2. Abramson HN, Ranaleben GE Jr (1960) Simulation of fuel sloshing characteristics in missile tanks by use of small models. ARS J 30(7):603–612
3. Abramson HN, Garza LR, Kana DD (1962) Liquid sloshing in compartmented cylindrical tanks. ARS J 326:978–980
4. Lindholm US, Kana DD, Abramson HN (1962) Breathing vibrations of a circular cylindrical shell with an internal liquid, J Aerospace Sci 29(9):1052–1059
5. Abramson HN (1963) The dynamics behavior of liquids in moving containers, Appl Mech Rev 16(7):501–506
6. Hutton RE (1964) Fluid particle motion during rotary sloshing, Trans ASME J Appl Mech 85:123
7. Abramson HN, Chu WH, Garza LR (1963) Liquid sloshing in spherical tanks, AIAA J Spacecraft Rockets 2(3):453–455
8. Abramson HN, Garza LR (1965) Some measurements of liquid frequencies and damping in compartmented cyindrical tanks. AIAA J Spacecraft Rockets 2(3):453–455
9. Dodge FT, Kana DD, Abramson HN (1965) Liquid surface oscillations in longitudinally excited rigid cylindrical containers, AIAA J 3(4):685–695
10. Abramson HN, Chu WH, Kana DD (1966) Some studies of nonlinear lateral sloshing in rigid containers, J Appl Mech 33:777–784
11. Bauer HF, Chang SS, Wang JTS (1971) Nonlinear liquid motion in a longitudinally excited container with elastic bottom, AIAA J 9(12):2333–2339
12. Faltinsen OM (1974) A nonlinear theory of sloshing in rectangular tanks, J Ship Res 18(4):224–241
13. Handbook of Fluid Dynamics (1987) Edited by Jpn Soc Fluid Dynamics, Maruzen, Tokyo, p 720
14. McLachlan NW (1961) Bessel functions for engineers, 2nd edn. Oxford, London
15. Sogabe K (1974) Seisan Kenkyu, University of Tokyo 26(9):353

16. Kimura A, Ohashi H (1978) Trans Jpn Soc Mech Eng Part 1 44:3024–3033
17. Kimura A, Ohashi H (1978) Trans Jpn Soc Mech Eng Part 1 44:3446–3454
18. Kato Y, Nakanishi K, Nozaki T, Suzuki K, Emi T (1982) Wave motion of metal bath in bottom blown converter, Tetsu-to-Hagane 68:1604
19. Kojima S, Yoshida M, Iwatani A, Take H (1987) Quantitative evaluation of vessel vibration on the top and bottom blown converter. Tech Rep Kawasaki Steel Co Ltd 19:1
20. Dubke M, Schwerdtfeger K (1990) Phenomena occurring during drainage of metallurgical vessels: Effect of stopper rod on vortex formation and development of surface waves, Ironmaking Steelmaking 17:184–192
21. Hammerschmidt P, Tacke K-H, Popper H, Weber L, Dubke M, Schwerdtfeger K (1984) Vortex formation during drainage of metallurgical vessels, Ironmaking Steelmaking 11:332
22. Swartz MP (1990) Sloshing waves formed in gas-agitated baths. Chem Eng Sci 45(7):1765–1777
23. Koch K, Roth C, Peter M (1996) Cold model investigations of fluid flow and mixing characteristics by bottom-blowing in discontinuous metallurgical reactors, ISIJ Int 36:850–853
24. Madarame H (1971) Investigation of machinery 23(8):1129–1134
25. Murthy GGK, Mehrorta SP, Ghosh A (1988) Experimental investigation of mixing phenomena in a gas stirred liquid bath, Metall Trans B 19:839
26. Handbook of Hydraulics and Pneumatics (1989) Jpn Hydraulics and Pneumatics Soc. Ohmu Pub. Co., Jpn. Hydraulics and Pneumatics Soc., p 51
27. Kirshner TM (1966) Fluid amplifier. McGraw-Hill, New York, pp 118–124
28. Sawyer RA (1963) Two-dimensional reattaching jet flows including the effects of curvature on entrainment. J Fluid Mech 17:481–498
29. Iguchi M, Hosohara S, Koga T, Yamaguchi R, Morita Z (1992) The swirl motion of vertical bubbling jet in a cylindrical vessel. Tetsu-to-Hagane 78:1778–1785
30. Iguchi M, Hosohara S, Koga T, Yamaguchi R, Morita Z (1993) The swirl motion of vertical bubbling jet in a cylindrical vessel. ISIJ Int 33:1037–1044
31. Tsai CP, Qian ZM (1986) Model study on shaking ladle. Trans ISIJ 26:139–144
32. Xie Y, Oeters F (1992) Experimental studies on the flow velocity of molten metals in a ladle model at centric gas blowing. Steel Res 63:93–104
33. Xie Y, Oeters F (1992) Experimental studies on the bath oscillation during gas blowing into liquids, part 2: measurements in a converter model. Steel Res 63:277–280
34. Xie Y, Oeters F (1992) Experimental studies on the bath oscillation during gas blowing into liquids. I. Measurements using a single nozzle, Steel Res 63:227–233
35. Iguchi M, Kawabata H, Demoto Y, Morita Z (1993) Cold model experiment for developing a new velocimeter applicable to molten metal. Tetsu-to-Hagane 79:1046–1052
36. Iguchi M, Itoh Y, Morita Z (1994) Transient phenomena and amplitude of swirl motion appearing in a bottom blown cylindrical bath with low depth. Tetsu-to-Hagane 80:189–194
37. Iguchi M, Hosohara S, Kondoh T, Itoh Y, Morita Z (1994) Effects of the swirl motion of bubbling jet on the transport phenomena in a bottom blown bath. ISIJ Int 34:330–337
38. Baker JG, Mikina SJ (1936) The calculation of dampers for systems subject to self-induced vibration, J Appl Mech 3A:121
39. Iguchi M, Sasaki K, Nakajima K, Kawabata H (1998) Coanda effect on bubble characteristics in a bubbling jet rising near the side wall of a cylindrical vessel. ISIJ Int 38:1297–1303
40. Iguchi M, Ueda H, Chihara T, Morita Z (1996) Model study of bubble and liquid-flow characteristics in a bottom blown bath under reduced pressure. Metall Mater Trans B 27:765–772
41. Iguchi M, Eguchi Y, Kawasaki A, Kitamura S, Naito K (1999) Swril motions in a cylindrical bath induced by gas injection under reduced pressure on surface. ISIJ Int 8:767–771
42. Iguchi M, Kondoh T, Uemura T, Yamamoto F, Morita Z (1994) Bubble and liquid flow characteristics in a cylindrical bath during swirl motion o bubbling jet. Exp Fluids 16:255–262
43. Sasaki K, Yoshida J, Sasaki Y, Kumagai T, Iguchi M (2000) In: Proc 19th Symp Multiphase Flow
44. Nakanishi K, Fujii T, Szekely J (1975) Possible relationship between energy dissipation and agitation in steel-processing operations, Ironmaking Steelmaking 2(3):193–199

45. Recent Developments in Steelmaking Technology Using Gas Injection (1984) In: Proc. 100th and 101th Nishiyama Memorial Lecture, ISIJ, Tokyo
46. Mazumdar D, Guthrie RIL (1995) The physical and mathematical modelling of gas stirred ladle systems. ISIJ Int 35(1):1–20
47. Asai S, Okamoto T, He JC, Muchi I (1983) Mixing time of refining vessels stirred by gas injection. Trans ISIJ 23:43–50
48. Shinha UP, McNallan RJ (1985) Mixing in ladles by vertical injection of gas and gas-particle jets—A water model study. Metall Trans B 16B:850–853
49. Stapurewicz T, Themelis NJ (1987) Mixing and mass transfer phenomena in bottom-injected gas–liquid reactors, Can Metall Q 26(2):123–128
50. Mietz J, Oeters F (1989) Model studies of mixing phenomena in stirred melts, Can Metall Q 28(1):19
51. Mazumdar D, Das S, Bajpayee S (1997) A comparative study of mixing phenomena in gas stirred ladle systems. ISIJ Int 37(2):194–196
52. Iguchi M, Nakamura K, Tsujino R (1998) Mixing time and fluid flow phenomena in liquids of varying kinematic viscosities agitated by bottom gas injection. Metal Mater Trans B 29: 569–575
53. Mazumdar D, Guthrie RIL (1999) Mixing times and correlations for gas stirred ladle systems, ISS Trans 9:89–96
54. Zhu M, Inomoto T, Sawada I, Hsiao T (1995) Fluid flow and mixing phenomena in the ladle stirred by argon through multi-tuyere. ISIJ Int 35:472–479
55. Zhu M, Sawada I, Yamasaki NY, Hsiao T (1996) Numerical simulation of three-dimensional fluid flow and mixing process in gas-stirred ladles. ISIJ Int 36:503–511
56. Schlichting H (1979) Boundary layer theory, 7th edn. McGraw-Hill, New York, pp 90–91
57. Takatsuka T, Kiuchi H, Iguchi M (2000) Mixing time of cylindrical bath agitated by gas injection. CAMP-ISIJ 13:82

# Chapter 6
# Slag–Metal Interaction

## 6.1 Shape and Size of Entrained Metal Layer

In the in-bath smelting reduction processes and the desulfurization processes, gas is injected into a molten metal bath covered with a thick slag layer. Understanding the mixing phenomena in the two layers and the behavior of molten metal droplets and slag droplets generated at the slag–metal interface is crucial for an improved understanding of the performance of the processes [1–3]. It is very difficult, however, to investigate these effects using the real processes at present, and accordingly, a number of model experiments have been performed by using water and oil to simulate molten metal and molten slag, respectively [4, 5]. The mixing phenomena and reverse emulsification of oil droplets in a water bath covered with a thin oil layer has been extensively investigated [6, 7]. On the contrary, very limited information exists on the spatial and temporal distributions of molten metal droplets in the thick slag layer and of slag droplets in the molten metal layer, even for water model experiments.

A number of water model experiments have been performed on the distributions of water droplets in a thick silicone oil layer using a two-needle electroresistivity probe [8]. The density ratio of the silicone oil to water was 0.93, which differs markedly from 0.4 to 0.5 in the real refining processes. Therefore, similarity condition may not be fully satisfied in these experiments.

Lin and Guthrie [9, 10] used an aqueous $ZnCl_2$ solution and silicone oil of different kinematic viscosities as models for molten metal and molten slag, respectively, in order to elucidate the mechanism of the generation of molten metal droplets. The density ratio was approximately 0.6, which was close to slag–steel density ratio in real processes. Later, Takashima and Iguchi [11] carried out model experiments using the same liquids as Lin and Guthrie. In this section, we describe experimental results on holdup of aqueous $ZnCl_2$ solution carried up by rising bubbles into the upper silicone oil layer.

M. Iguchi and O.J. Ilegbusi, *Modeling Multiphase Materials Processes: Gas-Liquid Systems*, DOI 10.1007/978-1-4419-7479-2_6,

### 6.1.1 Experiment

Figure 6.1 shows a schematic of the experimental apparatus. An aqueous $ZnCl_2$ solution was used to fill in a cylindrical vessel made of transparent acrylic resin to a prescribed level. The aqueous $ZnCl_2$ solution layer was covered with a silicone oil layer. Air was injected through a single-hole nozzle located on the center of the bottom of the vessel. The nozzle had an inner diameter, $d_{ni}$, of 2 mm, an outer diameter, $d_{no}$, of 4 mm, and a protrusion height $h_n$ of 4.5 mm. The air flow rate was adjusted with a regulator and a mass flow controller. The physical properties of the aqueous $ZnCl_2$ solution and the variety of silicone oils employed are listed in Table 6.1. The experimental conditions specified, for example, by the thickness of each layer and the air flow rate, are listed in Table 6.2.

An electroresistivity probe cannot be used to measure the holdup of aqueous $ZnCl_2$ solution droplets in silicone oil because of the low electrical conductivity of aqueous $ZnCl_2$ solution. A suction pipe method was therefore employed in which holdup was defined as the volumetric ratio of the solution to the total volume of a mixture of the solution and silicone oil. The inherent error in the suction pipe technique was determined by applying an electroresistivity probe and the suction pipe simultaneously to a water–silicone oil system [12]. It was found to be within $\pm 7\%$.

In what follows, the aqueous $ZnCl_2$ solution and silicone oil are simply referred to as molten metal and slag, respectively.

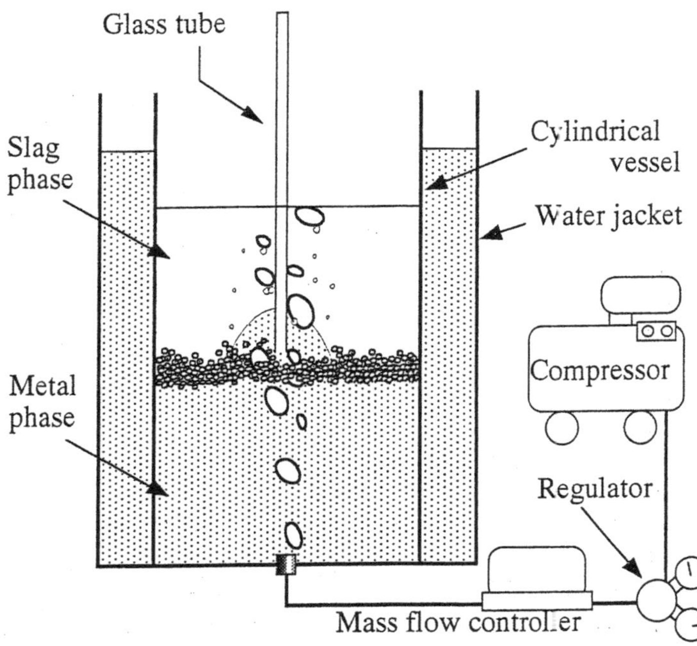


**Fig. 6.1** Experimental apparatus

**Table 6.1** Physical properties of liquids at 293 K

| Liquid | Density ($kg/m^3$) | Kinematic viscosity ($m^2/s$) | Viscosity (mPa s) |
|---|---|---|---|
| Silicone oil A | 818 | $1.0 \times 10^{-6}$ | 0.82 |
| Silicone oil B | 915 | $5.0 \times 10^{-6}$ | 4.6 |
| Silicone oil C | 943 | $10.0 \times 10^{-6}$ | 9.4 |
| Aqueous $ZnCl_2$ solution | 1,670 | $5.0 \times 10^{-6}$ | 8.4 |
| Water | 997 | $0.857 \times 10^{-6}$ | 0.854 |

**Table 6.2** Experimental conditions

| | Vessel diameter ($\times 10^{-2}$ m) | Thickness of metal layer ($\times 10^{-2}$ m) | Thickness of slag layer ($\times 10^{-2}$ m) | Gas flow rate ($\times 10^{-6}\ m^3\ s^{-1}$) |
|---|---|---|---|---|
| Vessel A | 20 | 6, 9, 12.5, 16 | 6, 9, 12.5, 16 | 41.4, 50, 60, 80 |
| Vessel B | 12.25 | 12.5 | 12.5 | 10, 20, 25, 30, 35, 41.4 |
| Vessel C | 7.5 | 12.5 | 12.5 | 10, 15, 20 |

Figure 6.2 shows a schematic of the interface between the molten metal and slag layers in the presence of molten metal droplets. The molten metal in the lower layer is carried up into the upper molten slag layer by bubbles passing through the interface. Such molten metal is often referred to as elevated molten metal. The mountain-like region occupied mainly by this metal is called elevated region. The interface between the elevated region and the surrounding molten slag layer is not stable, but the shape of the interface changes with respect to time in a complex manner due to the passage of rising bubbles.

A mixture of air, molten metal, and molten slag is sucked with a suction pipe. The pipe is 400 mm long and has inner and outer diameters of 4 and 6 mm, respectively. The suction volume of the mixture is approximately 8 ml in every experimental run. After a sufficiently long settling time, the three phases are separated as shown in the upper right part of Fig. 6.2. The length of each phase is denoted by $A$, $B$, or $C$. The gas holdup $\alpha$ and the molten metal holdup $\varepsilon$ are defined by

$$\alpha = \mathrm{C}/(A + B + C), \tag{6.1}$$

$$\varepsilon = \mathrm{A}/(A + B + C). \tag{6.2}$$

As the gas flow rate increases, several molten metal droplets are generated and accumulated above the interface between the two layers. For convenience, these droplets are called accumulated metal droplets to distinguish them from the elevated molten metal. Figure 6.3 shows that air, elevated molten metal, and accumulated molten metal droplets coexist in the suction pipe.

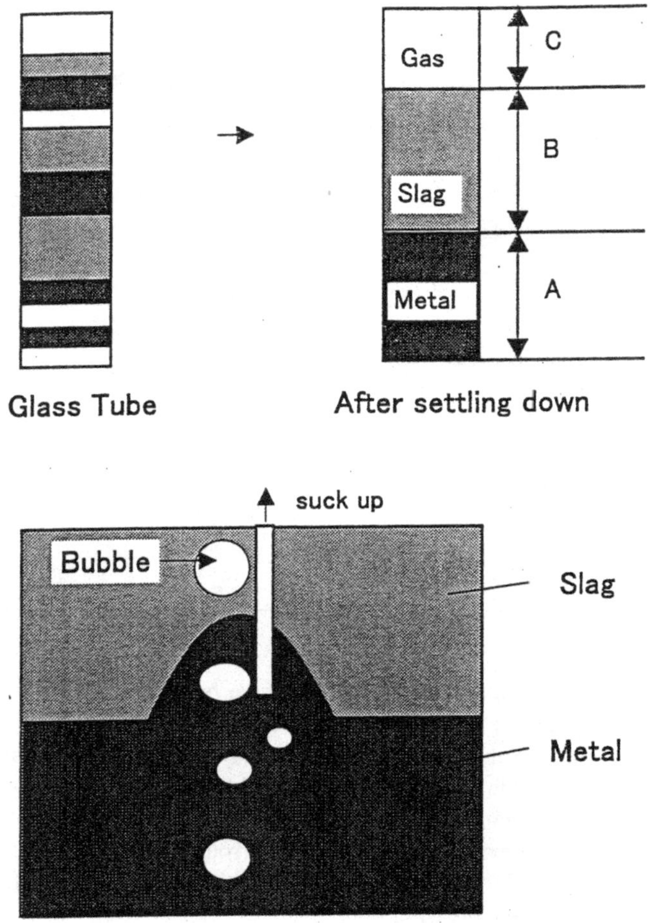


**Fig. 6.2** Metal behavior caused by bubbles through a slag–metal interface; lower gas flow rate

The molten metal holdup, $\varepsilon'$, including the elevated molten metal and accumulated molten metal droplets can be measured in a real situation by using the expression:

$$\varepsilon' = [A + \eta(r)B]/(A + B + C), \tag{6.3}$$

where $\eta(r)$ is the distribution function for the accumulated molten metal droplets in the elevated region. The introduction of this function is desired due to the difficulty of distinguishing between the elevated molten metal and accumulated molten metal droplets in the suction pipe.

The following two types of distribution functions are assumed.

$$\eta(r) = \beta r / r_{\mathrm{B}}, \tag{6.4}$$

$$\eta(r) = 0.30, \tag{6.5}$$

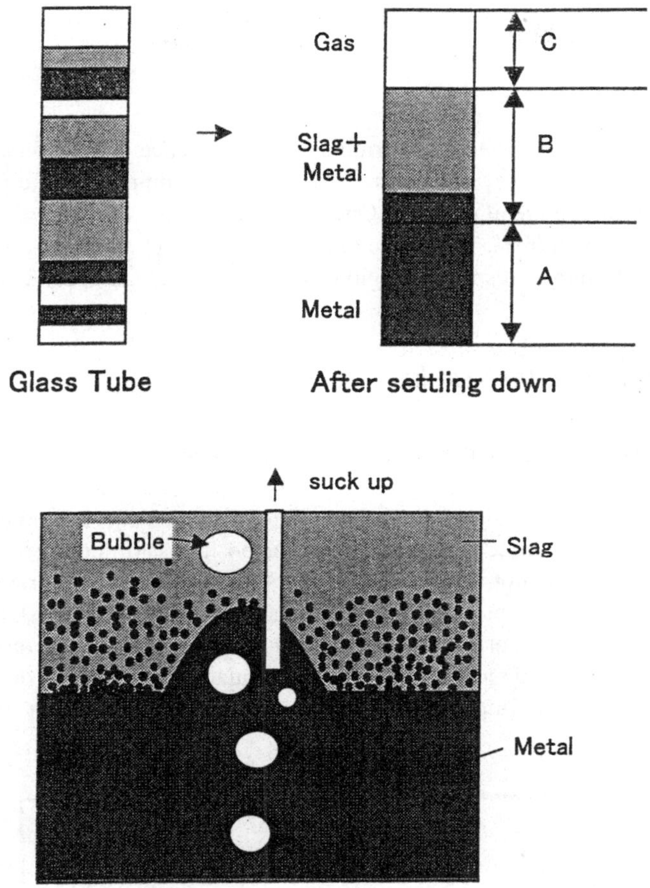


**Fig. 6.3** Metal behavior caused by bubbles through a slag–metal interface; higher gas flow rate

where $r_B$ is the radius of the bubble dispersion region in the elevated region shown in Fig. 6.3. Equation (6.4) implies that the holdup of accumulated molten metal droplets in the elevated region increases linearly with respect to the radial distance and then reaches the holdup of accumulated molten metal droplets outside the elevated region, $\beta$. On the other hand, (6.5) states that the holdup of accumulated molten metal droplets is uniform everywhere in the elevated region. A more detailed description of the distribution function given in a subsequent section.

The elevated molten metal holdup in the elevated region, $\varepsilon$, can be expressed in terms of $\alpha$, $\varepsilon'$ and $\beta r / r_B$, which is given by (6.4), as follows:

$$\begin{aligned} \varepsilon &= A/(A+B+C), \\ &= \left[\varepsilon' - (1-\alpha)\beta r/r_B\right] / (1 - \beta r/r_B). \end{aligned} \tag{6.6}$$

Meanwhile, the application of (6.5) for $\eta(r)$ yields

$$\varepsilon = [\varepsilon' - 0.3(1-\alpha)]/0.7. \tag{6.7}$$

In summary, (6.2) is used to determine $\varepsilon$ in the absence of accumulated molten metal droplets, while there are two methods for determining $\varepsilon$ in the presence of accumulated molten metal droplets. One is based on (6.6) with $r_B = 4.0\,\text{cm}$, and the other is based on (6.7). This value of $r_B$ was decided by observing the behavior of bubbles in the bubble dispersion region with the aid of a high-speed video camera.

## 6.1.2 Experimental Results

### 6.1.2.1 Total Holdup Distribution in the Slag Layer

Figure 6.4 shows the radial distributions of total molten metal holdup $\varepsilon'$ in the molten slag layer at three representative axial positions. It should be noted that the axial distance $z$ is not measured from the initial, horizontal interface between the two layers but from the actual horizontal interface after a steady state has been established. The actual horizontal interface is lower than the initial interface due to the existence of elevated molten metal and accumulated molten metal droplets. This actual interface is here simply called a horizontal interface between the two layers.

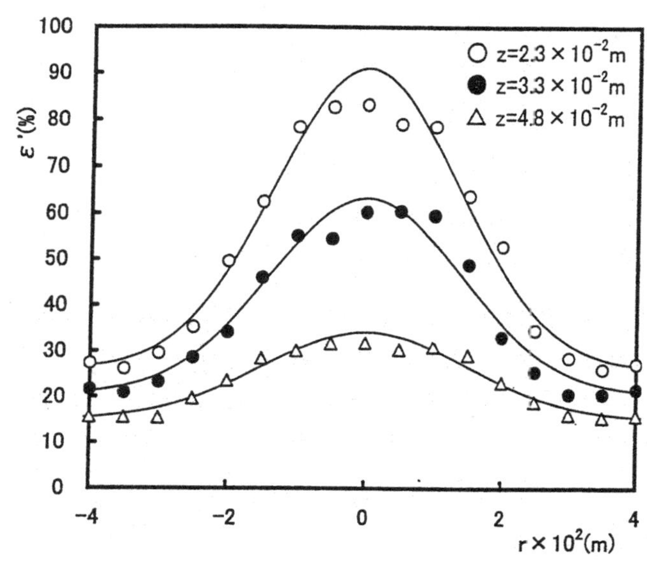


**Fig. 6.4** Radial distributions of total metal holdup in the slag layer. ($D = 12.25 \times 10^{-2}$ m, $Q_g = 25 \times 10^{-6}\,\text{m}^3/\text{s}$, $H_m = H_s = 12.5 \times 10^{-2}$ m)

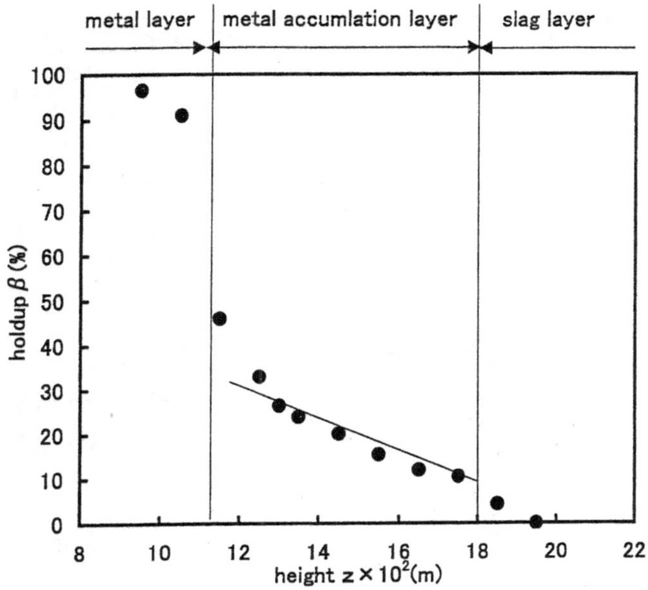


**Fig. 6.5** Accumulated metal holdup in the silicone oil layer ($D = 12.25 \times 10^{-2}$ m, $Q_g = 25 \times 10^{-6}$ m$^3$/s, $H_m = H_s = 12.5 \times 10^{-2}$ m)

At every axial position, the total metal holdup $\varepsilon'$ follows a Gaussian error curve. Also, $\varepsilon'$ decreases with an increase in $r$ and approaches a constant value denoted by $\beta$.

The holdup of accumulated molten metal droplets in the molten slag layer located outside the bubble dispersion region, and designated by $\beta$ as above, increases linearly toward the horizontal interface between the molten metal and slag layers, and then suddenly increases after $\beta$ exceeds 30% (see Fig. 6.5). On the other hand, in the absence of accumulated molten metal droplets, $\beta$ is at most 10% even just (say 1 cm) above the horizontal interface between the two layers. The contour lines of the total molten metal holdup, $\varepsilon'$, calculated from the data plotted in Fig. 6.4, are shown in Fig. 6.6.

#### 6.1.2.2 Horizontal Distribution of Elevated Molten Metal Holdup

Two types of distribution are assumed for the accumulated molten metal holdup, $\beta$, as illustrated in Figs. 6.7 and 6.8. In Fig. 6.7, contour lines of $\beta$ are parallel to the horizontal interface between the molten slag and metal layers. In this case, $\beta$ decreases as the radial distance decreases in the elevated region, and the distribution function expressed by (6.4) seems reasonable.

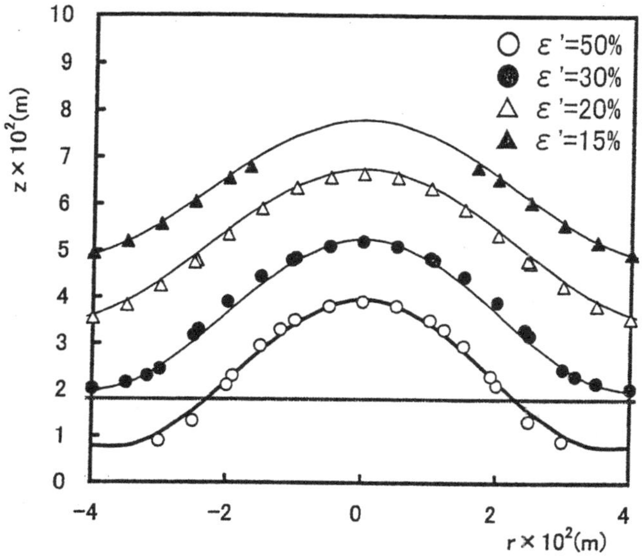


**Fig. 6.6** Contour lines for total metal holdup in the silicone oil layer ($D = 12.25 \times 10^{-2}$ m, $Q_g = 25 \times 10^{-6}\, m^3/s$, $H_m = H_s = 12.5 \times 10^{-2}$ m)

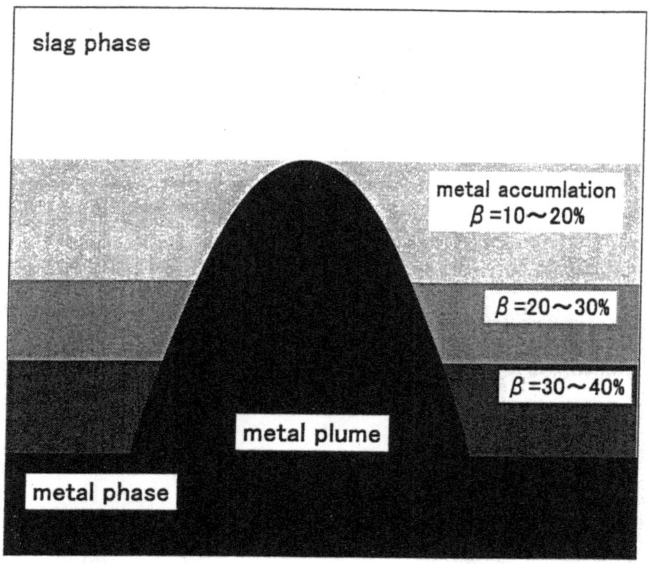


**Fig. 6.7** Schematic of metal plume and metal accumulation

Figure 6.8 shows that the contour lines of $\beta$ are parallel to the actual interface between the two layers. It is also considered constant everywhere in the elevated region. This result validates the use of (6.7) in such a situation.

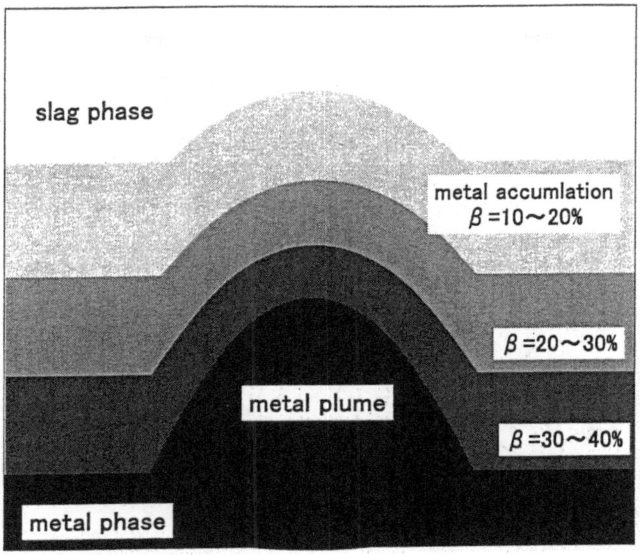


**Fig. 6.8** Schematic of metal plume and metal accumulation

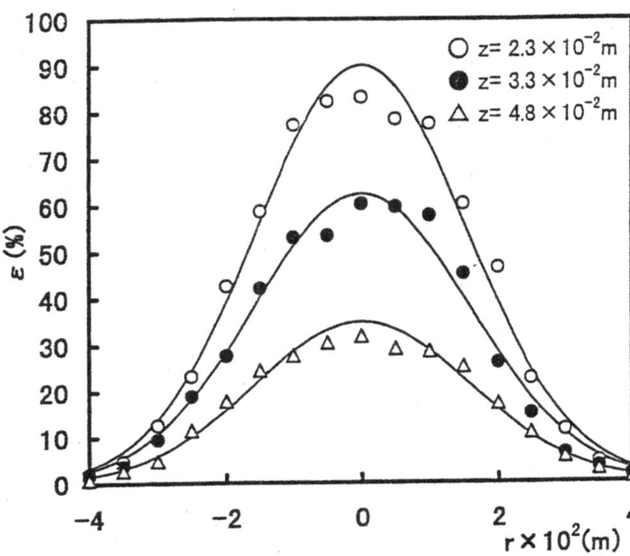


**Fig. 6.9** Radial distributions of metal holdup in the slag layer ($D = 12.25 \times 10^{-2}$ m, $Q_g = 25 \times 10^{-6}\,\text{m}^3/\text{s}$, $H_m = H_s = 12.5 \times 10^{-2}$ m)

Figure 6.9 shows the elevated molten metal holdup $\varepsilon$ calculated using the data on $\varepsilon'$ of Fig. 6.4 and (6.6). Also, Fig. 6.10 shows the distribution of $\varepsilon$ calculated by using the same data on $\varepsilon'$ and (6.7). In each figure, $\varepsilon$ follows a Gaussian error curve similar to the $\varepsilon'$ distribution.

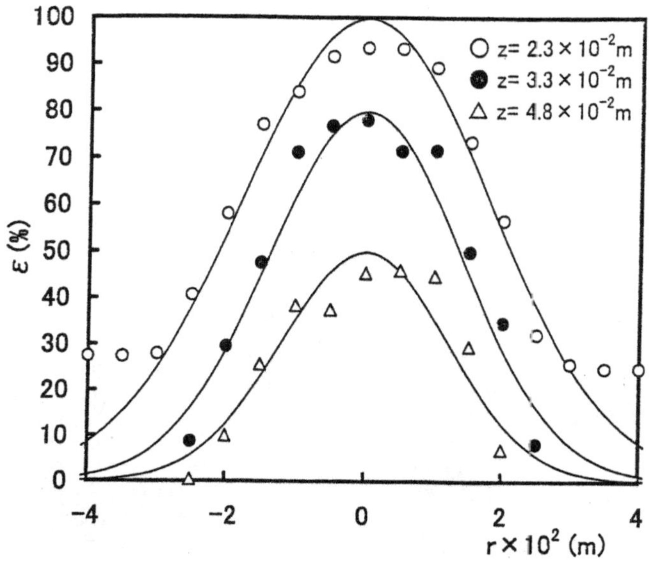


**Fig. 6.10** Radial distributions of metal holdup in the slag layer ($D = 12.25 \times 10^{-2}$ m, $Q_g = 25 \times 10^{-6}\,\text{m}^3/\text{s}$, $H_m = H_s = 12.5 \times 10^{-2}$ m)

The contour lines of elevated molten metal holdup $\varepsilon$ calculated from (6.6) are given in Fig. 6.11 and those calculated from (6.7) are shown in Fig. 6.12. Both figures indicate that the elevated region is parabolic in shape. When bubbles escape from the vicinity of the top of the elevated region, molten metal is engulfed in the wakes of the bubbles and lifted up into the upper slag layer. The filament-like molten metal consequently breaks up into many droplets.

#### 6.1.2.3 Height and Volume of Molten Metal in the Elevated Region

Definitions

The contour lines of $\varepsilon$ shown in Figs. 6.11 and 6.12 can be approximated by the following parabolic curve.

$$f(\varepsilon, r) = A(\varepsilon)r^2 + \Delta H(\varepsilon). \tag{6.8}$$

For the sake of simplicity, a part of the elevated region in which $\varepsilon$ is less than 10% is not taken into consideration. This is because the top of the elevated region fluctuates significantly with time due to the passage of rising bubbles. Consequently, the shape of the elevated region is described by (6.8) with $\varepsilon = 10\%$, and accordingly, $\Delta H(\varepsilon = 10\%)$ is regarded as the height of the elevated region and represented by $\Delta H_{\text{cl,mea}}$.

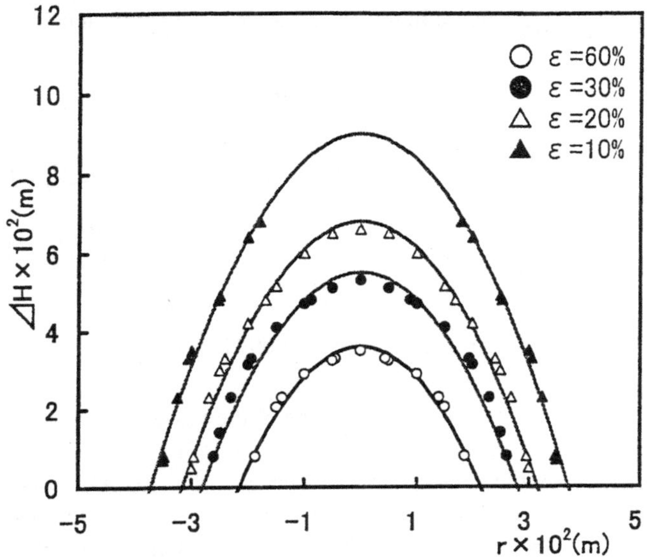


**Fig. 6.11** Contour lines for metal holdup in the slag layer ($D = 12.25 \times 10^{-2}$ m, $Q_g = 25 \times 10^{-6}\ m^3/s$, $H_m = H_s = 12.5 \times 10^{-2}$ m)

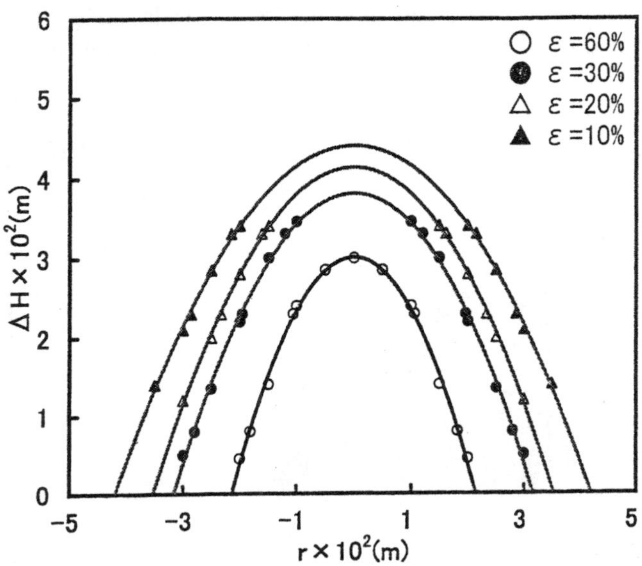


**Fig. 6.12** Contour lines for metal holdup in the slag layer ($D = 12.25 \times 10^{-2}$ m, $Q_g = 25 \times 10^{-6}\ m^3/s$, $H_m = H_s = 12.5 \times 10^{-2}$ m)

The volume of molten metal in the elevated region can be calculated from (6.9).

$$V_{\mathrm{m}} = \iint 2\pi r(r,z)\mathrm{d}r\ \mathrm{d}z/100. \tag{6.9}$$

A total area occupied by the molten metal at any height in the elevated region is denoted by $S_{\mathrm{m}}$, expressed by

$$S_{\mathrm{m}} = \int 2\pi r\varepsilon(r,z)\mathrm{d}r/100, \tag{6.10}$$

$$\varepsilon(r,z) = \varepsilon_{\mathrm{cl}}(z)\exp\left(-0.6931r^2/b_\varepsilon{}^2\right). \tag{6.11}$$

where $\varepsilon_{\mathrm{cl}}(z)$ is the value of $\varepsilon$ on the centerline of the vessel and $b\ \varepsilon$ is the half-value radius of the horizontal distribution of $\varepsilon$.

Figure 6.13 shows the relationship between $S_{\mathrm{m}}$ and the axial distance $z$. The area $S_{\mathrm{m}}$ decreases linearly with respect to $z$, and accordingly, the hatched region can be regarded as the volume of the elevated molten metal in the elevated region, designated by $V_{\mathrm{m,mea}}$.

Height of Elevated Region

Molten metal flow in the vertically upward direction is driven by rising bubbles in the elevated region as well as in the lower molten metal layer. If we neglect the

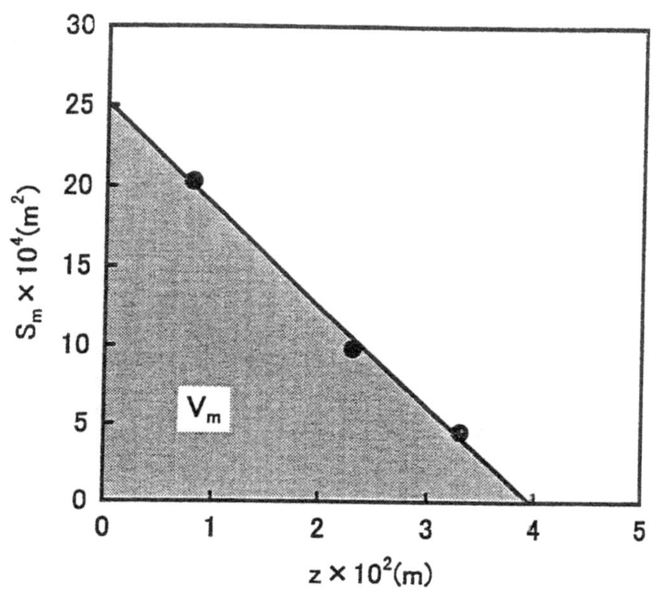


**Fig. 6.13** Calculation method of metal volume $V_{\mathrm{m}}$ lifted up into the slag layer ($D = 12.25 \times 10^{-2}$ m, $Q_{\mathrm{g}} = 25 \times 10^{-6}\ \mathrm{m}^3/\mathrm{s}$, $H_{\mathrm{m}} = H_{\mathrm{s}} = 12.5 \times 10^{-2}$ m)

buoyancy force acting on bubbles in the elevated region, the following relationship holds between the kinematic energy and potential energy of ascending molten metal.

$$(1/2)\rho_m \overline{u}_{m,cl}^2 = (\rho_m - \rho_s) g \Delta H_{cl}, \tag{6.12}$$

where $\Delta H_{cl}$ is the height of the elevated region and $\overline{u}_{m,cl}^2$ is the axial mean velocity of molten metal ascending along the centerline of the vessel. From (6.12) $\Delta H_{cl}$ is expressed by

$$\Delta H_{cl} = \rho_m \overline{u}_{m,cl}^2 / [2g(\rho_m - \rho_s)]. \tag{6.13}$$

This $\Delta H_{cl}$ is the calculated value of the height of the elevated region, $\Delta H_{cl,cal}$.

The measured values of $\Delta H_{cl}$, represented by $\Delta H_{cl,mea}$, are compared with $\Delta H_{cl,cal}$ in Figs. 6.14 and 6.15, using (6.6) and (6.7), respectively. The data on $\Delta H_{cl,mea}$ for a water–air system are also included in the figures. The dotted line in each figure represents the condition, $\Delta H_{cl,mea} = \Delta H_{cl,cal}$. All the measured values are much larger than $\Delta H_{cl,cal}$ due to the buoyancy force acting on bubbles passing through the elevated region.

The experimental data in both cases can be satisfactorily correlated, indicating that $\Delta H_{cl}$ is not a function of the kinematic viscosity of molten metal and slag and bath diameter, $D$.

The data in Fig. 6.14 are slightly larger than those in Fig. 6.15. This may be attributable to the existence of accumulated molten metal droplets in the elevated region. In (6.6), the accumulated molten metal droplets are assumed to be nonexistent at the centerline of the vessel while (6.7) assumes that there are droplets there.

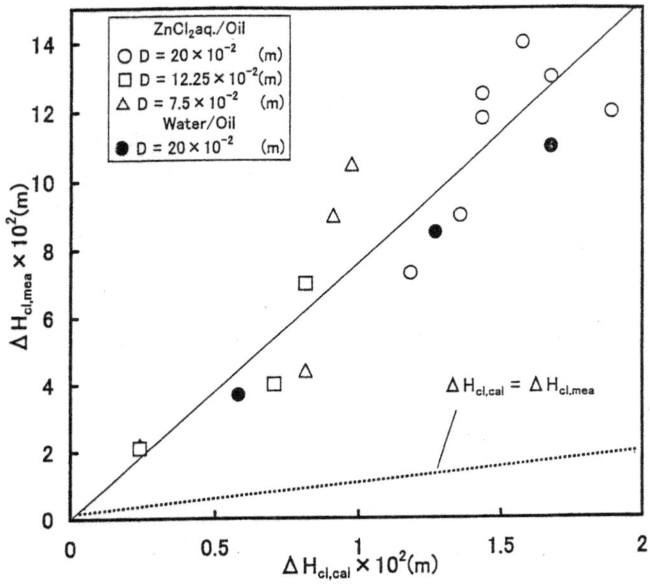


**Fig. 6.14** Relation between measured and calculated values of $\Delta H_{cl}$

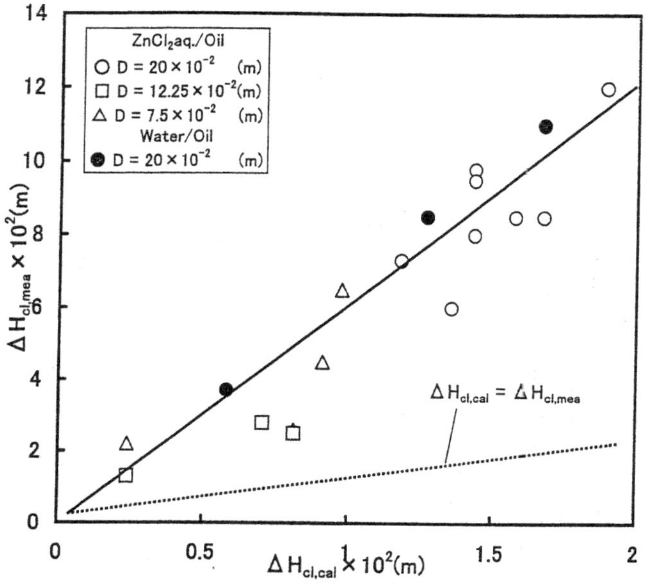


**Fig. 6.15** Relation between measured and calculated values of $\Delta H_{\rm cl}$

The following empirical relation, represented by the solid line, can be derived from $\Delta H_{\rm cl}$ on the basis of the data presented in Fig. 6.15.

$$\Delta H_{\rm cl} = 6\rho_{\rm m}\overline{u}_{\rm m,cl}{}^2/[2g(\rho_{\rm m} - \rho_{\rm s})]. \tag{6.14}$$

Volume of Elevated Molten Metal in the Elevated Region

If we neglect the volume of bubbles in the elevated region, the volume of molten metal brought up in the elevated region, $V_{\rm m}$, can be determined by integrating the height of the region with respect to the radial distance $r$, thus:

$$V_{\rm m} = \int 2\pi r \Delta H \,{\rm d}r, \tag{6.15}$$

where $R$ is the radius of the vessel, and $\Delta H$ is a function of $r$. The height $\Delta H$ can be derived from (6.12), giving:

$$\Delta H = 6\rho_{\rm m}\bar{u}_{\rm m}{}^2/[2g(\rho_{\rm m} - \rho_{\rm s})]. \tag{6.16}$$

From existing experimental investigations [13, 14], the mean velocity of molten metal flow approaching the interface between the molten slag and metal layers, $\overline{u}_{\rm m}$, is given by

$$\bar{u}_{\rm m} = \bar{u}_{\rm m,c,l}, \exp\left(-0.6931 r^2/b_{\rm u}{}^2\right). \tag{6.17}$$

where $b_u$ is the half-value radius of the radial distribution of $\overline{u}_m$. In the case of centered bottom blowing, $b_u$ is expressed by

$$b_u = 0.14H_m. \tag{6.18}$$

Most molten metal droplets are found to be generated in the circular region whose outer edge is located at approximately $r = 2b_u$.

Combining (6.16) and (6.17) gives the following expression for $V_m$:

$$V_m = 0.022\rho_m H_m{}^2 \bar{u}_{m,cl}{}^2/[g(\rho_m - \rho_s)]. \tag{6.19}$$

The value of $V_m$ calculated from (6.19) is denoted by $V_{m,cal}$. The experimental data, $V_{m,mea}$, obtained by assuming (6.6) and (6.7) are shown in Figs. 6.16 and 6.17, respectively. Equation (6.19), represented by the broken line, underestimates the volume of the molten metal droplets in the elevated region. The data on $V_m$ for both the real and air–water systems are better correlated by the method depicted in Fig. 6.17. The following empirical relation can thus be deduced with reference to Fig. 6.17.

$$V_m = 0.156\rho_m H_m{}^2 \bar{u}_{m,cl}{}^2/[g(\rho_m - \rho_s)]. \tag{6.20}$$

The difference between (6.19) and (6.20) is attributable to the existence of bubbles in the elevated region.

Equation (6.20) implies that the volume of molten metal droplets $V_m$ is independent of the kinematic viscosities of molten metal and slag as well as the diameter of the vessel.

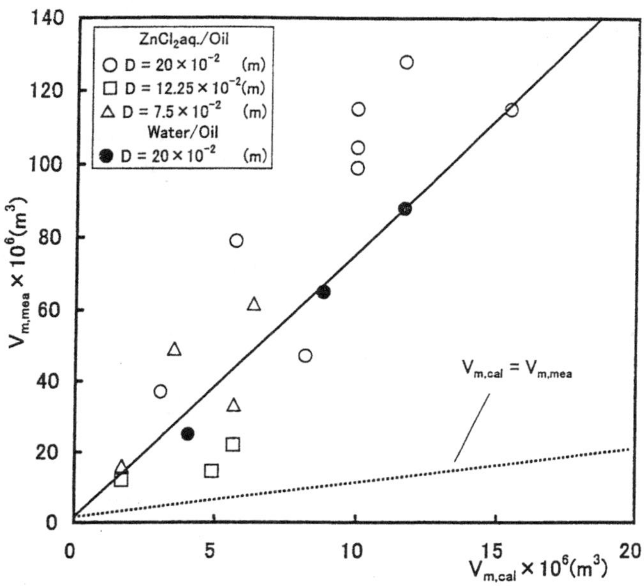


**Fig. 6.16** Relation between measured and calculated values of $V_m$

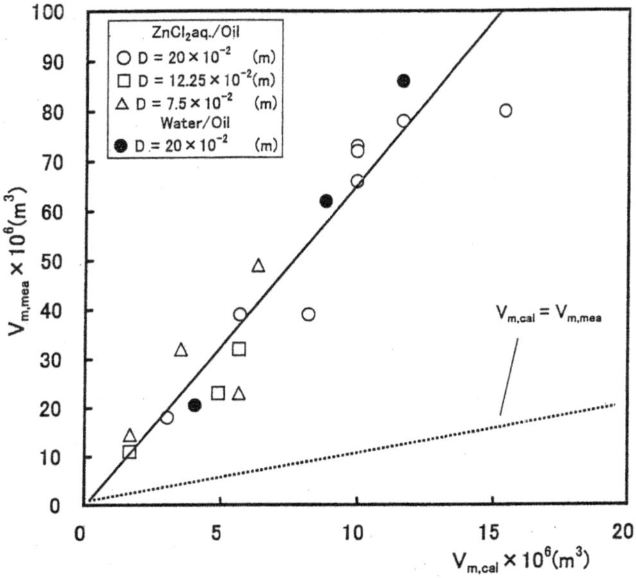


**Fig. 6.17** Relation between measured and calculated values of $V_m$

#### 6.1.2.4 Accumulated Molten Metal Droplets

The diameters of accumulated molten metal droplets were measured with the same suction pipe. The inner diameter of the suction pipe was 4 mm and the suction volume was 8 ml. The droplets were sucked with the suction pipe together with slag. The shape and size of each molten metal rising in the suction pipe was observed with a high-speed video camera, and the size was evaluated in terms of a volume-equivalent diameter.

Figure 6.18 shows a histogram of the diameters of molten metal droplets, $d_m$. The result nearly follows a logarithmic normal distribution. The mean droplet diameter falls between 1 and 1.5 mm.

Figure 6.19 shows the variation of $d_m$ with the gas flow rate $Q_g$. The mean diameter for the aqueous $ZnCl_2$ solution is much smaller than for water.

Asai [15] developed an inviscid theory for the generation of a slag droplet at the interface between molten slag and metal in the absence of bubbles based on the energy equation. The reported diameter, $d_{pc}$, is given by

$$d_{pc} = 2[3\sigma_{ms}/\{g(\rho_m - \rho_s)\}]^{1/2}. \qquad (6.21)$$

This equation may also be valid for predicting the diameter of a molten metal droplet.

It may be assumed that the mean diameter of molten metal droplets generated in the presence of bubbles is also predicted by (6.21). Furthermore, the following

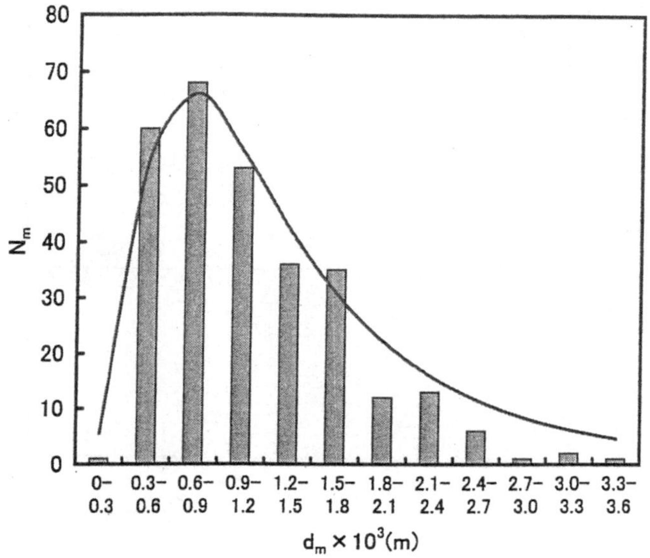


**Fig. 6.18** Histogram of diameters of metal droplets ($D = 12.25 \times 10^{-2}$ m, $Q_g = 25 \times 10^{-6}$ m$^3$/s, $H_m = H_s = 12.5 \times 10^{-2}$ m)

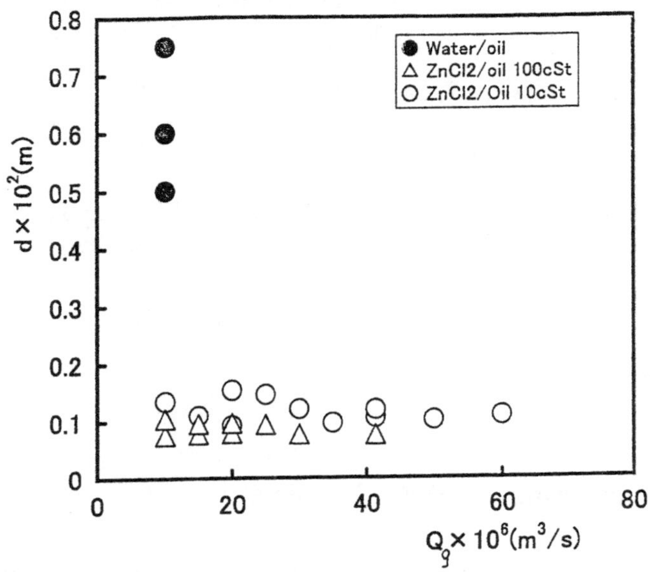


**Fig. 6.19** Relation between mean diameter of metal droplets and gas flow rate

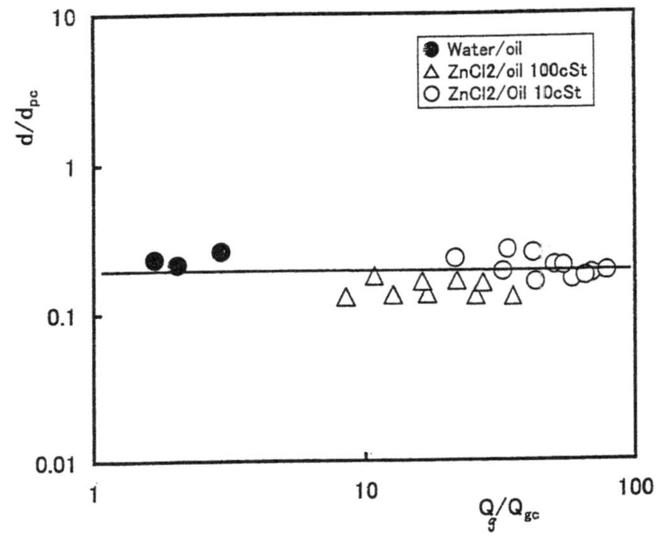


**Fig. 6.20** Relation between mean diameter of metal droplets and gas flow rate

empirical relation originally derived from the initiation of slag droplets or the onset of reverse emulsification in the presence of rising bubbles [16], may be used in order to nondimensionalize the gas flow rate $Q_g$, thus:

$$Q_{gc} = V^{4.52}(\nu_s/\nu_m)^{0.308}(H_s/D)^{-0.498}g^{-1.76}H_m{}^{-0.240}, \tag{6.22}$$

$$V = (\sigma_{ms}g/\rho_s)^{0.25}. \tag{6.23}$$

The comparison of $d_m/d_{pc}$ with $Q_g/Q_{gc}$ is given in Fig. 6.20. The mean droplet diameter is independent of the gas flow rate $Q_g$ and expressed by

$$d_m = 0.2d_{pc} = 0.69[\sigma_{ms}/\{g(\rho_m - \rho_s)\}]^{1/2}, \tag{6.24}$$

where $\sigma_{ms}$ is the interfacial tension.

## 6.2 Characteristics of Metal Droplets

In Sect. 6.1.2.4, we discussed the shape and size of molten metal carried into the upper silicone oil layer by rising bubbles, and the diameters of molten metal droplets accumulated at the slag metal interface. Evaluation of the birth rate, death rate, and lifetime of molten metal droplets is essential for practical applications. In this section, particular attention is paid to the behavior of small molten metal droplets in the silicone oil layer. The birth and death rates and lifetime of the molten metal droplets are defined, and empirical equations are introduced for these quantities.

### 6.2.1 Experiment

The experimental apparatus and fluids used as models for molten metal and slag are the same as those described in Sect. 6.2. When the air flow rate is relatively high, a number of small metal droplets that later accumulated at the slag–metal interface are generated, as shown schematically in Fig. 6.1. Droplet generation ceases below a certain critical air flow rate.

Figure 6.21 shows the change in the vertical position of the slag–metal interface with time. Displacement of the interface from its initial position at time $t$ is represented by

$\Delta H_m$ (cm). This parameter was measured with a video camera.

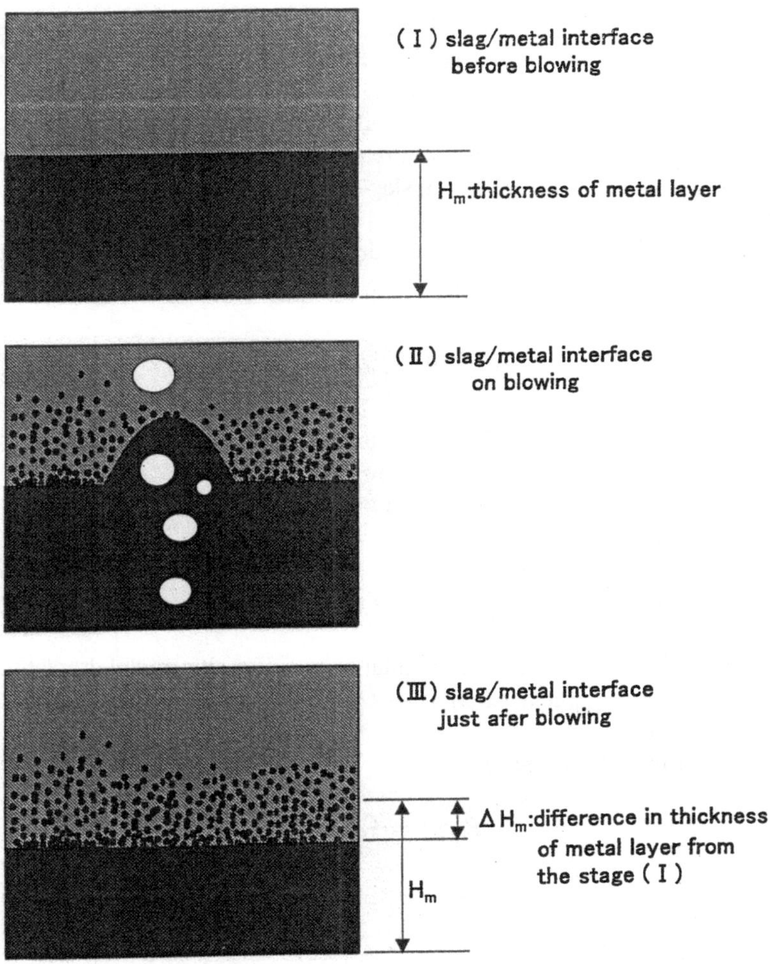


**Fig. 6.21** Change in the thickness of metal layer

The total volume of accumulated metal droplets, $V_d(t)$ ($cm^3$), can be expressed by [9]:

$$V_d(t) = \Delta H_m A, \quad (6.25)$$

where $A$ is the cross-sectional area of the vessel ($cm^2$). The death rate of the accumulated droplets, $R_D{}'$ ($cm^3/s$), after the stoppage of air injection, i.e., in the absence of rising bubbles can also be measured by observing a change in $\Delta H_m$.

## 6.2.2 *Experimental Results*

### 6.2.2.1 Mechanism of Metal Droplet Generation

Figure 6.22 shows that molten metal engulfed in the wake of a bubble is carried into the upper slag layer as the bubble passes through the slag–metal interface. Such filament-like melt breaks up, resulting in the generation of many small metal droplets in the upper slag layer. After a while, most of the metal droplets reach the horizontal slag–metal interface and then return to the lower metal layer. The metal droplets accumulated at the horizontal slag–metal interface until steady state is attained [9, 11].

The total volume of accumulated metal droplets at time $t$ in the presence of rising bubbles can be expressed by [9]

$$V_d(t) = R_B t - \int R_D(T) dt, \quad (6.26)$$

where $R_B$($cm^3/s$) and $R_D$($cm^3/s$) are the birth rate and death rate of molten metal droplets, respectively, and $R_B$ is assumed to be constant.

Differentiating (6.26) with respect to time $t$ gives the accumulating rate of molten metal droplets as follows:

$$dV_d(t)/dt = R_B - R_D(t). \quad (6.27)$$

Equation (6.27) states that the accumulating rate of molten metal droplets on the horizontal molten slag–metal interface is simply the difference between the birth rate and death rate.

Following Lin and Guthrie [9], the following relationship is further assumed.

$$R_D(t) = V_d(t)/\tau. \quad (6.28)$$

Substituting (6.28) into (6.27) and integrating the resulting equation with time under the initial condition

$$V_d(t) = 0 \text{ at } t = 0 \quad (6.29)$$

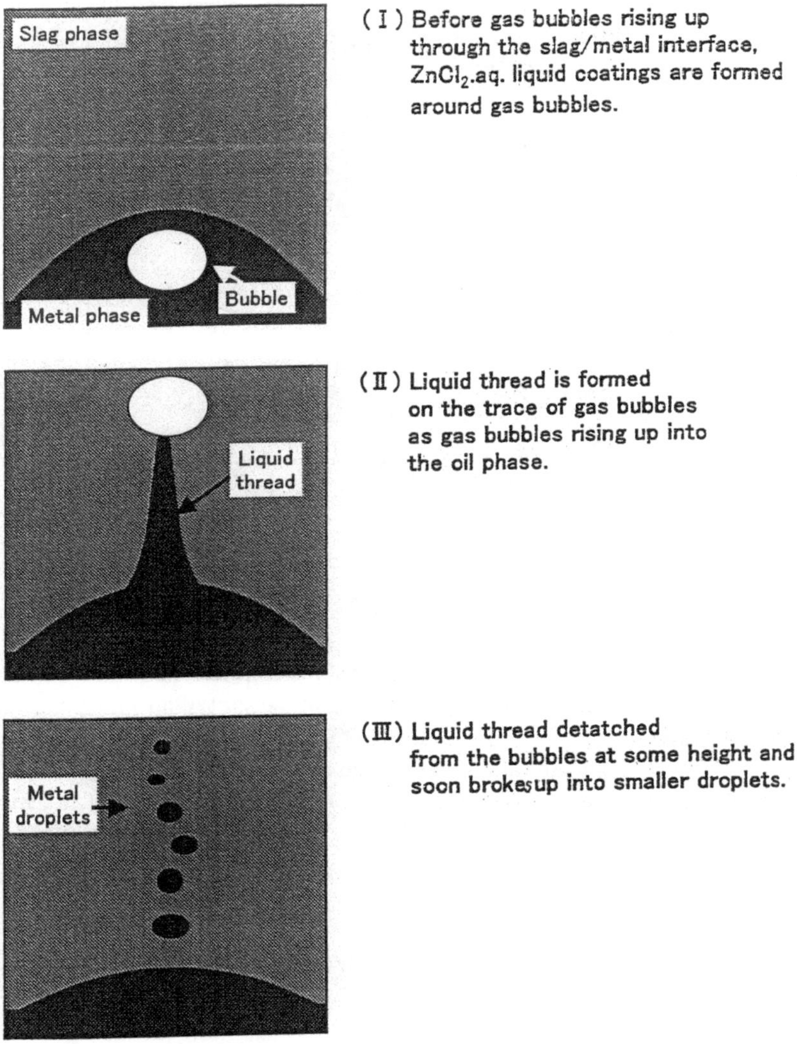


**Fig. 6.22** Droplet formation caused bubbles through a slag/metal interface

gives:

$$V_{\mathrm{d}}(t) = R_{\mathrm{B}}\tau[1 - \exp(-t/\tau)]. \tag{6.30}$$

If we designate the value of $V_{\mathrm{d}}(t)$ at $t = \infty$ by $V_\infty$ ($\mathrm{cm}^3$), we have

$$R_{\mathrm{B}} = V_\infty/\tau. \tag{6.31}$$

This equation implies that $\tau$(s) is the residence time of molten metal droplets in the slag layer at steady state. In other words, $\tau$ is regarded as the mean lifetime of molten metal droplets.

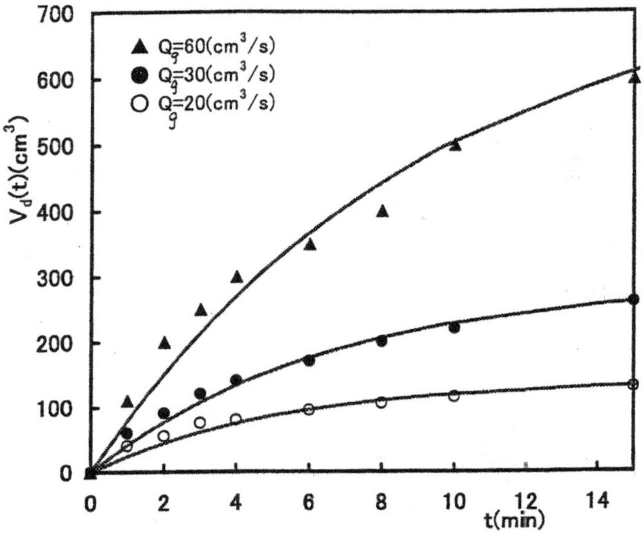


**Fig. 6.23** Effect of gas flow rate on the volume of droplets entrained within upper phase ($D = 12.25\,\text{cm}$, $\nu_s = 100\,\text{cSt}$)

Figure 6.23 shows the measured $V_d(t)$ for three different flow rates of 20, 30, and $60\,\text{cm}^3/\text{s}$. The origin of time, $t = 0$, was set at the start of air injection into the bath. The data for each air flow rate follow (6.30) and subsequently reach a constant value, $V_\infty$. That is, the death rate $R_D(t)$ increases with time and is finally balanced by the birth rate, $R_B$, as indicated by (6.27). The total volume of accumulated molten metal droplets, $V_\infty$, at the steady state, and the birth rate $R_B$ can be determined from this figure.

#### 6.2.2.2 Total Volume of Accumulated Molten Metal Droplets at Steady State, $V_\infty$

As shown in Fig. 6.24, $V_\infty$ increases with an increase in air flow rate $Q_g$ ($\text{cm}^3/\text{s}$). Also, $V_\infty$ is an increasing function of the vessel diameter $D$ (cm) and the kinematic viscosity of molten slag, $\nu_s$, (see Fig. 6.25). The effects of the thickness of molten metal layer, $H_m$ (cm), and of molten slag layer, $H_s$ (cm), on the total volume of molten metal droplets, $V_\infty$, are shown in Figs. 6.26 and 6.27, respectively. In Fig. 6.26, $V_\infty$ increases monotonically with the thickness of the molten oil layer. On the other hand, in Fig. 6.27 $V_\infty$ becomes independent of $H_m$ when $H_m$ exceeds a certain critical value. This value was found to be closely related to the boundary between the momentum-dominant and buoyancy-dominant regions [17].

Dimensional analysis can be performed using Buckingham's $\Pi$ theorem, assuming the following functional relationship:

$$V_\infty = f(Q_g, D, H_s, \rho_s, \Delta\rho, \sigma_{ms}, g, \mu_s, \mu_m), \tag{6.32}$$

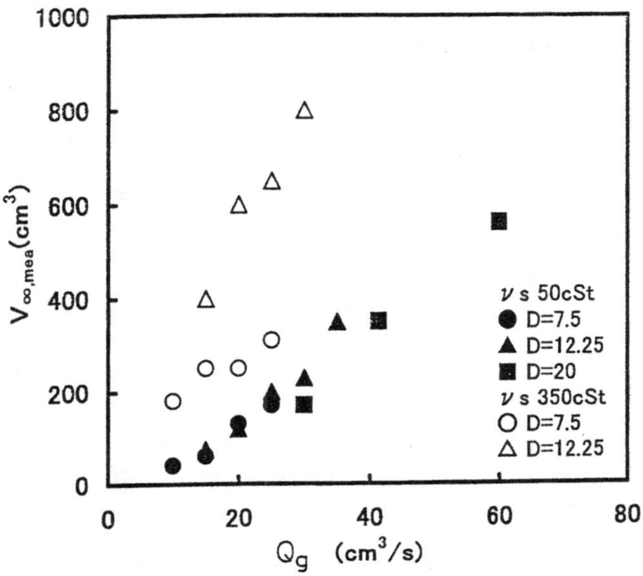


**Fig. 6.24** Effect of vessel diameter on the volume of droplets entrained within upper phase

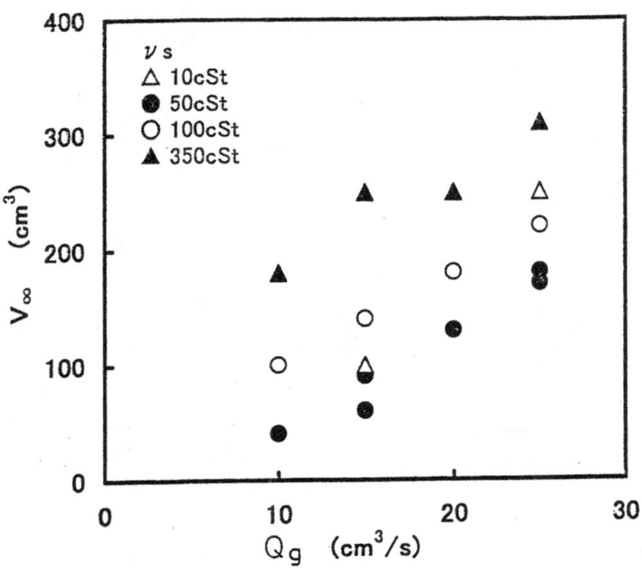


**Fig. 6.25** Effect of upper phase viscosity on the volume of droplets entrained within upper phase ($D = 7.5\,\text{cm}$)

where $\rho_s$ is the density of slag (g/cm$^3$), $\Delta\rho(= \rho_m - \rho_s)$ is the density difference (g/cm$^3$), $\rho_m$ is the density of melt (g/cm$^3$), $\sigma_{ms}$ is the interfacial tension (dyn/cm), $g$ is the acceleration due to gravity (cm/s$^2$), $\mu_s$ is the viscosity of slag (g/cms), and $\mu_m$ is the viscosity of molten metal (g/cms).

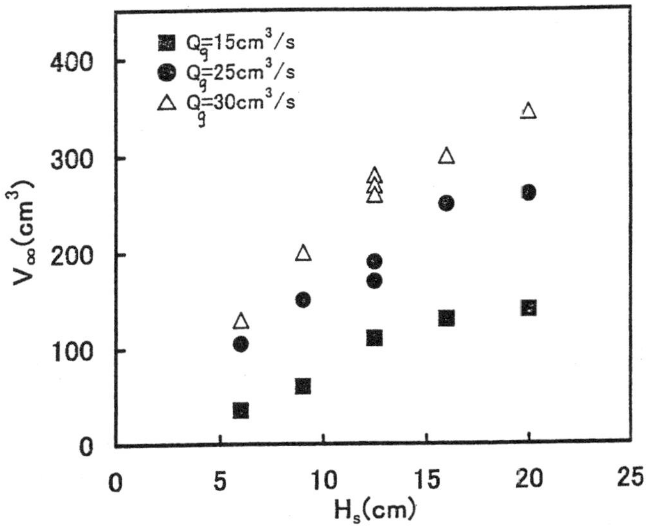


**Fig. 6.26** Effect of upper phase thickness on the volume of droplets entrained within upper phase ($D = 12.25\,\mathrm{cm},\ H_\mathrm{m} = 12.5\,\mathrm{cm}$)

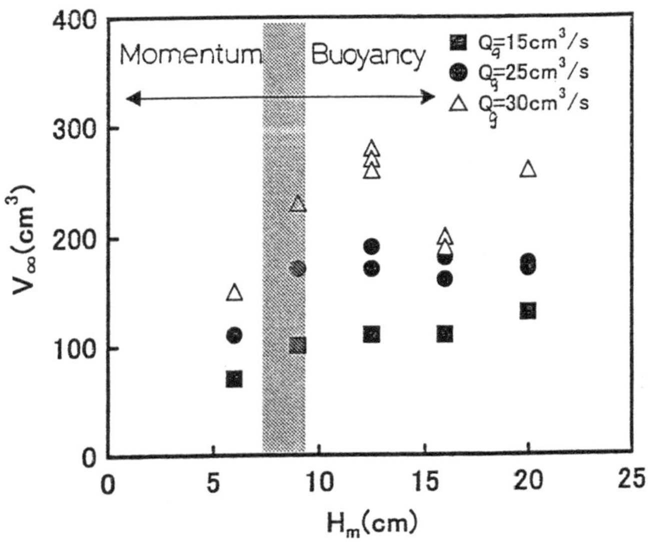


**Fig. 6.27** Effect of lower phase thickness on the volume of droplets entrained within upper phase ($D = 12.25\,\mathrm{cm},\ H_\mathrm{m} = 12.5\,\mathrm{cm}$))

Seven dimensionless parameters could be determined, thus:

$$\Pi 1 = V_\infty {\sigma_\mathrm{ms}}^{-.5} \Delta\rho^{1.5} g^{1.5}, \tag{6.33}$$

$$\Pi 2 = Q_\mathrm{g} {\sigma_\mathrm{ms}}^{-1.25} \Delta\rho^{1.25} g^{0.75}, \tag{6.34}$$

$$\Pi 3 = D {\sigma_\mathrm{ms}}^{-0.5} \Delta\rho^{0.5} g^{0.5}, \tag{6.35}$$

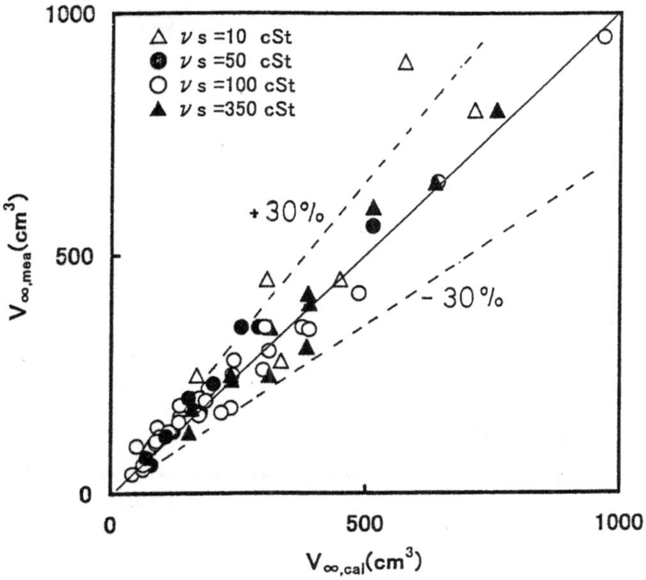


**Fig. 6.28** Relation between measured and calculated values of $V_\infty$

$$\Pi 4 = H_s \sigma_{ms}{}^{-0.75} \Delta\rho^{-0.25} g^{0.25}, \tag{6.36}$$

$$\Pi 5 = \rho_s{}^{-1} \Delta\rho, \tag{6.37}$$

$$\Pi 6 = \mu_s \sigma_{ms}{}^{-0.75} \Delta\rho^{-0.25} g^{0.25} \tag{6.38}$$

$$\Pi 7 = \mu_m \sigma_{ms}{}^{-0.75} \Delta\rho^{-0.25} g^{0.25}. \tag{6.39}$$

After some rearrangements, the following somewhat complicated empirical relation results:

$$V_\infty = 3.1 Q_g{}^{1.62} D^{-0.48} H_s \sigma_{ms}{}^{-0.785} \Delta\rho^{0.785} g^{-0.025} \times \left(Q_g{}^{-0.0165} D^{0.037} \sigma_{ms}{}^{0.0021} \Delta\rho^{0.138} g^{0.0061} \rho_s{}^{-0.14}\right)^{\mu s/\mu m} \tag{6.40}$$

Figure 6.28 shows a comparison between the measured $V_\infty$, designated by $V_{\infty,\text{mea}}$, and calculated values from (6.40), designated by $V_{\infty,\text{cal}}$. It is evident that the data can be approximated by (6.40) within a scatter of ±30%. Such a scatter is quite acceptable for this type of measurement.

#### 6.2.2.3 Birth Rate of Molten Metal Droplets

Lin and Guthrie [9] proposed the following empirical equation for the birth rate of molten metal droplets.

$$R_{B,LG} = 0.19 \mu_m \rho_s \mu_s{}^{-1/3} \Delta\rho^{-5/3} Q_g / d_B, \tag{6.41}$$

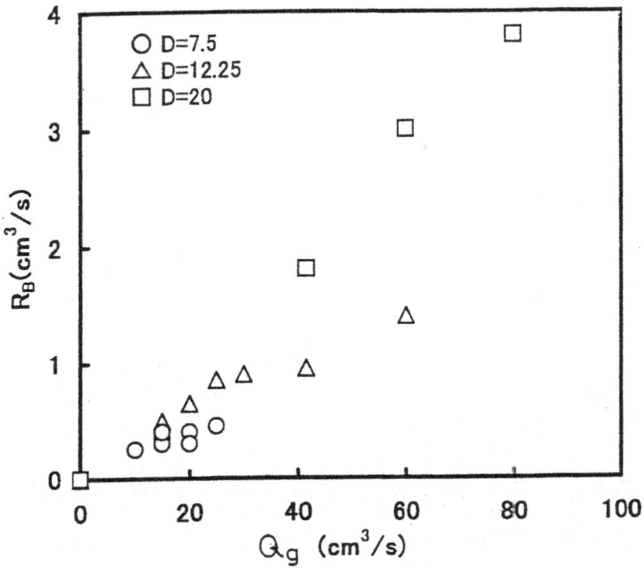


**Fig. 6.29** Effect of vessel diameter on the droplet birth rate ($\nu_s = 100\,\text{cSt}$)

where $d_B$(cm) is the mean diameter of bubbles passing through the slag–metal interface. Model experiments were performed using only one vessel with an inner diameter $D = 7.5\,\text{cm}$, and as such, the effect of $D$ is not considered in (6.41). In what follows, an empirical equation for $R_B$ is derived with reference to (6.41) and on the basis of experimental results for three vessels of different sizes.

Figure 6.29 demonstrates that $R_B$ increases monotonically with respect to air flow rate $Q_g$. In addition, $R_B$ is an increasing function of the vessel diameter $D$. As $D$ increases, the recirculating flow velocity in the molten metal and slag layers outside the bubble dispersion region becomes small. Accordingly, the wake of a bubble becomes more stable with an increase in $D$ and the volume of molten metal carried into the upper slag layer by the bubble becomes greater. This translates to an increase in $R_B$ with $D$.

Figure 6.30 shows that the measured values of $R_B$ are not dependent on the kinematic viscosity of slag except $\nu_s = 10\,\text{cSt}$ ($\text{cm}^2/\text{s}$). Specifically, the birth rate is proportional to $D^{1.25}$ and independent of the kinematic viscosity of slag, $\nu_s$, for $\nu_s > 10\,\text{cSt}$.

Based on these results, (6.41) was modified as follows:

$$R_B = 0.082{\mu_m}^{2/3}\rho_s\Delta\rho^{-5/3}Q_g/d_B(D/7.5)^{1.25}. \tag{6.42}$$

Figure 6.31 compares the measured values of $R_B$, designated by $R_{B,\text{mea}}$, with the values calculated from (6.42), designated by $R_{B,\text{cal}}$. Equation (6.42) can indeed approximate the measurement for $\nu_s = 50$–$350\,\text{cSt}$ within a scatter of $\pm 15\%$.

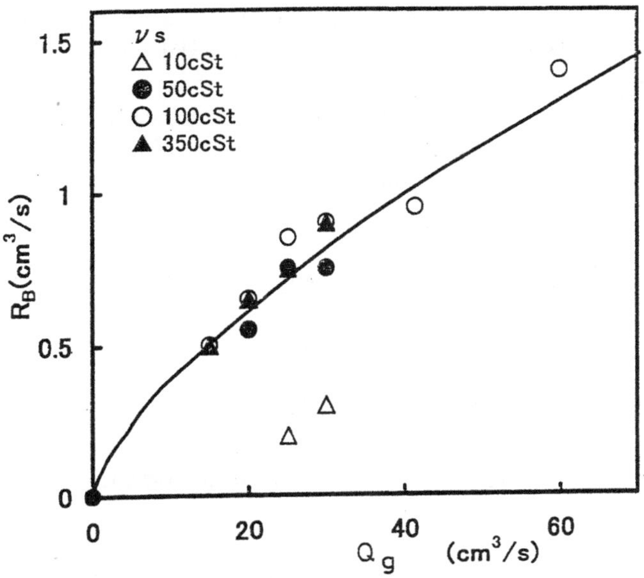


**Fig. 6.30** Effect of upper phase viscosity on the droplet birth rate ($D = 12.25\,\mathrm{cm}$)

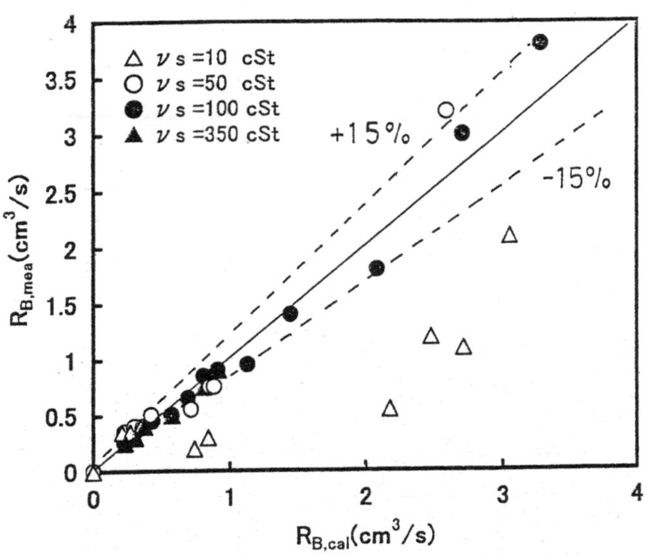


**Fig. 6.31** Relation between measured and calculated values of $R_B$

The discrepancy between the measurement and (6.42) for $\nu_s = 10\,\mathrm{cSt}$ can be explained as follows: According to Hetsroni [18], the wake of a bubble becomes turbulent when the bubble Reynolds number $R_B$, defined in terms of the relative velocity between bubbles and molten slag, mean bubble diameter and the kinematic viscosity of molten slag, exceeds a critical value of approximately 400. The Reynolds number, $Re_B$, for a bubble rising in the slag layer of $\nu_s = 10\,\mathrm{cSt}$ after passing through the slag–metal interface, is higher than 400, and hence, the wake becomes turbulent. Meanwhile, the wake of a bubble in the slag layer of $\nu_s \geq 50\,\mathrm{cSt}$ remains laminar because $Re_B$ is much smaller than 400. Molten metal is more easily carried into the upper slag layer when the wakes of bubbles are laminar than when they are turbulent.

#### 6.2.2.4 Lifetime of Molten Metal Droplet, $\tau$

Filament-like molten metal carried by bubbles into the upper slag layer breaks up into small droplets, as already illustrated in Fig. 6.22. The droplets accumulate on the horizontal slag–metal interface, and some of them subsequently return to the lower molten metal layer. Such a series of events is repeated at steady state. As can be inferred from (6.31), $\tau$ is regarded as the mean lifetime of the molten metal droplets.

Figure 6.32 compares measured values of $\tau$, represented by $\tau_{mea}$, and the calculated values based on (6.40) and (6.42), denoted by $\tau_{cal}$. The measurement and calculation are in good agreement within a scatter of $\pm 15\%$ except for $\nu_s = 10\,\mathrm{cSt}$. The mean lifetime becomes larger as the kinematic viscosity of slag increases. This is because the falling velocity of molten metal droplets in the upper slag layer becomes low with an increase in $\nu_s$.

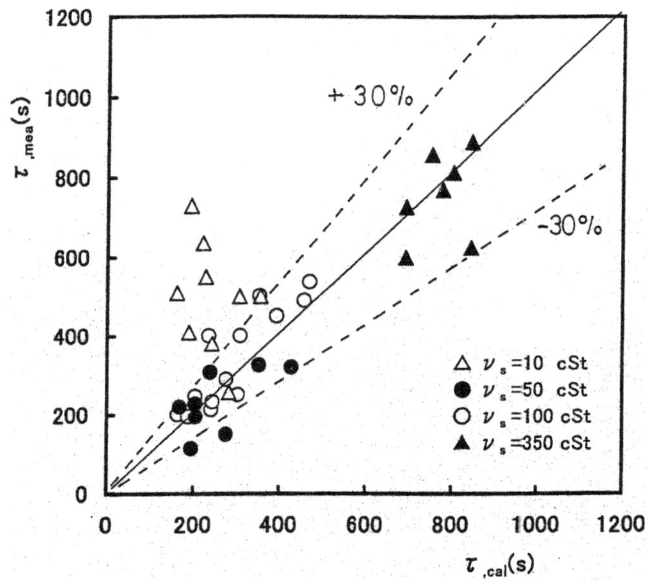


**Fig. 6.32** Relation between measured and calculated values of $\tau$

#### 6.2.2.5 Death Rate of Molten Metal Droplets After Stoppage of Gas Injection

The behavior of accumulated molten metal droplets after the stoppage of gas injection has been observed with a high-speed video camera. The generation of molten metal droplets in the central part of the vessel ceases and accumulated molten metal droplets return to the lower molten metal layer. With time, the position of the horizontal slag–metal interface elevates gradually and finally reaches its initial position before gas injection. We designate the difference between the actual position of the interface at time $t'$ measured from the stoppage of gas injection and the initial position, by $\Delta H_m{}'$. The total volume of molten metal droplets remaining in the upper slag layer, $V_d{}'(t')$, can be described by,

$$V_d{}'(t') = \Delta H_m{}' A. \tag{6.43}$$

Figure 6.33 indicates that $V_d{}'(t')$ (cm$^3$) decreases more rapidly with an increase in the vessel diameter $D$. A sudden decrease in $\Delta H_m{}'$ immediately after the stoppage of gas injection suggests that the molten metal in the elevated region appearing in the central part of the bath returns rapidly to the lower molten metal layer. After that, $V_d{}'(t')$ decreases linearly with time, $t'$, in every case and thus the death rate of accumulated molten metal droplets in the absence of rising bubbles is regarded as being constant.

Figure 6.34 shows the variation of measured values of death rate, $R_D{}'$ (cm$^3$/s) with the gas flow rate $Q_g$. The kinematic viscosity of slag, $\nu_s$ was chosen as a

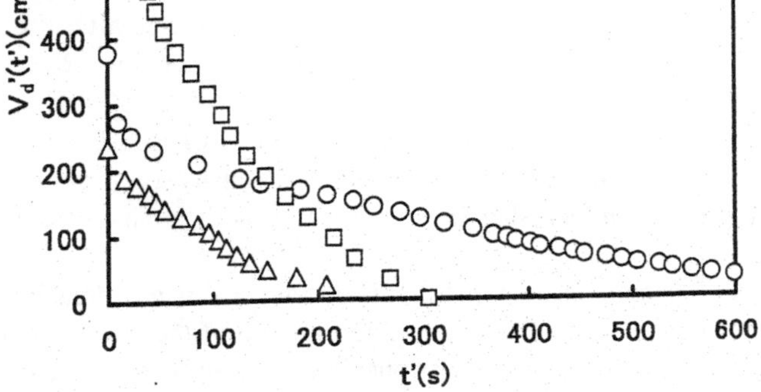


**Fig. 6.33** Effect of vessel diameter on the droplet death rate ($\nu_s = 100$ cSt)

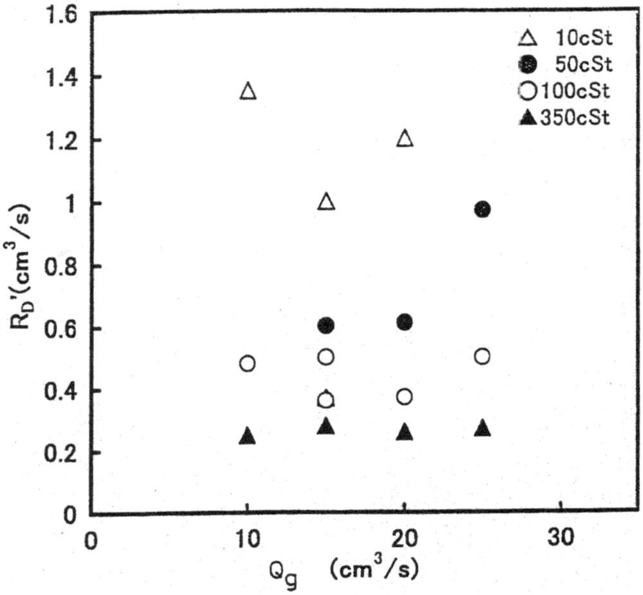


**Fig. 6.34** Effect of upper phase viscosity on the droplet death rate ($D = 7.5\,\text{cm}$)

parameter. It is evident that $R_D{}'$ decreases as $\nu_s$ increases. A dimensionless analysis was performed to derive an empirical equation for $R_D{}'$ by assuming

$$R_D{}' = f(D, \Delta\rho, \sigma_{ms}, \mu_s, g). \tag{6.44}$$

A comparison of the results of the analysis and the measured values yields the following empirical relation:

$$R_D{}' = 2.93 \times 10^{-3} D^{1.5} \mu_s{}^{-0.25} \Delta\rho^{-0.44} \sigma_{ms}{}^{0.69}. \tag{6.45}$$

Figure 6.35 compares the measured values of the death rate after the stoppage of gas injection, $R_D{}'_{,mea}$, and calculated values based on (6.45), $R_D{}'_{,cal}$. The measurement is approximated by (6.45) within a scatter of $\pm 30\%$. The lack of proportionality between $R_D{}'$ and the cross-sectional area of the vessel $A(= \pi D^2/4)$ can be explained by the fact that the number density of accumulated molten metal droplets in the upper slag layer decreases with an increase in the vessel diameter $D$.

At a glance, it seems curious that the death rate $R_D{}'$ is a decreasing function of the density difference $\Delta\rho$ and an increasing function of the interfacial tension $\sigma_{ms}$. The reason why such a functional relationship holds can be explained as follows:

The mean diameter of molten metal droplets, $d_m$, is expressed by (6.22). This diameter increases with $\sigma_{ms}$ and decreases with $\Delta\rho$. When the kinematic viscosity

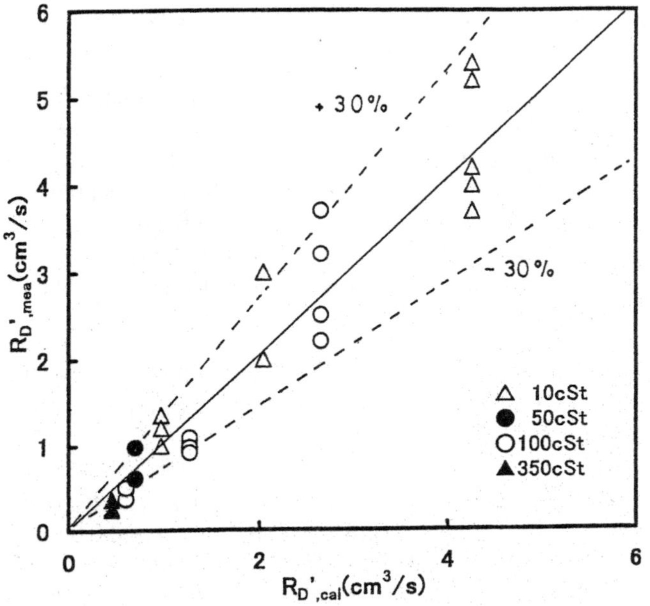


**Fig. 6.35** Relation between measured and calculated values of $R_D'$

of slag is high ($\nu_s \geq 10\,\text{cSt}$), a Reynolds number defined in terms of $d_m$, the falling velocity of molten metal droplets, and the kinematic viscosity of slag, $\nu_s$, is very small. Thus, the fluid dynamic drag on each molten metal droplet nearly follows Stokes's law. Consequently, the falling velocity of the droplets decreases as the mean diameter decreases, validating (6.45).

## 6.3 Summary

### *6.3.1 Shape and Size of Entrained Metal Layer*

- The horizontal distribution of the total molten metal holdup, $\varepsilon'$, in the elevated region can be approximated by a Gaussian distribution.
- The shape of the elevated region is parabolic. The maximum height of the elevated region appearing on the centerline of the bath, $\Delta H_{cl}$, is expressed by (6.14).
- The volume of the elevated molten metal in the elevated region is approximated by (6.20).
- The mean diameter of molten metal droplets, $d_m$, in the slag layer is expressed by (6.24).

### 6.3.2 *Characteristics of Metal Droplets*

Molten metal in the lower layer is lifted up by bubbles passing through the slag–metal interface into the upper slag layer and forms elevated, i.e., mountain-like region in the central part of the vessel. At a certain distance above the elevated region, filament-like molten metal in the wakes of the bubbles breaks up into small molten metal droplets. These droplets accumulate on the horizontal slag–metal interface with time, and some of them return to the lower molten metal layer. Such a series of events was observed with a high-speed video camera. The main findings can be summarized as follows:

(a) The total volume of accumulated molten metal droplets above the slag–metal interface in steady state can be approximated by the empirical relation in (6.40) within a scatter of $\pm 30\%$.
(b) The birth rate of accumulated molten metal droplets can be predicted by the empirical relation in (6.42) within a scatter of $\pm 15\%$. After steady state has been established, the death rate $R_D(t)$ becomes equal to the birth rate $R_B$.
(c) The empirical relation in (6.45) is proposed for the death rate $R_D{}'$ of accumulated molten metal droplets after the cessation of gas injection. This relation predicts $R_D{}'$ to within a scatter of $\pm 30\%$.

## References

1. Zaidi A, Sohn HY (1995) Measurement and correlation of drop-size distribution in liquid-liquid emulsions formed by high-velocity bottom gas injection. ISIJ Int 35:234–241
2. Iwamasa PK, Fruehan RJ (1996) Separation of metal droplets from slag. ISIJ Int 36:1319–1327
3. Matsuo M, Katayama H, Ibaraki T, Yamauchi M, Kanemoto M, Ogawa T (1996) The characteristic of a thick slag layer and mechanism of reduction of iron ore in the smelting reduction. Tetsu-to-Hagane 82:725–730
4. Hartland S (1967) The coalescence of a liquid drop at a liquid-liquid interface. Part I: Drop shape, Trans Inst Chem Eng 45:T97
5. Iguchi M, Nozawa K, Morita Z (1991) Bubble characteristics in the momentum region of air-water vertical bubbling jet. ISIJ Int 31:952–959
6. Ilegbusi OJ, Iguchi M, Nakajima K, Sano M, Sakamoto M (1998) Modeling mean flow and turbulence characteristics in gas-agitated bath with top layer. Metall Mater Trans B 29B: 211–222
7. Iguchi M, Nakatani T, Ueda H (1997) Model study of turbulence structure in a bottom blown bath with top slag using conditional sampling. Metall Mater Trans B 28B:87–94
8. Iguchi M, Sasaki K, Nakamura K, Takahashi K (1997) Model study on bubble and liquid flow characteristics in a bottom blown bath with a thick slag layer. CAMP-ISIJ 10:915
9. Lin Z-H, Guthrie RIL (1994) Modeling of metallurgical emulsions. Metall Mater Trans B 25B:855–864
10. Lin Z-H (1997) The Modelling of emulsification, slag foaming and auoy addition behaviour in intensively stirred metaUurgical reactors, Ph.D. Dissertation, McGill University, Montreal
11. Takashima S, Iguchi M (2000) Metal droplet holdup in the thick slag layer subjected to bottom gas injection. Tetsu-to-Hagane 86:217–224
12. Masuda H (1990) Jpn Soc Multiphase Flow, p 19

13. Castello-Branco MASC, Schwerdtfeger K (1994) Large-scale measurements of the physical characteristics of round vertical bubble plumes in liquids. Metall Mater Trans B 25B:359–371
14. Iguchi M, Ueda H, Uemura T (1995) Bubble and liquid flow characteristics in a vertical bubbling jet. Int J Multiphase Flow 21:861–873
15. Asai S (1984) 100th and 101th Nishiyama Memorial Lecture, ISIJ Tokyo, p 67
16. Iguchi M, Sumida Y, Okada R, Morita Z (1993) Evaluation of the critical gas flow rate using water model for the entrapment of slag into a metal bath subject to gas injection. Tetsu-to-Hagane 79:569–575
17. Iguchi M, Nozawa K, Tomida Y, Morita Z (1991) Bubble characteristics in the buoyancy region of air-Water vertical bubbling jet. Tetsu-to-Hagane 77:1426–1433
18. Hetsroni G (1989) Particles-turbulence interaction. Int J Multiphase Flow 15:735–746

# Chapter 7
# Surface Flow Control

## 7.1 Overview

Recently, optimum control of mixing in current refining processes accompanied by bottom gas injection has become increasingly important because the metallurgical reactions in the bath proceed at different rates and in different sites with a lapse of time. The intensity of mixing is commonly represented by the mixing time $T_m$ [1–8]. Measuring the mixing time in the bath of real metallurgical reactors is difficult. Therefore, it is usually predicted on the basis of water model experiments using electric conductivity sensor and dilute aqueous KCl solution as tracer. The mixing time is known to be influenced by operating variables, such as the bath diameter $D$, bath depth $H_L$, the location of bottom nozzle, and gas flow rate $Q_g$ [8]. However, only $Q_g$ can be easily controlled during processing. Even the effect of the gas flow rate $Q_g$ on the mixing time is relatively weak.

When swirl motion occurs in a bath contained in a cylindrical vessel, the mixing time is considerably shorter than a situation in the swirl motion that is stopped by bringing a circular cylinder into contact with the bath surface. This result indicates that the mixing time in the bath can be changed drastically by controlling the surface flow.

As is widely known, mixing in a bath is governed mainly by large-scale recirculation and turbulent motion. The former is characterized by the mean velocity components in the three directions, while the latter is characterized by the root-mean-square (rms) values of the three turbulence components and the Reynolds shear stresses. Desirable mixing condition would be realized when the two kinds of motions are produced together. Unfortunately, these motions on the mixing time in a bath subjected to surface flow control are poorly understood. This chapter discusses these effects with reference to experiments in which three types of boundary conditions are imposed on the surface of a water bath stirred by bottom gas injection.

M. Iguchi and O.J. Ilegbusi, *Modeling Multiphase Materials Processes: Gas-Liquid Systems*, DOI 10.1007/978-1-4419-7479-2_7,

## 7.2 Experiment

### 7.2.1 Experimental Apparatus and Procedure

Figure 7.1 shows a schematic of the experimental apparatus. The transparent cylindrical vessel made of acrylic resin had an inner diameter $D$ of $200 \times 10^{-3}$ m and a height $H$ of $390 \times 10^{-3}$ m. The vessel was filled with water to a prescribed depth $H_L$ and air was injected using a compressor through a central single-hole bottom nozzle of inner diameter $d_{ni} = 2 \times 10^{-3}$ m. The gas flow rate $Q_g$ was adjusted with a mass flow controller, but the experiments were carried out mainly for $Q_g = 120 \times 10^{-6}\,m^3/s$. The cylindrical test vessel was placed inside another transparent vessel of a square cross-section. Water was used to fill the space between the two vessels in order to reduce parallax effect on the accuracy of LDV measurements.

The test vessel was mounted on a three-dimensional traversing device. The axial, radial, and tangential coordinates were $z$, $r$, and $\theta$, respectively. The corresponding velocity components of water flow in the bath were $u$, $v$, and $w$. The origin of the coordinate system was placed at the center of the nozzle exit, as shown in Fig. 7.1. These velocity components were measured by making use of a two-channel laser Doppler velocimeter [10, 11].

The mixing time was measured by a conventional technique based on the detection of a change in the electric conductivity of water induced by the addition of dilute aqueous KCl solution. The well-known 95% criterion was used to determine the mixing time [8, 9].

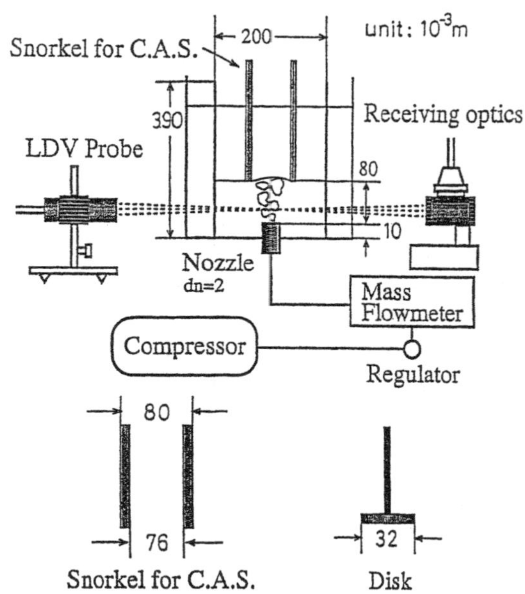


**Fig. 7.1** Experimental apparatus

### 7.2.2 Boundary Conditions on Bath Surface

When the gas flow rate $Q_g$ was $120 \times 10^{-6}\,m^3/s$ and the top of the bath was free, there arose a swirl motion similar to the rotary sloshing typical of a bath in a cylindrical vessel subjected to external, horizontal, or vertical forced oscillation. This motion was termed the first type of swirl motion [12].

The swirl motion could be stopped in one of the two ways, namely, by bringing a circular cylinder with an inner diameter of $78 \times 10^{-3}$ m and an outer diameter of $80 \times 10^{-3}$ m into contact with the bath surface or by bringing a flat circular disk with a diameter of $32 \times 10^{-3}$ m and a thickness of $5 \times 10^{-3}$ m into contact with the bath surface. The value of $32 \times 10^{-3}$ m was the minimum disk diameter with which the swirl motion could be stopped completely for the experimental conditions investigated. The first option, using circular cylinder is referred to as the CAS model, while the second, utilizing the circular disk, is termed the disk model.

### 7.2.3 Data Processing

The axial, radial, and tangential velocity components, $(u_i,\ v_i,\ w_i)$, were decomposed into the mean, $(\overline{u}, \overline{v}, \overline{w})$, and turbulence components, $(u'_i,\ v'_i,\ w'_i)$, as follows:

$$\overline{u} = \sum u_i/Nu, \tag{7.1}$$

$$\overline{v} = \sum v_i/Nv, \tag{7.2}$$

$$\overline{w} = \sum w_i/Nw, \tag{7.3}$$

$$u'_i = u_i - \overline{u}, \tag{7.4}$$

$$v'_i = v_i - \overline{v}, \tag{7.5}$$

$$w'_i = w_i - \overline{w}, \tag{7.6}$$

where the subscript $i$ designates the $i$th digitized datum, the overbar denotes the ensemble averaged value, and *Nu*, *Nv*, and *Nw* are the numbers of the axial, radial, and tangential velocity data, respectively. The *Nu*, *Nv*, and *Nw* values were chosen to be larger than $1 \times 10^4$ so that the velocity data could be analyzed with sufficient accuracy [13].

The rms values of the three turbulence components were calculated from

$$u'_{rms} = \left[\sum u_i'^2/Nu\right], \tag{7.7}$$

$$v'_{rms} = \left[\sum v_i'^2/Nv\right], \tag{7.8}$$

$$w'_{rms} = \left[\sum u_i'^2/Nu\right]^{1/2}, \tag{7.9}$$

The Reynolds shear stress $\overline{u'v'}$ was calculated from

$$\overline{u'v'} = \sum u'_i v'_i / Nuv, \tag{7.10}$$

where $Nuv$ is the number of a set of the axial and radial velocity data obtained simultaneously. It should be noted that the Reynolds shear stress, i.e., apparent shear stress due to turbulent motions is a useful measure of the intensity of local turbulent mixing in the bath.

## 7.3 Experimental Results

### *7.3.1 Mixing Time*

The results of the mixing time $T_m$ are presented against the radial sensor position $R_s$ in Fig. 7.2. The bath depth $H_L$ is $100 \times 10^{-3}$ m, and the gas flow rate $Q_g$ is $120 \times 10^{-6}\,m^3/s$. The electric conductivity sensor is placed at $z = 50 \times 10^{-3}$ m. This sensor position is designated by Hs in Fig. 7.2.

In the presence of the swirl motion, the mixing time $T_m$ is approximately 15 s, while it is approximately 100 s when the swirl motion is stopped by making use of

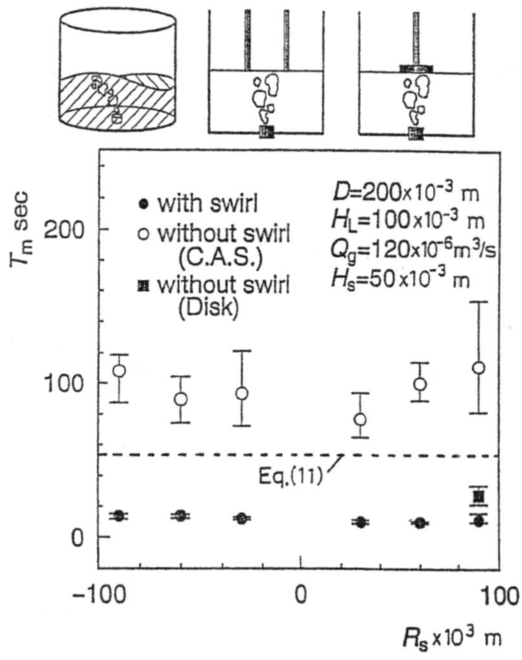


Fig. 7.2 Mixing time values under three boundary conditions imposed on the bath surface

the CAS model. In effect, the CAS model prolongs the mixing time by about one order of magnitude over the value in the presence of swirl motion. On the other hand, the mixing time for the disk model is about 20 s, which is slightly longer than that in the presence of the swirl motion.

The first type of swirl motion always appears for $Q_g = 120 \times 10^{-6}\,m^3/s$ in the bath of $D = 200 \times 10^{-3}$ m and $H_L = 100 \times 10^{-3}$ m whenever the free surface is present. However, we can imagine a situation in which the swirl motion is absent even in the presence of free surface. In this imaginary case, $T_m$ can be calculated from the following empirical relation [13]:

$$T_m = 1,200 Q_g^{-0.47} D^{1.97} H_L^{-1} \nu_L^{0.47}. \tag{7.11}$$

Equation (7.11) yields $T_m = 53.1$ s under the experimental condition considered. This value is much longer than those obtained in the presence of swirl motion and in the disk model but much shorter than that in the CAS model. Thus, when the swirl motion is absent in a real bath under certain blowing conditions, the mixing time in the bath may be significantly shortened by using the disk model.

## 7.3.2 *Fluid Flow Phenomena*

### 7.3.2.1 Mean Velocity Components

CAS Model

All velocity measurements were carried out for a bath depth of $H_L = 80 \times 10^{-3}$ m although the mixing time measurements were done for $H_L = 100 \times 10^{-3}$ m. This difference did not cause any problem because the flow patterns for $H_L = 80 \times 10^{-3}$ m and $100 \times 10^{-3}$ m were nearly the same.

The radial distributions of the axial and radial mean velocity components, $\overline{u}$ and $\overline{v}$, are shown in Figs. 7.3 and 7.4, respectively. The $\overline{u}$ distribution indicates that water moves upward for $r < 20 \times 10^{-3}$ m and downward for $20 \times 10^{-3}\,m < r < 40 \times 10^{-3}$ m. The broken line in each figure denotes the location of the outer edge of the circular cylinder. Outside this region ($r > 40 \times 10^{-3}$ m), $\overline{u}$ and $\overline{v}$ nearly vanish. In other words, a dead water region develops for $r > 40 \times 10^{-3}$ m.

For the purpose of a better understanding of the flow field in the bath with the CAS model, the resultant velocity vectors for $\overline{u}$ and $\overline{v}$ are presented in Fig. 7.5. Only in the inner region of $r < 40 \times 10^{-3}$ m, is violent vertical water motion evident. Thus, the CAS model almost completely suppresses the recirculation motion in the radial region outside the outer edge of the circular cylinder. This is mainly responsible for the long mixing time obtained with the CAS model.

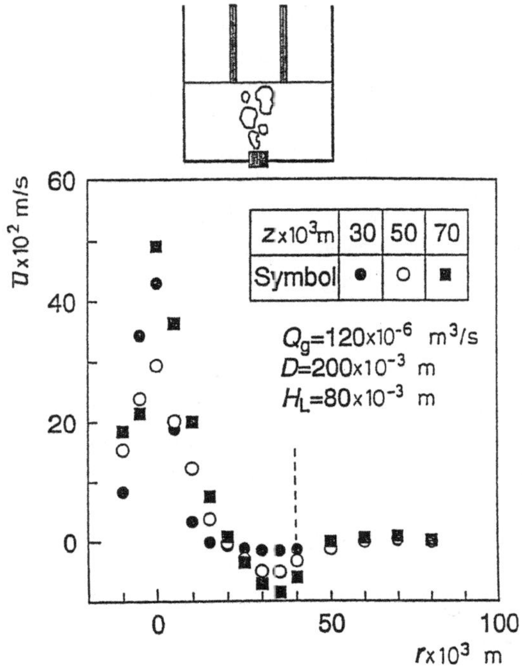


**Fig. 7.3** Axial mean velocity $u$ in the bath in CAS model

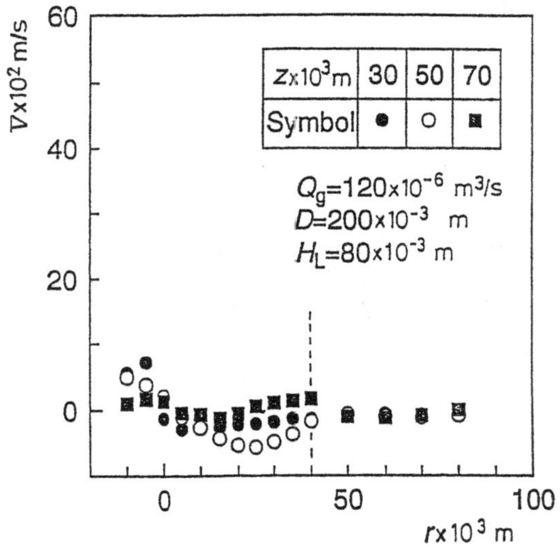


**Fig. 7.4** Radial mean velocity $v$ in the bath in CAS model

Disk Model

The radial distributions of the axial and radial mean velocity components, $\overline{u}$ and $\overline{v}$, are shown in Figs. 7.6 and 7.7, respectively. Unlike the results in the CAS model, the measured $\overline{u}$ and $\overline{v}$ values never diminish in the radial region outside the outer edge of the disk. The magnitudes of $\overline{u}$ and $\overline{v}$ are approximately the same in this region.

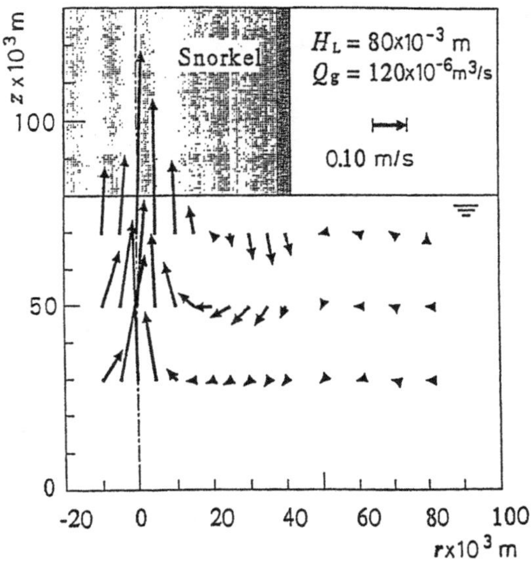


**Fig. 7.5** Resultant velocity vectors of $u$ and $v$ in CAS model

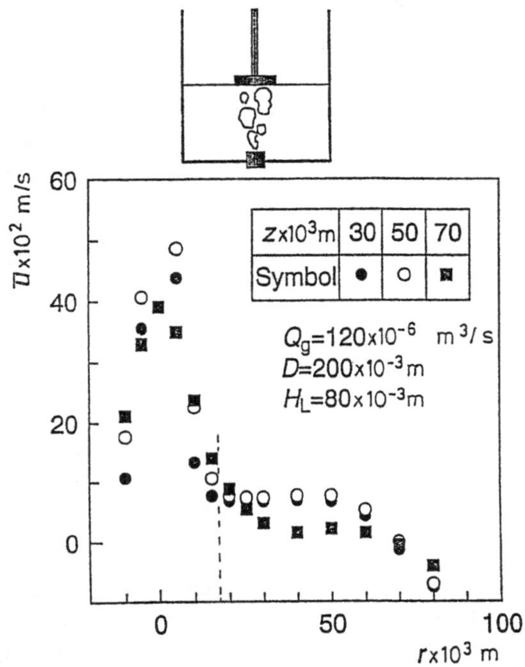


**Fig. 7.6** Axial mean velocity $u$ in the bath in disk model

The radial mean velocity $\bar{v}$ exhibits a relatively high positive value beneath the bath surface ($z = 70 \times 10^{-3}$ m), indicating the existence of a strong outward-moving surface flow. Near the bottom of the bath ($z = 30 \times 10^{-3}$ m), $\bar{v}$ becomes negative. This result implies the existence of a strong inward-moving bottom flow.

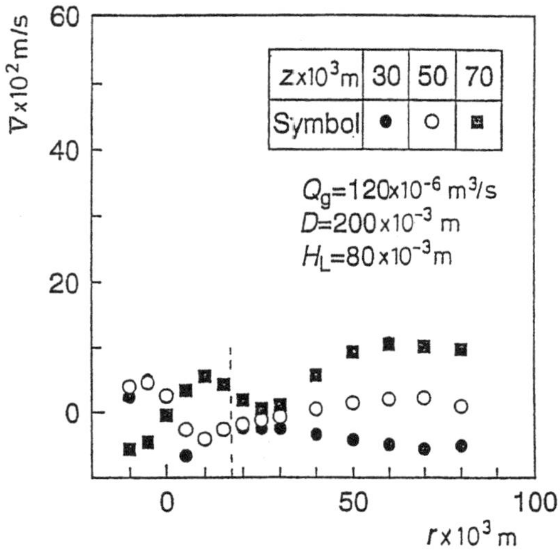


**Fig. 7.7** Radial mean velocity $v$ in the bath in disk model

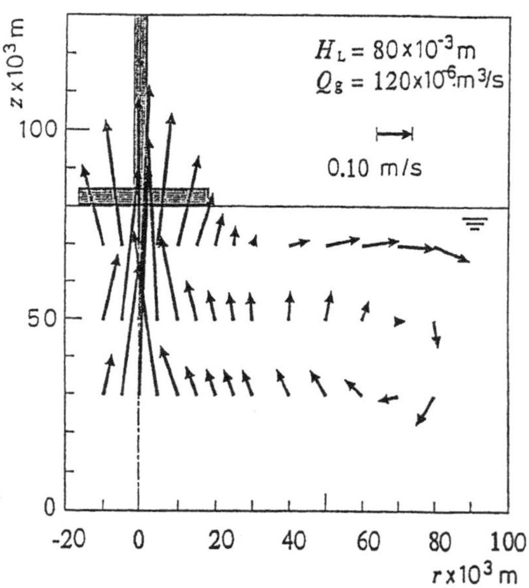


**Fig. 7.8** Resultant velocity vectors of $u$ and $v$ in disk model

The resultant velocity vectors of $\overline{u}$ and $\overline{v}$ are plotted in Fig. 7.8. Due to the outward moving surface flow induced by the disk, a strong recirculation flow develops outside the bubbling jet region. Hence, mixing in the bath is much stronger than the CAS model.

### Free Surface

The velocity vectors in the three directions are reproduced from Iguchi et al. [9] and replotted in Fig. 7.9. In the CAS and disk models, the tangential mean velocity com-

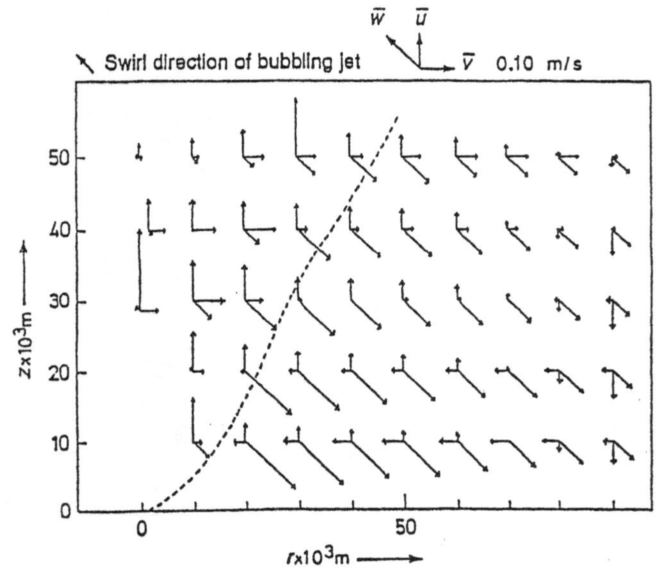


**Fig. 7.9** Axial, radial, and tangential velocity values in the presence of swirl motion

ponent $w$ is negligible, whereas in the presence of the swirl motion it is comparable to the axial and radial mean velocity components, $\overline{u}$ and $\overline{v}$.

The resultant vectors of $\overline{u}$ and $\overline{v}$ are shown in Fig. 7.10. The recirculation flow is clearly evident outside the bubbling jet region. Such an intense swirl motion strongly enhances the mixing, and consequently shortens the mixing time.

#### 7.3.2.2 Root-Mean-Square Turbulence Components and Reynolds Shear Stress

Mixing in the bath is closely associated with the turbulence characteristics in addition to the mean velocity components presented above. For further understanding of the mixing condition, information is given below on the rms values of the axial and radial turbulence components, $u'_{\mathrm{rms}}$ and $v'_{\mathrm{rms}}$, and the Reynolds shear stress $\overline{u'v'}$ at three representative axial positions, $z = 30, 50$, and $70 \times 10^{-3}$ m. In the presence of swirl motion, however, velocity measurements were impossible at $z = 70 \times 10^{-3}$ m due to the wave motion of the bath surface.

(a) $z = 30 \times 10^{-3}$ m (near the bottom wall)

Figures 7.11 and 7.12 show the rms values of the axial and radial turbulence components, $u'_{\mathrm{rms}}$ and $v'_{\mathrm{rms}}$, and the Reynolds shear stress $\overline{u'v'}$ at $z = 30 \times 10^{-3}$ m. This axial position is located near the bottom wall. The measured $u'_{\mathrm{rms}}$, $v'_{\mathrm{rms}}$, and $\overline{u'v'}$ values in the CAS model are approximately equal to the values in the disk model. These characteristics become very small outside the bubbling jet region,

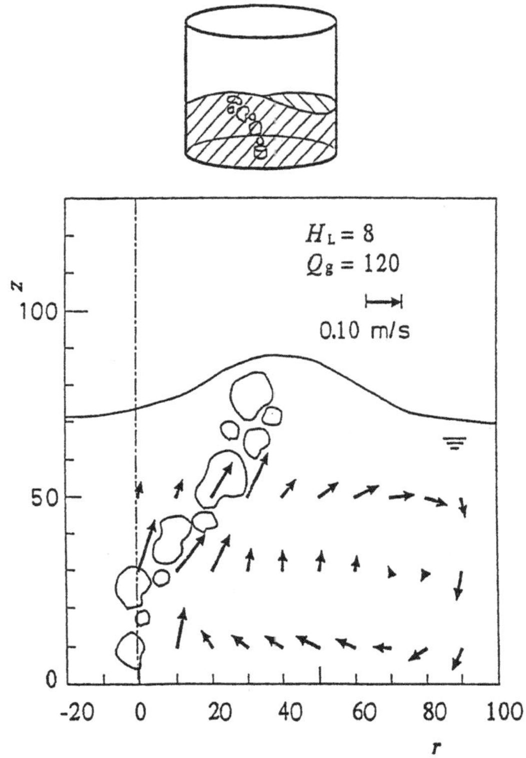


**Fig. 7.10** Resultant velocity vectors of $u$ and $v$ in the presence of swirl motion

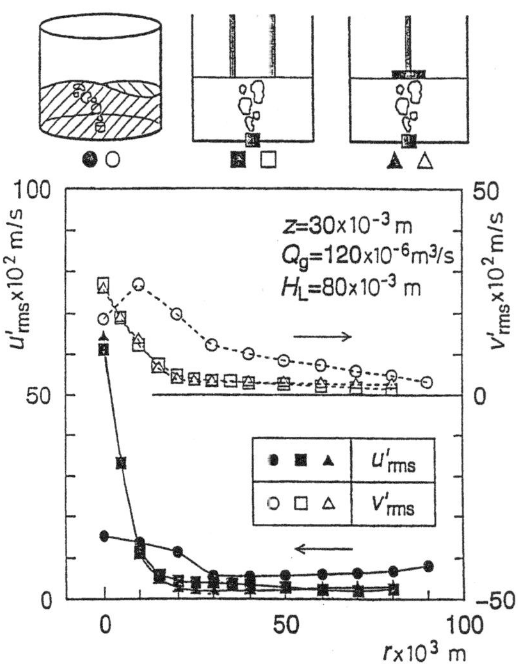


**Fig. 7.11** The rms values of axial and radial turbulence components at $z = 30 \times 10^{-3}$ m under three boundary conditions

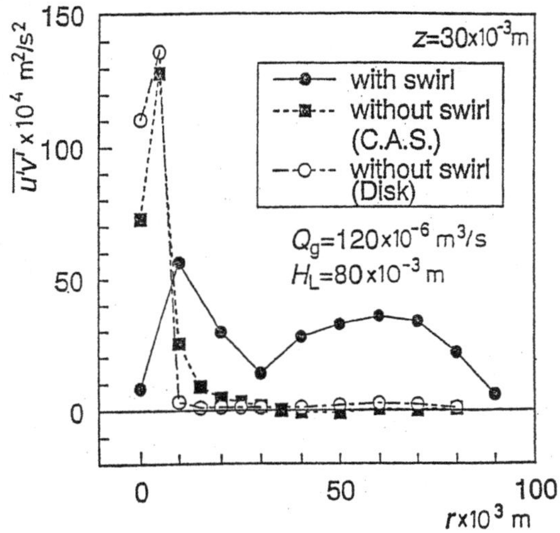


**Fig. 7.12** Reynolds shear stress values at $z = 30 \times 10^{-3}$ m under three boundary conditions

which is in the region where the axial mean velocity $\overline{u}$ is negative. On the other hand, in the presence of swirl motion, the rms values of the two turbulence components and the Reynolds shear stress become slightly smaller in the bubbling jet region than those in the CAS and disk models but much larger outside the bubbling jet region. A peak appears in the Reynolds shear stress distribution. This peak is closely associated with the swirl motion of the bubbling jet [9].

(b) $z = 50 \times 10^{-3}$ m (central part of the bath)

Figure 7.13 shows the rms values of the axial and radial turbulence components, $u'_{\mathrm{rms}}$, $v'_{\mathrm{rms}}$, at $z = 50 \times 10^{-3}$ m. The results follow a similar trend to that observed at $z = 30 \times 10^{-3}$ m.

Near the centerline of the bath in Fig. 7.14, the Reynolds shear stress in the disk model is smaller than that in the CAS model. This is because the rising motion of bubbles is suppressed at this axial position by the circular disk. In the presence of the swirl motion, the Reynolds shear stress is much larger than those obtained both in the CAS and disk models. The radial position at which the Reynolds shear stress exhibits a peak shifts from $r = 20 \times 10^{-3}$ m at $z = 30 \times 10^{-3}$ m to $r = 30 \times 10^{-3}$ m at $z = 50 \times 10^{-3}$ m.

Such high rms and Reynolds shear stress values in the presence of swirl motion also contribute strongly to the shortening of the mixing time.

(c) $z = 70 \times 10^{-3}$ m (beneath the bath surface)

Figures 7.15 and 7.16 show the rms values of the axial and radial turbulence components and the Reynolds shear stress at $z = 70 \times 10^{-3}$ m in the CAS and disk models. The measured $u'_{\mathrm{rms}}$, $v'_{\mathrm{rms}}$, and $\overline{u'v'}$ in the two models are approximately the same. Recalling that the mixing time in the disk model is much shorter than that in the CAS model, it can be concluded that the difference between the mixing times in the disk and CAS models is mainly due to the existence of a large-scale recirculation motion in the bath.

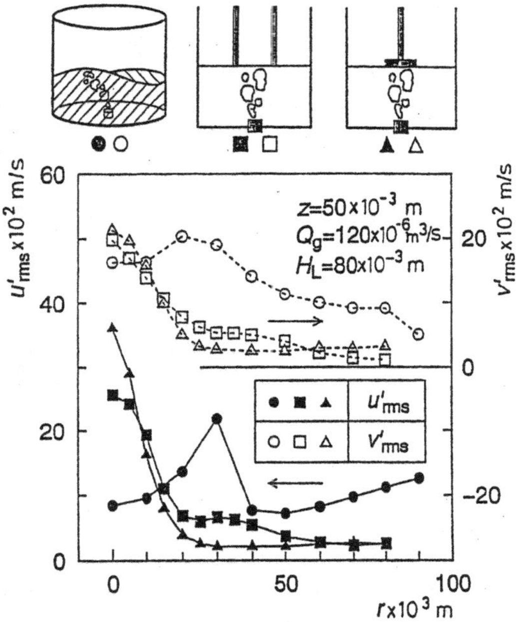


**Fig. 7.13** The rms values of axial and radial turbulence components at $z = 50 \times 10^{-3}$ m under three boundary conditions

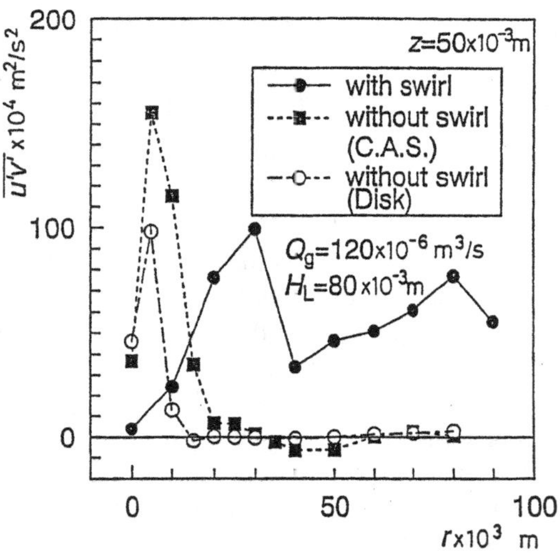


**Fig. 7.14** Reynolds shear stress values at $z = 50 \times 10^{-3}$ m under three boundary conditions

## 7.4 Conclusions

(a) When swirl motion is stopped by bringing a circular cylinder into contact with the bath surface, an intense fluid flow region is localized in the radial region inside the outer edge of the circular cylinder ($r < 40 \times 10^{-3}$ m), whereas a dead water region develops outside the outer edge ($r > 40 \times 10^{-3}$ m). This is

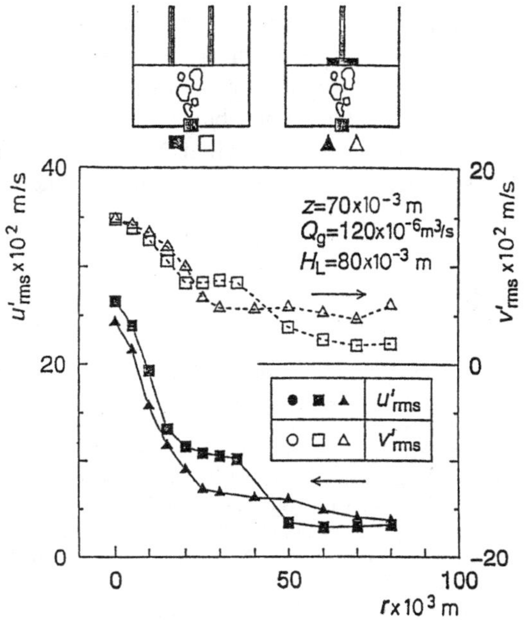


**Fig. 7.15** The rms values of axial and radial turbulence components at $z = 70 \times 10^{-3}$ m under two boundary conditions

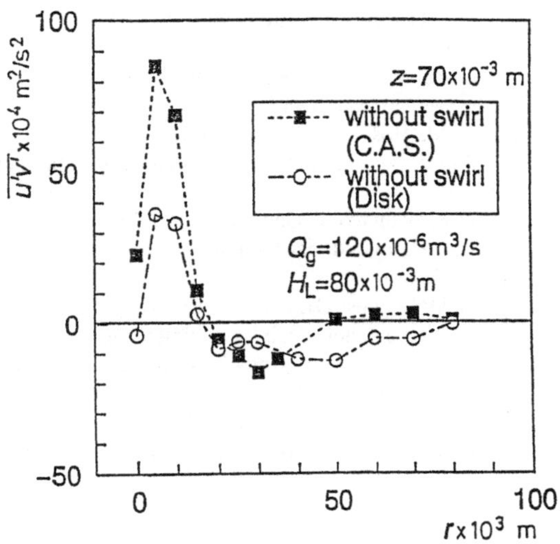


**Fig. 7.16** Reynolds shear stress values at $z = 70 \times 10^{-3}$ m under two boundary conditions

due to the fact that the surface flow moving outward is drastically suppressed by the circular cylinder and accordingly, the recirculation flow is nearly suppressed completely. Due to the existence of this dead water region, the mixing time becomes very long in the CAS model.

(b) When swirl motion is stopped by using a circular disk, the surface flow moving outward becomes much stronger than that in the CAS model, and hence, the recirculation flow also becomes more intense. As a result, the mixing time in the disk model is much shorter than the CAS model, but slightly longer than the value obtained in the presence of swirl motion. The latter result can be explained by the fact that the tangential water flow associated with the swirl motion of the bubbling jet also contributes to the mixing in the bath.

(c) These results collectively indicate that mixing in the bottom blown bath can be significantly influenced by controlling the surface flow.

## References


1. Nakanishi K, Fujii T, Szekely J (1975) Possible relation between energy dissipation and agitation in steel processing operations, Ironmak Steelmak 2:193–197
2. Nakanishi K, Saito K, Nozaki T, Kato Y, Suzuki K, Emi T (1982) Physical and metallurgical characteristics of combined blowing processes, Proc Steelmak Conf 65:101
3. Asai S, Okamoto T, He JC, Muchi I (1983) Mixing time of refining vessels stirred by gas injection. Trans Iron Steel Inst Jpn 23:43–50
4. Sinha UP, McNallan MJ (1985) Mixing in ladles by vertical injection of gas and gas-particle jets-a water model study. Metall Trans 16B:850–853
5. Stapurewicz T, Themelis NJ (1987) Mixing and mass transfer phenomena in bottom-injected gas-liquid reactors, Can Metall Quart 26:123–128
6. Krishnamurthy GG, Mehrotra SP, Ghosh A (1988) Experimental investigation of mixing phenomena in a gas stirred liquid bath. Metall Trans 19B:839–850
7. Mietz J, Oeters F (1989) Flow field and mixing with eccentric gas stirring. Steel Res 60: 387–394
8. Mazumdar D, Guthrie RIL (1995) The physical and mathematical modelling of gas stirred ladle systems. Iron Steel Inst Jpn Int 35:1–20
9. Iguchi M, Hosohara S, Kondoh T, Itoh Y, Morita Z (1994) Effects of the swirl motion of bubbling jet on the transport phenomena in a bottom blown bath. Iron Steel Inst Jpn Int 34: 330–337
10. Iguchi M, Kondoh T, Uemura T (1994) Simultaneous measurement of liquid and bubble velocities in a cylindrical bath subject to centric bottom gas injection. Int J Multiphase Flow 20:753–762
11. Iguchi M, Ilegbusi OJ, Ueda H, Kuranaga T, Morita Z (1996) Water model experiment on the liquid flow behavior in a bottom blown bath with top layer. Metall Mater Trans B 27:35–41
12. Iguchi M, Hosohara S, Koga T, Yamaguchi R, Morita Z (1993) The swirl motion of vertical bubbling jet in a cylindrical vessel. Iron Steel Inst Jpn Int 33:1037–1044
13. Iguchi M, Nakamura K, Tsujino R (1998) Mixing time and fluid flow phenomena in liquids of varying kinematic viscosities agitated by bottom gas injection. Metall Mater Trans B 29: 569–575

# Chapter 8
# Two-Phase Flow in Continuous Casting

## 8.1 Flow Characteristics

### *8.1.1 Overview*

In the current continuous casting processes, Ar gas is injected into a submerged entrance nozzle (SEN) to prevent clogging of $Al_2O_3$ to the inner surface of the SEN [1–6]. The gas disintegrates into small bubbles of varying diameters as it issues out of the SEN. This occurs as a result of the intense shear forces exerted by steel on the gas. Large bubbles rise toward the meniscus due to buoyancy and are subsequently removed from the mold, while smaller bubbles are carried deep into the mold. The small bubbles are trapped in the steel, causing pinhole defects [4, 7, 8]. Considerable effort has been devoted to removing these small bubbles in addition to suppressing the entrainment of mold powder into the steel [9–12]. These efforts have resulted in improvements in the shape and size of the nozzle [13, 14] and the application of electromagnetic braking [2, 4, 15, 16]. These notwithstanding, pinhole defects have not been successfully and completely eliminated from castings.

Understanding the behavior of steel–Ar two-phase jet and steel flow around the jet is essential for the design of effective methods of removing small bubbles. Numerical techniques have been developed to predict such a flow field [1, 2, 11, 12]. Although the accuracy of numerical prediction is improving, the underlying physics is not fully developed and the techniques are still relatively complicated to use in a routine manner. A simplified method is therefore highly desirable.

In this section, a description is provided of the use of water–air two-phase jet to model steel–Ar two-phase system. The horizontal and vertical velocity components are measured with a two-channel laser Doppler velocimeter [17]. Empirical relations are given for the mean values and the root-mean-square values of the turbulence components. Predictions of the trajectory of the two-phase jet are also discussed.

M. Iguchi and O.J. Ilegbusi, *Modeling Multiphase Materials Processes: Gas-Liquid Systems*, DOI 10.1007/978-1-4419-7479-2_8,

## *8.1.2 Experiment*

### 8.1.2.1 Experimental Apparatus

Figure 8.1 shows a schematic of the experimental apparatus. Water and air were supplied with a pump and a compressor, respectively. The test section had a length of 30.0 cm, a width of 15.0 cm, and a depth of 50.0 cm. The origin of the Cartesian coordinate system $(x, y, z)$ was placed at the center of the outlet of a pipe having an inner diameter $d_{ni}$ of 0.9 cm. The pipe outlet was arranged to flush with the narrow face. The water and air were mixed in the pipe at a position of 10.0 cm upstream of the outlet. The water flow rate was varied from 2.5 to 7.5 l/min (41.7–125 $cm^3/s$) while the air flow rate $Q_g$ was increased from 4 to 24 $cm^3/s$. The ratio of the flow rate of air to water was determined with reference to the actual operating conditions as is presented later. There was no attachment of water–air two-phase jet to one of the wide faces due to the Coanda effect [18] for the experimental conditions investigated.

### 8.1.2.2 Dimensional Analysis

Details of the actual continuous casting processes and model experiments are given in Table 8.1. Also, included are the following three dimensionless parameters for single-phase liquid flow (*Re*, *Fr*, *We*) and one dimensionless parameter for liquid–gas two-phase flow ($r_v$):

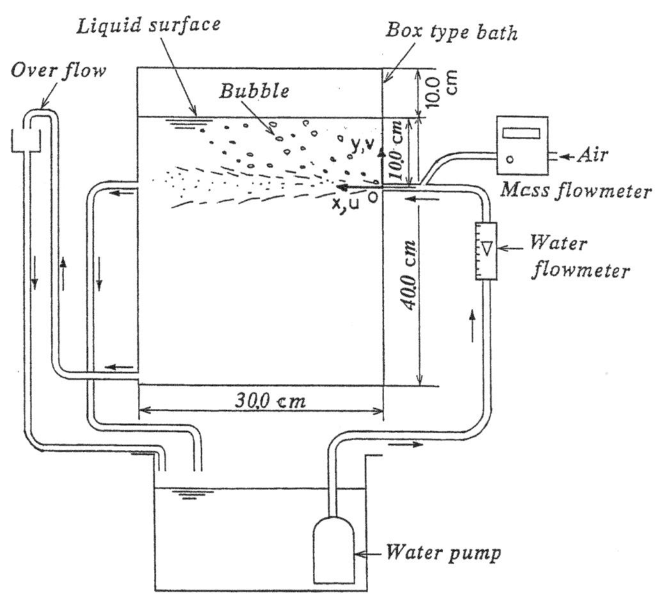


**Fig. 8.1** Experimental apparatus

**Table 8.1** Comparison of process parameters between model and actual continuous casting mold

| | Actual | Model |
|---|---|---|
| *Liquid* | *Molten steel* | *Water* |
| Liquid temperature | 1,833–1,853 (K) | 293 (K) |
| Liquid density | 7,000–7,200 ($kg/m^3$) | 1,000 ($kg/m^3$) |
| *Gas* | *Argon* | *Air* |
| Gas temperature[a] | 293 (injection point)–1,853 (nozzle exit) (K) | 293 (K) |
| Gas density | 1.6–0.25 ($kg/m^3$) | 1.2 ($kg/m^3$) |
| Mold size | | |
| Narrow face | 0.23–0.27 (m) | 0.125 (m) |
| Wide face | 0.85–2.3 (m) | 0.300 (m) |
| Nozzle shape | Square (two ports) | Round |
| Nozzle size | (0.07 × 0.07)–(0.09 × 0.09) (m) | $\phi$ 0.009 (m) |
| Liquid flow rate | 3.0–5.0 (ton/min) | 2.5–7.5 (kg/min) |
| Liquid velocity at nozzle outlet | 0.4–1.5 (m/s) | 0.65–1.96 (m/s) |
| Gas flow rate | 10–70 (L/min) | 4–24 ($cm^3/s$) |
| Gas velocity at nozzle outlet | 0.020–0.24 (m/s) | 0.062–0.38 (m/s) |
| Bubble diameter of argon trapped in steel product | 30–300 ($\mu$m) | – |
| Reynolds number *Re* for single-phase flow[b] | 30,000–200,000 | 5,000–16,000 |
| Froude number *Fr* for single-phase flow | 0.4–2 | 2–7 |
| Weber number *We* for single-phase flow | 55–1,000 | 50–500 |
| Velocity ratio $r_v$ for two-phase flow | 0.6–15 | 3.0–40 |

[a]Argon gas is injected into the submerged entrance nozzle at a room temperature. It expands descending in the nozzle due to heat transfer. Estimation of the temperature of the argon gas at the nozzle outlet is very difficult

[b]Steel jet and water jet are turbulent. When a jet is turbulent, its behavior is known to be hardly dependent on the Reynolds number

$$Re = u_0 d_{ni}/\nu_L \tag{8.1}$$

$$Fr = u_0/(g d_{ni})^{1/2} \tag{8.2}$$

$$We = \rho_L d_{ni} {u_0}^2/\sigma \tag{8.3}$$

$$r_v = u_0/[(g Q_g/(d_{ni} u_0))^{1/2} \tag{8.4}$$

In the above relations, Re is the Reynolds number, $u_0$ is the liquid flow velocity at the nozzle outlet, $d_{ni}$ is the inner diameter of the nozzle, $\nu_L$ is the kinematic viscosity of liquid, *Fr* is the Froude number, $g$ is the acceleration due to gravity, $\rho_L$ is the density of liquid, *We* is the Weber number, $\sigma$ is the surface tension of liquid, $r_v$ is the velocity ratio defined subsequently, and $Q_g$ is the gas flow rate.

The argon gas flow rate is much smaller than the molten steel flow rate. Accordingly, as a first step, experimental conditions in the water model experiment were determined by referring to the aforementioned three dimensionless parameters describing the liquid flow in the actual mold. Selected values of the four dimensionless parameters in the water model study were nearly the same as in the actual processes, as shown in Table 8.1. Consequently, the results of this study are considered to be useful for actual continuous casting mold redesign and process modification.

Although only isothermal systems are considered in this section, the hydrodynamics in a real system may be influenced by thermal gradients in the mold. The temperature difference between the molten steel in the mold and the issuing steel jet is at most 20 K, and the bulk modulus of steel is quite small. Therefore, natural convection force due to temperature stratification is small compared to the inertia force of the steel jet and the buoyancy force acting on argon bubbles in the jet.

### 8.1.2.3 Experimental Procedure

Photographs of the water–air two-phase jets thus generated were taken with a still camera, from which the diameter of bubbles was determined. The velocity components of water flow in the $x$, $y$, and $z$ directions are $u$, $v$, and $w$, respectively. For convenience, the $x$ and $y$ directions are referred to as the axial and vertical directions, respectively. Measurements of the $u$ and $v$ components were made at the $z = 0$ plane with a two-channel laser Doppler velocimeter, as shown in Fig. 8.2. The uncertainty in these measurements was less than 3%. The farthest axial position from the origin, $x = 21.0$ cm, was chosen because the flow field upstream of this position was hardly affected by the flow issuing out of a pipe placed at the opposite narrow face. The water–air two-phase jet was bent upward due to the buoyancy force acting on air bubbles. The centerline of the jet is represented with a thick solid line in Fig. 8.2, and the deflection from the $x$ axis is represented by $y_c$.

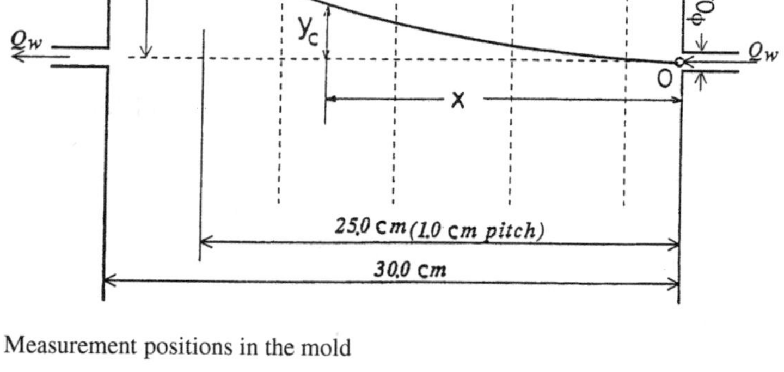


**Fig. 8.2** Measurement positions in the mold

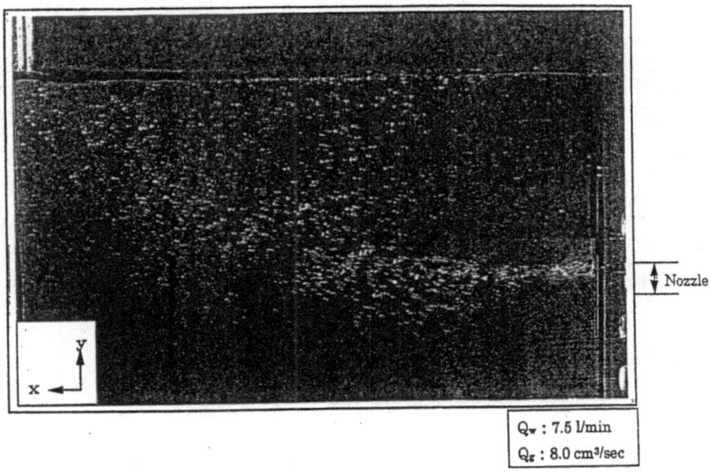


**Fig. 8.3** Photograph of a water–air two-phase jet

## 8.1.3 Experimental Results

### 8.1.3.1 Dispersion of Bubbles and Mean Bubble Diameter

A photograph of the water–air two-phase jet is shown in Fig. 8.3. The white patches in the figure represent bubbles. The upward bending of the jet is quite evident. The maximum bubble diameter is about 0.5 cm, which is approximately within the range of Ar bubble in real continuous casting molds.

### 8.1.3.2 Mean Velocities and Root-Mean-Square Turbulence Components

Figure 8.4 shows the axial and vertical mean velocity components of water, $\overline{u}$ and $\overline{v}$, measured along the $x$ axis. The value of $\overline{u}$ decreases monotonically in the $x$ direction for every gas flow rate. In particular, in the absence of gas flow, $\overline{u}$ is inversely proportional to x, as widely known for single-phase jets. This result is discussed in a subsequent section. Meanwhile, $\overline{v}$ for the two-phase jets becomes positive on the $x$ axis, implying that upward water flow is induced by the bubbles.

The root-mean-square axial turbulence component, $u'_{\text{rms}}$, is normalized by the axial mean velocity of single-phase water jet on the $x$ axis, $\overline{u}_{\text{m,sw}}$, in Fig. 8.5. The subscripts m and sw denote maximum value and a single-phase water jet, respectively. In the horizontal region $x < 10\,\text{cm}$, $u'_{\text{rms}}/\overline{u}_{\text{m,sw}}$ is strongly influenced by the existence of bubbles, and the maximum value is approximately equal to the bubbling jet value of 0.5 [17]. In the region $x > 10\,\text{cm}$, $u'_{\text{rms}}/\overline{u}_{\text{m,sw}}$ for the two-phase jet is slightly larger than that of the single-phase jet and remains approximately constant

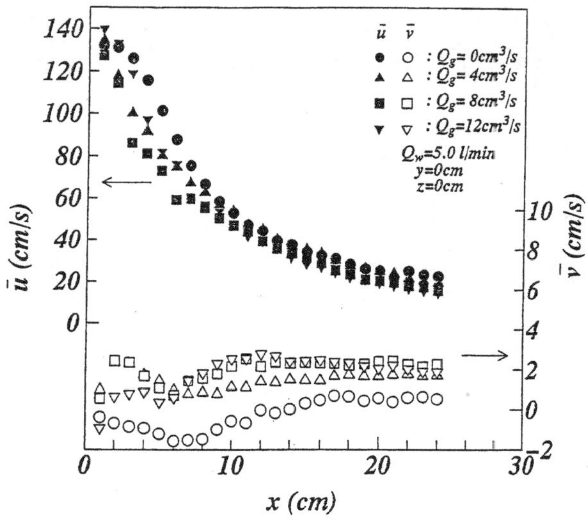


**Fig. 8.4** The axial and vertical mean velocities on the $x$ axis

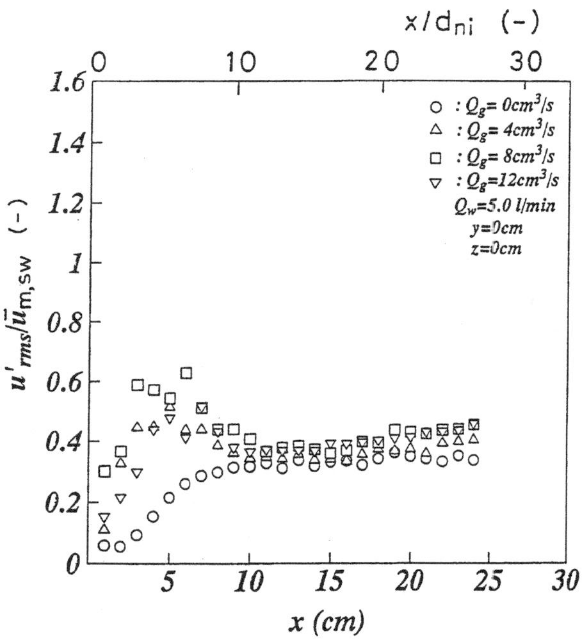


**Fig. 8.5** The root-mean-square value of axial turbulence component on the $x$ axis

in the axial direction. The measured turbulence intensity in the single-phase water jet is approximately 0.3. The dependence of $u'_{\text{rms}}/\bar{u}_{\text{m,sw}}$ on the air flow rate $Q_{\text{g}}$ is quite weak.

#### 8.1.3.3 Vertical Distribution of Axial Mean Velocity

Figure 8.6 shows the vertical distribution of the axial mean velocity $\overline{u}$ normalized by the centerline value for the single-phase water jet ($Q_g = 0\,\mathrm{cm}^3/\mathrm{s}$), $\overline{u}_{m,sw}$. The vertical distance $y$ is nondimensionalized by the half-value radius for the single-phase jet, $b_{u,sw}$. The measured velocity $\overline{u}$ for the single-phase jet can be approximated by a normal distribution, represented by a solid line in the figure. The vertical position at which a peak occurs in the $\overline{u}$ distribution for the two-phase jets shifts upward as the gas flow rate increases, whereas the magnitude of the peak slightly decreases as $Q_g$ increases. Such an upward shift in the peak position is due to the buoyancy force acting on bubbles in the water–air two-phase jets.

#### 8.1.3.4 Vertical Distribution of Root-Mean-Square Turbulence Components

The root-mean-square values of the axial and radial turbulence components, $u'_{rms}$ and $v'_{rms}$, are normalized by the axial mean velocity on the centerline of the single-phase water jet, $\overline{u}_{m,sw}$, and plotted in Figs. 8.7 and 8.8, respectively. The vertical distance $y$ is nondimensionalized in the same manner as in Fig. 8.6. The measured values of $u'_{rms}/\overline{u}_{m,sw}$ and $v'_{rms}/\overline{u}_{m,sw}$ for the single-phase water jet are close to the data of Wygnanski and Fiedler [19]. Below the $x$ axis ($y < 0$) both $u'_{rms}/\overline{u}_{m,sw}$ and $v'_{rms}/\overline{u}_{m,sw}$ are nearly independent of the gas flow rate while above the $x$ axis ($y \geq 0$) they increase slightly with an increase in $Q_g$. This is because turbulence is generated in the wake of the bubbles [20], and the number of bubbles increases as $Q_g$ increases.

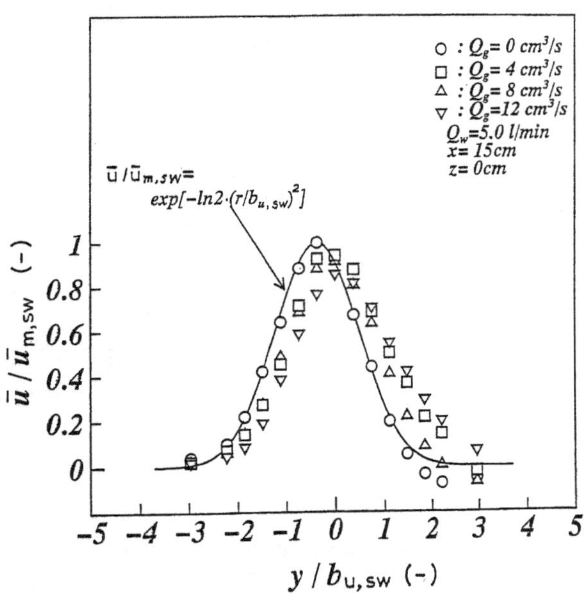


**Fig. 8.6** Vertical distributions of axial mean velocity

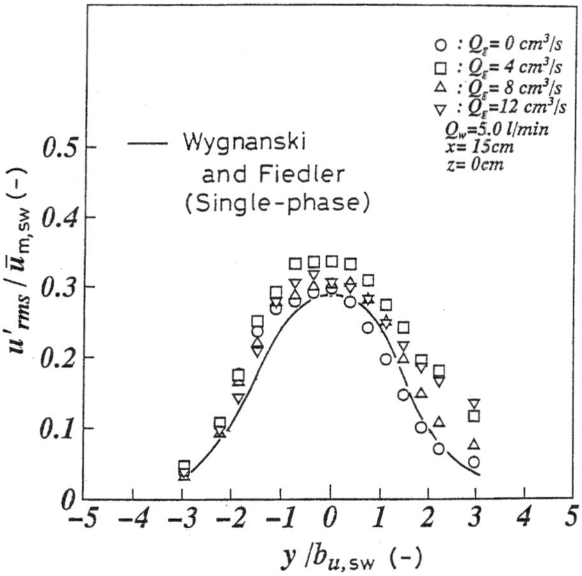


**Fig. 8.7** Vertical distributions of the root-mean-square value of axial turbulence component

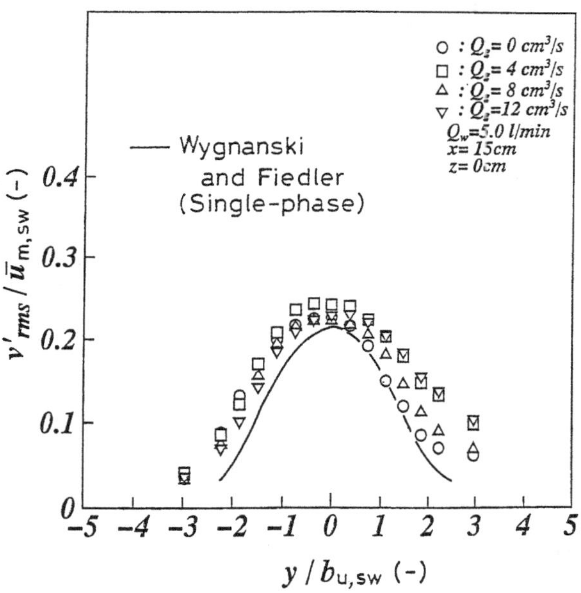


**Fig. 8.8** Vertical distributions of the root-mean-square value of vertical turbulence component

Figure 8.9 shows the vertical distribution of nondimensionalized Reynolds shear stress against $y/b_{u,sw}$. Similar to $u'_{rms}/\overline{u}_{m,sw}$ and $v'_{rms}/\overline{u}_{m,sw}$, the distribution of Reynolds shear stress is hardly influenced by the air flow rate for $y < 0$.

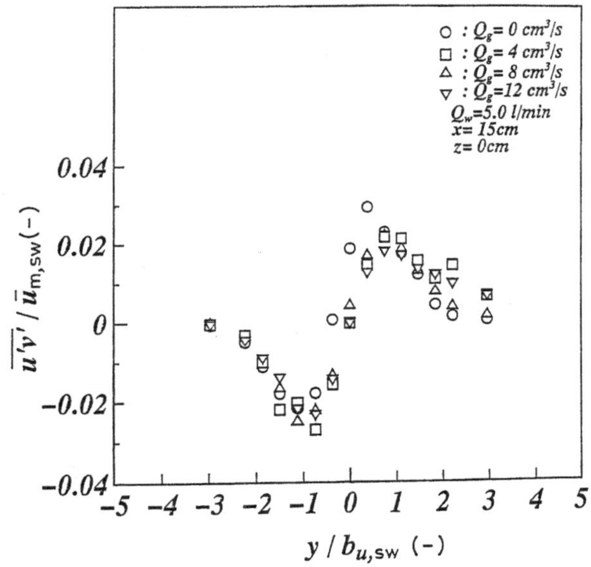


**Fig. 8.9** Vertical distributions of Reynolds shear stress

Above the $x$ axis ($y \geq 0$) the position at which $\overline{u'v'}/\overline{u}_{\mathrm{m,sw}}{}^2$ exhibits a peak shifts slightly upward as $Q_\mathrm{g}$ increases, but the distribution of $\overline{u'v'}/\overline{u}_{\mathrm{m,sw}}{}^2$ is not sensitive to $Q_\mathrm{g}$.

#### 8.1.3.5 Empirical Relations for Mean Velocity Components

(a) *Mechanism of upward deflection of water–air two-phase jet from the x axis*

The behavior of the water–air two-phase jets described in this section appears to be quite similar to a single-phase jet subjected to cross-flow, as briefly explained below. Consider a single-phase water jet issuing out of a pipe placed horizontally. If there exists a uniform cross-flow of water directed upward around the jet, the jet is bent in the cross-flow direction, i.e., in the vertical direction due to the drag force acting on the jet, as illustrated in Fig. 8.10a. The deflection of the jet in the vertical direction can be described by introducing a velocity ratio defined $u_0/U_\mathrm{crs}$ [21, 22] where $u_0$ is the velocity of water jet at the pipe outlet and $U_\mathrm{crs}$ is the cross-flow velocity of water flow.

Concerning the horizontal water–air two-phase jets, upward moving water flow is also induced by the buoyancy force acting on the bubbles (see Fig. 8.10b). In a similar manner to the single-phase water jet deflected by a cross-flow, a velocity ratio is introduced characterizing the upward moving water flow in the two-phase jet.

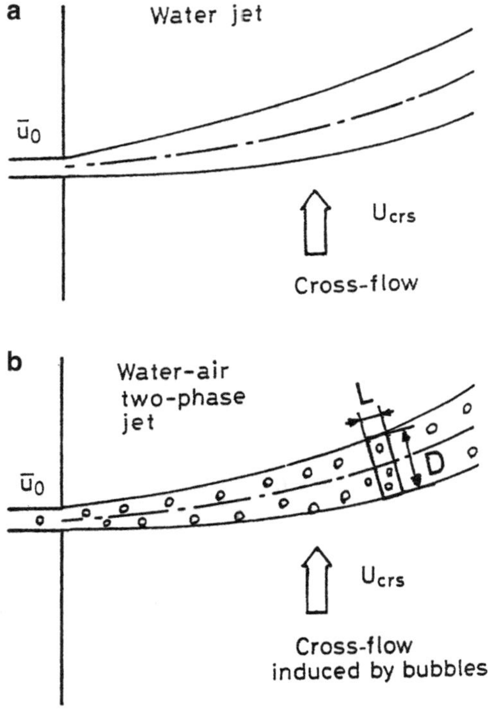


**Fig. 8.10** Schematic illustration of (**a**) water jet and (**b**) water–air two-phase jet subjected to cross-flow

(b) *Velocity scale for cross-flow induced by buoyancy force on bubbles*

Using the analysis of Themelis et al. [23] for horizontal bubbling jets as a base, consider a control volume shown in Fig. 8.11b having diameter $D$ and length L. The following four assumptions are introduced.

1. Gas holdup α in the control volume is very low.
2. The deflection of water–air two-phase jet is very small; $y_c/x \ll 1$.
3. Accordingly, the vertical mean velocity component of water flow, $\bar{v}$, is much smaller than the axial mean velocity component $\bar{u}$.
4. The mean bubble velocity $\overline{u_B}$ is equal to the axial mean velocity of water $\bar{u}$ in the control volume.
5. The buoyancy force acting on bubbles in the control volume is balanced by the drag force acting on the control volume.

Based on assumptions (2) and (4), the relationship between the buoyancy force and the drag force is expressed by [24]

$$(\rho_w - \rho_g) L g Q_g / \overline{u_B} = C_D D L \rho_m {U_{crs}}^2 / 2 \tag{8.5}$$

$$\rho_m = (1 - \alpha)\rho_w + \alpha \rho_g \tag{8.6}$$

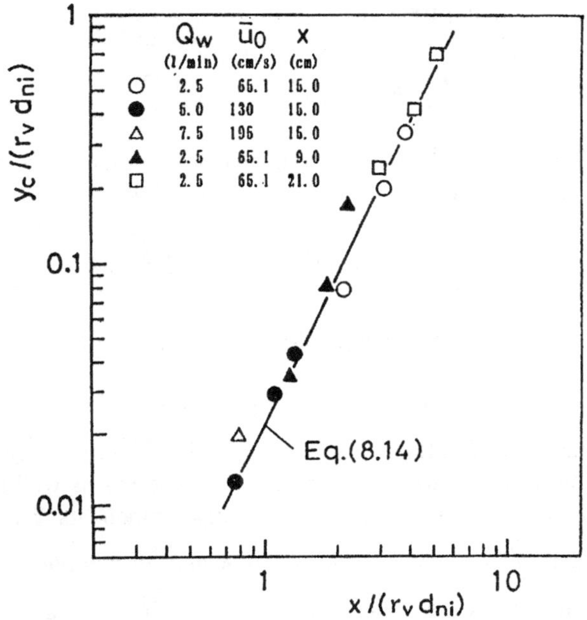


**Fig. 8.11** Correlation of the vertical deflection of water–air two-phase jet

where $\rho_w$ is the density of water, $\rho_g$ is the density of air ($\rho_g \ll \rho_w$), $g$ is the acceleration due to gravity, $C_D$ is the drag coefficient, $\rho_m$ is the mean density of water and air mixtures, and $U_{crs}$ is the cross-flow velocity.

Assumption (1) states that $\rho_m$ is nearly equal to $\rho_w$, hence, $U_{crs}$ reduces to

$$U_{crs} = [2gQ_g/(D\overline{u_B})]^{1/2} \tag{8.7}$$

In order to evaluate $\overline{u_B}$, we further assume that the momentum of fluid in the axial direction is preserved, as proposed by Crowe and Riesebieter [22, 24]. The momentum of gas is negligibly small compared to that of water, and hence, the following equation can be derived.

$$\rho_w {u_0}^2(\pi {d_{ni}}^2/4) = \rho_w \overline{u}^2(\pi D^2/4) \tag{8.8}$$

where $u_0$ is the velocity of water flow at the outlet of the pipe. Accordingly, $\overline{u}$ is given by

$$\overline{u} = (d_{ni}/D)u_0 \tag{8.9}$$

From assumption (3) $\overline{u_B}$ is expressed by

$$\overline{u_B} = (d_{ni}/D)u_0 \tag{8.10}$$

Combination of (8.8) with (8.10) gives

$$U_{\mathrm{crs}} = [2gQ_{\mathrm{g}}/(C_{\mathrm{D}}d_{\mathrm{ni}}u_{\mathrm{O}})]^{1/2} \tag{8.11}$$

The drag coefficient $C_{\mathrm{D}}$ is further assumed to be constant, and the cross-flow velocity $U_{\mathrm{crs}}$ is rewritten thus:

$$U_{\mathrm{crs}} = k[gQ_{\mathrm{g}}/(C_{\mathrm{D}}d_{\mathrm{ni}}u_{\mathrm{O}})]^{1/2} \tag{8.12}$$

where $k$ is a constant. The following velocity ratio $r_{\mathrm{v}}$, therefore, is introduced by putting $k = 1$:

$$r_{\mathrm{v}} = u_{\mathrm{O}}/U_{\mathrm{crs}} = u_{\mathrm{O}}/[gQ_{\mathrm{g}}/(d_{\mathrm{ni}}u_{\mathrm{O}})]^{1/2} \tag{8.13}$$

In a similar manner to Pratte and Baines [21, 22], this velocity ratio is used to correlate the upward deflection of water–air two-phase jets as well as the vertical distributions of the axial and vertical mean velocity components, $\overline{u}$ and $\overline{v}$.

(c) *Empirical relation for upward deflection of water–air two-phase jets*

Figure 8.11 illustrates the deflection of the centerline of water–air two-phase jets from the $x$ axis. All the measured values of $y_{\mathrm{c}}$ are satisfactorily correlated and approximated by the solid line, expressed by:

$$y_{\mathrm{c}}/(r_{\mathrm{v}}d_{\mathrm{ni}}) = 0.020\{x/(r_{\mathrm{v}}d_{\mathrm{ni}})\}^{1.8} \tag{8.14}$$

This empirical relation is valid for

$$4 < r_{\mathrm{v}} < 24 \tag{8.15}$$

$$0.7 < x/(r_{\mathrm{v}}d_{\mathrm{ni}}) < 5 \tag{8.16}$$

It should be noted that the upward deflection $y_{\mathrm{c}}$ is zero in a single-phase water jet. Accordingly, a value of zero should be substituted for $r_{\mathrm{v}} > 24$.

(d) *Empirical relations for vertical distributions of $\overline{u}$ and $\overline{v}$*

The vertical distribution of $\overline{u}$ can be described if the maximum value $\overline{u}_{\mathrm{m}}$ and the half-value radius $b_{\mathrm{u}}$ are known. This section focuses on $\overline{u}$ and only the maximum value of $\overline{v}$ is discussed.

### Maximum Values of Velocity Components

- Single-phase water jet ($Q_{\mathrm{g}} = 0\,\mathrm{cm}^3/\mathrm{s}$)

The maximum value of the axial mean velocity component, $\overline{u}_{\mathrm{m,sw}}$, for the single-phase water jet, which occurs at the axis of the jet, is nondimensionalized by the water velocity at the pipe outlet, $u_{\mathrm{O}}$. The results are plotted against $x/d_{\mathrm{ni}}$ in Fig. 8.12.

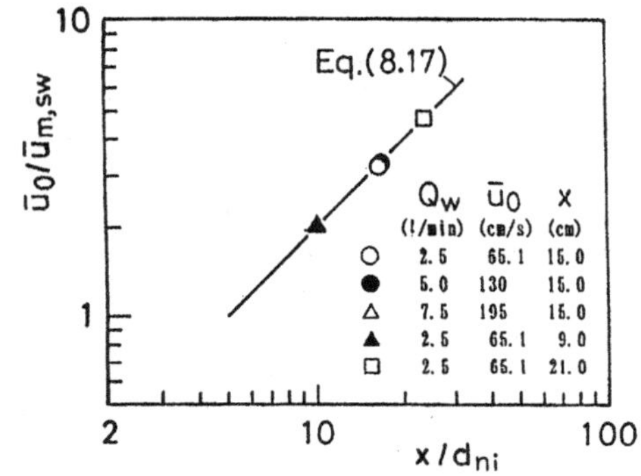


**Fig. 8.12** Correlation of the maximum value of axial mean velocity in single-phase water jet

The distribution of $\bar{u}_{\mathrm{m,sw}}$ follows the widely known inverse proportionality to the axial distance $x$ [22, 25] and can be expressed as:

$$u_0/\overline{u}_{\mathrm{m,sw}} = 0.20x/d_{\mathrm{ni}} \tag{8.17}$$

$$5 < x/d_{\mathrm{ni}} < 30 \tag{8.18}$$

On the other hand, the maximum value of the vertical mean velocity component, $\overline{v}_{\mathrm{m,sw}}$ does not appear on the axis of the jet but at a certain distance from the $x$ axis. The relationship between $\overline{v}_{\mathrm{m,sw}}$ and $\overline{u}_{\mathrm{m,sw}}$ can be approximated by [22]:

$$\overline{v}_{\mathrm{m,sw}}/\overline{u}_{\mathrm{m,sw}} = 0.019 \tag{8.19}$$

- Water–air two-phase jet

The measured values of $\overline{u}_{\mathrm{m}}$ are nondimensionalized by $u_0$ and $r_{\mathrm{v}}$, and plotted against $x/(r_{\mathrm{v}}d_{\mathrm{ni}})$ in Fig. 8.13. The solid line, drawn through the mean of the measured values, is expressed by:

$$u_0/(\overline{u}_{\mathrm{m}}r_{\mathrm{v}}) = 0.22\{x/(r_{\mathrm{v}}d_{\mathrm{ni}})\}^{1.25} \tag{8.20}$$

The applicable range of (8.20) is given by (8.15) and (8.16). Equation (8.17) should be used for a velocity ratio $r_{\mathrm{v}}$ over 24.

Figure 8.14 illustrates the maximum value of the vertical velocity component for water–air two-phase jets. The measured values can be correlated in the same manner as that for $\overline{u}_{\mathrm{m}}$. The solid line indicates the following empirical relation:

$$u_0/(\overline{v}_{\mathrm{m}}r_{\mathrm{v}}) = 2.5\{x/(r_{\mathrm{v}}d_{\mathrm{in}})\}^{0.80} \tag{8.21}$$

$$4 < r_{\mathrm{v}} < 24 \tag{8.15}$$

$$0.15 < x/(r_{\mathrm{v}}d_{\mathrm{ni}}) < 5 \tag{8.22}$$

Equation (8.19) should be used for $r_{\mathrm{v}} > 24$.

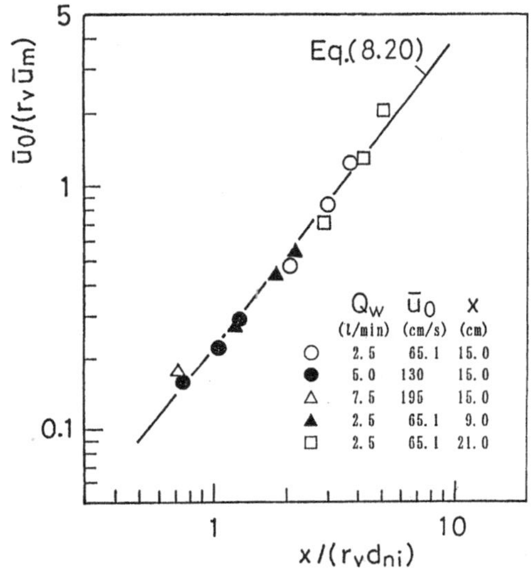


**Fig. 8.13** Correlation of the maximum value of axial mean velocity in water–air two-phase jet

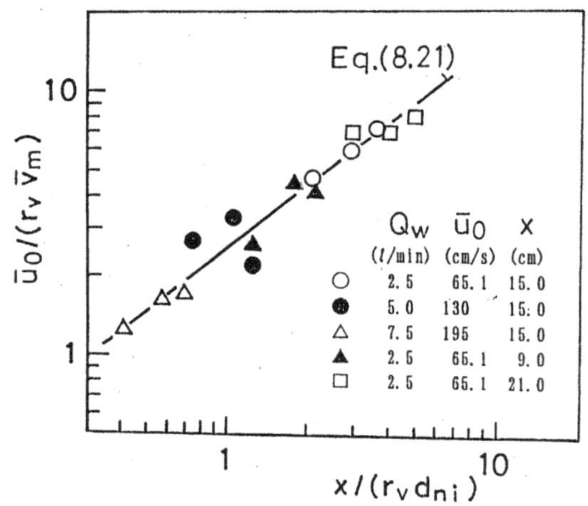


**Fig. 8.14** Correlation of the maximum value of vertical mean velocity in water–air two-phase jet

Half-Value Radius

The half-value radius of the vertical distribution of the axial mean velocity component is denoted by $b_u$. The measured values shown in Fig. 8.15 can all be approximated by the solid line of the form:

$$b_u/(r_v d_{ni}) = 0.12\{x/(r_v d_{ni})\}^{0.92} \tag{8.23}$$

The applicable range of (8.23) is given by (8.15) and (8.22). When $r_v$ is greater than 24, the following empirical relation, which was originally derived for single-phase jets [22], should be used:

$$b_u = 0.10x \tag{8.24}$$

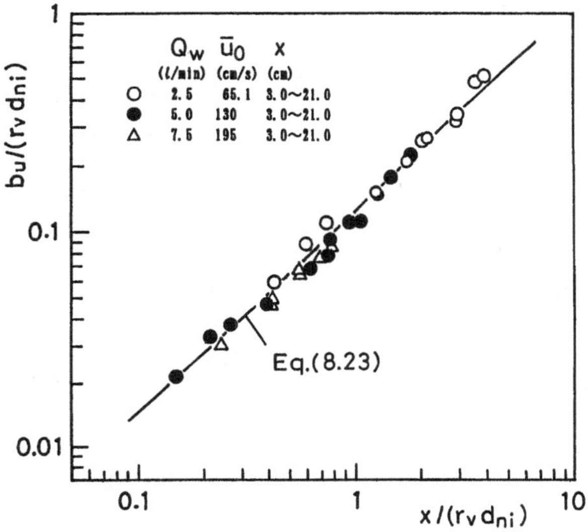


**Fig. 8.15** Correlation of the half-value radius of the vertical distribution of axial mean velocity in water–air two-phase jet

## *8.1.4 Summary*

The velocity measurements for horizontal water–air two-phase jets have been described. These results are used to understand the behavior of molten steel jet issuing out of the SEN subjected to Ar gas injection in a continuous casting mold. The main findings can be summarized as follows.

- Deflection of water–air two-phase jet

A water–air two-phase jet generated with a horizontally placed pipe was deflected upward due to the buoyancy force acting on bubbles in the jet. A velocity ratio, $r_v$, is introduced to correlate the behavior of the two-phase jet by referring to single-phase jets subjected to cross-flow. An empirical equation, (8.14), is proposed for the deflection of the centerline of the jet from the horizontal plane.

- Axial and vertical mean velocity components

Empirical relations are derived for the maximum values of the axial and vertical velocity components, $\overline{u}_m$ and $\overline{v}_m$, as well as the half-value radius of the axial velocity component, $b_u$. The distributions of $\overline{u}_m$, $\overline{v}_m$, and $b_u$ could be predicted by the empirical relations in (8.20), (8.21), and (8.23) when the velocity ratio $r_v$ falls between 4 and 24. When $r_v$ is greater than 24, empirical relations originally derived for single-phase jets should be used to predict the three quantities in the water–air two-phase jets.

- Axial and vertical turbulence components

The effects of the gas flow rate $Q_g$ on the axial and vertical turbulence components are relatively small under the experimental conditions described. The root-mean-square values of these turbulence components agree approximately with those for single-phase water jets.

As a first step toward understanding horizontal water–air two-phase jets, only the water flow characteristics in the jet are considered here. The behavior of small bubbles, which is responsible for pinhole defects, has not been discussed.

The velocity ratio $r_v$ is not a function of the physical properties of liquid and gas, as can be seen from (8.13). Accordingly, the aforementioned empirical relations describing the behavior of horizontal water–air two-phase jets appear to be applicable to molten steel–Ar two-phase jets as long as the argon gas flow rate is much smaller than the molten steel flow rate.

## 8.2 Mold Powder Entrapment

### *8.2.1 Overview*

Figure 8.16 shows a schematic of the current continuous casting mold practice. The entrapment of mold powder into molten steel strongly influences the quality of the steel product. The mechanism of entrapment has therefore been extensively investigated [26–33]. Considerable effort has also been devoted to preventing it [34–36]. The following three types of entrapment phenomena have been suggested [28]:

(a) Karman's vortex streets formed behind the immersion nozzle entrap the mold powder placed on the meniscus of the mold [27–29, 31, 33].
(b) High shear stress between the reversing molten steel flow reflected from the narrow face and mold powder, induces shear flow instability to cause entrapment of the mold powder [27, 30, 32, 33].
(c) Argon gas injected into the immersion nozzle to prevent clogging of alumina on to the inner wall of the nozzle disturbs the molten steel/mold powder interface and causes the mold powder entrapment.

Unfortunately, there is yet no clear explanation of the critical condition for each of the above entrapment phenomena to occur in real processes. Among these phenomena, the second has been most actively investigated, probably because of the ease of carrying out model experiments [27, 30, 32] The description in this section similarly focuses on this mode of powder entrapment.

A number of researchers have considered steady flow in the mold and conducted model experiments using water and silicone oil. The flow field in the mold can be illustrated schematically as in Fig. 8.17a. The silicone oil is pushed away from the narrow face of the mold by the reversing flow reflected from the face [27]. Silicone oil droplets are pulled from the silicone oil layer into the water layer in the presence of high shear stress. Such a situation may not be realistic in the real processes

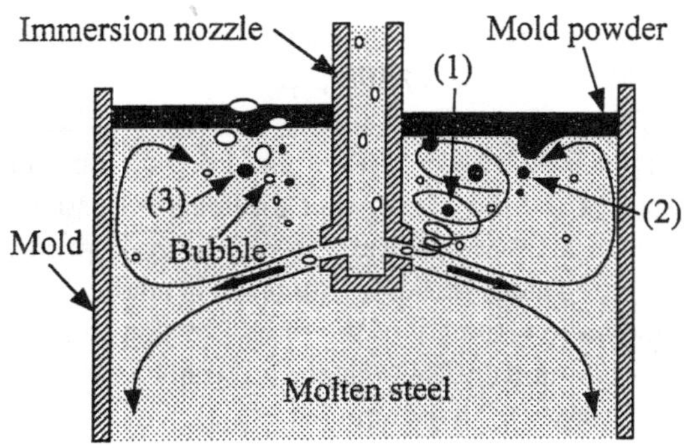


**Fig. 8.16** Schematic of mold powder entrapment

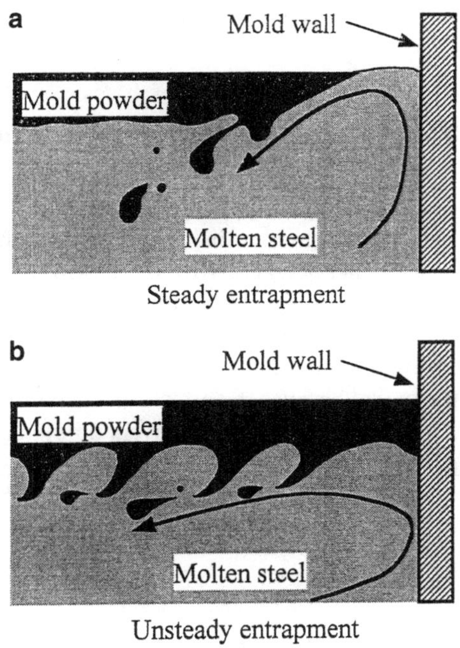


**Fig. 8.17** Entrapment of mold powder due to reversing flow. (**a**) Steady entrapment, (**b**) Unsteady entrapment

because when the mold powder is pushed away from the narrow face, it cannot be supplied between the solidifying steel and the oscillating mold. Mold powder is known to be supplied continuously even when mold powder entrapment occurs.

It is therefore more plausible to assume that the mold powder entrapment occurs under unsteady conditions, as shown in Fig. 8.17b. Existing studies on the slab casting mold have shown that the steel flow in the mold is essentially unsteady due to uneven flows discharging out of the ports of the immersion nozzle [37–39]. Mold powder entrapment under unsteady flow conditions seems to be closely associated with the Kelvin–Helmholtz instability, hereinafter abbreviated as KHI [40,41]. Some researchers have suggested that KHI is one of the causes of the onset of mold powder entrapment and that a critical velocity difference can be predicted by the original KHI theory or an infinitely extending interface [33,42]. However, it is not clear whether the original KHI theory is applicable to a finite interface between mold powder and molten steel.

The original KHI theory [40, 41] implies that the critical velocity difference responsible for the onset of mold powder entrapment is influenced by the interfacial tension and the densities of molten steel and mold powder. In addition to these parameters, the kinematic viscosity of mold powder must be taken into consideration. As a first step, particular attention is paid, in the description here, to the effect of the kinematic viscosity of the mold powder on the entrapment, and model experiments carried out using salt water and silicone oils.

In order to determine the critical velocity difference for the initiation of KHI, it is necessary to measure the velocities of salt water flow and silicone oil flow near the silicone oil/salt water interface. A particle imaging velocimetry (PIV) technique is described as it is considered the most adequate for this purpose.

## *8.2.2 Experimental Apparatus and Procedure*

The experimental apparatus is schematically shown in Fig. 8.18. The vessel made of transparent acrylic resin had an inner length of 1,000 mm, an inner width of 100 mm, and an inner height of 70 mm. These dimensions were decided with reference to the current continuous slab casting molds of steel in practice. Specifically, the model was approximately half the size of real slab casting molds. Only the geometrical similitude for the inner length and the inner width were taken into consideration in the model because the effects of the physical properties of upper and lower liquids and the depths of the two layers on the onset of the KHI could be evaluated from existing correlations. Salt water and silicone oils with different physical properties

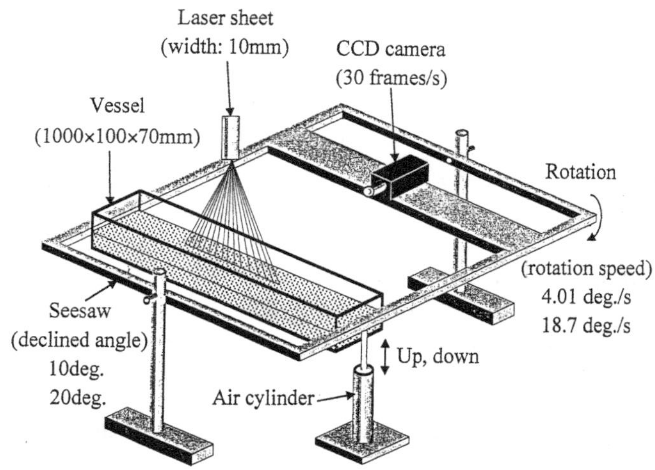


**Fig. 8.18** Schematic of apparatus for mold powder entrapment

**Table 8.2** Physical properties of salt water and silicone oil at 298 K

| Liquid | Kinematic viscosity $\nu_1$, $\nu_2$ (mm²/s, cSt) | Density $\rho_1$, $\rho_2$ (kg/m³) | Interfacial tension on $\sigma_{12}$ (mN/m) |
|---|---|---|---|
| Salt water | 1.0 | 1,013 | |
| Silicone oil 2 | 2.0 | 873 | 52.7 |
| Silicone oil 10 | 10 | 935 | 52.7 |
| Silicone oil 50 | 50 | 960 | 52.7 |
| Silicone oil 100 | 100 | 965 | 53.0 |
| Silicone oil 350 | 350 | 970 | 53.0 |

were used. The temperatures of the two liquids in the vessel were 298 K. The physical properties of the model liquids at this temperature are listed in Table 8.2. The actual mold powder layer did not melt completely, and only the lower part was in the liquid phase. Hence, the upper part was assumed to behave like a solid wall. The upper end of the silicone oil layer therefore was covered with a flat solid wall in order to model the nonmelted mold powder. The depth ratio of the salt water layer to the silicone oil layer was varied from 1:1 to 5:1.

Solid particles having diameters ranging from 75 to 150 μm were dispersed in the salt water layer as tracers for PIV measurements. The density of the particles was 1,013 kg/m³, and accordingly, the density of the salt water was adjusted to the same value. Also, particles having the same density as silicone oil 10 were used as tracers for the measurement of flow velocity in the oil. Each silicone oil type was tagged with a number. For example, silicone oil 10 implies that the kinematic viscosity of the oil is 10 mm²/s. Adequate tracer particles are available for silicone oil 10 and 100, but nothing exists at present for the other types.

A laser sheet having a width of 10 mm was used for the illumination of the flow field to make the movements of tracer particles visible. After confirming the particles dispersed in the still salt water layer and silicone oil layer did not move, the vessel was declined 10 or 20° to the horizontal at a rotation velocity of 0.070 or 0.327 rad/s, using an air cylinder. A CCD camera mounted on the frame of the vessel was used for recording the movements of the particles at 30 frames/s. A velocity difference for the onset of shear flow instability was measured with PIV. The cross-correlation method was used to process the video images.

The velocity of silicone oil flow near the interface, $V_1$, is not usually measured as described above, and accordingly, it was calculated from the following equation of continuity:

$$V_1 = -(H_2/H_1)\,V_2 \tag{8.25}$$

where $H_1$ is the depth of the silicone oil layer, $H_2$ is the depth of the salt water layer, and $V_2$ is the velocity of salt water flow near the interface, as can be seen in Fig. 8.19.

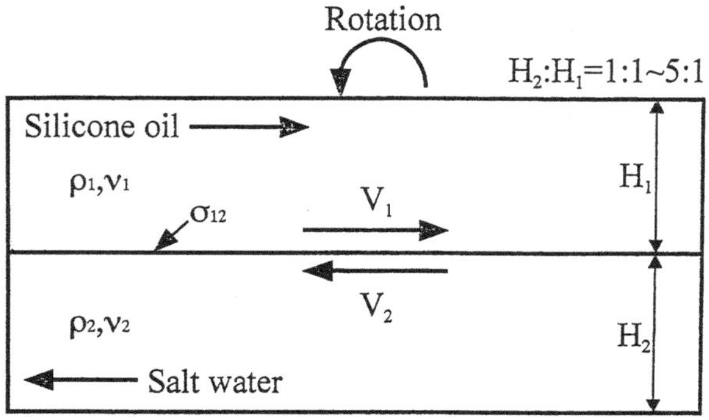


**Fig. 8.19** Symbols used in this study

## *8.2.3 Some Aspects of Kelvin–Helmholtz Instability*

### 8.2.3.1 Critical Velocity Difference for the Onset of Kelvin–Helmholtz Instability

The original theory on KHI was concerned with a shear flow instability occurring at an interface between two horizontal liquid layers extending infinitely. A critical velocity difference for the onset of the KHI is expressed by [40, 41]

$$(V_{1\mathrm{cr}} - V_{2\mathrm{cr}})^4 = 4\sigma_{12} g\,(\rho_2 - \rho_1)\,(\rho_1 + \rho_2)^2 \,/\, \left({\rho_1}^2 {\rho_2}^2\right) \qquad (8.26)$$

where $\rho_1$ is the density of the upper liquid, $\rho_2$ is the density of the lower liquid, $\sigma_{12}$ is the interfacial tension, and the subscript cr denotes a critical value. This equation cannot be applied directly to the flow filed considered here because the flow field in the actual mold is confined by the vertical walls of the mold and nonmelted mold powder. However, (8.26) provides a milestone for understanding the influential parameters for the entrapment of mold powder.

According to (8.26), the interfacial tension, $\sigma_{12}$, and density difference, $\Delta\rho(= \rho_2 - \rho_1)$ appear to play an important role on the entrapment of mold powder. Although the kinematic viscosity of mold powder is not included in the equation, it varies over a wide range in the actual processes depending on the chemical compositions of molten steel and casting speed. Thus, mold powder having a higher kinematic viscosity is expected to suppress the KHI. The effect is therefore considered of kinematic viscosity of mold powder on the onset of mold powder entrapment in addition to the two parameters, $\sigma_{12}$ and $\Delta\rho$.

Existing KHI theories for fluids contained in a vessel of rectangular cross-section are briefly reviewed (see Fig. 8.20). A critical velocity difference for the onset of KHI in the flow field illustrated in Fig. 8.20a was given by Milne-Thompson [43] in the form:

$$(V_{1\mathrm{cr}} - V_{2\mathrm{cr}})^2 = (\rho_2 - \rho_1)\,g\,(H_1/\rho_1 + H_2/\rho_2) \qquad (8.27)$$

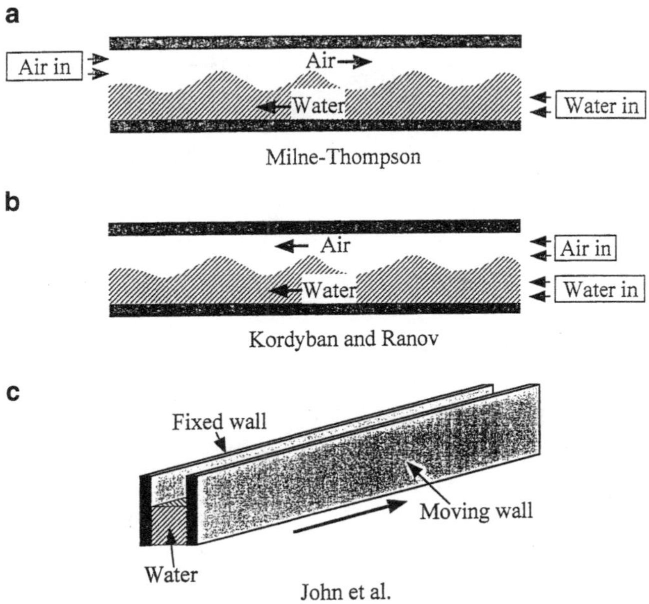


**Fig. 8.20** Models for Kelvin–Helmholtz instability. (**a**) Milne-Thompson, (**b**) Kordyban and Ranov, (**c**) John et al.

where $g$ is the acceleration due to gravity, and $H_1$ and $H_2$ are the silicon oil layer depth and salt water layer depth, respectively. In deriving (8.27), it was assumed that the effect of surface tension is negligible and that $KH_1 \ll 1$ and $KH_2 \ll 1$, where $K(= 2\pi/\lambda)$ is the wave number.

Kordyban and Ranov [44] considered the flow field shown in Fig. 8.20b and derived the following equation for the critical velocity difference:

$$(V_{1\mathrm{cr}} - V_{2\mathrm{cr}})^2 = \frac{\rho_2}{\rho_1} \times \frac{g}{K} \times \frac{1}{\coth(KH_1 - 0.9) + 0.45\coth^2(KH_1 - 0.9)} \tag{8.28}$$

where $\lambda$ is the wavelength. In addition, John et al. [45] considered the flow field shown in Fig. 8.20c and proposed

$$(V_{1\mathrm{cr}} - V_{2\mathrm{cr}})^2 = \frac{\rho_1 + \rho_2}{\rho_1\rho_2}\left((\rho_2 - \rho_1)\frac{g}{K} + \sigma_{12}K\right) \tag{8.29}$$

This flow field is somewhat different from that considered here, Fig. 8.17b.

#### 8.2.3.2 Wavelength and Amplitude of Instability Wave

The following equation has been compared with measured values of the wavelength $\lambda$ [42,45]:

$$\lambda = 2/\mathrm{K} \tag{8.30}$$

$$K = \left[(\rho_2 - \rho_1)\, g\sigma^{-1}\right]^{1/2} \tag{8.31}$$

where $K$ is the wave number. The amplitude of instability waves, $A_\mathrm{m}$, is given by [46]

$$A_\mathrm{m} = 1/K \tag{8.32}$$

## 8.2.4 *Experimental Results*

### 8.2.4.1 Visualized Flow Field and Velocity Vectors

Figures 8.21 and 8.22 show visualized images taken by a CCD camera from the side and top of the experimental apparatus. Instability does not appear near the end walls of the vessel due to end effect. In Fig. 8.21, a silicone oil droplet can be seen after the rupture of instability waves. It is evident in Fig. 8.22 that the instability initiates over the entire horizontal cross-section. Figure 8.23 shows velocity vectors of salt water flow around the silicone oil/salt water interface just after the onset of KHI. Figure 8.24 shows some examples of the histories of the salt water flow velocity measured with the PIV. The mean velocity ranges from 0 to 32 cm/s, and the acceleration ranges from 0 to 20 $\mathrm{cm/s^2}$. It can be concluded that PIV is adequate for the measurement of such an unsteady velocity field.

### 8.2.4.2 Critical Velocity Difference for the Onset of KHI

Figure 8.25 shows the velocity distribution in the vertical cross-section for silicone oil 10. Both the distributions of salt water flow and silicone oil flow are approximately uniform at every time of measurement. On the other hand, the velocity distributions for silicone oil 100 shown in Fig. 8.26 are not uniform because of its high kinematic viscosity.

Figures 8.27 and 8.28 show the critical velocity differences describing the onset of KHI for silicone oil 10 and 100, respectively. The results shown with open circles were determined by using both the measured values of $V_{1\mathrm{cr}}$ and $V_{2\mathrm{cr}}$, while open triangles were determined by using the measured values of $V_{2\mathrm{cr}}$ and calculated value of $V_{1\mathrm{cr}}$ based on (8.25). For every salt water layer depth, the two sets of results (i.e., the two symbols) agree with each other in Fig. 8.27. As the kinematic viscosity of silicone oil was increased from 10 to 100 $\mathrm{mm^2/s}$, a discrepancy appears between the two symbols as seen in Fig. 8.28, but it is limited to about 15%. A number of researchers [27, 30, 32] have also observed a critical velocity difference of approximately 20 cm/s for water–silicone oil systems shown in Fig. 8.17a.

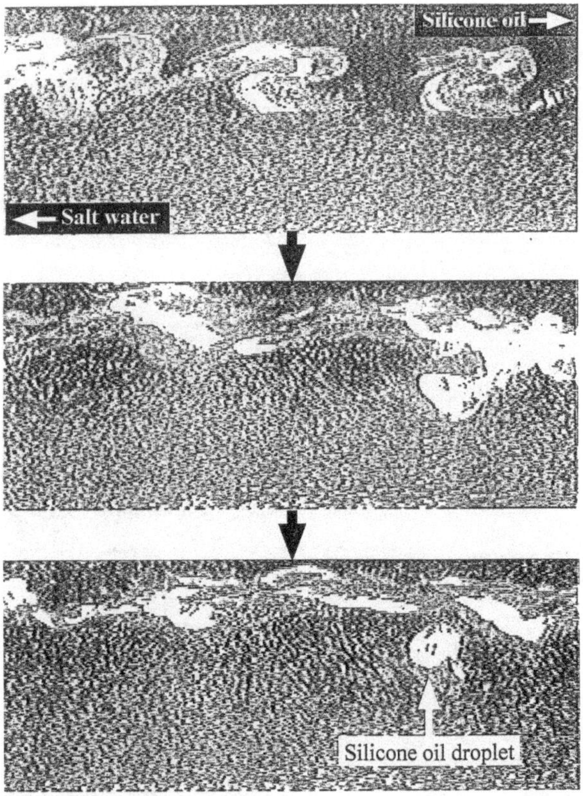


**Fig. 8.21** Entrapment of silicone oil droplet observed from side of the vessel (silicone oil 2, depth ratio ($H_2 : H_1$): 2:1, decline angle: 20°, rotation speed: 4.01 deg./s)

In the following, data on $V_{2cr}$ are presented for comparison with the calculations. Data on $V_{1cr}$ are not considered as they were obtained only for silicone oils 10 and 100.

#### 8.2.4.3 Comparison of Measured and Calculated Critical Salt Water Flow Velocity

For the system utilizing silicone oil 2, the critical salt water velocity, $V_{2cr}$, was essentially independent of the inclination angle and the rotation speed of the vessel, as shown in Table 8.3. That is, the acceleration of the salt water flow did not affect the critical salt water velocity. Accordingly, the mean value of the measured critical salt water flow velocities was calculated and presented in Table 8.4 and Fig. 8.29 for different depth ratios. The following relationship was assumed in calculating (8.27)–(8.29):

$$V_{1cr} = -V_{2cr} \tag{8.33}$$

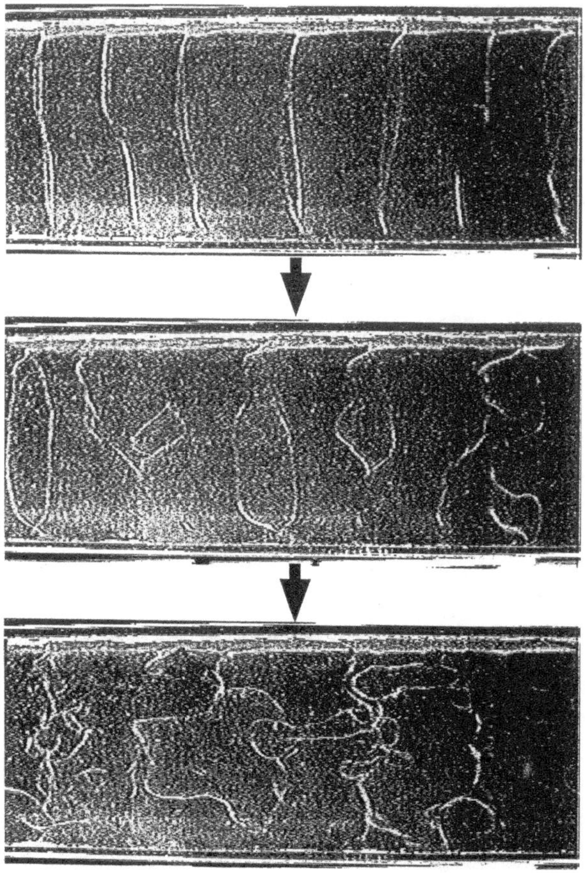

**Fig. 8.22** Entrapment of silicone oil droplet observed from top of the vessel (silicone oil 2, depth ratio ($H_2 : H_1$)1:1, decline angle: 20°, rotation speed: 18.7 deg./s)

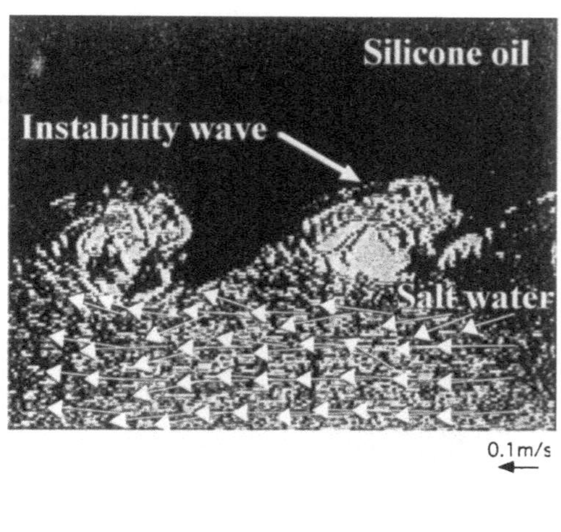


**Fig. 8.23** Velocity vectors measured with PIV (silicone oil 2, depth ratio ($H_2 : H_1$): 1:1, decline angle: 20°, rotation speed: 18.7 deg./s)

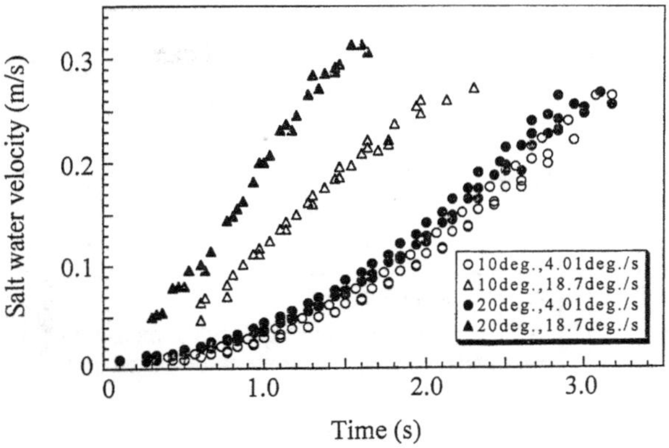


**Fig. 8.24** Histories of salt water velocity near interface (silicone oil 2, depth ratio ($H_2$ : $H_1$): 1:1)

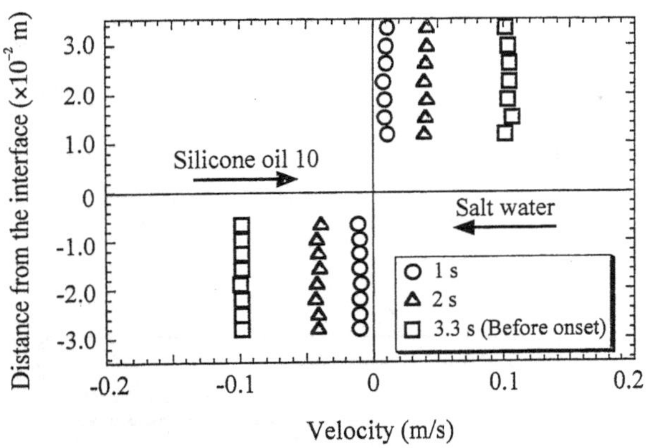


**Fig. 8.25** Change in velocity distribution (silicone oil 10, depth ratio ($H_2$ : $H_1$): 1:1, decline angle: 10°, rotation speed: 4.01 deg./s)

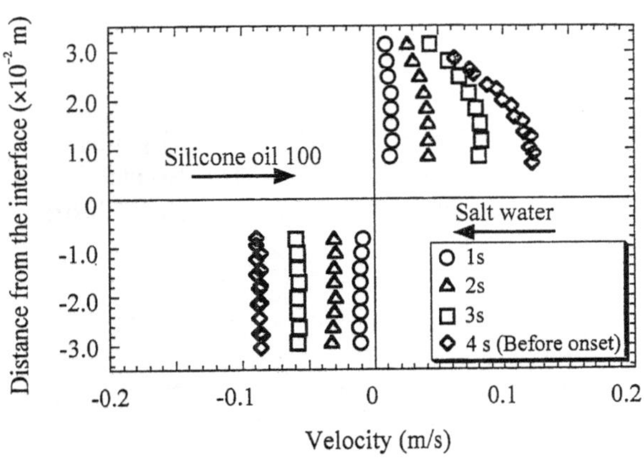


**Fig. 8.26** Change in velocity distribution (silicone oil 100, depth ratio ($H_2$ : $H_1$): 1:1, decline angle: 10°, rotation speed: 4.01 deg./s)

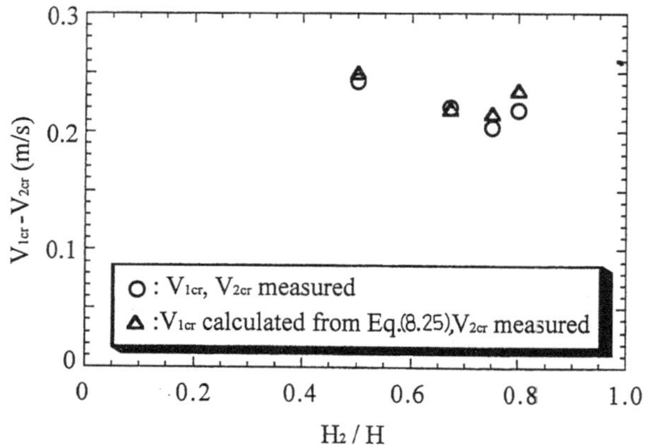


**Fig. 8.27** Critical velocity difference as a function of salt water depth (silicone oil 10, decline angle: 20°, rotation speed: 18.7 deg./s)

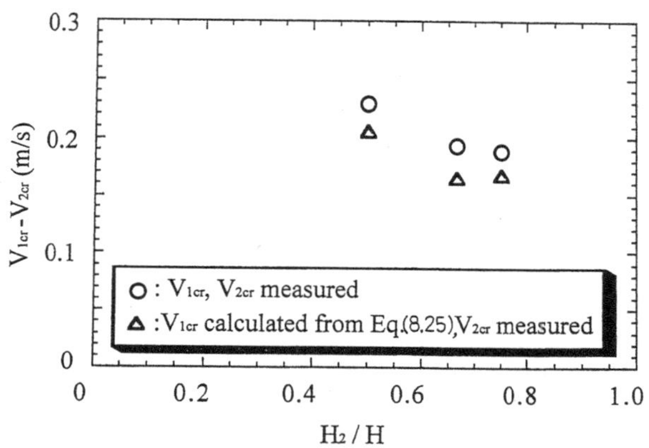


**Fig. 8.28** Critical velocity difference as a function of salt water depth (silicone oil 100, decline angle: 20°, rotation speed: 18.7 deg./s)

Also, in calculating (8.28) and (8.29), $K$ was evaluated by substituting the measured values of $\lambda$. A relationship proposed by Asai [47] is also included in Fig. 8.29. This relations was originally derived for the droplet formation illustrated in Fig. 8.17a, and expressed as,

$$(V_{1cr} - V_{2cr})^4 = 48\sigma_{12}g(\rho_2 - \rho_1)/{\rho_1}^2 \tag{8.34}$$

The mean value of the critical salt water velocity is satisfactorily approximated by (8.27) for the salt water depth ratio $H_2/H_1$ up to around 0.75.

There is a good agreement in Fig. 8.30 between the measured values of critical salt water velocity and (8.27) for silicone oil 10. Equations (8.26) and (8.28) are not suitable for correlating the measured data here. Equation (8.27) may therefore be

**Table 8.3** Critical velocities of salt water layer, $|V_{2cr}|$ (silicone oil 2) ($\times 10^{-2}$m/s)

| | $\lvert V_{2cr}\rvert$ measured with PIV ($\times 10^{-2}$ m/s) | | | |
|---|---|---|---|---|
| | Decline angle: 10° | | Decline angle: 20° | |
| $H_2 : H_1$ | 4.01 deg./s | 18.7 deg./s | 4.01 deg./s | 18.7 deg./s |
| 1:1 | 16 | 16 | 16 | 16 |

**Table 8.4** Critical velocities of salt water layer, $|V_{2cr}|$ (silicone oil 2) ($\times 10^{-2}$ m/s)

| $H_2 : H_1$ | $\lvert V_{2cr}\rvert$ (PIV) | Original (8.26) | (8.27) | (8.28) | (8.29) | (8.34) |
|---|---|---|---|---|---|---|
| 1:1 | 16 | 9.5 | 16 | 11 | 9.5 | 13 |
| 2:1 | 10 | 6.4 | 11 | 7.5 | 6.3 | 7.2 |
| 3:1 | 7.1 | 4.8 | 7.9 | 5.2 | 4.6 | 4.9 |
| 4:1 | 4.3 | 3.8 | 6.3 | 4.0 | 3.8 | 3.8 |
| 5:1 | 3.7 | 3.2 | 5.2 | 3.1 | 3.2 | 3.1 |

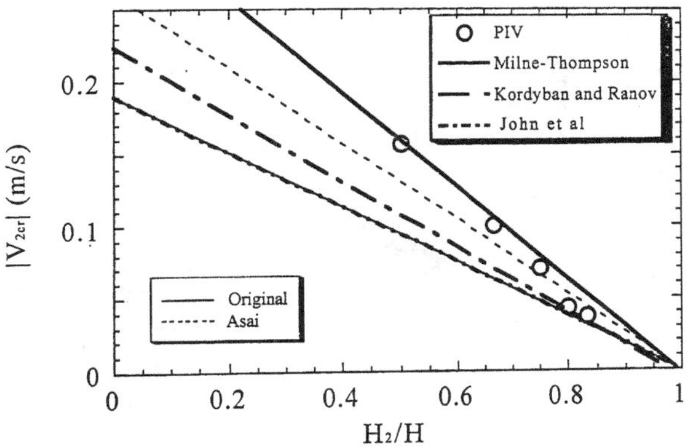


**Fig. 8.29** Critical velocities of salt water layer: silicone oil 2

considered most appropriate for predicting the critical salt water velocity, although $KH_1$ and $KH_2$ are on the order of unity, and hence, the conditions $KH_1 \ll 1$ and $KH_2 \ll 1$ are not satisfied.

The agreement between measurement and (8.27) is not satisfactory for depth ratios of 0.8 and 0.833 in Fig. 8.29. In addition, for silicone oils 50, 100, and 350 the salt water attached to the top wall as the instability grew for the same depth ratios. That is, when the amplitude of the KHI is larger than $H_1$, the interfacial instability is disturbed by the top wall.

Figure 8.31 demonstrates that all the measured values of the critical salt water velocity can be predicted by (8.27) regardless of the kinematic viscosity of silicone oil. Yamasaki et al. [30] and Komai et al. [42] also observed that the critical velocity difference is a weak function of the kinematic viscosity of silicone oil.

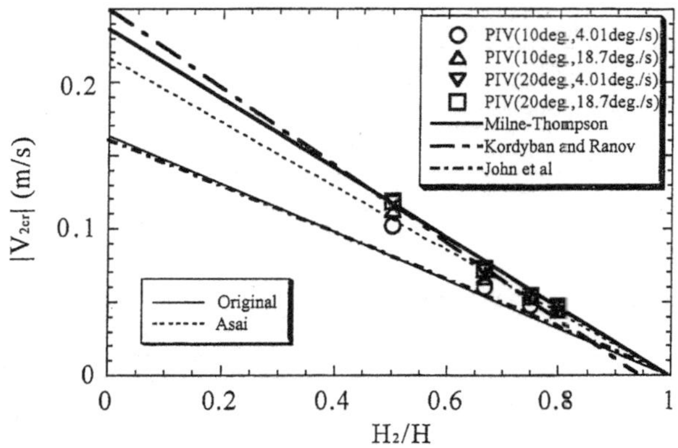


**Fig. 8.30** Critical velocities of salt water layer: silicone oil 10

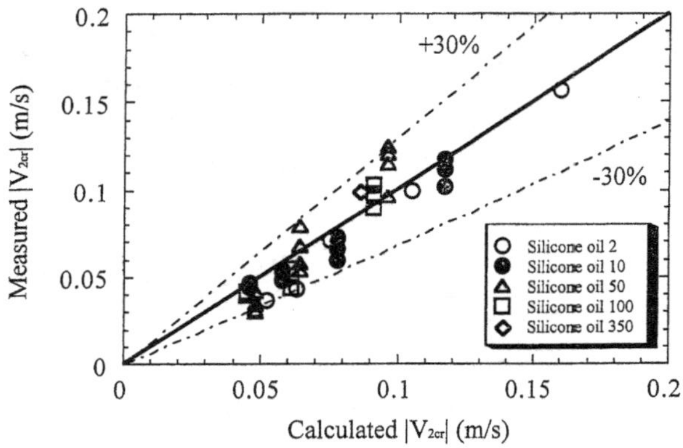


**Fig. 8.31** Comparison between calculated $V_{2cr}$ from Milne-Thomson equation and measured $|V_{2cr}|$

#### 8.2.4.4 Wavelength and Amplitude of KHI

Figure 8.32 shows that the wavelength decreases with an increase in the salt water layer depth. When the kinematic viscosity of silicone oil is low, say $2\,\text{mm}^2/\text{s}$, the measured values of wavelength $\lambda$ can be approximated by (8.30). As the kinematic viscosity of silicone oil increased, the wavelength increased (see Table 8.5), and (8.30) overestimated the $\lambda$. Yamasaki et al. [30] observed that the diameter of a liquid paraffin droplet became large with an increase in the kinematic viscosity, which is consistent with the above finding on the wavelength behavior.

The amplitude of KHI is a function of the depth ratio $H_2/H_1$ and the kinematic viscosity of silicone oil, as shown in Table 8.6. Figure 8.33 shows that (8.32) cannot predict the amplitude.

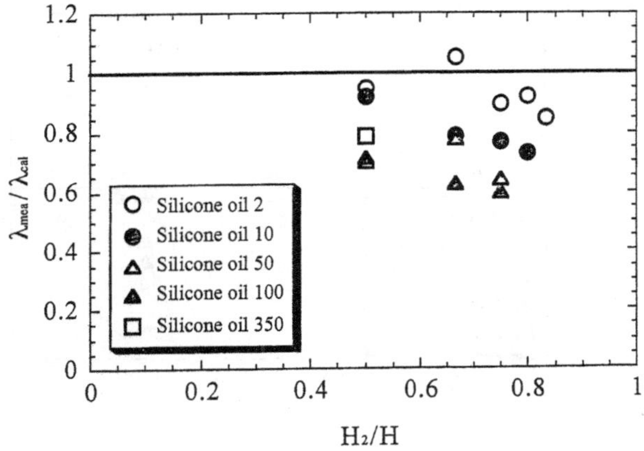


**Fig. 8.32** Wave length $\lambda$

**Table 8.5** Wavelength ($\times 10^{-2}$ m)

| $H_2 : H_1$ | Silicone oil 2 | 10 | 50 | 100 | 350 |
|---|---|---|---|---|---|
| 1:1 | 3.7 | 4.8 | 4.5 | 4.8 | 5.6 |
| 2:1 | 4.1 | 4.1 | 5.0 | 4.2 | |
| 3:1 | 3.5 | 4.0 | 4.1 | 4.0 | |
| 4:1 | 3.6 | 3.8 | | | |
| 5:1 | 3.3 | | | | |

**Table 8.6** Amplitude ($\times 10^{-2}$ m)

| $H_2 : H_1$ | Silicone oil 2 | 10 | 50 | 100 | 350 |
|---|---|---|---|---|---|
| 1:1 | 1.0 | 1.2 | 1.1 | 1.1 | 1.0 |
| 2:1 | 1.0 | 0.96 | 1.1 | 0.80 | |
| 3:1 | 1.0 | 0.80 | 0.70 | 0.60 | |
| 4:1 | 1.0 | 0.79 | | | |
| 5:1 | 0.90 | | | | |

#### 8.2.4.5 KHI-Induced Mold Powder Entrapment in Continuous Casting Mold

A simple scaling analysis can be used to assess the possibility of mold powder entrapment due to KHI. Assume that the depth of the melted mold powder layer is 0.05 m and that the depth of the molten steel layer is equal to the thickness of the reversing molten steel flow, say 0.10 m. Substituting $\rho_1 = 3{,}000\,\mathrm{kg/m^3}$, $\rho_2 = 7{,}000\,\mathrm{kg/m^3}$, $H_1 = 0.05$ m, and $H_2 = 0.10$ m into (8.27), the critical velocity difference ($V_{1cr} - V_{2cr}$) becomes approximately 1 m/s. If $H_1$ is changed from 0.05 to 0.02 m and $H_2$ is changed from 0.1 to 0.05 m, $V_{1cr} - V_{2cr}$ becomes approximately 0.6 m/s. Such a steel flow velocity fluctuation frequently occurs in the real continuous casting mold [37–39]. Thus it can be concluded that the possibility of mold powder entrapment due to KHI is considerably high, as long as the onset of KHI is based on (8.27).

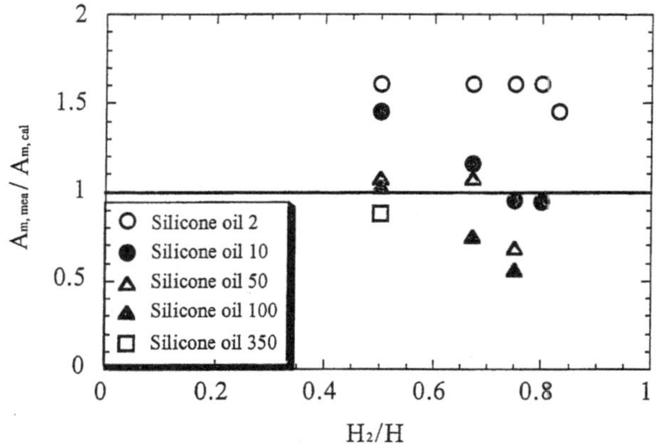


**Fig. 8.33** Amplitude $A_{\mathrm{m}}$

It should, however, be noted that the interfacial tension between mold powder and molten steel is much higher than that between silicone oil and salt water. Further investigation is necessary to provide a clear explanation of the onset of KHI in real continuous casting molds.

### *8.2.5 Summary*

When the thickness of the silicone oil layer is larger than the amplitude of waves caused by shear flow instability at the silicone oil/salt water interface, the critical salt water velocity is independent of the kinematic viscosity of the oil, $\nu_1$, in the range of $\nu_1$ from 2 to 350 $\mathrm{mm^2/s}$ and satisfactorily predicted by an analytical solution, (8.27), proposed by Milne-Thompson.

The wavelength $\lambda$ increases with an increase in the kinematic viscosity of silicone oil. The analytical equation for $\lambda$ (8.30) agrees with the measured values of $\lambda$ for $\nu_1 = 2\,\mathrm{mm^2/s}$, but overestimates $\lambda$ as $\nu_1$ exceeds $10\,\mathrm{mm^2/s}$. The amplitude of the interfacial instability is also measured.

## References

1. Thomas BG, Huang X, Sussman RC (1994) Simulation of argon gas flow effects in a continuous slab caster. Metall Mater Trans B 25:527–547
2. Kubo N, Ishii T, Kubota J, Suzuki M, Nakata M (1997) Two-phase flow simulation of molten steel and argon gas in continuous casting mold. CAMP ISIJ 10:758
3. Tanaka S, Zeze M, Tsujino R, Hirata M, Mukumoto I (1994) Laboratory rolling test of slab samples with artificial defects. CAMP ISIJ 7:284
4. (1994) 19th Committee of JSPS: inclusions in steel. JSPS, Tokyo

5. Sasai K, Mizukami Y (1994) Crack formation mechanism of gas injection type immersion nozzle. CAMP ISIJ 7:24
6. Inoue T, Yasunaka H, Matsumoto H, Kobayashi T, Kimura M (1994) Examination about formation mechanism of pinhole defect on cast slab for ultra low carbon steel. CAMP ISIJ 7:282
7. Abbel G, Damen W, de Gendt G, Tiekink W (1996) Argon bubbles in slabs. ISIJ Int 36(suppl):s.219
8. Kasai N, Watanabe Y, Kajiwara K, Toyoda M (1997) Mechanism of fine bubbles entrapment beneath the surface of continuously cast slabs. Tetsu-to-Hagane 83:24–29
9. Kasai N, Kawasaki M, Hanazaki K (1990) Effect of the molten steel flow on powder trapping in the continuous casting mould. CAMP-ISIJ 3:1114–1116
10. Teshima T, Kubota J, Suzuki M, Ozawa K, Masaoka T, Miyashita S (1993) Influence of casting conditions on molten steel flow in continuous casting mold at high speed casting of slabs. Tetsu-to-Hagane 79:576–582
11. Takatani K, Nakai K, Kasahara N, Watanabe T, Nakajima H (1990) Analysis of heat transfer and fluid flow in the continuous casting mold with electromagnetic brake. Tech Rep Sumitomo Metal Ind Ltd 42:11–18
12. Ho Y-H, Chen C-H, Hwang W-S (1994) Analysis of molten steel flow in slab continuous caster mold. ISIJ Int 34:255–264
13. Tokunaga T, Miyake T, Morishita M, Ebina K (1994) Optimization of submerged entrty nozzle for continuous caster. CAMP-ISIJ 7:306
14. Yokoya S, Takagi S, Souma H, Iguchi M, Asako Y, Hara S (1998) Prevention of air suction from the contact-part between sliding gate and immersion nozzle. ISIJ Int 38:1346–1352
15. Nakanishi K (1996) Japanese state of the art continuous casting process. ISIJ Int 36(suppl):S14–S17
16. Ayata K, Miyazawa K, Uesugi H, Takeuchi E, Mori H, Tozawa K (1999) Outline of national project on application of electromagnetic force to continuous casting of steel. CAMP-ISIJ 12:49–50
17. Iguchi M, Ueda H, Uemura T (1995) Bubble and liquid flow characteristics in a vertical bubbling jet. Int J Multiphase Flow 21:861–873
18. Iguchi M, Sasaki K, Nakajima K, Kawabata H (1998) Coanda effect on bubble characteristics in a bubbling jet rising near the side wall of a cylindrical vessel. ISIJ Int 38:1297–1303
19. Wygnanski I, Fiedler H (1969) Some measurements in the self-preserving jet. J Fluid Mech 38:577–612
20. Hetsroni G (1989) Particles-turbulence interaction. Int J Multiphase Flow 15:735–746
21. Pratte BD, Baines WD (1967) Profiles of the round turbulent jet in a cross flow. Proc ASCE J Hydraul Div 92:53
22. Rajaratnam N (1981) Turbulent jet (trans: Nomura Y). Morikita Publishing Co. Ltd, Tokyo, p 178, p 34
23. Themelis NJ, Tarasof P, Szekely J (1969) Gas-Liquid momentum transfer in a copper converter, Trans Metall Soc AIME 245:2425–2433
24. Crowe CT, Riesebieter H (1967) An analytical and experimental study of jet deflection in a crossflow. Fluid Dynamics of Roter and Fan Supported Aircraft at Subsonic Speeds, AGARD reprints, Paris, NATO
25. Ishigaki H (1982) Studies on the properties of turbulent jets : 1st report, constant density jet and turbulence model. Trans. Jpn. Soc. Mech. Eng. 49:1692–1700
26. Japanese Society for the Promotion of Science (JSPS) (1994) Recent development in studies of non-metallic inclusions in steel, JSPS, p 87
27. Kasai N, Kawasaki M, Hanazaki K, Sakashita T (1990) CAMP-ISIJ 3:1114
28. Tanaka H, Kuwatori H, Nishihara R (1992) Analysis of continuous casting powder entrapping pheno mena by water-model experiments. Tetsu-to-Hagane 78:761–766
29. Yamaoka Y, Suzuki M, Murakami K (1993) Powder trapping in continuous casting mold by water model. CAMP-ISIJ 6:279
30. Yamasaki T, Hasegawa H, Huji T, Imoto T (1997) Fundamental research of flux entrainment behavior on the meniscus. CAMP-ISIJ 10:236

31. Watanabe K, Nakata M, Suzuki M, Shiomi T (1998) The effect of mold powder properties on powder entrapment. CAMP-ISIJ 11:861
32. Katai T, Otsuka Y, Shigematsu K, Imoto T, Iiboshi H (1998) Critical condition of mould flux entrapment with fluctuation of sliding nozzle gate. CAMP-ISIJ 11:861
33. Miyake T (1999) Characteristics of molten slag entrainment in the steel making process, Jpn J Multiphase Flow 13:2
34. Yokoya S, Westhoff R, Asako Y, Hara S, Szekely J (1994) Numerical analysis of immersion nozzle outlet flow pattern through using swirling flow in continuous casting. Tetsu-to-Hagane 80:759–764
35. Santis MD, Ferretti A (1996) Thermo-fluid-dynamics modelling of the solidification process and behaviour of non-metallic inclusions in the continuous casting slabs. ISIJ Int 36:673–680
36. Yokoya S, Takagi S, Souma H, Iguchi M, Asako Y, Hara S (1998) Removal of inclusion through bubble curtain created by swirl motion in submerged entry nozzle. ISIJ Int 38:1086–1092
37. Hatta N, Fukui T, Nakajima Y, Ikeda J (1996) Nonsteady behaviour of free surface configuration and velocity distribution in a flow field of molten steel in a mold. Tetsu-to-Hagane 82:285–290
38. Tozawa K, Idogawa S, Sorimachi K (1995) Investigation of periodical change of molten steel flow in continuous casting mold. CAMP-ISIJ 8:1114
39. Gupta D, Chakraborty S, Lahiri AK (1997) Asymmetry and oscillation of the fluid flow pattern in a continuous casting mould: a water model study. ISIJ Int 37:654–658
40. Landau LD, Lifshitz EM (1970) Fluid mechanics, 2nd edn (trans: Takeuchi H). Tokyo Tosho Co. Ltd., Tokyo, p 264
41. The Japan society of fluid mechanics (1998) Handbook of fluid mechanics, Maruzen, 2nd edn., p 166
42. Komai T, Saeki Y, Tsutsumi N (1987) JSPS, 19th Committee, 19-10801, Solidification-426
43. Milne LM-Thompson (1968) Theoretical hydrodynamics. Macmillan, London
44. Kordyban ES, Ranov T (1970) Mechanism of slug formation in horizontal two-phase flow, Trans ASME, J Basic Eng 92:857–864
45. Walker JS, Talmage G, Brown SH, Sondergaard NA (1993) Kelvin–Helmholtz instability of couette flow between vertical walls with a free surface. Phys Fluids A 5:1466–1472
46. Mishima K, Ishii M (1980) Theoretical prediction of onset of horizontal slug flow. J Fluid Eng 102:441–445
47. Asai S (1984) 100th and 101st Nishiyama memorial lecture. ISIJ, Tokyo, p 67

# Chapter 9
# Modeling Gas–Liquid Flow in Metallurgical Operations

## 9.1 Overview

This chapter discusses the mathematical models and solution techniques usually employed in gas–liquid flow in metallurgical applications. Three approaches are generally used for modeling gas–liquid flows. The first approach considers the gas–liquid mixture as a single phase with variable density and solves a single set of transport equations. The second approach solves separate transport equations for the liquid and gas phases. The third method quantifies the mixing and flow characteristics by employing an energy balance for the system. These approaches are reviewed in the following section.

## 9.2 Review of Modeling Methods

There are numerous physical systems of industrial importance, where a gas bubble stream agitates a melt or liquid phase contained in a vessel. Gas injection is largely employed in the steelmaking industry, but it also has important applications in non-ferrous industry such as copper conversion and lead-making process, in chemical engineering (bubble column), and in environmental engineering such as reduction of temperature stratification in a lake. Although in most of these systems the ultimate objective is to effect a chemical change in the liquid phase, knowledge of the fluid flow phenomena has to be a component of any effort aimed at providing a good quantitative representation of these systems.

Due to the need for cleaner steels and higher productivity, secondary refining or ladle metallurgy has grown dramatically in the past two decades. The ladle metallurgical process can vary from simple for inclusion removal, to more complex systems such as desulfurization, dephosphorization, and alloying. In order to enhance refining processes in the steelmaking industry, gas injection into melts has been widely employed as a means to generate a turbulent recirculating flow to promote mixing. Gas stirring technique involves the introduction of gas into the system through one or a number of nozzles situated at the base of the ladle. The resulting column of

M. Iguchi and O.J. Ilegbusi, *Modeling Multiphase Materials Processes: Gas-Liquid Systems*, DOI 10.1007/978-1-4419-7479-2_9,

bubbles rises due to the buoyancy and leaves the reactor through the free surface at the top. The gas rising through the bulk liquid performs mixing, and thus enhances reaction rates, homogenizes chemical compositions, removes particulates and eliminates temperature stratification through generation of a circulating turbulent motion. The momentum interaction between the gas and melt induces a toroidal recirculation zone, which facilitates metallurgical processes such as degassing and desulfurization. Through careful choice of the dispersed phase, a chemical reaction may be promoted with the impurities in the native steel, thereby enhancing their flotation to the surface to form a slag. The other advantages of ladle refining by gas stirring are that its capital cost is low and it provides good refining efficiency through faster approach to equilibrium by enhancing mass transfer rate between the slag and metal.

Mechanical mixing at high temperature is impractical. On the contrary, electromagnetic mixing has disadvantages of being more expensive than gas mixing and does not provide sufficient mixing power. Therefore, gas mixing appears to be the best compromise for the steelmaking process among the various competing methods.

A schematic representation of a gas-agitated system is show in Fig. 9.1. The system consists of a water pool with a top layer of dissimilar fluid of smaller density. This arrangement simulates a real steelmaking process with a top slag cover. The system is agitated by gas injected through a nozzle at the bottom.

The size and shape of rising bubbles significantly affect the flow characteristics, and hence, mixing in a ladle. If the bubbles are spherical in shape and small in size, they slip through the liquid causing only local movement of the liquid. Such movement of the liquid does not result in bulk motion of the bath; hence mixing in the bath depends mainly on molecular diffusion. However, if the bubbles are large in size and irregular in shape, they entrain the surrounding liquid as they rise in the vessel. Thus the liquid–gas mixture rises from the nozzle tip to the bath surface.

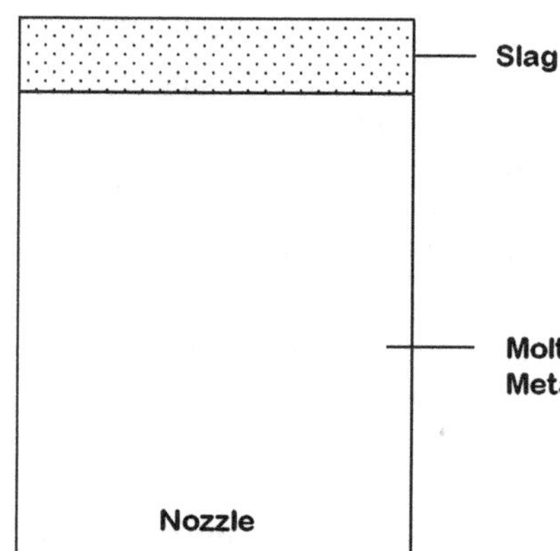


**Fig. 9.1** Schematic sketch of a gas injection system

This two-phase mixture termed "plume" rises to the bath surface and subsequently flows down into the bath along the vessel walls, causing gross circulation of the liquid known as "recirculatory flow."

The most critical parameter that characterizes the mixing and plume behavior when the gas is injected through a submerged nozzle is the gas velocity at the nozzle tip as a function of gas flow rate and bath height. Sahai and Guthrie [1, 2] investigated the hydrodynamics of a body of fluid contained in a cylindrical vessel stirred by central gas injection from the bottom of the vessel by employing both water modeling and mathematical modeling techniques. The plume velocity was found to increase with the gas flow rate according to a one-third power, increase with the depth of the fluid according to a one-fourth power, and decrease with the vessel radius according to a one-third power.

A major problem with the gas-stirring method of fluid mixing and impurity removal is the re-entrainment of the slag into the molten steel. This phenomenon is generally associated with the mixing and flow characteristics at the slag/metal interface and has attracted the interest of many researchers in recent years [3–8].

A number of model experiments have been conducted to investigate the mass transfer at the slag/metal interface [3–5], the critical condition and mechanism for occurrence of two types of emulsification [6, 9–12], and the liquid flow characteristics in molten metal baths [12, 13]. Water is generally employed in model experiments since its viscosity is of the same order of magnitude as that of molten steel, and bubble rise behavior in water and metals is quite similar. Iguchi et al. [11] used water-model experiments to establish an empirical correlation between the critical conditions and gas flow rate, geometry of vessel, and physical properties of the slag and metal. Lin and Guthrie [14] investigated the effect of density ratio between the two layers on the emulsification. It was found that the emulsification is dominated by the entrainment of lower (denser) layer into upper layer at large density ratios, while inverse emulsion of top layer droplets occurs in the lower layer at small density ratios.

Mazumdar et al. [15] carried out extensive water modeling to investigate the influence of baffle design on the hydrodynamics performance of tundishes. Three different designs of steelmaking tundishes were considered, including a two-strand slab caster, a six-strand billet caster, and a five-strand, skewed, delta-shaped tundish. It was demonstrated that the tundish performance was significantly affected by an increase in the number of strands, symmetry of tundish design, and flow operating conditions. The experimental results also indicated that changing the characteristics of the baffle design could lead to significant performance improvement.

The gas phase distribution and plume shape are important parameters in the quantification of mixing in the system. Tacke et al. [16] performed experiments and observed that gas distribution in the bath was a function of gas and liquid properties as well as gas flow rate. Castillejos and Brimacombe [17] reported time-averaged gas fraction maps for air–water system, demonstrating symmetry about the central axis and a decrease in gas fraction from the nozzle tip to the bath surface. In a subsequent study, Castillejos and Brimacombe [18] experimentally measured the plume shape and consistency in a water pool agitated by an injected air steam.

Iguchi et al. [10] measured the velocity and turbulence field in a similar system. Ilegbusi and Szekely [19] successfully predicted the plume shapes using a two-phase mathematical model, and the results compared favorably with the data of Castillejos and Brimacombe [17].

The shape of the two-phase plume has been discussed in detail by Murty and co-workers [20]. Based on extensive photographic studies, it was concluded that the plume shape was approximately that of a truncated cone. The plume profile and its dimensions depended significantly on the various operating variables [16, 20, 21].

Szekely et al. [22] assumed this region (gas–liquid) to be a solid core, having only an upward velocity component, and used the measured velocity as a boundary condition at the interface between the core and liquid region. DebRoy et al. [23] considered the region to be liquid with a variable density determined by the gas volume fraction. The gas volume fraction, being the source of the buoyancy, was calculated as the ratio of gas flow rate to the overall gas–liquid volume flow rate in the plume region, with the assumption of no slip between the liquid and gas phases.

Most of the early work on the physical and mathematical modeling of gas-stirred ladles was concerned with fluid flow and mixing times in the melt [1, 2, 23, 24]. Mixing is generally characterized by $\tau_m$, which is defined as the time taken by the melt in a vessel to reach a state of chemical and/or thermal homogeneity. Since the definition of perfect homogeneity is somewhat subjective, the mixing time is traditionally defined as the time taken by the tracer concentration at the detection point to reach about 95% of the fully mixed values. Researchers have attempted to correlate the mixing time, $\tau_m$ with the rate of energy input per unit volume of the bath $\varepsilon_m$ in the form of the following general relationship:

$$\tau_m = c\varepsilon_m^{-n} \tag{9.1}$$

where $c$ and $n$ are empirical constants. The issue of the relative contributions of bulk convection and turbulent diffusion to mixing in gas-stirred systems was addressed by Mazumdar and Guthrie [25]. Their approach was to first ignore the contribution of mixing in solving the governing differential equations, followed by a similar analysis in which the convection terms were retained and eddy diffusion terms were neglected.

Chung and Lange [26] studied convective diffusion and dispersion in steel melts and the relative contributions of bulk convection and eddy diffusion to the Peclet number. It was found that for Peclet numbers less than five the mixing time was determined by turbulent diffusion, while for increasing Peclet number, the contribution of the bulk convection to mixing increased due to the decrease in the effective diffusion distance.

In a gas-stirred ladle, the turbulence caused by a gas/liquid plume rising through the bath in the upper portion of the vessel and the circulating motion in the bath facilitate the metallurgical reactions at the interface between two immiscible liquid phases by giving high mass transfer coefficient and interfacial area. The mass transfer parameter $K$ is the only parameter typically measured in the experiments on liquid phase mass transfer because of the difficulty in measuring the interfacial area between the slag and metal. Due to slag/metal mixing, the interfacial area is generally greater than the planar area between the two phases. Asai et al. [27] comprehensively reviewed the previous work on the effect of gas flow rate on the mass transfer rate between two immiscible liquids. The results of such studies were summarized by the correlation $K \propto Q^n$, where $n$ is an empirical exponent and $Q$ is the gas flowrate.

The gas–liquid flow is also important in the continuous casting process, and relatively little work has been reported on two-phase flow in the continuous casting mold. Argon gas is sometimes employed at several stops in the continuous casting process (ladle, tundish, and mold) to enhance the mixing, to help prevent nozzle clogging, and to promote the flotation of solid inclusion particles from the liquid steel. It usually enters the continuous casting mold after injection into the submerged entry nozzle (SEN) and eventually escapes from the liquid steel surface through the mold flux powder layer. Besho et al. [28] compared the calculated flow patterns, gas volume fraction, and inclusion distribution in a full-scale water model with the experimental measurements and observations. Bubble dispersion could be controlled by nozzle submergence depth and nozzle geometry such as nominal part angle. In a similar study, Andrzejewsky et al. [29] found that carefully controlled argon injection and submergence depth were able to improve flow in a wide mold and even reduce surface flow velocity and level fluctuations.

Fundamental studies on heat transfer occurring in ladles have been few in contrast to the considerable body of work that exists on the fluid flow phenomena. It is well known that the real system is non-isothermal because the heat losses through various bounding surfaces of the ladle and the top of the melt result in thermal stratification.

Szekely and Lee [30] studied the effect of slag thickness on heat losses by conduction and radiation through the top surface. In this study, the top part of an initially molten slag gradually solidifies, while the slag in contact with the hot metal remains molten. The heat loss was found to decrease as the slag thickness increased from 0.61 to about 5 cm. There was no significant heat loss beyond a thickness of 5 cm for the operating conditions investigated.

Charaborty and Sahai [31] modeled the transient fluid flow and heat transfer in melts that have been stirred with inert gas in a ladle. Two different heat loss conditions were imposed at the free surface of the melt to reflect the insulating effect of slag of varying thickness. In one case, the top free surface was assumed to be perfectly insulated, representing the use of a thick slag layer. An appreciable heat loss through the top was considered in the second case to simulate the effect of a slag layer of insufficient thickness. It was found that significant temperature stratification occurred in the melt being held in the ladle with the insulating slag layer. Use of a thin slag resulted essentially in the reverse situation. The bulk of the melt in the ladle was well mixed due to the strong buoyancy-driven convection currents and resulted in temperature homogenization of the melt.

High-temperature ladle metallurgical processes primarily involve three modes of gas injection, namely, top, side, and bottom injection. The principal aim of all these injection techniques is ultimately to provide improved mixing conditions leading to homogenization of temperature and chemical compositions. Effective mixing depends on the scale of the convective velocities in the liquid and the breakup of eddies into smaller and smaller ones due to turbulence. Generally, mixing is considered to be more effective when gas is injected from the bottom of the vessel.

In recent years, mathematical modeling studies of the process dynamics and associated efficiency of liquid metal processing operations have been wide spread. This is expected since the visual opacity of high temperatures and the relatively large sizes of metal processing units place significant limitations on direct experimental observations. Consequently, mathematical models particularly coupled with physical (water) models appear to be the most viable means of analyzing the complex transport phenomena such as fluid flow, dispersion, and mixing in typical liquid metal processing operations. Numerous investigations addressing fundamental and applied aspects of ladle processing operations have been reported [1, 2, 7, 8, 32–38]. These investigations have led to an improved understanding of gas–liquid interactions in such systems and have provided a better insight into their overall process dynamics.

Analysis of the motion of the gas bubbles can be conducted either in the Eulerian or Lagrangian frame of reference. In the Eulerian frame of reference, the problem is formulated in terms of partial differential equations which describe the balances of mass and momentum, while in the latter approach, the trajectories of individual bubbles are tracked by solving ordinary differential equations in time. The Lagrangian method has distinct advantages over the Eulerian method in terms of simplicity of formulation, ability to accommodate complicated exchange process, computer memory requirements, and computational efforts.

## 9.3 Mathematical Models

The mathematical models describing fluid motion in gas-stirred ladle systems can be broadly classified into three groups:

(a) The quasi-single-phase models based on the continuum approach [39] in which the gas–liquid mixture in the upwelling plume is considered to rise like a homogenous fluid and a single set of transport equations is solved for the two-phase mixture.
(b) Two-fluid models [7, 8, 40, 41] in which separate continuity and momentum conservation equations are considered for each individual phase. In essence, the two-fluid models are based on the concept of unequal phase velocities. According to transport equation solved for gas phase, two-phase models can be classified into two groups, namely, Eulerian–Eulerian and Eulerian–Lagrangian models. Details of the models will be reviewed in Sect. 9.3.1.
(c) Models based on energy balance.

### 9.3.1 Quasi-Single-Fluid (Momentum Balance) Models

In the quasi-single-fluid models, the rising gas–liquid mixture is treated like a homogenous fluid of reduced density, and one set of continuity and momentum equations is solved for the two-phase mixture. The quasi-single-phase modeling technique has been relatively more popular and has been used extensively by Szekely and co-workers [34,42,43] and Guthrie and co-workers [1,2,15,25,44–47]. The $k-\varepsilon$ model is often used to represent turbulence. In most applications, the void fraction distribution is assumed a priori rather than being solved for, and this limits the predictive capability of these models.

Although the models based on momentum balance are conceptually sound, the predictions are not always satisfactory and fail to account for many observed phenomena. This is attributed to the various assumptions made in simplifying the model and the characterization of a large number of empirical parameters contained in the model equations. It may be noted further that these models are valid only in the turbulent region and fail close to the walls of the vessel.

Mazumdar and Guthrie [25] numerically and experimentally investigated gas-stirred baths using quasi-single models and performed an experiment involving bubbling air into the bottom of a water bath. The experimental setup was a small-scale model of a steel-processing ladle. In order to obtain agreement with the velocity measurements, adjustments were made to the empirical constants in the standard $k-\varepsilon$ turbulence model.

The formulation of models based on this approach basically involves setting up the equations of motion in continuity, together with appropriate boundary conditions. The general forms of these equations used by various investigators [44, 48] are given below using cylindrical polar coordinates, and assuming axisymmetric conditions.

*Continuity equation*:

$$\frac{\partial}{\partial z}(r\,\rho u) + \frac{\partial}{\partial r}(r\,\rho v) = 0, \tag{9.2}$$

*Axial momentum*:

$$\frac{\partial(\rho u)}{\partial z} + \frac{\partial(\rho uu)}{\partial z} + \frac{\partial(\rho ruv)}{r\partial r} = \frac{\partial}{\partial z}\left(\mu_{\text{eff}}\frac{\partial u}{\partial z}\right) + \frac{\partial}{r\partial r}\left(r\mu_{\text{eff}}\frac{\partial u}{\partial r}\right) - \frac{\partial p}{\partial z}$$
$$+\frac{\partial}{\partial z}\left(\mu_{\text{eff}}\frac{\partial u}{\partial z}\right) + \frac{\partial}{r\partial r}\left(r\mu_{\text{eff}}\frac{\partial v}{\partial r}\right) + \rho_{\text{l}}g\alpha_{\text{G}} + \rho_{\text{l}}g\beta_{\text{l}}(T - T_{\text{ref}}). \tag{9.3}$$

*Radial momentum*:

$$\frac{\partial(\rho u)}{\partial z} + \frac{\partial(\rho uv)}{\partial z} + \frac{\partial(\rho rvv)}{r\partial r} = \frac{\partial}{\partial z}\left(\mu_{\text{eff}}\frac{\partial v}{\partial z}\right) + \frac{\partial}{r\partial r}\left(r\mu_{\text{eff}}\frac{\partial v}{\partial r}\right) - \frac{\partial p}{\partial r}$$
$$+\frac{\partial}{\partial z}\left(\mu_{\text{eff}}\frac{\partial u}{\partial r}\right) + \frac{\partial}{r\partial r}\left(r\mu_{\text{eff}}\frac{\partial v}{\partial r}\right) - 2\mu_{\text{eff}}\frac{v}{r^2}. \tag{9.4}$$

*Energy equation*:

$$\frac{\partial(\rho T)}{\partial z}+\frac{\partial(\rho uT)}{\partial z}+\frac{\partial(\rho rT)}{r\partial r}=\frac{\partial}{\partial z}\left(\Gamma_{\text{eff}}\frac{\partial T}{\partial z}\right)+\frac{\partial}{r\partial r}\left(r\Gamma_{\text{eff}}\frac{\partial T}{\partial r}\right). \tag{9.5}$$

In (9.2)–(9.5), $u$ and $v$ represent velocity components in the axial ($z$) and radial ($r$) directions, respectively. The gas volume fraction $\alpha_G$ appears in the axial direction momentum balance equation. For example, $\rho g\,\alpha_G$ is the buoyancy force per unit volume through which the free convection effect is incorporated. The effective viscosity $\mu_{\text{eff}}$ appears in both the axial and radial direction momentum conservation equations and accommodates the effect of momentum transfer caused by the turbulence phenomena.

The velocity and temperature profiles are obtained from solution of the model equations together with the appropriate boundary conditions. By integrating velocity profiles over the entire region outside the plume, it is possible to calculate the average recirculatory flow rate.

The transport equations (9.2)–(9.5) do not form a closed set because $\alpha_G$, the gas volume fraction, and $\mu_{\text{eff}}$, the effective viscosity, embodied in the momentum equations require prior specification. The model used in the literature to calculate $\alpha$ and $\mu_{\text{eff}}$ are reviewed in Sect. 9.3.1.1.

Using a quasi single model, Mazumdar and Guthrie [44] calculated flow field and investigated the effects of process parameters on the mixing in gas-stirred ladles, and also performed an experiment to compare the numerical predictions. Figure 9.2 shows the calculated and measured velocity profile obtained by Mazumdar and Guthrie [44] in the ladle when gas injection is from the middle of the system. The calculated flow field is qualitatively in agreement with the experimental data.

#### 9.3.1.1 Two-Phase Zone Modeling

There is a large body of work on the characterization of the two-phase zone (size, phase distribution, bubble size, and liquid and gas velocities). In earlier studies, the two-phase region was treated as a solid body moving with a defined velocity which then induced the fluid motion through friction on the surrounding liquid [22]. In latter treatments, it was recognized that fluid motion was induced mainly by density differences between the two-phase plume and surrounding liquid [48, 49]. Subsequently, the two-phase region has been treated as a single phase with a variable density. The gas concentration within the plume has been calculated theoretically with no-slip models [23] and drift flux models [48]. The gas concentration variation with radius has been prescribed as a uniform or as an arbitrary function. Sahai and Guthrie [1,2] considered a simple assumption of uniform gas fraction over the entire plume. The size of the plume has been defined based on visual observation and its geometry has been generally considered to be cylindrical or at an imposed angle.

In general, the geometry of the two-phase region is usually determined empirically [1, 2, 9]. In contrast, gas volume fractions within the central two-phase region

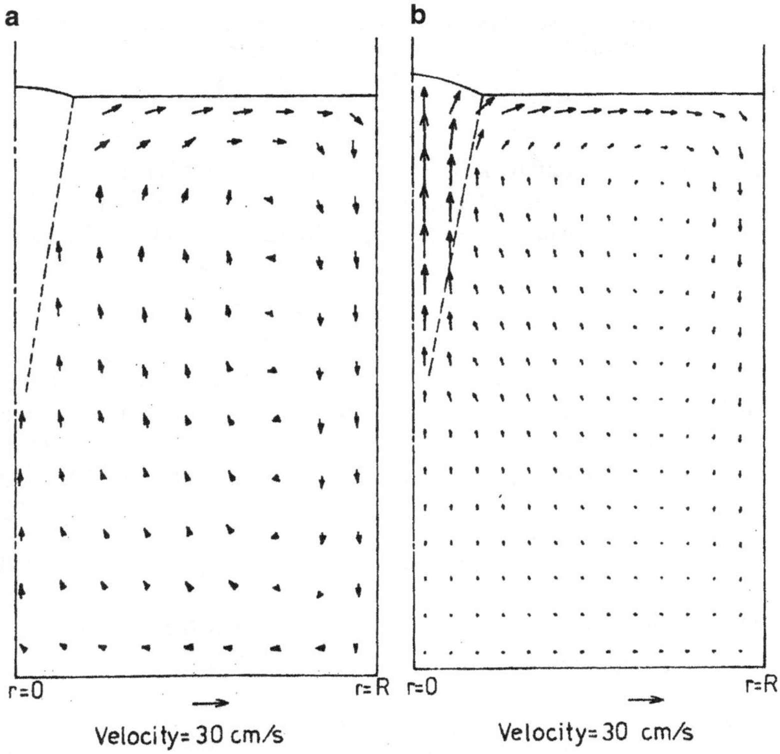


**Fig. 9.2** Comparison of predicted velocity field (**b**) with experimentally measured velocity field (**a**) [44]

are generally derived theoretically through considerations of relevant continuity principles, and various methods are applied to find this region.

To estimate the gas volume fraction in the two-phase plume region, knowledge of the average rise velocity of the gas–liquid mixture is required and this can be estimated using one of many currently available procedures. From such rise velocities and known physical dimensions of the two-phase plume, the gas voidage $\alpha_G$ can be readily obtained. The value of $\alpha_G$ calculated in this manner corresponds to zero or no slippage and can be subsequently incorporated in the buoyancy term $\rho g\,\alpha_G$, in the momentum equation [1,2,9].

Szekely and co-workers [50] derived an ordinary differential equation for the average rise velocity in the two-phase region by applying a momentum balance over a control volume in the two-phase region, expressed as;

$$\frac{\mathrm{d}U_z}{\mathrm{d}z} + \frac{U_z}{z} = \frac{\pi Q_q g}{2(\tan x)^2 U_z^2 z^2}, \tag{9.6}$$

in where, $U_z$ is the mixture rise velocity at any axial station, $z$. The boundary condition applicable to this equation is

$$\text{at } z = h_O,\ U_z = U_O. \tag{9.7}$$

Szekely and co-workers used this equation to determine gas void fraction; thus,

$$\alpha_{\mathrm{G}}(z) = \frac{Q_{\mathrm{g}} - \pi r_{\mathrm{c}}^2 \alpha_{\mathrm{G}}(1-\alpha_{\mathrm{G}})U_{\mathrm{s}}}{2\pi \int_0^{r_c} U_z r \, \mathrm{d}r}, \tag{9.8}$$

where $U_{\mathrm{s}}$ is the slip velocity typically considered to be equivalent to the rise velocity of a characteristic single bubble in the quiescent liquid. It is deduced from the following equation which uses the knowledge of the average bubble diameter in the system:

$$U_{\mathrm{s}} \cong U_{\mathrm{B}} = 1 - 0,8 \left( \frac{g d_{\mathrm{B}}}{2} \right)^{1/2}. \tag{9.9}$$

Sahai and Guthrie [1,2] applied an alternative concept to estimate gas volume fraction distribution in the rising plume. Through detailed considerations of relevant hydrodynamic principles, the authors suggested a simple, yet effective formula for the estimation of average rise velocity of the upwelling mixture in the two-phase zone, expressed as;

$$U_{\mathrm{p}} = 4.4 \frac{Q^{1/3} L^{1/4}}{R^{1/3}}. \tag{9.10}$$

Using $U_{\mathrm{p}}$, the gas volume fraction within the plume has been derived from the volume continuity principle as

$$\alpha_{Gav} = \frac{Q \frac{L}{U_{\mathrm{p}}}}{\alpha r_{av}^2 L} \tag{9.11}$$

in where $r_{av}$ refers to the radius of the plume. This equation does not consider bubble slippage and the gas volume fraction is assumed to be spatially uniform over the entire volume of the two-phase plume.

A conceptually different and a new approach to estimate the gas volume fraction within the plume has been proposed by Zhang et al. [51]. This model does not require prior specification of the two-phase domain. Furthermore, the gas volume fraction within the flow domain is deduced from the following empirical relation:

$$\alpha_{\mathrm{G}}(r,z) = \alpha_{\mathrm{m}} \exp\left[ -\frac{r^2}{C_{\mathrm{g}}^2 (H - Z + h_0)^2} \right] \tag{9.12}$$

in where $\alpha_{\mathrm{G,m}}$ represents the void fraction at the axis of symmetry.

Castillejos et al. [52,53] solved the quasi-single-model equations considering that the ladle is occupied by a single-phase fluid with spatially variable density. Gas void fraction is prescribed with an empirical model developed from the experimental data of Castillejos and Brimacombe [18,52,53]; thus:

$$\frac{\alpha_{\mathrm{G}}}{\alpha_{\mathrm{G,max}}} = \exp\left[ -0,7 \left( \frac{r}{r_{\mathrm{max}/2}} \right)^{2.4} \right], \tag{9.13}$$

where $\alpha_{G,max}$ is the maximum void fraction given by

$$\alpha_{G,max} = 81.5\left[\left(\frac{9d_0^5}{Q_0^2}\right)^{0.26}\left(\frac{\rho_L}{\rho_{go}}\right)^{0.13}\left(\frac{z}{d_0}\right)^{0.94}\right]^{-0.1} \tag{9.14}$$

When a parameter $N$ (defined subsequently in 9.18) is less than 1.35, then

$$\alpha_{G,max} = 106.9\left[\left(\frac{9d_0^5}{Q_0^2}\right)^{0.26}\left(\frac{\rho_L}{\rho_{go}}\right)^{0.13}\left(\frac{z}{d_0}\right)^{0.94}\right]^{-0.1} \tag{9.15}$$

When $N \geq 1.35$,

$$\Gamma\alpha_{max/2}\left(\frac{9}{Q_0^2}\right)^{1/5} = 0.275\left[\left(\frac{9d_0^5}{Q_0^2}\right)^{0.155}\left(\frac{\rho_L}{\rho_{go}}\right)^{0.11}\left(\frac{z}{d_0}\right)^{0.91}\right]^{0.51}. \tag{9.16}$$

The density of mixture fluid or equivalent fluid is calculated from the lever rule; thus,

$$\rho = (1-\alpha_G)\rho_L + \alpha_G\rho_g. \tag{9.17}$$

The parameter $N$ in the above equations is expressed as

$$N = \left(\frac{gd_0^5}{Q_0^2}\right)^{0.26}\left(\frac{\rho_L}{\rho_G}\right)^{0.13}\left(\frac{z}{d_0}\right)^{0.94}. \tag{9.18}$$

DebRoy et al. [23] suggested the following relationship assuming no-slip between the continuous and gas phases

$$\alpha_{G,ave} = \frac{1}{2\pi}\frac{Q}{\int_0^{r_c} rU_z \mathrm{d}z}. \tag{9.19}$$

An alternative model suggested by Grevet et al. [50] called drift–flux model which allows partial slip between the phases gives

$$\alpha_{G,ave} = \frac{1}{2\pi}\frac{Q - \pi r_c^2\alpha_G(1-\alpha_G)U_\infty}{\int_0^{r_c} rU_z \mathrm{d}z} \tag{9.20}$$

Woo et al. [34] compared the performance of drift–flux, no-slip model and experimentally obtained model for void fraction. One of the representative results is presented in Fig. 9.3, showing the radial variation of the absolute value of mean velocity at $z/H = 0.3$ and 0.98, respectively, with $H$ being the height of the bath. The predictions from the drift-flux model and the correlation agree satisfactorily

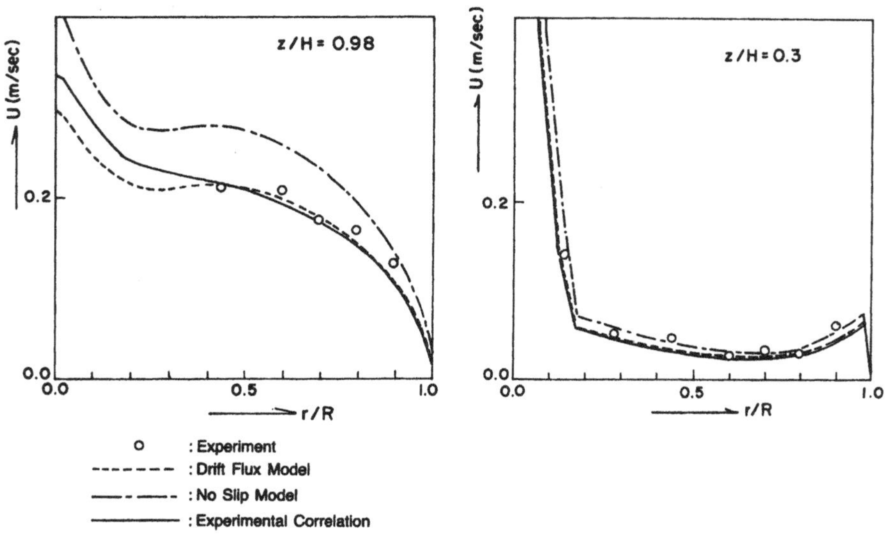


**Fig. 9.3** Comparison between predictions with three different models for the two-phase region and experimental data [34]

with experiments; however, the no-slip model predicts a relatively higher mean velocity. This result implies that the no-slip model predicts a higher value of void fraction. In addition, the no-slip model seriously deviates from experiments near the free surface where the flow is primarily driven by the strong convective outflow from the two-phase region.

### 9.3.1.2 Turbulence Modeling

In general, the overall flow field that may be generated by agitating a fluid contained in a cylindrical vessel by an axisymmetrically introduced gas jet or gas bubble stream may be adequately represented by using a variety of turbulence models [54]. The principal mechanism of momentum transfer in these systems is associated with convection rather than with the diffusive transport mechanism. Therefore, it follows therefore that predictions regarding the velocity fields are not expected to be very sensitive to the particular turbulence model chosen. However, if we were to consider these systems as reactors, then the knowledge of turbulence characteristics such as the turbulent energy dissipation and the like is of crucial importance because of its role in determining the rates of various processes that may occur in these systems. As an example, the rate of turbulent energy dissipation will have a major effect in determining the mass transfer coefficient between the fluid and suspended solid particles [55]. Furthermore, the turbulence levels in the system should have a significant effect in determining mass transfer rates between the fluid and the walls, and hence in affecting the erosion rates of the walls in case of high-temperature melts.

To characterize the effective turbulent viscosity of liquid in the bath, two models, namely, differential models and algebraic models, have frequently been used. The differential models also can also be categorized into two. In the first group, the effective viscosity $\mu_{\text{eff}}$ is determined by solving a differential equation, which expresses the conservation of turbulent energy coupled with a prescribed length scale. Szekely and co-workers [22, 42] used this approach in their earlier models. However, it had been realized that this model is not valid for recirculatory turbulent flow [42, 48].

The second group is called the two-equation $k-\varepsilon$ model. The model developed originally by Launder and Spalding [56] basically involves two partial differential equations representing the conservation of kinetic energy and the dissipation rate of turbulent energy [1, 2, 7, 8, 40, 52, 53, 57].

Szekely and Asai [58] were among the first to model bubble-driven flows in ladles, using a $k-\varepsilon$ turbulence model and assuming that the bubbles were contained in a cylindrical region of given diameter. The boundary conditions at the bubble column–liquid interface were taken from measurements. Their work gave a qualitative but not quantitative agreement with the measurements. Sahai and Guthrie [1, 2] developed a more elaborate model based on the $k-\varepsilon$ model and treated the two-phase region with a phenomenologic model.

Sahai and Guthrie [1] used the phenomenological approach to analyze fluid flow in a ladle and employed the two-equation $k-\varepsilon$ turbulent flow model with the existing quasi-single phase calculation procedure [2] and obtained the spatial distribution of turbulent kinetic energy which is important in identifying the location of (a) maximum mixing and (b) introduction of refining agents. Grevet et al. [50] performed experiments using a laser Doppler anemometer to measure rms velocity components and Reynolds stress in the liquid region and compared measured data with predictions obtained using the $k-\varepsilon$ model.

## $k-\varepsilon$ Model

The transport equations for the $k-\varepsilon$ model proposed by Jones and Launder [59] for the turbulence kinetic energy and turbulence energy dissipation in transient form can be expressed as follows:

*Turbulence kinetic energy*

$$\frac{\partial}{\partial z}\left(\rho_{\mathrm{L}}\alpha_{\mathrm{L}}w_{\mathrm{L}}k\right)+\frac{1}{r}\frac{\partial}{\partial r}\left(r\rho_{\mathrm{L}}\alpha_{\mathrm{L}}v_{\mathrm{L}}k\right)=\frac{1}{r}\frac{\partial}{\partial r}\left(r\alpha_{\mathrm{L}}\Gamma_{\mathrm{L}}\frac{\partial k}{\partial r}\right)+S_k. \tag{9.21}$$

*Turbulence energy dissipation*

$$\frac{\partial}{\partial z}\left(\rho_{\mathrm{L}}\alpha_{\mathrm{L}}w_{\mathrm{L}}\varepsilon\right)+\frac{1}{r}\frac{\partial}{\partial r}\left(r\rho_{\mathrm{L}}\alpha_{\mathrm{L}}v_{\mathrm{L}}\varepsilon\right)=\frac{1}{r}\frac{\partial}{\partial r}\left(r\alpha_{\mathrm{L}}\Gamma_{\varepsilon}\frac{\partial \varepsilon}{\partial r}\right)+S_{\varepsilon}. \tag{9.22}$$

$\Gamma_k$ and $\Gamma_\varepsilon$ are the diffusion coefficients and are expressed as:

$$\Gamma_k = \mu_l + \frac{\mu_t}{\sigma_k}, \tag{9.23}$$

$$\Gamma_\varepsilon = \mu_l + \frac{\mu_t}{\sigma_\varepsilon}, \tag{9.24}$$

where $\sigma_k$ and $\sigma_\varepsilon$ are the Schmidt numbers for $k$ and $\varepsilon$, respectively. $S_k$ and $S_\varepsilon$ are source terms are given as

$$S_k = \rho_L \alpha_L (G_k - \varepsilon) + \alpha_L G_{kb}, \tag{9.25}$$

$$S_\varepsilon = \rho_L \alpha_L \frac{\varepsilon}{k} (C_L G_k - C_2 \varepsilon) + \alpha_L c_L G_{kb} \frac{\varepsilon}{k}. \tag{9.26}$$

$G_k$ is the rate of production of turbulent energy, expressed as

$$G_k = \mu_t \left\{ \left( \frac{\partial w_1}{\partial r} + \frac{\partial v_1}{\partial z} \right)^2 + 2 \left[ \left( \frac{\partial w_1}{\partial z} \right)^2 + \left( \frac{\partial v_1}{\partial r} \right)^2 \right] \right\}. \tag{9.27}$$

The second terms on the right hand side in (9.25) and (9.26) are the turbulence production due to the motion of bubbles. Lopez de Bertodano [60] proposed the following equations for $G_{kb}$;

$$G_{kb} = 0.75 \frac{c_b c_d \rho_L \alpha_L \alpha_G}{d_b} |u_r|^3 . \tag{9.28}$$

The effective viscosity is then calculated from

$$\mu_{eff} = \mu_e + \mu_t = \mu_e + \frac{C_d \rho k^2}{\varepsilon}, \tag{9.29}$$

where $\mu_e$ is the molecular viscosity. The effective diffusivity is deduced from

$$\Gamma_{eff} = \frac{\mu_e}{\sigma_T} + \frac{\mu_t}{\sigma_{T,t}} \tag{9.30}$$

Although the $k - \varepsilon$ turbulence model has been applied extensively [1, 2, 44, 61] for modeling fluid flow in gas-stirred ladle systems, some fluid model studies [50] indicate that the $k - \varepsilon$ model cannot accurately simulate the distribution of various turbulence parameters in the gas-stirred system. Despite this, it is demonstrated that the $k - \varepsilon$ model has been reasonably successful in predicting the bulk liquid flows as these are largely dominated by inertial rather than turbulence viscous forces. It is important to note here that the inadequacy of the $k - \varepsilon$ model to simulate turbulence in the gas-stirred reactors has been attributed to the quasi-single-phase modeling technique [40], since exact two-phase computational procedures have been shown to produce fairly accurate estimates of turbulence parameters in the system.

Using the $k - \varepsilon$ turbulence model, Grevet et al. [39, 50] showed that turbulence was fairly isotropic in the bulky liquid region except in the vicinity of the solid surface. However, in the two-phase plume region, it was observed that the theoretically predicted turbulent velocity obtained by the $k - \varepsilon$ model fell between the axial and radial turbulent velocities. The $k - \varepsilon$ modeling results also exhibit a large discrepancy in predicting turbulent characteristics and thus needed to be refined or replaced by a more sophisticated model for predicting the structure of turbulence. Mazumdar and Guthrie [44] compared the mean velocity predictions of the $k - \varepsilon$ model and another simpler model (namely, the bulk effective viscosity model) and found that there is no significant difference between the two models in predicting the velocity field in the continuous phase. The result confirms Grevet's arguments that the principal mechanism of transfer phenomena in these systems is associated with convection transport.

### Effective Viscosity Model

In gas-stirred ladle systems, an average effective viscosity model [23, 43, 44] or a differential model of turbulence (i.e. $k - \varepsilon$ model) is typically employed to calculate the distribution of flow variables and turbulence parameters within the systems. It has often been argued that since such ladle flows are dominated by inertial (rather than turbulence viscous) forces, the effective viscosity value is likely to play only a secondary role in affecting the overall generated flow patterns. Thus, despite their simplicity, algebraic models of effective viscosity have been reasonably successful for realistic prediction of ladle hydrodynamics.

DebRoy et al. [23] and Szekely et al. [43] were among the first to propose computational schemes for ladle flows based on a bulk effective viscosity model. The Pun–Spalding formula [54, 56] was adopted wherein the effective viscosity was expressed as

$$\mu_e = KD^{2/3}L^{-1/3}\rho_L^{2/3}\left(mU_0^2\right)^{1/3}. \tag{9.31}$$

Sahai and Guthrie [1, 2] identified the difficulties with (9.31) based on the data obtained from the $k - \varepsilon$ model and suggested the following correlation:

$$\mu_e = C\rho_L L(Q(1-\alpha_G)9/D)^{1/3}, \tag{9.32}$$

where $C$ is the proportionality constant, starting from the following general definition of turbulent viscosity,

$$\mu_t = C_\mu \rho k^2/\varepsilon. \tag{9.33}$$

Mazumdar [62] developed an effective viscosity ignoring the laminar contribution. Assuming that energy-containing eddies are isotropic, the following expression for turbulence kinetic energy was proposed,

$$k = \frac{3}{2}u'^2, \tag{9.34}$$

where $u'$ is the fluctuating velocity component. Taking the average $u'$, to be proportional to the mean speed of liquid recirculation, $u' = c_1\overline{u}$, turbulence kinetic energy becomes

$$k = \frac{3}{2}c_1^2\overline{u}^2. \tag{9.35}$$

The turbulence energy dissipation rate $\varepsilon$ can also be deduced once the specific rate of energy input to the gas-stirred system is known. Considering an efficiency factor $\eta$ governing the generation and dissipation of turbulence kinetic energy, $\varepsilon$ can be readily estimated from the following relationship:

$$\varepsilon = \eta\,\dot{\varepsilon}_{\mathrm{m}}, \tag{9.36}$$

where $\dot{\varepsilon}_{\mathrm{m}}$ is the specific rate of energy input to the system, which is equal to $\rho\varphi/\pi R^2 (= \rho_{\mathrm{L}} g \alpha_{\mathrm{G}} L/\rho_{\mathrm{L}}\ \pi R^2 L)$. By substituting these terms into the effective viscosity expression, the following relation can be obtained:

$$\mu_{\mathrm{e}} = \left(\frac{g\pi C_\mu C_1^4}{4\eta g}\right)\left(\frac{\rho_{\mathrm{L}} R^2 \overline{U}^4}{\alpha_{\mathrm{G}}}\right). \tag{9.37}$$

The performance of different algebraic and differential turbulence models on the calculated flow field has been compared [48]. The effective viscosities predicted close to the symmetry axis and the top surface are considerably larger with differential models than algebraic models. Such results lead to a comparatively higher entrainment of the liquid in the plume and consequently, to higher velocities in the low-velocity region of the ladle (outer bottom region). Algebraic models are simpler and entail lower computational costs: however, due to the recirculatory nature of the flow, transport of the turbulence properties occurs so that the use of differential model is desirable.

Reynolds Stress Model

The Reynolds stress model involves solving the transport equations for the individual stresses $\left(u_i' u_j'\right)$. These transport equations can be derived from the momentum equations. They contain a triple-order velocity correlation and a pressure velocity correlation that must be modeled to obtain closure. The transport equations for the Reynolds stresses can be expressed in the general form

$$\begin{aligned}
&\frac{\partial}{\partial t}\left\langle u_i' u_j'\right\rangle + u_k \frac{\partial\left\langle u_i' u_j'\right\rangle}{\partial x_k} \\
&= -\frac{\partial}{\partial x_k}\left[\left\langle u_i' u_j' u_k'\right\rangle + \left\langle \frac{P}{\rho}\left(\delta_{kj} u_i' + \delta_{ik} u_j'\right)\right\rangle - \upsilon \frac{\partial}{\partial x_k}\left\langle u_i' u_j'\right\rangle\right] \\
&\quad - \left[\left\langle u_j' u_k'\right\rangle \frac{\partial u_j}{\partial x_k} + \left\langle u_j' u_k'\right\rangle \frac{\partial u_i}{\partial x_k}\right] + \left\langle \frac{P}{\rho}\left[\frac{\partial u_i'}{\partial x_j} + \frac{\partial u_j'}{\partial x_i}\right]\right\rangle - 2\upsilon\left\langle \frac{\partial u_i'}{\partial x_k}\frac{\partial u_j'}{\partial x_k}\right\rangle
\end{aligned} \tag{9.38}$$

where the terms on the right-hand side represent diffusive transport, production, pressure–strain, and dissipation terms, respectively. Due to complexity of the above (9.38), Park and Yang [38] used the following simplified form:

$$\frac{\partial\left\langle u_i' u_j'\right\rangle}{\partial t} + \langle u_k\rangle \frac{\partial\left\langle u_i' u_j'\right\rangle}{\partial x_k} = \frac{\partial}{\partial x_k}\left(\frac{\upsilon_t}{\sigma_k}\frac{\partial\left\langle u_i' u_j'\right\rangle}{\partial x_k}\right) + P_{ij} + \Phi_{ij} - \varepsilon_{ij} + R_{ij}, \quad (9.39)$$

where $P_{ij}$ represents the stress production rate, $\phi_{ij}$ is a source/sink term due to the pressure strain, $\varepsilon_{ij}$ is the viscous dissipation, and $R_{ij}$ is the rotational term.

The stress production rate can be expressed as

$$P_{ij} = -\left(\left\langle u_i' u_k'\right\rangle \frac{\partial u_j}{\partial x_k} + \left\langle u_j' u_k'\right\rangle \frac{\partial u_i}{\partial x_k}\right). \quad (9.40)$$

The pressure strain is calculated from

$$\Phi_{ij} = -C_3 \frac{\varepsilon}{k}\left(\left\langle u_j' u_j'\right\rangle - \frac{2}{3}\delta_{ij}k\right) + C_4\left(P_{ij} - \frac{2}{3}\delta_{ij}P\right). \quad (9.41)$$

A comparison of flow field and turbulence characteristics obtained with the Reynolds stress model and the $k - \varepsilon$ model is presented in a subsequent section "Momentum Interaction Between the Phases".

## 9.3.2 *Two-Fluid Model*

In the two-fluid models, separate continuity and momentum equations are derived for the liquid and gas phases with an appropriate interface mass and momentum transfer relations. The two-fluid method is potentially capable of producing more accurate solutions, since extra relations (specifying Interphase slip, for example) are included in the model. This method has been widely used in nuclear reactor thermal hydraulics following the pioneering work by Harlow and Amsden. Most of these applications have ignored turbulence or simulated it by means of constant effective viscosity.

The two-phase flow formulation can also be categorized into two groups, namely, the Eulerian–Eulerian models and Eulerian–Lagrangian model. In the Eulerian–Eulerian model, a set of transport equations is solved for the gas phase while in the Eulerian–Lagrangian model, the gas phase is treated as particles, which interact with the continuous phase.

### 9.3.2.1 Eulerian–Eulerian Model

In the Eulerian–Eulerian models, separate transport equations are solved for each phase. Ilegbusi et al. [7] used such flow model to calculate the flow characteristics of an argon-stirred ladle. The predicted mean flow and turbulence parameters compared quite well with the experimental data of Iguchi et al. [63].

Figure 9.1 illustrates a typical gas-stirred ladle [7]. The two phases are separated by sharp (but flexible) boundaries and are immiscible. There are thus two averaged densities and velocities. The phases are assumed to share space in proportion to their existence probabilities or volume fraction so as to satisfy the total continuity relation:

$$\alpha_{\mathrm{L}} + \alpha_{\mathrm{G}} = 1, \tag{9.42}$$

where $\alpha_{\mathrm{L}}$ and $\alpha_{\mathrm{G}}$ represent the volume fractions of the liquid and gas, respectively. Turbulence is assumed to be a property of the liquid phase, and dispersion of bubbles is accounted for through phase mass diffusion.

Within this framework, the governing equations can be expressed as follows.

*Mass conservation*

$$\frac{\partial}{\partial x}(\rho_i \alpha_i u_i) + \frac{1}{r}\frac{\partial}{\partial r}(r\rho_i \alpha_i v_i) = \frac{1}{r}\frac{\partial}{\partial r}\left(rD\frac{\partial f_i}{\partial r}\right) \tag{9.43}$$

where the subscript represents the phase with $i = L$ for the liquid phase and $i = G$ for the gaseous phase. The parameter $D$ associated with this term is assumed to represent an effective diffusion coefficient similar to turbulent eddy viscosity and is here assumed to be $D = \mu_{\mathrm{eff}}$, with $\mu_{\mathrm{eff}}$ being the effective viscosity, which represents a sum of the molecular and turbulent components as presented below.

*Axial momentum conservation (x-direction)*

$$\frac{\partial}{\partial x}\left(\rho_i \alpha_i u_i^2\right) + \frac{1}{r}\frac{\partial}{\partial r}(r\rho_i \alpha_i u_i v_i) = -\alpha_i \frac{\partial P}{\partial x} + F_i(u_j - u_i) + \frac{1}{r}\frac{\partial}{\partial r}\left(r\alpha_i \mu_{\mathrm{eff}}\frac{\partial u_i}{\partial r}\right) + \frac{1}{r}\frac{\partial}{\partial r}\left(rDu_i\frac{\partial \alpha_i}{\partial r}\right). \tag{9.44}$$

*Radial momentum conservation (r-direction)*

$$\frac{\partial}{\partial x}(\rho_i \alpha_i u_i v_i) + \frac{1}{r}\frac{\partial}{\partial r}\left(r\rho_i \alpha_i v_i^2\right) = -\alpha_i \frac{\partial P}{\partial r} + F_i(v_j - v_i) + \frac{1}{r}\frac{\partial}{\partial r}\left(r\alpha_i \mu_{\mathrm{eff}}\frac{\partial v_i}{\partial r}\right) + \frac{1}{r}\frac{\partial}{\partial r}\left(rDv_i\frac{\partial \alpha_i}{\partial r}\right). \tag{9.45}$$

*Energy equation*

$$\frac{\partial}{\partial x}(\rho_i \alpha_i u_i c_{p,i} T_i) + \frac{1}{r}\frac{\partial}{\partial r}(r\rho_i \alpha_i v_i c_{p,i} T) = \frac{1}{r}\frac{\partial}{\partial r}\left(r\alpha_i k_{\mathrm{eff}}\frac{\partial T_i}{\partial r}\right) + \frac{\partial}{\partial z}\left(\alpha_i k_{\mathrm{eff}}\frac{\partial T_i}{\partial z}\right) + Q_i(T_1 - T_2), \tag{9.46}$$

where $P$ is the pressure which is assumed to be shared by the phases, $F_i$ is the interface friction term and represents the momentum exchange between phases, and $Q_i$ in the energy equation is the interface heat transfer term.

*Drag force between the phases*

The momentum transfer between the phases occurs on the basis of the drag function. Ilegbusi et al. [8] employed the following formulation for momentum exchange between the two phases:

$$F = 0.75 \frac{c_d \rho_1 \alpha_L \alpha_G}{d_b} |u_r| , \tag{9.47}$$

where $u_r$ is the slip velocity vector between the two phases and $c_d$ is the drag coefficient. There are extensive works on the drag coefficient in the literature. For example, the "Dirty water" model of Kuo and Wallis [64] expresses the drag coefficient as

$$c_d = \left\{ \begin{array}{ll} 6.3/Re_b^{0.385} & Re_b > 100,\ We \le 8 \\ 2.67 & Re_b > 100,\ We > 8 \\ We/3.0 & Re_b > 2065.1/We^{2.6} \end{array} \right\},$$

where $Re_b$ is the Reynolds number based on the gas bubble diameter,

$$Re_b = \frac{\rho_1 |u_r| d_b}{\mu_1}, \tag{9.48}$$

and *We* is the Weber number defined as

$$We = \frac{\rho_1 |u_r|^2 d_b}{\gamma}, \tag{9.49}$$

where $\gamma$ is the interfacial tension between the phases.

The parameter $Q_i$ is the interphase heat transfer term expressed as

$$Q_i = C_p C_h F_i \tag{9.50}$$

where $C_p$ is the mean specific heat and $C_h$ is an empirical constant.

Using Eulerian–Eulerian two-phase model Ilegbusi et al. [8] investigated flow characteristics in a gas-agitated water bath. The water bath comprises a water pool and a light fluid above it. The system represents molten steel and a slag layer. Figures 9.4 and 9.5 present representative predictions of Ilegbusi et al. [8] compared with the experimental data of Iguchi et al. [13].

Figure 9.4 shows the predicted mean axial velocity along the axis of bath when silicon oil is used as the top layer. The results show that the velocity decays over a relatively short distance above the nozzle. Figure 9.4 as well as others from Ilegbusi et al. [8] not presented here for brevity shows that the decay length is independent of gas flow rate, indicating that the mixing in the system may be dominated by the turbulence modulation resulting from the migration of bubbles through the liquid.

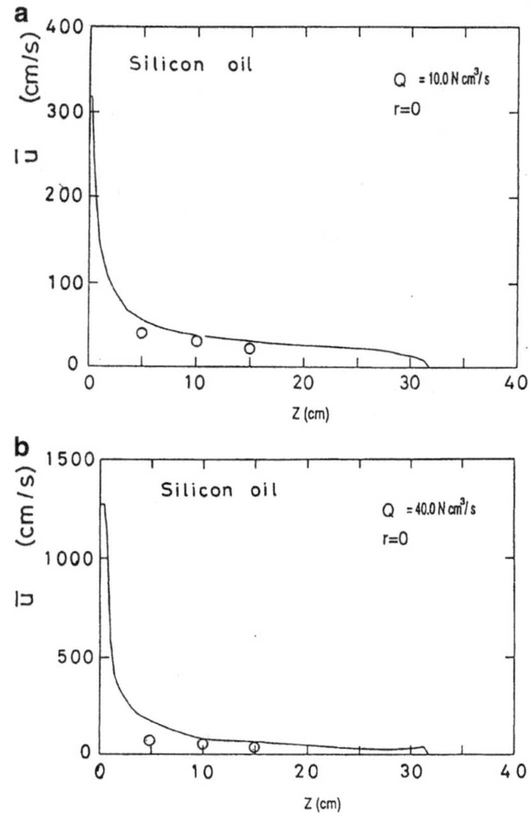


**Fig. 9.4** Predicted and measured axial mean velocity for water/silicon system: (**a**) $Q = 10\,\mathrm{Ncm^3/s}$ and (**b**) $40\,\mathrm{Ncm^3/s}$ [8]

#### 9.3.2.2 Eulerian–Lagrangian model

The Lagrangian–Eulerian approach was first introduced by Johansen and Boysan [33]. In this method, the trajectories of a steady stream of bubbles are computed in a Lagrangian field, while the phenomena in liquid motion are determined by means of an Eulerian scheme. This method is different from quasi-single phase calculation in the sense that the dispersion of bubbles is coupled to the mean and turbulent flow fields. This coupling implies that the movement and spread of the plume are not known a priori. This approach is very promising for asymmetrical as well as symmetrical gas injection problems. The Lagrangian–Eulerian approach was employed by Mazumdar and Guthrie [47] and Neifer et al. [65] in model analyses. Mazumdar and Guthrie investigated two-phase regions taking into account the effect of a free surface on the flow field, while Neifer et al. simulated the distribution of concentrations of alloy elements and temperature fields.

Park and Yang [38] analyzed mixing in two types of gas-stirred ladle systems namely, a conventional cylindrical vessel and a through-flow configuration, by employing a Eulerian–Lagrangian two-phase model. The performance of two turbulent models, the $k-\varepsilon$ model and the Reynolds stress models, were compared.

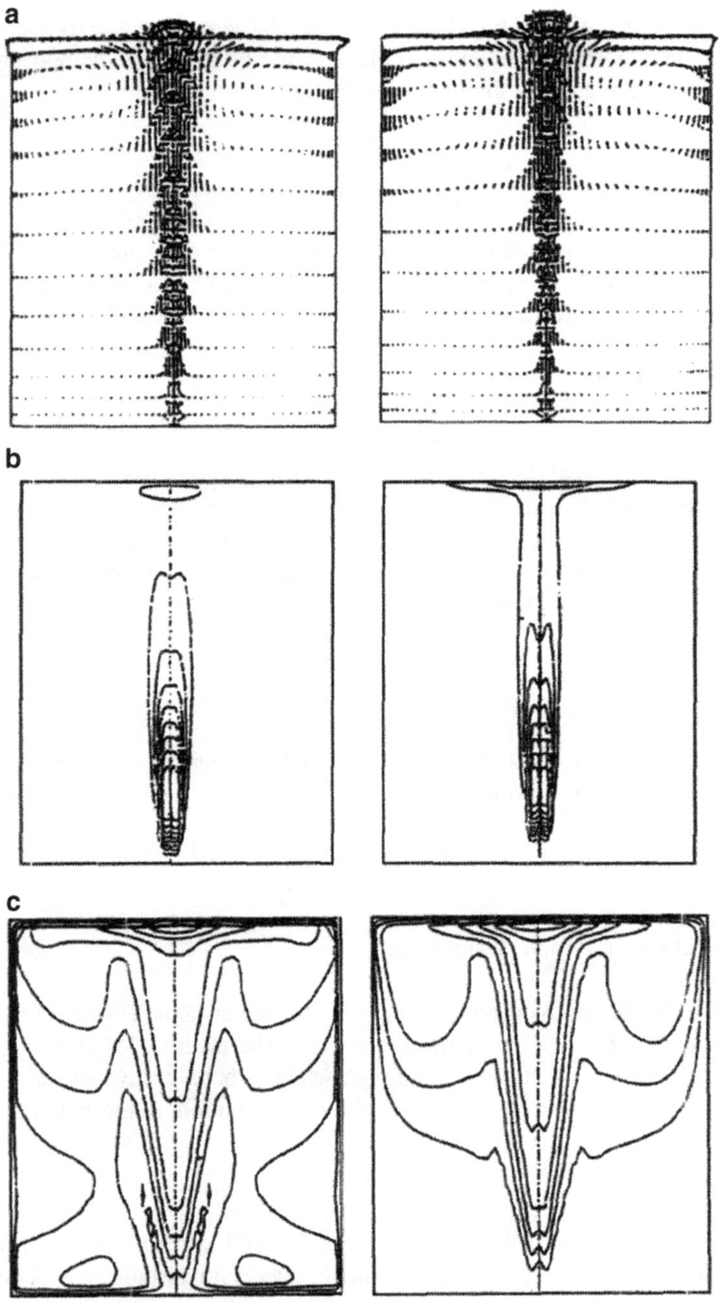


**Fig. 9.5** Comparison of results predicted by the $k - \varepsilon$ and Reynolds stress models: (**a**) velocity vector; (**b**) turbulent eddy dissipation; and (**c**) effective viscosity [38]

The transport equation for the continuous phase or Eulerian section of the transport equations is very similar to those in the quasi-single method presented previously and, therefore, will not be repeated here.

#### Gas-Phase Equations

The trajectory of a dispersed-phase bubble can be predicted by integrating the force balance written in the Lagrangian reference frame [33, 38]. This force balance equates the bubble inertia with the forces acting on the bubble and can be written as

$$\frac{\mathrm{d}u_\mathrm{b}}{\mathrm{d}t} = \frac{18\mu_\mathrm{l}}{\rho_\mathrm{b} D_\mathrm{b}^2}\frac{C_\mathrm{D} R_\mathrm{e}}{24}(u_\mathrm{l} - u_\mathrm{b}) + g_x \frac{(\rho_\mathrm{b} - \rho_\mathrm{l})}{\rho_\mathrm{b}} + \frac{1}{2}\frac{\rho_\mathrm{l}}{\rho_\mathrm{b}}\frac{\mathrm{d}}{\mathrm{d}x}(u_\mathrm{l} - u_\mathrm{b}) + \left(\frac{\rho_\mathrm{l}}{\rho_\mathrm{b}}\right) u_\mathrm{l}\frac{\partial u_\mathrm{l}}{\partial x}, \tag{9.51}$$

where $R_\mathrm{e} = \rho_\mathrm{L} D_\mathrm{b} \frac{|u_\mathrm{l} - u_\mathrm{b}|}{\mu_\mathrm{e}}$ is the Reynolds number, $c_\mathrm{d}$ is empirical drag coefficient, and $g_x$ is the gravitational acceleration. The term on the right-hand side represents the drag, gravitational acceleration, added mass, and pressure effects, respectively. The above equation is supplemented by the simple kinematics relationship which defines the trajectories of bubbles

$$\frac{\mathrm{d}x}{\mathrm{d}t} = u_\mathrm{b}. \tag{9.52}$$

The drag coefficient $c_\mathrm{d}$ is a function of the relative Reynolds number and can be expressed in the general form

$$c_\mathrm{d} = a_1 + \frac{a_2}{Re} + \frac{a_3}{Re^2}. \tag{9.53}$$

#### Turbulent Effect on Bubble Trajectories

A stochastic bubble-tracking method incorporates the instantaneous gas flow velocity, for example, $\hat{u} = \overline{u} + u'$, in the $x$ direction. The magnitudes of $u'$, $v'$, and $w'$ during the lifetime of a fluid eddy through which a bubble traverses are sampled assuming that they obey a Gaussian probability distribution. In the $x$ direction,

$$u' = \xi\sqrt{u'^2}, \tag{9.54}$$

where $\xi$ is a normally distributed random number and $\sqrt{u'^2}$ is the local rms value of the velocity fluctuations. Since the kinetic energy of turbulence is known for turbulent flow calculations, the magnitudes of the rms fluctuating component can be obtained (assuming isotropy) as

$$\sqrt{u'^2} = \sqrt{v'^2} = \sqrt{w'^2} = \sqrt{2\frac{k}{\varepsilon}}. \tag{9.55}$$

The value of the random number $\xi$ is applied for the characteristic lifetime of the eddy, defined as

$$\tau = \frac{C_\mu^{3/4}}{\sqrt{2}}\frac{k}{\varepsilon}. \tag{9.56}$$

Momentum Interaction Between the Phases

It is impossible to track trajectories of all the bubbles in the solution field. Therefore, statistically adequate number of bubbles is chosen. Void fraction distribution can be obtained using the residence time distribution, which is the sum of the residence times of all bubbles visiting a particular control volume, expressed as

$$\alpha_g = \frac{Q}{N\Delta V}\sum_{m=1}^{n} t_{R,m} \tag{9.57}$$

where $Q$ is the volumetric flowrate of the gas, $n$ is the number of bubbles that visit that control volume, and $t_R$ is the residence time, and $N$ is the number of nozzles.

The momentum interaction term $F_i$ in the momentum equation can be calculated using the fact that the drag force experienced by the bubbles acts with equal magnitude but in the opposite direction of the liquid:

$$F_i = \frac{Q}{N\Delta V}\sum_{m=1}^{n}\int_0^{t_{R,m}} \frac{3}{4}\frac{\mu_l}{d_b^2} C_D Re(V_i - U_i)\mathrm{d}t. \tag{9.58}$$

The numerical results of Park and Yang [38] obtained using the Eulerian–Lagrangian two-phase model are presented in Figs. 9.5 and 9.6 in which the performance of $k-\varepsilon$ and Reynolds stress turbulence models is compared. Figure 9.5 compares the velocity field, turbulent eddy dissipation, dispersed-phase mass concentration, and effective viscosity using both models. The results from both models considered are generally similar. The velocity field exhibits a jet-like character in the vicinity of the symmetrical axis. Recirculation zones form near the top left and right corners that are almost stagnant, implying poorly mixed regions.

Assuming isotropy (inherent in the $k-\varepsilon$ model) for the anisotropic turbulence in the system could result in an overestimation of the turbulent kinetic energy due to the fact that the radial turbulent velocity is less than the axial turbulent velocity. Thus, the $k-\varepsilon$ model can be expected to overpredict the results of the Reynolds stress model.

Figure 9.6 compares the predictions of Park and Yang and the experimental results of Johansen et al. [33, 66]. Both models agree well with the measured data except for the axial velocity at relatively low height. This discrepancy may be attributed to the fact that the models neglected of bubble break up and coalescence in the region near the air injection nozzle, due to a lack of accurate information.

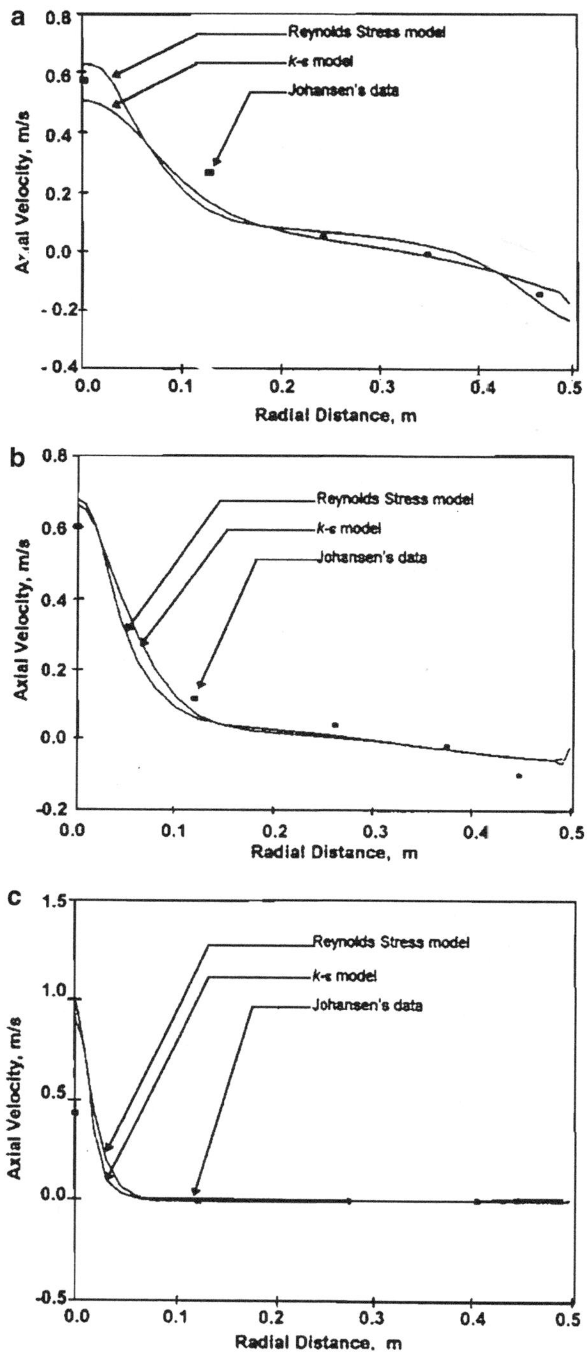


**Fig. 9.6** Comparison of axial velocity with experimental data at locations (**a**) $Z^* = 0.9$; (**b**) $Z^* = 0.5$; and (**c**) $Z^* = 0.1$ [38]

### 9.3.3 Mathematical Models Based on Energy Balance

Mathematical models based on the energy balance approach involve equating the steady-state energy input rate to the bath ($E_i$) to the rate of energy losses due to dissipation ($E_D$). The manner in which $E_i^1$ and $E_D^1$ have been characterized and quantified by various investigators has resulted in different models. For example, Sahai and Guthrie [1, 2] assumed that the net force on the liquid was the product of the total number of bubbles rising within the ascending plume and the drag force exerted by each bubble. The energy dissipation losses due to bath turbulence were estimated using the $k - \varepsilon$ model.

One of the main drawbacks of almost all models in this category is that the gas kinetic energy contribution to input energy rate is assumed to be negligible. This assumption is perhaps based on experimental investigations of Lehrer [67] who estimated that only 6 pct of the total gas kinetic energy is contributed toward the total energy input. However, this estimate does not seem to be universally correct. In a later study, employing a water bath and an orifice, Haida and Brimacombe [68] estimated this contribution to be around 15 pct. In a recent investigation, Murthy et al. [20] examined this aspect in detail and demonstrated that the contribution of kinetic energy to the mixing process in fact depends on operating conditions, and under certain situations specifically pertaining to small nozzle diameters, large bath heights, and large gas flow rates, almost the whole of the gas kinetic energy contributes to the mixing process.

*Energy balance*

The rate of energy input to the bath ($\dot{E}_i = V_L\dot{\varepsilon}$), which is actually contributes to the mixing process, is the sum of the rate of energy input due to buoyancy effect of rising bubble ($\dot{E}_b = V_L\dot{\varepsilon}_b$) and a fraction of the rate of total kinetic energy associated with the gas at the nozzle exit ($\dot{E}_k = V_L\dot{\varepsilon}_k$). Thus,

$$\varepsilon = \varepsilon_b + \varepsilon_k, \tag{9.59}$$

where $V_L$ is the bath volume, and $\varepsilon_b$, $\dot{\varepsilon}$, and $\dot{\varepsilon}_k$ correspond to the rate of energy input per unit bath volume.

Buoyancy contribution of energy input $\varepsilon_b$ can be expressed as

$$\dot{\varepsilon}_b = \frac{\dot{E}_b}{V_L} = \frac{4\dot{Q}P_{atm}T_L}{298\,2\pi D^2 H}\ln\left(1 + \frac{\rho_L g H}{P_{atm}}\right) \tag{9.60}$$

and

$$\dot{\varepsilon} = \frac{\dot{E}_i}{V_L} = \frac{8\rho_g\dot{Q}^3}{\pi^2 dn^4 V_L}, \tag{9.61}$$

where $T_L$ is the bath temperature. The input energy to the bath is usually assumed to be dissipated or lost in the following ways:

(a) Energy loss due to liquid circulation ($\dot{E}_c$), which occurs during downward flow, presumably into the turbulent eddies, and ultimately by viscous dissipation;
(b) Energy loss due to bubble slip, which is otherwise known as energy dissipation in the wakes behind the bubbles ($\dot{E}_s$)
(c) Energy loss due to net bubble break up; and
(d) Energy loss due to viscous drag at the walls.

Adopting the approach used by Bhavaraju et al. [69] in modeling bubble columns, Sano and Mori [70] and Murty et al. [71] assumed the rate of energy dissipation to be equal to the sum of the rate of energy dissipation associated with liquid circulation and that due to bubble slip.

The net energy dissipation due to bubble break up and from bubble coalescence is generally assumed to be negligible [72]. The amount of viscous drag at the walls depends on liquid viscosity. For low viscosity liquids such as molten metals and water, the energy dissipation due to viscous drag at the walls can be neglected. Thus at steady state;

$$\dot{E}_i = \dot{E}_c + \dot{E}_s. \tag{9.62}$$

The rate of energy loss due to liquid circulation in the bath ($\dot{E}_c$) is calculated as the rate of energy associated with rising liquid in the plume ($\dot{E}_{LIP}$) minus the rate of kinetic energy associated with liquid flowing downward outside of the plume ($\dot{E}_{LOP}$). The rate of energy associated with the rising liquid in the plume can be expressed as

$$\dot{E}_{LIP} = \frac{1}{2}[\rho_L A_p(1-\phi)U_{LP}]U_{LP}{}^2, \tag{9.63}$$

where $A_p$ is average cross-sectional area of the plume which is given as

$$A_p = \frac{\pi}{4}d_c{}^2. \tag{9.64}$$

The parameter $\phi$ in (9.63) is the average gas holdup in the plume and is defined as

$$\phi = \frac{\dot{Q}_M(h/U_{LP})}{A_p h} = \frac{\dot{Q}_M}{A_p U_{LP}}. \tag{9.65}$$

This implies that $\phi$ = total volume of gas in the plume volume/volume of the plume.

Finally,

$$\dot{E}_c = \frac{1}{2}\rho_2 A_p(1-\phi)U_{LP}^3\left[1 - \frac{A_p^2}{A_{op}^2}(1-\phi)^2\right]. \tag{9.66}$$

*Energy loss due to bubble slip*

Energy dissipation due to the bubble slip is generally expressed as

$$\dot{E}_{\mathrm{s}} = \frac{U_{\mathrm{s}}}{U_{\mathrm{s}} + U_{\mathrm{LP}}} \dot{E}_{\mathrm{i}}, \tag{9.67}$$

where $U_{\mathrm{s}}$ is the bubble slip velocity, calculated using empirical correlations [58]. This approach, however, is not very satisfactory since it fails to satisfy the necessary condition of $U_{\mathrm{s}}/U_{\mathrm{LP}}$ being negligible for the entrainment processes to be effective. Murty et al. [71] used the following formulation for energy loss considering the rate of energy required to lift the plume as a rigid mass to height $h(E_{\mathrm{mp}})$ and the rate of energy required to lift both the liquid and the gas as separate entities inside the plume height $h(E_{\mathrm{LGP}})$; thus;

$$\dot{E}_{\mathrm{s}} = (1-\phi)(\rho_{\mathrm{L}} - \rho_{\mathrm{G}})(\dot{Q} - A_{\mathrm{p}}\phi U_{\mathrm{lp}})gh. \tag{9.68}$$

Therefore, $\dot{E}_{\mathrm{i}}$ becomes;

$$\dot{E}_{\mathrm{i}} = \frac{1}{2}\rho_{\mathrm{L}}(1-\phi)A_{\mathrm{p}}U_{\mathrm{LP}}^{3}\left[1 - \frac{A_{\mathrm{p}}^{2}}{A_{\mathrm{op}}^{2}}(1-\phi)\right] + (1-\phi)(\rho_{\mathrm{L}} - \rho_{\mathrm{G}})gh(\dot{Q} - A_{\mathrm{p}}\phi U_{\mathrm{LP}}) \tag{9.69}$$

with the supplementary relation;

$$Q_{\mathrm{M}} - A_{\mathrm{p}}\dot{U}_{\mathrm{LP}}\phi = 0. \tag{9.70}$$

Equations (9.69) and (9.70) mathematically describe the mixing under consideration. The two unknowns $\phi$ and $U_{\mathrm{LP}}$ are determined based on the process parameters.

## 9.4 Boundary Conditions

Since the ladles are usually cylindrical, the governing equations are generally solved for half of the system due to the symmetry at the axis. Initially fluid in the bath is assumed to be at rest and has a uniform temperature, which can be mathematically expressed as

$$u = v = 0,\ T = T_{\mathrm{in}} \tag{9.71}$$

Bernard et al. [73] assigned small initial values for $k$ and $\varepsilon$ to set the initial eddy viscosity roughly equal to the molecular viscosity of water ($10^{-6}\ \mathrm{m^2/s}$).

There is no flux across the symmetry plane such that

$$\frac{\partial u}{\partial r} = 0,\ \frac{\partial v}{\partial r} = 0,\ \frac{\partial k}{\partial r} = 0,\ \frac{\partial \varepsilon}{\partial r} = 0,\ V = 0,\ \frac{\partial T}{\partial r} = 0,\ \frac{\partial \alpha_{\mathrm{L}}}{\partial r} = 0,\ \frac{\partial \alpha_{\mathrm{G}}}{\partial r} = 0. \tag{9.72}$$

At the free surface, generally zero stresses are assumed; therefore, the axial gradient of tangential velocity, temperature and turbulence kinetic energy and dissipation rate are zero. Thus,

$$\text{at } z = H : \frac{\partial V}{\partial z} = 0, \frac{\partial T}{\partial z} = 0, \frac{\partial k}{\partial z} = 0, \frac{\partial \varepsilon}{\partial z} = 0. \tag{9.73}$$

The top surface is usually assumed to be flat; therefore, axial velocity of liquid is set to zero [8, 19, 40, 52];

$$u = 0. \tag{9.74}$$

Gas is allowed to leave the bath at the rate at which it arrives at the top slab of cells. To prevent accumulation of gas at the top slab of cells, Schawarz and Turner [40] and Ilegbusi et al. [19] used the following boundary condition:

$$\frac{\partial \alpha_g}{\partial z} = 0. \tag{9.75}$$

The walls are assumed to be solid and impermeable so that flow velocities are zero. No slip conditions are employed at the boundary walls. These wall boundary conditions are implemented using the near-wall approach of Launder and Spalding [54] for the momentum and scalar transport equations. The method assumes that near the wall, Couette flow prevails and the velocity profile obeys the universal logarithmic law. Therefore, the log-law is generally employed to calculate the velocity component parallel to the wall; thus

$$\frac{u_1}{u_\tau} = \left(\frac{1}{\kappa}\right) \ln y^+ + 5.4, \tag{9.76}$$

where $\kappa(= 0.435)$ is the von-Karman constant, $u_\tau$ is the shear velocity, and $y^+$ is the dimensionless distance of the node from the wall, expressed as

$$y^+ = \frac{y u_\tau}{\mu_1}, \tag{9.77}$$

$$u_\tau = \left(\frac{\tau_w}{\rho_1}\right)^{0.5}, \tag{9.78}$$

where $\tau_w$ is the wall sheer stress.

The turbulence parameters $k$ and $\varepsilon$ near the wall are calculated from the following relations:

$$k = 4.2 u_\tau^2, \tag{9.79}$$

$$\varepsilon = \frac{u_\tau^3}{\kappa y}. \tag{9.80}$$

Quasi-single models typically set the axial velocity at the inlet to the gas punch velocity; thus,

$$\text{at } z = 0 : u = U_o, 0 < r < r_n \tag{9.81}$$

Two-phase models typically assume that only the gas enters the bath at the nozzle such that,

$$f_l = 0, f_g = 1, u_g = U_o, \tag{9.82}$$

where $f_l$ and $f_g$ represent the liquid and gas fractions, respectively.

Assuming a 10-pct turbulence intensity generated by the incoming gas bubbles, Schwarz and Turner [40] and Ilegbusi et al. [8] used the following boundary condition for the $k$ and $\varepsilon$ at the nozzle:

$$k = 0.01U_0^2. \tag{9.83}$$

$$\varepsilon = \frac{47k^{1.5}}{r_n}. \tag{9.84}$$

## 9.5 Numerical Solution

Integration of the governing partial differential equations presented in Sect. 9.3 over a control volume leads to an algebraic equation of the form;

$$A_P\phi_P = A_N\phi_N + A_S\phi_S + A_E\phi_E + A_Eq_E + A_W\phi_W + S_C, \tag{9.85}$$

where $\phi$ is a generic variable and the $A$s are the convection and diffusion coefficients, representing the combined influences of convection and diffusion processes. To obtain these coefficients (from convection and diffusion fluxes), several methods can be employed such as upwind, exponential, and finite difference hybrid [1, 2, 52, 56, 57]. Although these schemes incorporate various assumptions (for example, the upwind method ignores diffusion effects), Balaji and Mazumdar [57] found that the differentiation schemes have a practically negligible influence on the overall accuracy of the predicted results in gas–liquid flow in ladle operations. Sometimes diffusive and convective terms are discretized using separate methods. For example, Bernard et al. [73] used straightforward central differencing scheme for the diffusive terms, and the skew-upwind differencing scheme for the convective fluxes.

In single-fluid methods, the system of discretized equations is usually solved with the SIMPLE algorithm of Patankar and Spalding (quasi). This method is available in most commercial codes such as PHOENICS and FLUENT. Sahai and Guthrie [1, 2] used a modified version of the SIMPLE method which uses GALA (gas and liquid analyzer) algorithm. In this method, the physical properties of the fluid mixture in a cell in the two-phase region (plume) are averaged on a volumetric basis. This requires replacing the more conventional mass continuity equation with a volume continuity equation. In this way, the volume rather than mass of fluids entering a fluid volume element is taken to equal the volume of fluids flowing out. This approach is necessary because in two-phase flow problems, there often is a significant discrepancy between the total in- and out-flow of mass to a cell.

In multi-fluid models, there is a separate solution field for each phase. Transported quantities interact via interphase terms. The pressure in both phases is assumed to be the same within a computational cell. Field equations for each phase are weighted with the volume fraction of that phase. The model is solved using the inter-phase slip algorithm (IPSA) embodied in the PHOENICS computational code with modifications for the interfacial parameters and other source terms. This approach and its derivatives have been adopted in other commercial computational codes.

In the Eulerian–Lagrangian methods, Eulerian parts of the equations are generally solved with methods described previously for the quasi-single-fluid models. The bubble trajectory equations are solved by direct integration [33, 74] of (9.51) and (9.52). The solution procedure employed by Johansen and Boysan to solve the Eulerian–Lagrangian model [33] can be summarized as follows:

(a) In the first step, calculate the trajectories of bubbles assuming velocities are zero.
(b) Calculate the distribution of force exerted by liquid on the bubbles, the void fraction and turbulence production due to the bubbles.
(c) Solve continuous phase momentum and continuity equations and obtain correct velocities and pressure.
(d) Solve for turbulence kinetic energy and dissipation rate, and obtain turbulence viscosities.
(e) Return to step (a) for the next iteration and continue this procedure until converged results are obtained [75].

## References

1. Sahai Y, Guthrie RIL (1982) Hydrodynamics of gas stirred melts: Part II. Axisymmetric flows. Metall Trans B 13B:203–211
2. Sahai Y, Guthrie RIL (1982) Hydrodynamics of gas stirred melts: Part I. Metall Trans 13B: 193–211
3. Kim SH, Fruehan RJ (1987) Physical modeling of liquid/liquid mass transfer in gas stirred ladles. Metall Trans B 18B:381–390
4. Mietz J, Schneider S, Oeters F (1991) Emulsification and mass transfer in ladle metallurgy. Steel Res 62:10
5. Murthy GGK, Mehrotra SP (1992) Mixing in liquid baths by gas injection. Iron-Making Steel-Making 19(5):377–389
6. Gerlach F, Frohberg MG (1993) Mass transfer in a bottom blowing cold model converter. Steel Res 64(1):7–14
7. Ilegbusi OJ, Szekely J, Iguchı M, Takeuchi H, Morita ZI (1993) A comparison of experimentally measured and theoretically calculated velocity fields in a water model of an argon stirred ladle. ISIJ Int 33:474–478

8. Ilegbusi OJ, Iguchi M, Nakajima K, Sano M, Sakamoto M (1998) Modeling mean flow and turbulence characteristics in gas-agitated bath with top layer. Metall Mater Trans B 29B: 211–222
9. Mazumdar Nakajima HD, Guthrie RIL (1988) Possible roles of upper slag phases on the fluid dynamics of gas stirred ladles. Metall Trans 19B:507/11
10. Iguchi M, Demoto Y, Sugawara N, Morita Z (1992) Behavior of Hg-air vertical bubbling jets in a cylindirical vessel. ISIJ Int 32:998–1005
11. Iguchi M, Sumida Y, Okada R, Morita Z (1994) Evaluation of critical gas flow rate for the entrapment of slag using a water model. Iron Steel Inst Jpn Int 34:164
12. Iguchi M, Okita K, Nakatani T, Kasai N (1997) Structure of turbulent round bubbling jet generated by premixed gas and liquid injection. Int J Multiphase Flow 23:249–262
13. Iguchi M, Ilegbusi OJ, Ueda H, Kuranaga T, Morita Z (1996) Water model experiment on the liquid flow behavior in a bottom blown bath with top layer. Metall Mater Trans B 27B:35–41
14. Lin Z, Guthrie RIL (1994) Modeling of metallurgical emulsions. Metall Mater Trans B 25B:855–864
15. Mazumdar D, Yamanoglu G, Guthrie RIL (1997) Hydrodynamic performance of steelmaking Tundish systems: A comparative study of three different Tundish designs. Steel Res 68(7): 293–300
16. Tacke KH, Schubert HG, Weber DJ, Schwerdtfeger K (1985) Characteristcs of round vertical gas bubble jet. Metall. Trans 16B:263–275
17. Castillejos EAH, Brimacombe JK (1986) SCANINJECT IV, Part 1, No. 16, 4th Conf. on Injection Metallurgy, MEFOS, Lulea, Sweden, paper no. 16.1
18. Castillejos EAH, Brimacombe JK (1987) Local properties of turbulent air water plumes in vertically injected jets. Metall Trans B 18B:649–658; 659–671
19. Ilegbusi OJ, Szekely J (1987) Melt stratification in ladles. Trans ISIJ 27:563–569
20. Murthy GGK, Ghosh A, Mehrotra SP (1988) Characterization of two-phase axisymmetric plume in a gas stirred liquid bath – a water model study. Metall Trans B 19B:885–892
21. Oryall GN, Brimacombe JK (1976) The physical behavior of a gas jet injected horizontally into liquid metal. Metal Trans 7B:391–403
22. Szekely J, Wang HJ, Kiser KM (1976) Flow pattern velocity an turbulence energy measurements and predictions in a water model of argon-stirred ladle. Metall Trans B 7B:287–295
23. DebRoy T, Mazumdar AK, Spalding DB (1978) Numerical prediction of recirculation flows with free convection encountered in gas-agitated reactors. Appl Math Modell 2:146–150
24. Guthrie RIL (1992) Engineering in process metallurgy. Oxford University Press, New York, p 528.
25. Mazumdar D, Guthrie RIL (1986) Mixing models for gas stirred metallurgical reactors. Metall Trans B 17B:725–733
26. Chung SH, Lange KW (1988) Convective diffusion and dispersion of additions in steel melts. Ironmaking Steelmaking 15:244–256
27. Asai S, Okamoto T, He J, Muchi I (1983) Mixing time of refining vessels stirred by gas injection. Trans ISIJ 23:43–50
28. Bessho N, Yoda R, Yamasaki H (1991) Numerical analysis of fluid flow in continuous casting mold by a bubble dispersion model. Iron Steelmaker 18(4):39–44
29. Andrzejewski P, Kohler KU, Pluschkeli W (1992) Model investigations on the fluid flow in continuous casting molds on wide dimensions. Steel Res 3:242–246
30. Szekely J, Lee RG (1968) The effect of slag thickness on heat loss from ladles holding molten steel. Trans Metall Soc AIME 242:961–965
31. Chagraborty S, Sahai Y (1992) Effect of slag on heat loss and liquid steel flow in ladles before and during teeming to a continuous casting tundish. Metall Trans B 23B:135–151
32. Johansen ST, Boysan F, Ayers WH (1987) Mathematical modeling of bubble driven flows in metallurgical processes. Appl Sci Res 44:197–207
33. Johansen ST, Boysan F (1988) Fluid dynamics in bubbled stirred ladles, Part II: mathematical modeling. Metall Trans B 19B:755–764
34. Woo JS, Szekely J, Castillejos E AH, Brimacombe JK (1990) A study on the mathematical modeling of turbulent recirculating flow in gas-stirred ladles. Metall Trans B 21B:269–277

35. Davidson MR (1990) Numerical calculations of two-phase flow in a liquid bath with bottom injection: The central plume. Appl Math Modell 14:67–76
36. Mazumdar D, Guthrie RIL (1995) The physical and mathematical modeling of gas stirred ladle systems. ISIJ Int 35:1–20
37. Pan SM, Ho YH, Hwang WS (1997) Three-dimensional fluid flow model for gas stirred ladles. J Mater Eng Perform 6:625–635
38. Park HJ, Yang WJ (1997) Turbulent two-phase mixing in gas-stirred ladle systems for continuous casting applications. Numer Heat Transf A 31:493–515
39. Grevet JH, Szekely J, El-Kaddah N (1981) An experimental and theoretical study of gas bubble driven circulation systems. Int J Heat Mass Transfer 25:487–497
40. Schwarz MP, Turner WJ (1988) Applicability of the standard k-ε turbulence model to gas-stirred baths. Appl Math Modell 12:273–279
41. Turkoglu H, Farouk B (1991) Mixing time and liquid circulation rate in steelmaking ladles with vertical gas. ISIJ Int 31:1371–1380
42. Szekely J, Dilawari AH, Metz R (1979) The mathematical and physical modeling of turbulent recirculating flows. Metall Trans B 10B:33–41
43. Szekely J, El-Kaddah NE, Grevet JH (1980) Second international conference on injection metallurgy "SCANINJECT II". Jernkonteret Sweden 5:1–32
44. Mazumdar D, Guthrie RIL (1985) Hydrodynamic modeling of some gas injection procedure in ladle metallurgy operations. Metall Trans B 16B:83–90
45. Mazumdar D, Guthrie RIL (1993) On the mathematical models and solutions of gas stirred ladle systems. Appl Math Modell 17:255–262
46. Mazumdar D, Guthrie RIL (1993) On the numerical calculation and non-dimensional representation of velocity fields in bubbles-stirred ladle systems. Steel Res 64(6):286–291
47. Mazumdar D, Guthrie RIL (1994) An assessment of a two-phase calculation procedure for hydrodynamic modeling of submerged gas injection in ladles. ISIJ Int 34(5):384–392
48. DebRoy T, Mazumdar AK (1981) Predicting fluid flow in gas stirred systems. J Met 33(11):42–47
49. McKelliget JW, Cross M, Gibson RD, Brimacombe JK (1981) In: Spalding DB, Afgan NH (eds) Symp on heat and mass transfer in metallurgical systems. Hemisphere Publishing Corp., New York, NY, pp 349–372
50. Grevet JH, Szekely J, El-Kaddah N (1984) Melting rates in turbulent recirculating flow systems. Int J Heat Mass Transf 27:1116/1120
51. Zhang J, Du S, Wie S (1985) Flow field in a bath agitated by symmetrically placed impinging gas jet and submerged gas stream. Ironmaking Steelmaking 12:249/255
52. Castillejos EAH, Brimacombe JK (1989) Physical characteristics of gas jets injected vertically upward into liquid metal. Metall Trans B 20B:595–601
53. Castillejos EAH, Salcudean ME, Brimacome JK (1989) Fluid flow and bath temperature desertification in gas-stirred ladles. Metall Trans B 20B:603–611
54. Launder BE, Spalding DB (1972) Mathematical models of turbulence. Academic Press, London
55. Middleman S (1965) Mass transfer from particles in agitated systems applications of the Kolmogoroff theory. A I Ch E Jl 11:750–752
56. Launder BE, Spalding DB (1974) The numerical computation of turbulent flows. Comput Methods Appl Mech Eng 3:269–289
57. Balaji D, Mazumdar D (1991) Numerical computation of flow in gas-stirred ladle systems. Steel Res 62(1):16–24
58. Szekely J, Asai S (1975) Trans ISIJ 15:271
59. Jones WP, Launder BE (1972) Int J Heat Transf 15:301–314
60. Lopez de Bertodano M, Lee SJ, Lahey RT, Drew DA (1990) The prediction of 2-phase turbulence and phase distribution phenomena using a Reynolds stress model. J Fluid Eng 112:107–114
61. El-Kaddah N, Szekely J (1981) Mathematical model for desulphrization kinetics in argon-stirred ladles. Ironmaking Steelmaking (6):269–278

62. Mazumdar D (1989) On effective viscosity models for gas-stirred ladle systems. Metall Trans B 20B:967–969
63. Iguchi M, Nozawa K, Tomida H, Morita Z (1991) Tetsu-toHegane 77:840–852
64. Kuo JT, Wallis GB (1988) Int J Multiphase Flow 14:545
65. Neifer M, Rödi S, Sucker D (1993) Investigation on the fluid dynamic and thermal process control in ladles. Steel Res 64(1):54–62
66. Johansen ST, Robertson DGC, Woje K, Engh TA (1988) Fluid dynamics in bubble stirred ladles: Part I. Experiments. Metall Trans B 19B:745–755
67. Lehrer LH (1968) I & EC Process Des Dev 7:226–239
68. Haida O, Brimacombe JK (1983) Third int. conf. injection metallurgy, Lulea, Sweden, vol. 1, pp 5:1–5:15
69. Bhavaraju SM, Russel TWF, Blanch HW (1978) AIChE J 24:454–466
70. Sano M, Mori K (1983) Trans ISIJ 23:169–175
71. Murthy GGK, Ghosh A, Mehrotra SP (1989) Mathematical modeling of mixing phenomena in a gas stirred liquid bath. Metall Trans B 20B:53–59
72. Coshi JB, Sharma MM (1979) Trans Inst Chem Eng 54:244–251
73. Bernard RS, Maier RS, Falvey HT (2000) A simple computational model for bubble plumes. Appl Math Model 24:215–233
74. Boysan F, Ayers WH, Swithenbank J, Pan Z (1982) Three-Dimensional model of spray combustion in gas turbine combustors. J Energy 6:368–375
75. Sano M, Makino Y, Ozawa Y, Mori K (1986) Behavior of gas jet and plume in liquid metal. Trans ISIJ 26:298–304

# Chapter 10
# Numerical Modeling of Multiphase Flows in Materials Processing

## 10.1 Overview

This chapter discusses the numerical approaches to solving multiphase flow in materials processing. Chapter 9 already provided basic introduction to the modeling of gas–liquid phenomena. The focus in this chapter is liquid–solid systems which is prevalent in continuous casting operations. The general approach to numerical modeling is presented with reference to alloy solidification in order to provide additional examples of modeling complex multiphase systems. Such a system is complicated by the existence of an intermediate "mushy" zone due to phase transformation. Three approaches are generally used for modeling such multiphase systems namely, the continuum mixture models, two-phase models, and multiregion models. Control volume methods are usually employed for discretization of the transport equations in continuum mixture models and two-phase models, while the finite element methods (FEMs) are preferred in continuum mixture models. Multi-domain models usually involve the reduction of the partial differential equations to ordinary differential equations which are solved either analytically or numerically.

In general, solving multiphase problems such as alloy solidification involves calculating the heat transfer, fluid flow, and mass diffusion in the solid, liquid, and mushy regions. Because of the complexity of alloy solidification very few analytical solutions have been reported [1, 2]. Thus, in the past two decades or so, a great deal of research has been done to develop numerical methods and algorithms and more recently, specialist codes to model the key thermal and flow components during solidification and melting processes.

In modeling phase-change problems, it is always important to recognize the multiphase nature of the system. Typically the system will consist of a fully solid phase, a solid/liquid mushy phase, and a fully liquid phase. The two-phase mushy region could contain free-floating equiaxed grains. Suitable heat transfer equations that account for the two-phase nature could be based on the two-fluid models. The numerical techniques employed could thus be classified into the two-region and single-region methods.

In two-region methods, two independent sets of conservation equations are derived for the liquid and solid regions and appropriate boundary conditions are

M. Iguchi and O.J. Ilegbusi, *Modeling Multiphase Materials Processes: Gas-Liquid Systems*, DOI 10.1007/978-1-4419-7479-2_10,

employed at the liquid–solid interface to couple the equations. Such a numerical formulation produces smooth isotherms and streamlines in the vicinity of the liquid–solid interface. However, this interface is often irregular with time in most practical phase -change problems. Under this situation, both the liquid and solid phases may have irregular domains with Dirichlet boundary conditions at the interface. Hence an algebraic coordinate transformation [3–5] or boundary-fitted curvilinear coordinate [6] is needed for each of the liquid and solid regions. This complicates the use of the two-region methods. Therefore, some simplification involving geometric regularity of the interface is generally needed [7]. In addition, the governing equations based on a two-region method will become singular when either the liquid or solid phase does not exist. To solve the associated numerical difficulties, a number of single-region methods have been developed.

The fixed domain method does not deal with the particularized forms of energy and mass conservation principle for each region. Rather, the method considers the entire domain including all regions together. Thus, since the total domain does not change with time, the method is termed "single-region method."

Although a number of models have been proposed based on the variable domain method, the single-region method remains the most popular. This is because of its ease of implementation and, unlike the variable domain method it can be readily extended to multidimensional problems. In addition, alloy solidification models can be conveniently developed using the single-region method.

Studies on solidification modeling have been largely directed towards macroscopic phenomena. A variety of numerical techniques have been used for such modeling studies. Among these are the finite difference method (FDM) with or without the alternate direction implicit (ADI) time-stepping scheme, the FEM, the boundary element method (BEM), the direct finite difference method (DFDM), and the control volume element (VFM) method.

However, two major routes are generally followed to develop practical algorithms and codes to model casting processes: finite element (FE) and control/finite volume (C/FV) methods. FE methods were originally developed to solve deformation processes; the ability to address heat transfer, phase change, and fluid-flow processes emerged sometime later. Conversely, C/FV methods were initially used for conduction heat transfer, then convection driven flow and finally, free surface flows as in mold filling. Both approaches have now been developed to the point where they are able to represent the important macroscopic solidification behavior involved in casting. These numerical methods are described below.

## 10.2 Control Volume-Based Finite Difference Method

### *10.2.1 Continuum Mixture Model*

Bennon and Incropera [8–10] and Voller and coworkers [11–14] developed a continuum mixture model for the prediction of macroscopic transport behavior for a class of binary phase-change systems. The model is based on the integration of semi empirical laws and microscopic/atomic descriptions of transport behavior coupled with the principles of classical mixture theory. The continuum mixture model is similar to the standard single-phase flow equations and hence easier to compute than the complete two-phase flow equations. In this model, the evolution of the latent heat during the solidification process is embedded in the local enthalpy change rate $\partial H/\partial t$. Voller and coworkers separated the latent heat from the sensible heat such that the sensible heat becomes a continuous function across the liquid–solid interface. This treatment allowed the well-known SIMPLE algorithm [15] to be implemented in the enthalpy formulation. Based on the enthalpy method developed by Voller and coworkers, Brent et al. [16] proposed the enthalpy-porosity technique by treating the mushy zone as a porous medium.

It is important to recognize that each of the continuum equations remains valid in all regions (solid, liquid, and mushy) of the phase-change domain. Thus the continuum formulation eliminates the need to prescribe complex internal interfacial boundary conditions. Considerations need to be given only to boundary conditions on the external domain surface. In addition, because explicit tracking of internal boundaries between the solid, liquid, and mushy regions is unnecessary, the need for coordinate mapping, a quasi-steady approximation or numerical remeshing is eliminated. Hence, the complex binary phase change problem assumes many of the features of a strongly coupled single-phase problem.

Bennon and Incropera [8] employed an implicit control volume-based finite difference SIMPLER [15] scheme. The salient features of this methodology are: (1) the governing equations are discretized using a control volume approach with upwind differencing for the convective terms, (2) a staggered grid arrangement is employed so that the velocities are computed at the faces of the control volumes, (3) the velocity pressure coupling is handled by a SIMPLE type algorithm, (4) the solution marches in time using a fully implicit formulation and (5) at each matching time step, iterations are performed to account for the coupling between the various equations. The detailed derivation of this method is given by Patankar [15]. Here we consider only its application to binary phase change problems.

Figure 10.1 illustrates a typical solidified control volume in two-dimensional Cartesian coordinate system. To integrate the conservation equations, a staggered grid is employed in which the velocity component ($U_{\mathrm{e}}$, $V_{\mathrm{w}}$, $V_{\mathrm{n}}$, and $V_{\mathrm{s}}$) are defined at the control–volume interfaces (e, w, n, and s) while other dependent variables are defined at the center of the control volumes. The resulting discretized equations are solved with the SIMPLE algorithm. The equation governing the transport of mass, momentum, and heat can be written in the generalized form:

$$\frac{\mathrm{d}\,(\rho\phi)}{\mathrm{d}t} + \nabla u\phi = \nabla\,(\Gamma\nabla\phi) + S_{\phi}, \tag{10.1}$$

where, $\phi$ is a generic flow variable such as velocity component, enthalpy, and species mass fraction.

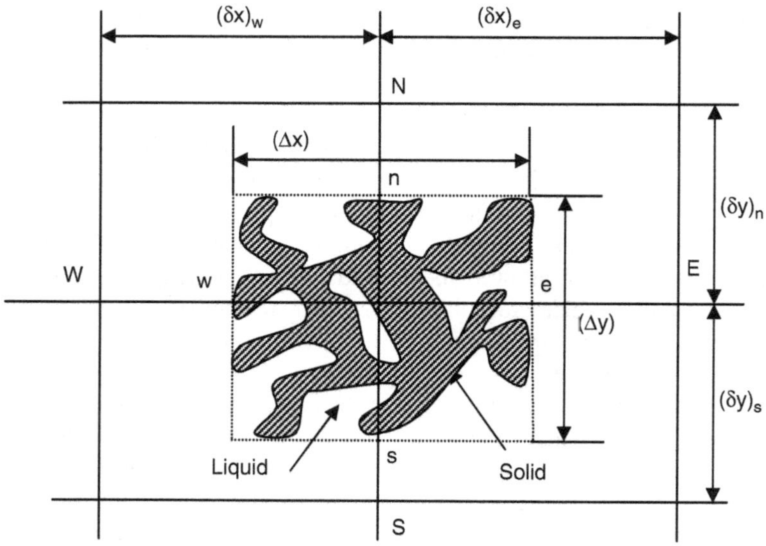


**Fig. 10.1** A partially solidified control volume and velocity storage locations

Integrating the generalized governing equation over this control volume and applying an upwind-differencing scheme, the discretized form of (10.1) becomes;

$$\frac{a_{\mathrm{p}}}{\alpha}\phi_{m,\mathrm{p}} = a_{\mathrm{E}}\phi_{m,\mathrm{E}} + a_{\mathrm{W}}\phi_{m,\mathrm{W}} + a_{\mathrm{N}}\phi_{m,\mathrm{N}} + a_{\mathrm{S}}\phi_{m,\mathrm{S}} + b + (1-\alpha)\frac{a_{\mathrm{p}}}{\alpha}\phi^{*}_{m,\mathrm{p}}, \quad (10.2)$$

where

$$a_{\mathrm{E}} = D_{\mathrm{E}} + (-F_{m,\mathrm{e}}, O), \quad (10.3)$$

$$a_{\mathrm{W}} = D_{\mathrm{W}} + (-F_{m,\mathrm{W}}, O), \quad (10.4)$$

$$a_{\mathrm{N}} = D_{\mathrm{N}} + (-F_{m,\mathrm{N}}, O), \quad (10.5)$$

$$a_{\mathrm{S}} = D_{\mathrm{S}} + (-F_{m,\mathrm{S}}, O), \quad (10.6)$$

$$a_P^0 = \frac{\rho_P^0 \Delta x \Delta y}{\Delta t}, \quad (10.7)$$

$$b = S_{\mathrm{C}}\Delta x \Delta y + a_P^0 \phi_{m,P}^0, \quad (10.8)$$

$$a_P = a_{\mathrm{E}} + a_{\mathrm{W}} + a_{\mathrm{N}} + a_{\mathrm{S}} + a_P^0 + S_P \Delta x \Delta y. \quad (10.9)$$

In (10.3)–(10.6) $D$ and $F$ represent, respectively, the diffusion and convection coefficients defined typically thus,

$$D_{\mathrm{e}} = \frac{\Gamma_{\mathrm{e}} A_{\mathrm{e}}}{(\delta x)_e}, \quad (10.10)$$

$$F_{m,\mathrm{e}} = (\rho f_m U_m A)_{\mathrm{e}}. \quad (10.11)$$

The other diffusion and convection coefficients are similarly defined.

The parameters $S_C$ in (10.8) represents the constant part of the linearized source term $S_\phi$ and $S_P$ in (10.8) is the coefficient of $\phi_p$. Under-relaxation and iteration within a time step is implied by (10.2).

The main problem with the continuum model is the handling of the source terms in the species conservation equation. Bennon and Incropera [8] expressed the advective source term in the general form;

$$S = -\nabla[\rho(V - V_s)(\phi_1 - \phi)], \tag{10.12}$$

where, $\phi$ specifically represents the solute composition, $C$. If the source term is linearized and integrated over the control volume, it may be expressed as:

$$\bar{S} = \int_w^e \int_s^n S_\phi \mathrm{d}x\mathrm{d}y = (S_c + S_P\phi_P)\Delta x \Delta y, \tag{10.13}$$

where, $S_C$ is the constant part of the source term that is independent of $\phi$.

Prior to integration, Bennon and Incropera [8] noted that the advection source term could be expressed as:

$$S = \nabla(\rho V \phi) - \nabla(\rho_s g_s V_s \phi_s) - \nabla(\rho_1 g_1 V_1 \phi_1), \tag{10.14}$$

where, $g$ is mass fraction, and subscript s and l represent solid and liquid phases, respectively.

Integration of the terms in (10.12) yields

$$\overline{S} = \overline{S}_m - \left(\overline{S}_s + \overline{S}_1\right), \tag{10.15}$$

where,

$$\overline{S}_m = F_e\phi_e^* - F_w\phi_w^* + F_n\phi_n^* - F_s\phi_s^*, \tag{10.16}$$

$$\overline{S}_k = F_{k,e}\phi_{k,e}^* - F_{k,w}\phi_{k,w}^* + F_{k,n}\phi_{k,n}^* - F_{k,s}\phi_{k,s}^*, \quad k = s, 1, \tag{10.17}$$

and $\phi^*$ is the value of $\phi$ at the previous iteration. The coefficients in (10.14) represent the total mass flow rates across each control volume interface (e, w, n, and s) and are defined as:

$$F_e = (\rho u A)_e, \quad F_w = (\rho u A)_w, \quad F_n = (\rho u A)_n, \quad F_s = (\rho u A)_s, \tag{10.18}$$

while the coefficients in (10.15) represent the mass flow rates of phase $k$ across the control volume interface and are defined as:

$$F_{k,e} = (\rho_k g_k u_k A_k)_e, \quad F_{k,w} = (\rho_k g_k u_k A_k)_w, \tag{10.19}$$

$$F_{k,n} = (\rho_k g_k u_k A_k)_n, \quad F_{k,s} = (\rho_k g_k u_k A_k)_s. \tag{10.20}$$

Mass conservation requires that,

$$f_{\mathrm{i}} = f_{\mathrm{s,i}} + f_{\mathrm{l,i}} \; (\mathrm{i} = \mathrm{e},\ \mathrm{w},\ \mathrm{n},\ \text{or s}). \tag{10.21}$$

Employing the upwinding scheme such that the interfacial values of $\phi$ are specified as the upwind nodal values yields the following discretized forms of $\overline{S}_m$ and $\overline{S}_k$:

$$\begin{aligned}\overline{S}_m &= [(F_{\mathrm{e}}) + (F_{\mathrm{w}}) + (F_{\mathrm{n}}) + (-F_{\mathrm{s}}) + (-F_{\mathrm{s}})]\, \Phi_P^* - (-F_{\mathrm{e}})\, \Phi_{\mathrm{E}}^* \\ &\quad - (F_{\mathrm{w}})\, \Phi_{\mathrm{W}}^* - (F_{\mathrm{n}})\, \Phi_{\mathrm{N}}^* - (F_{\mathrm{s}})\, \Phi_{\mathrm{S}}^*, \end{aligned} \tag{10.22}$$

$$\begin{aligned}\overline{S}_k &= \left[\left(F_{k,\mathrm{e}}\right) + \left(F_{k,\mathrm{w}}\right) + \left(F_{k,\mathrm{n}}\right) + \left(-F_{k,\mathrm{s}}\right)\right] \Phi_{k,P}^* - \left(-F_{k,\mathrm{e}}\right) \Phi_{k,\mathrm{E}}^* \\ &\quad - \left(F_{k,\mathrm{w}}\right) \Phi_{k,\mathrm{W}}^* - \left(F_{k,\mathrm{n}}\right) \Phi_{k,\mathrm{N}}^* - \left(F_{k,\mathrm{s}}\right) \Phi_{k,\mathrm{S}}^*, \end{aligned} \tag{10.23}$$

$(k = s, 1)$.

Amberg [17] and Shahani et al. [18] used a similar approach, but calculated temperature from the phase diagram equation using the relation,

$$\frac{\partial f}{\partial t} = \frac{1}{\varepsilon}(T - T_{\mathrm{L}}(c_{\mathrm{p}})), \tag{10.24}$$

instead of

$$T = T_{\mathrm{L}}(c_1) = T_o - \Gamma c_{\mathrm{p}}, \tag{10.25}$$

used by Bennon and Incropera [10], and Prakash and Voller [19].

Amberg [17] and Shahani et al. [18] used second-order accurate discretization of spatial derivatives. The velocity field was solved from the momentum equation using the well-known pressure-correction method. The nonlinear convective terms were discretized in a special manner to allow a reasonably large time step and to suppress non-physical oscillations in the solution. The diffusive term was treated implicitly also to avoid a severe limitation of the time-step size.

The difference between Amberg's method and that used by Bennon and Incropera [8], and Beckermann and Viscanta [20] is mainly that (10.24) is used instead of (10.25) which allows the system to be advanced to a new time level explicitly. Bennon and Incropera, and Beckermann and Viscanta used a completely implicit method where temperature and melt fraction etc. at the new time level were obtained simultaneously as solutions to a nonlinear system of equations. Their treatment allows larger time steps than the explicit method, but each successive step is more costly. It is not clear which of the two methods is most economical.

Since Amberg and coworkers used a rather explicit scheme, the overall stability was governed by a Courant number $Co = V_{\max}\mathrm{d}t/\mathrm{d}y$, $V_{\max}$ being the maximum velocity in the spatial stepwise domain.

The continuum mixture model and the control volume-based FDM are the predominant approaches employed by most researchers with only slight modifications. Those include the works of Christenson et al. [21], Voller et al. [13], Prescott and Incropera [22], Yoo and Viskanta [23], Shahani et al. [18], Chiang and Tsai [24], Chen and Tsai [25], Prescott et al. [26], Diao and Tsai [27], Schneider and

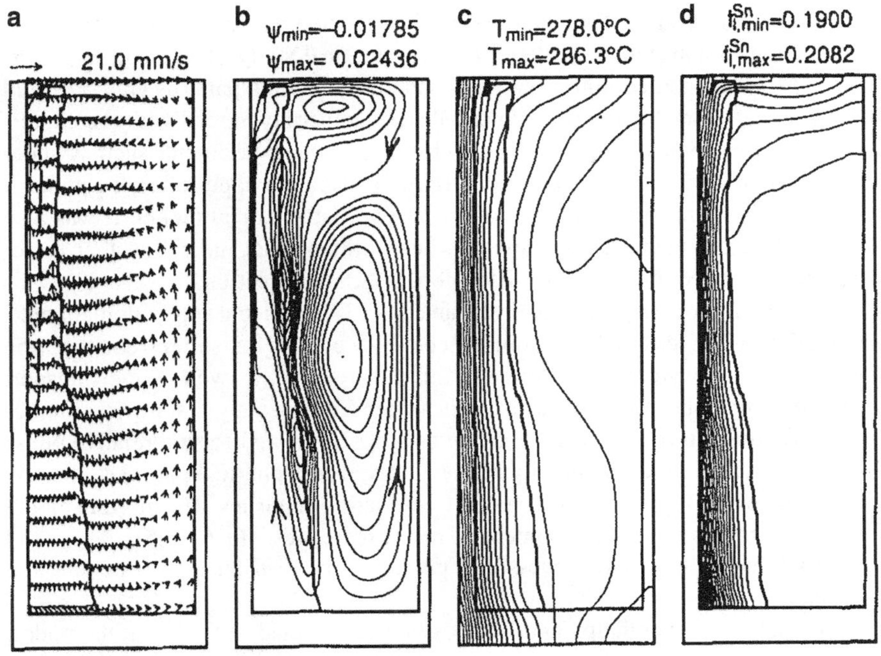


**Fig. 10.2** Convection conditions after 175 s of cooling (**a**) velocity vectors, (**b**) streamlines, (**c**) isotherms, and (**d**) liquid isocomps [35]

Beckermann [28], Ilegbusi and Mat [29, 30], Rady and Nada [31], Vreeman et al. [32], Mat and Ilegbusi [33]. For example, Prescott and Incropera [22] used a power law-differencing scheme. Raw and Lee [34] considered the solid phase as a liquid with an infinite viscosity and applied a weighting function to handle the viscosity change in the mushy region. Chen and Tsai [25] used an implicit control volume-based finite difference procedure with the SIMPLEC algorithm. The domain change due to solidification shrinkage was handled by a front tracking method.

Prescott and Incropera [26, 35] applied continuum mixture model to solidification of Pb–% 19 Sn alloy in an axisymmetric cylindrical mold made of stainless steel, cooled from its outer vertical wall. A typical result of Prescott and Incropera is given in Fig. 10.2. The figure shows the velocity vectors, streamlines, isotherms, and liquid isocomposition lines respectively after 175 s of solidification. Due to the phase equilibrium requirements, Sn is rejected from the mushy region during the solidification. Since rejected solute (Sn) is lighter than the solvent (Pb), an opposing solutal buoyancy flow is generated in the mold to the flow generated from temperature gradients between the mold surface and the center (Fig. 10.2a). Fluid exchange occurs between the mushy and melt regions, and the three small recirculation cells formed at the interface causes the formation of preferred flow path for interdendritic fluid. These channels lead to the formation of segregates which are solute deficient zones in the solidified parts.

In a subsequent study, Prescott et al. [26] conducted an experiment to study the solidification in a system considered in the numerical study of Prescott and Incropera [35]. Estimated cooling curves and macrosegregation patterns are compared with the experimental data. The numerical study generally agreed with the experiment. However, some of the phenomena observed in the experiments such as under cooling, recalescence and solid particle transport were not captured in the numerical study, due to the inherent assumptions in the mathematical model. It was also observed that the macrosegregation was three-dimensional indicating that three-dimensional convection patterns existed during the solidification.

Krane and Incropera [36], and Vreeman et al. [32] applied a single-domain mixture model of DC casting that simultaneously accounts for solute redistribution due to the fluid convection in the melt and mushy zone, as well as the transport of free-floating solid phase dendrites. These dendrites occupy a slurry region between the liquid slurry regions. Vreeman et al. [32] allowed for the relative phase motion with a solid–liquid velocity region, derived for spherical particles, thereby accounting for the settling of the heavier, free-floating dendrites. The mixture model vas used to simulate the development of macrosegregation in Al–4.5 wt% Cu and Al–6.0 wt%Mg billets, and successfully predicted the distribution of the alloying elements.

Since the mixture enthalpy and composition are defined explicitly at the nodes, Vreeman et al. [32] used the solid volume fraction, $f_s$, at each mode to determine whether the solid phase in a particular control volume forms a rigid dendrite structure $(f_s \langle f_{s,p})$. With the morphology of the solid phase defined by the solid volume fraction at each node and the velocity components defined at the control volume interfaces, the momentum equations for a rigid mushy zone are applied to the interfaces of control volumes whose solid volume fraction is greater than or equal to the packing fraction, while the slurry momentum equations are applied at the remaining control volume interfaces. In the liquid and slurry regions for which the solid volume fraction is less than the designated packing fraction $f_{s.p}$ at which free-floating dendrites are assumed to coalesce to form a rigid dendrite structure, a modified version of Ni and Incropera's slurry momentum equation is employed. In the remaining regions, the momentum equations originally developed by Bennon and Incropera [8], and modified by Prescott and Incropera [35] are employed to account for interdendritic fluid flow in a solid dendritic structure translating at the casting speed.

The solidification problems involve simultaneously solving advection–diffusion equations, namely, those associated with the solid and liquid phases. Hence, the adoption of any advection–diffusion scheme should be based on the Péclet number defined for the phases thus:

$$Pe_k = \frac{\rho u_k \delta_x}{\Gamma_x}, \tag{10.26}$$

in which subscript $k$ represents the phases, $\delta_x$ is the length of control volume in $x$ direction, and $\Gamma$ is the appropriate diffusion coefficient.

The influence that upwinding has on the numerical predictions therefore depends not only on the selected numerical mesh, but also on the phase thermophysical properties and velocities. In general, for solid+liquid system, the Péclet numbers associated with phase momentum and species transport are sufficiently large to justify the use of upwinding. However, although always providing physically realistic results, upwinding can overpredict diffusion energy transport, particularly for systems of high-conductivity phases such as metals.

Another issue related to the numerical solution of the continuum equations is that of false diffusion. False diffusion exists when flows are oblique to the orthogonal mesh and is most severe for large Péclet numbers. It is not a consequence of upwinding, but rather the treatment of variables as locally one-dimensional. The influence of false diffusion can be minimized by adopting fine computation meshes. For binary solid–liquid systems, false diffusion is generally most severe in the species conservation equation since local mass transfer Péclet numbers are large and real Fickian diffusion is small.

Most of the existing single-region methods produce zigzag isotherms and streamline in the vicinity of a liquid–solid interface [17]. This numerical error has been proven to arise from improper handling of the evolution (or absorption) of the latent heat. This may also be responsible for the poor solution convergence rate encountered in the use of the existing single-region methods.

### *10.2.2 Two-Fluid Models*

Typically, a phase-change system is treated as a single continuum i.e. there is no explicit differentiation between the solid and liquid phases. Numerical algorithms based on such single-domain approaches are by now well established and have produced useful and insightful models of solidification system. Some authors, however, have recognized that continuum approaches often lack a comprehensive treatment of the general nature of the systems. Recently full multiple-phase models have been proposed by Beckermann and coworkers [37–45].

A general solidification involves a spatial region over which both solid and liquid coexist in a so-called mushy region. Often the scale required for the resolution of the solid–liquid interface is several orders smaller than the typical cell size used in a discrete numerical solution of the governing macroscopic transport equations. A powerful modeling concept in this solution is the representative volume element (REV) introduced to the solidification modeling community by Ni and Beckermann [37]. Typically, REV is selected to include a representative and uniform sampling of the mushy region such that local scale solidification processes can be described by variables averaged over the REV.

The essential part of a solidification model is the definition of a set of consistent variables that can describe the state of the solidification systems. Four classes of variables can be identified as described below.

1. *Microscopic*. These variables vary at each point in the REV: In many solidification systems heat transfer and mass transport in the liquid phase are relatively rapid and it is often reasonable to assume the solid concentration of the variables.

The development of macroscopic equations that involve solid concentrations use the intrinsic volume average over the solid phase in REV, defined as,

$$\langle C_{\mathrm{s}}^{k} \rangle = \frac{1}{g} \int_{0}^{g} C_{\mathrm{s}}^{k} \mathrm{d}\Gamma, \tag{10.27}$$

where, $g$ is the solid volume fraction in the REV and the superscript $k$ is a marker for the $k$th component of an $m$ component system.

2. *Macroscopic-mixture*. These variables describe the state of the REV. They are the main variables in the mixture macroscopic transport equations. Appropriate definitions are, Mixture enthalpy:

$$|\rho \mathrm{H}| = \rho g_{shs} + (1-g)\,\rho_{\mathrm{l}} h_{\mathrm{l}}. \tag{10.28}$$

Mixture concentration (for a given solute component $k = 1, 2$)

$$[\rho C]^{k} = g\rho_{\mathrm{s}} \langle C_{\mathrm{s}}^{k} \rangle^{\mathrm{s}} + (1-g)\,\rho_{\mathrm{l}} c_{\mathrm{l}}^{k}. \tag{10.29}$$

3. *Mesoscopic*. These variables are uniform in the REV if heat and liquid mass transport are relatively rapid. The mesoscopic variables are: temperature $T$, liquid solute concentrations $C_{\mathrm{l}}^{k}$, and solid phase enthalpy,

$$h_{\mathrm{s}} = C_{\mathrm{s}} T, \tag{10.30}$$

where, $C_{\mathrm{s}}$ is a specific heat term which in general can be a function of temperature and concentration, and liquid phase enthalpy

$$h_{\mathrm{l}} = C_{\mathrm{l}} T + \Delta \mathrm{H}. \tag{10.31}$$

4. *Flow*. If the solid and liquid flows are uniform over the REV then the equations of motion for the solid and liquid phases can be developed in terms of a solid velocity $U_{\mathrm{s}}$ and liquid velocity $U_{\mathrm{l}}$. With these variables, a mixture velocity $u$ is defined thus:

$$\rho u = g\rho_{\mathrm{s}} u_{\mathrm{s}} + (1-g)\,\rho_{\mathrm{l}} u_{\mathrm{l}}. \tag{10.32}$$

In many models, however, it is convenient to work with the volume flow rate of the liquid phase $U(u, v)$. The interdendritic volume flow

$$U = (1-g)\,u_{\mathrm{l}} \tag{10.33}$$

is thus the velocity variable.

A numerical solution of the conservation equations will result in the current values for the mixture enthalpy $[\rho H]$ and mixture solute concentration $[\rho C]$ at each node of the domain.

Coupling models are used to determine the model fields of the solid-fraction, $g$; average solute concentration in the solid, $\langle C_s^k \rangle^s$; liquid solute concentration, $C_l^k$;and temperature $T$.

The model equations presented in Sect. 10.2.1 for multiphase system with solidification may be expressed in the following general form:

$$\frac{\partial \left( f_k \rho_k \psi_k \right)}{\partial t} + \nabla \left( f_k \rho_k v_k \psi_k \right) = \nabla \left( f_k \Gamma_k \nabla \psi_k \right) + S_k, \tag{10.34}$$

where $\Gamma_k$ stands for the diffusion coefficient of a general transfer field $\psi_k$ and $S_k$ represents all the source terms for the transport of $\psi_k$, including the interfacial interaction terms among the phases.

Beckermann and coworkers [42, 43] discretized (10.29) using a control volume-based FDM, in which the transient term is treated by a fully implicit scheme. The resulting algebraic equation in cartesian coordinates can be expressed as:

$$\left( \frac{f^0 g^0 \Delta x \Delta y}{\Delta t} + a_{\mathrm{E}} + a_{\mathrm{W}} + a_{\mathrm{S}} + a_{\mathrm{N}} + S_P \right) \psi_P$$
$$= \frac{f^0 g^0 \Delta x \Delta y}{\Delta t} \psi_P^0 + a_{\mathrm{E}} \psi_{\mathrm{E}} + a_{\mathrm{W}} \psi_{\mathrm{W}} + a_{\mathrm{S}} \psi_{\mathrm{S}} + a_{\mathrm{N}} \psi_{\mathrm{N}} + S_{\mathrm{C}}, \tag{10.35}$$

where, the superscript 0 denotes the old values of $f$, $g$, and $\chi_P$ from the previous time step at the central point $P$. The $a$'s are the diffusional convection coefficients and subscript E, W, S, and N denote the nodes neighboring the center node $P$, respectively. The upwind scheme is used for the advection terms.

The flow fields of the various phases in a multiphase system interact with each other in numerous ways e.g. through the constraint that the phases have to share the same volume at a time and through common pressure terms. These difficulties have been adequately resolved by the so-called IPSA (Interphase Slip Algorithm) algorithm, developed by Spalding [46]. IPSA is a basic feature of the PHOENICS commercial code [47]. Therefore, Beckermann and coworkers modified the existing multiphase model in PHOENICS to handle the solidification and then solved the discretization equations iteratively in the following order:

1. Guess the phase-change rate $\Gamma_s$ from the previous time step.
2. Compute the flow fields $v_s$ and $v_f$ from the continuity and momentum equations.
3. Compute the volume averaged concentration fields $C_s$ and $C_f$ from the species conservation equations.
4. Compute the temperature field $T$ from mixture energy equation.
5. Calculate the liquid temperature from $C_f$
6. Determine the nucleation rate according to the nucleation model and calculate the nuclei number.
7. Calculate the phase change rate $\Gamma_s$ from the interface species balance.
8. Feed the newly obtained $\Gamma_s$ back to step 2.
9. Repeat steps (a)–(i) until convergence of all fields is achieved before advancing to the next time step.

In the multiphase model for dendritic solidification the phase-change rate represents a critical parameter reflecting the coupling between the microscopic submodel for grain nucleation and growth, and the macroscopic model for transport phenomena. The temperature, concentration, and flow fields predicted from the macroscopic model are used to calculate the nucleation rate, grain growth rate and hence, the phase-change rate during the solidification. The phase-change rate so estimated in turn exerts a significant influence on momentum transfer through changing the phase volume fractions, energy transfer through releasing the latent heat as well as on species transport through rejection or incorporation of species at the solid/liquid interface. Furthermore, in solidification problems, due to the presence of multiple length scales, the time scales at which various physical phenomena evolve are significantly disparate. For example, the grain fraction is dictated by the dendrite tip growth, which is a much faster physical process than the macroscopic transport phenomena. Accurate modeling of microstructure hence requires a very fine time step. However, a large time step is desirable for the macroscopic conservation equations. Thévoz et al. [48] proposed a multiple time-step coupling scheme to resolve this problem.

The advantages of the multiple time-step scheme described above are: (1) it enables accurate modeling of both macro and micro behaviors, which is crucially important for multiscale modeling of dendritic alloys solidification; (2) it is computationally efficient, because the microscopic phenomena that must be resolved by very fine time steps are treated locally, while the macroscopic phenomena involving the dimension effect are simulated using a fairly large time step; (3) it is a stable and fast convergent solution algorithm because the source terms in the macroscopic conservation equations are obtained by carrying out integration of several microscopic phenomena, in which any single ill-behaved step would be smeared out in the integration.

Wang and Beckermann [42], and Beckermann and Ni [45] used a fixed arm spacing overlooks coarsening phenomena. In reality, it would be expected that over the solidification, the morphology of the REV, characterized by the secondary arm spacing, $\lambda$ will increase in size. This phenomenon will dilute, and one way of accounting for the coarsening, which has a theoretical basis is to enhance the diffusion coefficient. Voller and Beckermann [49] suggest that a suitable additive enhancement is $D^{+} = \frac{0.1\lambda^2}{4t_f}$ where $t_f$ is solidification time.

Using two-phase model Beckermann and Ni [45] modeled globolutic solidification of Al–4% Cu alloy in a rectangular cavity. One of the vertical boundary walls is cooled with a constant heat flux and the other walls are assumed to be adiabatic. Beckermann and Ni considered three alternative situations. The first simulation considered the solid phase as stationary. The second and third alternatives consider both solid and liquid motion, but with different nucleation rates. Selected representative results of Beckermann and Ni [45] are presented in Figs. 10.3 and 10.4.

Figure 10.3 shows the velocity field and the extent of the mushy region calculated in the second case at $t = 10, 30, 50$, and 70 s, respectively. In this case, nucleation rate was taken as $\dot{n} = 10^{11}\ 1/\text{m}^3\text{s}$. In this figure (Fig. 10.3a), solid contours concentrate at the lower parts of the cavity indicating that much of the solid particle moves

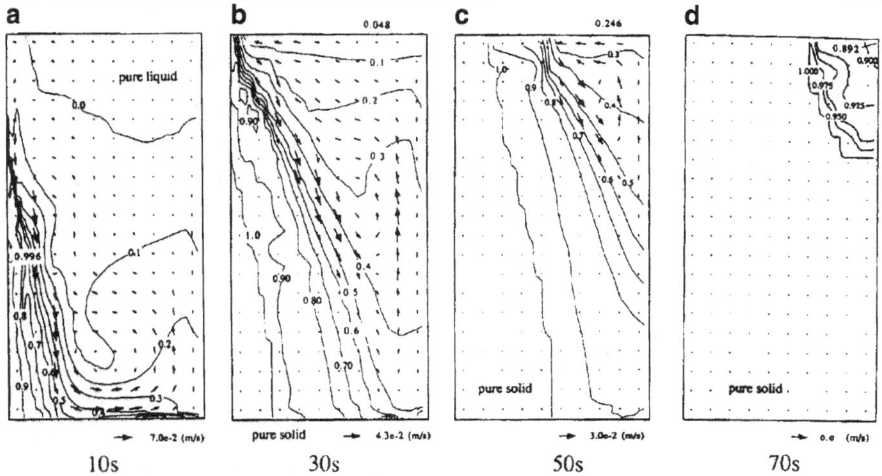


**Fig. 10.3** Liquid velocity vectors and solid volume fraction contours at four instances for simulation 2 (moving solid, $n = 10^{11}\ 1/\mathrm{m}^3\mathrm{s}$) [45]. Note that solid fraction contours are plotted on the velocity vectors in each case

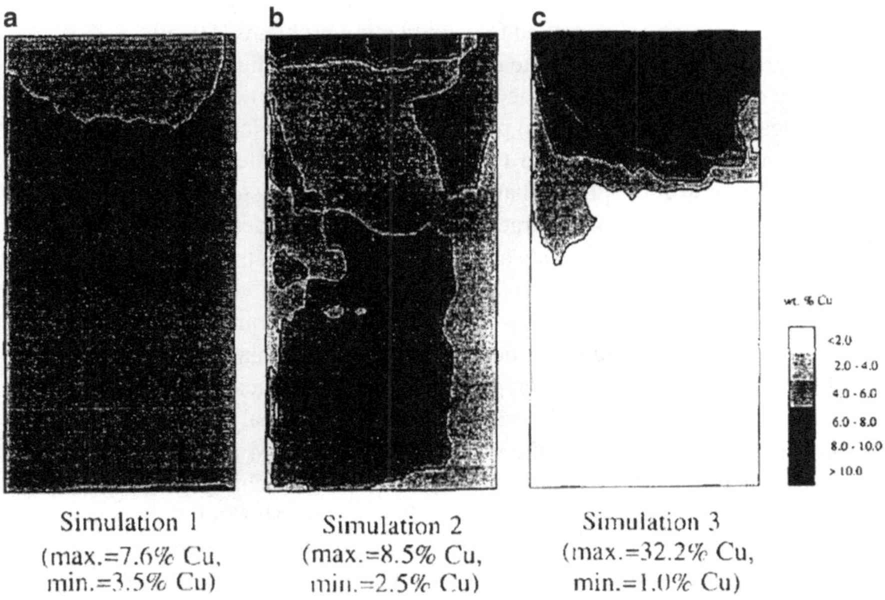


**Fig. 10.4** Comparison of final macrosegregation patterns for three alloy concentrations [45]

downwards due to gravity and counter clockwise circulation. Since after $f_s = 0.67$, the solid particles are assumed to form a solid structure, the velocity at the bottom left of the cavity is very small, while smaller particles are advected to the right side by the counter clockwise circulation. Since solid particles move downwards, the last solidified part in the cavity is the upper right region (Fig. 10.3d).

The effect of solid motion and nucleation rate on macrosegregation is given in Fig. 10.4 in which the results of case 1, case 2, and case 3 are presented in Fig. 10.4a–c, respectively. The final macrosegregation pattern is significantly affected from the motion and nucleation rate. In the first case, downward motion of solute-rich melt causes the bottom part of the cavity to be positively segregated, while in the second and third cases, the sedimentation of solute-poor grain results in a negative segregation at the bottom part of the cavity.

## 10.3 The Finite Element Method

The FEM has been an attractive method for analyzing heat and mass transport, in solidification processing of materials. This is because of its ability to handle problems with complex geometries and inherent nonlinear properties arising out of dependence of these properties on the field variables as well as the presence of nonlinear boundary conditions. The FEM is popular for two main reasons. First, it is able to reproduce using fewer nodes than does FDM, the shape of complex domains. Second, large commercial or public domain codes have already been developed. The FEM has been successfully applied to solidification problems [50–59].

Early attempts at applying finite element analysis to solidification problems focused only on heat conduction. The most important phenomena taken into account are the release of latent heat due to phase change. If this is incorporated in the governing equations as a variation in the specific heat of material, it is evident that there occurs a jump at the phase-change temperature in the specific heat curve. This is analogous to the peak of a Dirac delta function. In order that this peak is not missed in the analysis, an alternate averaging procedure on the smoother enthalpy–temperature curve was suggested [60].

In order to include the effect of free convection in the liquid and mushy regions, the whole set of coupled momentum, energy, and mass balance equations have to be solved simultaneously. However, poor computational efficiency of the simultaneous solution approach commonly used in the early FEM analysis of incompressible fluids is a major impediment in the widespread use of FEM in computational fluid dynamics. Therefore, special sophisticated acceleration techniques and algorithms have been developed to speed up FEM calculations. Most of these techniques originate from finite difference calculations. To decouple the continuity and momentum equations for incompressible fluids, Chorin [61] developed the fractional step method. The momentum is first solved disregarding the pressure gradient term. Then, the provisional velocity field obtained is corrected by taking into account the pressure contribution through the enforcement of the incompressibility requirement.

Applying the splitting techniques to discretize the diffusion and convection terms, the momentum and energy equations are split according to physical processes, and contribution from each of these processes is calculated separately in the time integration procedure. Ramaswamy and Jue [62] reported successful use of one such technique where the convective terms of the momentum and energy equations are integrated in time using the second-order Adams Bashforth scheme whereas the diffusive terms are calculated using the backward Euler scheme.

The formulation for FEM is more complicated, but it is not significantly different from the FDM. Although the specific heat (or mass) matrix is nondiagonal in the standard formulation, it can be lumped when using explicit schemes. This lumping technique which introduces a numerical diffusion similar to the numerical viscosity encountered in upwinding schemes of convection calculations, is only slightly less accurate than the exact integration method. Furthermore, it can remove some inconsistencies in the calculated temperature field for example, when two media of different temperature are suddenly brought into thermal contact. Although no ADI algorithm exists in FEM, the super time step technique can partly remove the stability criterion of explicit schemes. The DFDM technique of Ohnaka [63] combines the formulation simplicity of FDM with the enmeshment facility of FEM. The local heat balance is written in a way that is similar to that in FDM. It has been compared with the FEM and ADI–FDM techniques by Thomas et al. [64].

In the finite element solution procedure, the nodal values of quantities are interpolated to approximate the velocity, temperature and pressure in the domain thus:

$$U_i(\overline{x},t) = \phi^{\mathrm{T}}(\overline{x})\,\overline{U}_i(t)\,, \tag{10.36}$$

$$P_i(\overline{x},t) = \psi^{\mathrm{T}}(\overline{x})\,\overline{P}_i(t)\,, \tag{10.37}$$

$$T_i(\overline{x},t) = \theta^{\mathrm{T}}(\overline{x})\,\overline{T}_i(t)\,, \tag{10.38}$$

$$C_i(\overline{x},t) = \Omega^{\mathrm{T}}(\overline{x})\,\overline{C}_i(t)\,, \tag{10.39}$$

where, $\overline{U}_i$, $\overline{P}_i$, $\overline{T}_i$, $\overline{C}_i$ are the column vectors, of the element nodal point velocities, pressure, temperature, and concentration, respectively.

To discretize the equations, the Galerkin's method is usually employed. In the Galerkin's procedure, the second-order diffusion terms in the momentum and energy equations and the pressure term are reduced to first-order terms and a surface integral, by the application of temperature dependent properties (enthalpy or specific heat). It is readily applied to coarse grids with large time step. The temperature is held constant until all the latent heat associated with the nodal volume is completely released. This ensures that no latent heat is lost, but this approach has been found to be unstable in some problems [65].

Felicelli et al. [51] applied continuum mixture equations similar to Beckermann and coworkers to solve the governing equations by the FEM, using a generality function, the Petrov–Galerkin formulation. Using the Galerkin's procedure and the above approximations, the following matrix form of the governing equations ensues.

$$\{C\}\{U\} = [0]\,, \tag{10.40}$$

$$[N]\frac{\mathrm{d}u}{\mathrm{d}t} + [A(U)]\{U\} + [K(T,U)]\{U\} - [C]\{P\} + [G(T)]\{T\} = \{F(T,T_{\mathrm{ref}})\}\,, \tag{10.41}$$

where the various matrices are derived from the element shape functions and their derivatives.

The main thrust in the finite element formulation of the problems is to avoid the separated characterization of the phase-change interface and instead to include it through the variations in the material properties [65–67]. The progress of the interface can then be viewed using the appropriate isothermals in the solution domain.

The other class of formulations of the FEM is based on the definition of an effective specific heat. This results in the inclusion of the latent heat effect in the capacitance matrix. There are a number of ways in which this can be provided for. Each of these methods makes use of an enthalpy temperature curve, for example.

The effective specific heat at every integration point is computed according to the following relationship,

$$C_P = \sqrt{\frac{\nabla H \nabla H}{\nabla T \nabla T}}, \tag{10.42}$$

or in terms of the finite element formulation as follows:

$$C_P = \sqrt{\frac{\{H(T)[B]\{H(T)\}\}}{\{T\}[B]\{T\}}}, \tag{10.43}$$

where, $\{T\}$ and $\{H(T)\}$ are the vectors of the nodal temperatures and enthalpies and $[B]$ is the matrix formed from the derivatives of the element shape function. This method incorporates the variation of enthalpy in each element. It performs better than the shape method even on coarse meshes [68].

The finite element formulations described above result in a set of nonlinear time dependent matrix equations. In general, various iterative solution procedures can be used [69]. But the choice of a solution procedure is highly problem-dependent and involves frequent trade-offs between computational efficiency and available high-speed storage. Among the most popular and computationally efficient methods is the Newton–Raphson method in which the entire set of global unknown is solved together.

For the solution of the transient problem, it is necessary to have a robust time-stepping algorithm. The matrix equations result in a first-order equation in time. For the solution of this problem a three-level unconditionally stable scheme has been proposed [70]. One describes the three-level scheme class of time-step algorithm in the two-level scheme, which can vary between explicit, and implicit solution strategies [67]. The explicit scheme requires no matrix inversion, but the time step is limited by stability consideration. On the other hand, the implicit method is unconditionally stable but involves matrix inversion.

The numerical algorithms used in FE formulation for phase change problems can be categorized into fixed and moving grid methods. A typical fixed grid method is the enthalpy method where an enthalpy function is introduced and the discontinuity at the solid/liquid interface is rounded off in an averaging process. The solid/liquid interface boundary is obtained from the calculated temperature field. This round-off process may decrease the accuracy of the final results as discussed by Pardo and Weckman [71].

Felicelli et al. [54] used a fixed finite element algorithm to calculate macrosegregation and the formation of channels and freckles in Pb–Sn alloys. They assumed that the mushy zone is a porous medium with an isotropic permeability and considered superficial velocity components for the fluid velocities in the mushy region. The superficial velocities are

$$u = \phi u_{\mathrm{p}}, \ w = \phi w_{\mathrm{pa}}, \tag{10.44}$$

where, $u_{\mathrm{p}}$ and $w_{\mathrm{p}}$ is the components of the pore velocity and $\phi$ is the liquid fraction.

Following Marshall et al. [72], Heinrich and Marshall [73], Heinrich and Yu [74], Ganesan and Poirier [75], Nandapurkar et al. [52], and Felicelli et al. [54] employed a computational algorithm, which uses a standard penalty function formulation. The advantage of penalty function procedure is that the pressure is eliminated in the momentum equation. The elimination is done using the following pseudo-constitutive relation;

$$P = P_{\mathrm{s}} - \lambda \left( \frac{\partial U}{\partial x} + \frac{\partial W}{\partial z} \right), \tag{10.45}$$

where, $P_{\mathrm{s}}$ is the hydrostatic pressure and $\lambda$ is a large penalty parameter. As a result, the pressure is not calculated but if needed can be recovered a posteriori by solving a Poisson equation.

Felicelli et al. [54] used a bilinear Lagrangian isoparametric element to discretize the transport equations. The convective terms are dealt with using a Petrov–Galerkin formulation in which the weighting function is perturbed in the convective term. The perturbed weighting function is expressed as:

$$W_i^k = N_i + \alpha_k P_i \quad k = 1, 2, 3, \tag{10.46}$$

where $N_i$ is the shape function corresponding to node $i$, and $k = 1, 2$ corresponds to $u$-momentum equation and $v$-momentum equation, respectively.

Using the above terms the weak form of the momentum equations are obtained. Then the convective terms are treated explicitly and diffusive terms are evaluated implicitly. All variables are interpolated using the bilinear isoparametric shape functions for a generic function $f(x, z)$ as;

$$f(x,z) = \sum_i N_i(x,z) f_i. \tag{10.47}$$

Felicelli et al. [57] extended the model to three-dimensional problems and applied it to solidification of a Pb–Sn alloy in a parallelepiped and in a cylinder. The transport equations were discretized and integrated in time using a FEM based on the bilinear Lagrangian isoparametric element. A Petrov–Galerkin technique was used to treat the convective terms of the transport equations. The momentum equations were solved with Galerkin's/least-square method, using equal-order bilinear interpolation for both velocity and pressure. A conjugate gradient algorithm with diagonal preconditioning and sparse format storage was employed in the solution of the algebraic equations. The finite element mesh system used by Felicelli et al. [57] is shown in Fig. 10.5. This system employs 12,360 hexahedral elements and 13,733 nodes.

A representative result of Felicelli et al. [57] is shown in Fig. 10.6. Figure 10.6a shows the calculated fluid fraction contours up to $z = 0.005$ m, and the velocity vector. Four preformed flow channels developed in the system due to remelting. Consider a differential volume element within the mushy zone that has a fraction liquid $\phi_1$. Its temperature and liquid concentration are given by the point $(T_1, Ce_1)$ in the liquidus line. Remelting occurs in this element when $\frac{\partial \phi}{\partial t} > 0$ without necessarily achieving a fraction liquid of 1. In order for this to happen either the temperature or the concentration of the element must increase (assuming $k < 1$). Solute-rich liquid flows upward within the channel into the all-liquid region, gradually losing solute along its path, impinges on the top surface and comes down to feed back depleted solute into the channels. The solute-depleted liquid entering the neighborhood of the channel mouth produces a volcano in this region.

The estimated mixture concentration contours are presented in Fig. 10.6b. The solute-rich regions are formed along the channel paths. The plumes that emerge from the channels are evident which are richer in solute than the surrounding liquid.

Felicelli et al. [57] also performed a two-dimensional simulation on a vertical cross section of the three-dimensional casting. The velocities obtained in the 3D calculation were higher than those in the 2D calculation, indicating more intense convection in 3D than 2D. The higher velocities resulted in more segregation in the casting. However, the channel sizes were found to be the same in both calculations.

Nigro et al. [59] introduced a special integration method based on temperature model in order to avoid some of the difficulties introduced by the discontinuities presented in most of the fixed grid phase-change formulation. In this way it is possible to get exact Jacobian matrix of the Newton scheme retaining the quadratic convergence rate. This idea was originally presented as a discontinuous integration scheme by Steven [76] to solve Poisson equation with discontinuous coefficients and by Crivelli and Idelsohn [50]. Fachinotti et al. [77] adopted this technique to solve the conductive and/or convective heat equation with isothermal or mushy phase change oriented to continuous casting process.

The application of the phasewise discontinuous integration method for the solution of thermal phase change problems consists of dividing the element integration domain into several element sub-domains according to the distribution of phases within the element. The nodal solid fraction of each element is computed for each iteration. Before computing the contribution of each element, a special routine identifies the corresponding case according to the nodal solid fraction. To do this it is necessary to compute the intersection coordinates at each triangle edge with the solidus and liquidus line coming from the thermodynamic equilibrium diagram.

Another variation is the moving or deforming FEM method. In the deforming FEM method, finite element nodes are moving with respect to time, to adapt to the

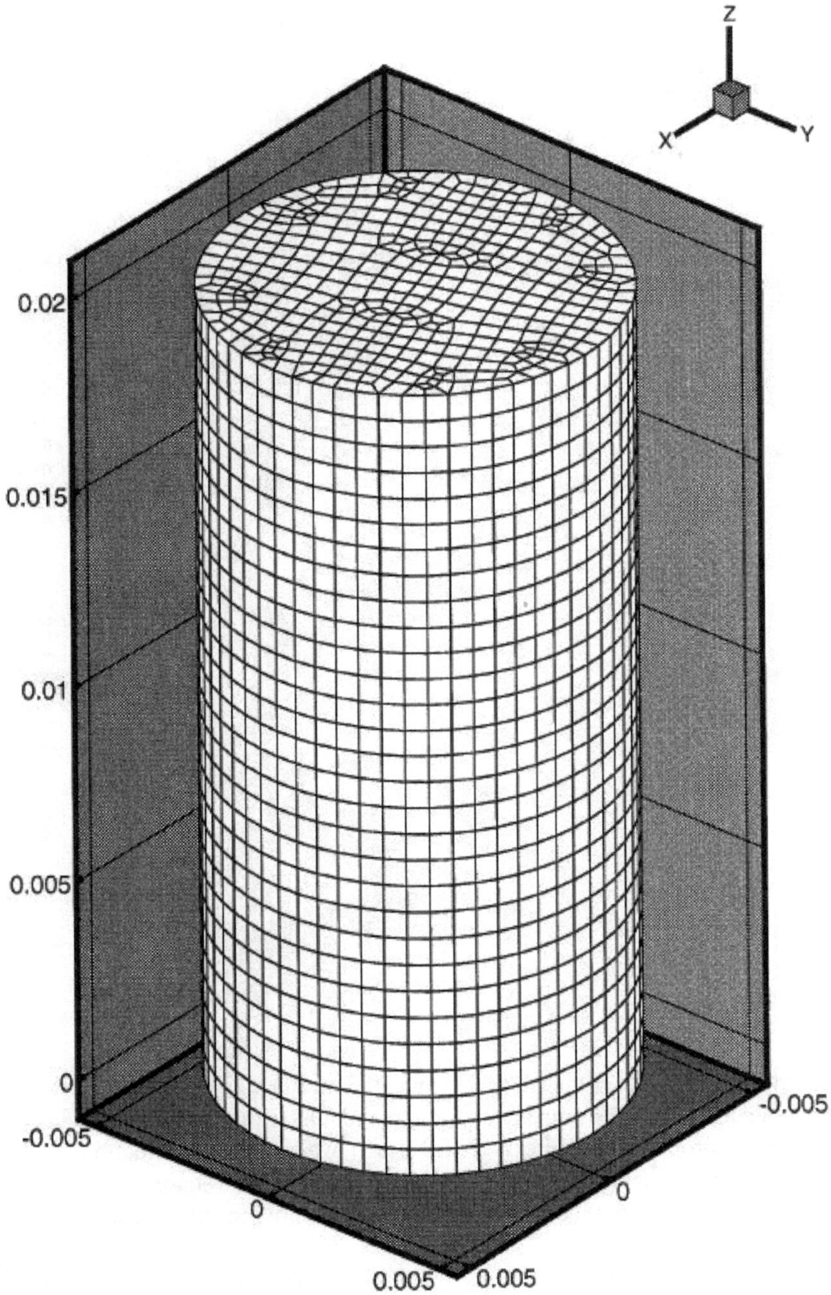


**Fig. 10.5** Finite element mesh used for calculations [57]

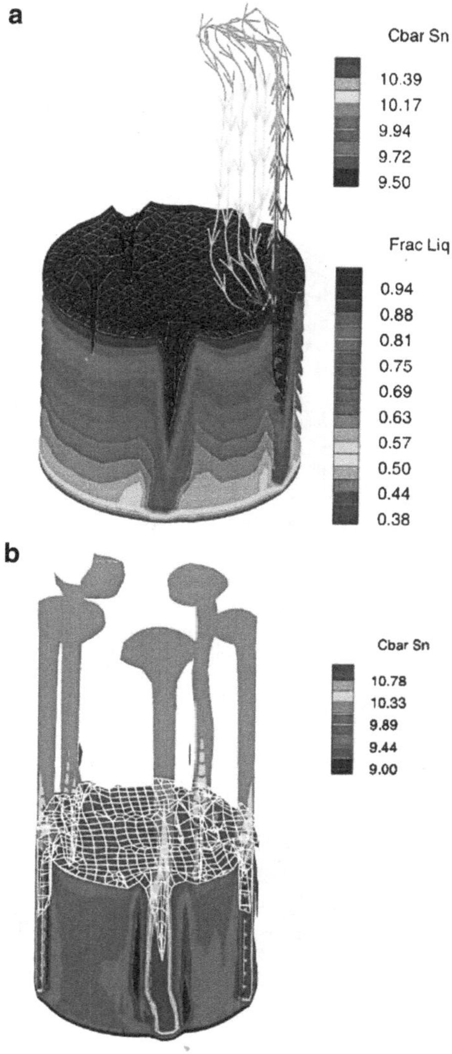


**Fig. 10.6** Solidification of Pb–10 wt%Sn in a cylinder at 10 min. (**a**) Isosurfaces of volume fraction of liquid and streamtraces emerging from a channel, (**b**) isosurfaces of mixture concentration of Sn [57]

motion of the solid/liquid interface. Because the interface is continuously tracked, the accuracy of the final results is much improved. One example of the moving grid method is that proposed by Miller and Miller [78], where Burger's equation is solved with more finite element nodes moving and concentrating in a sharp transition layer. Generally, fixed grid methods offer simplicity and moving grid methods provide accuracy.

The moving or deforming FEM method has been demonstrated successfully for modeling the solidification of pure materials by Lynch and O'Neill [79],and Zabaras and Ruan [80, 81]. In this method, time dependent finite element interpolation functions are introduced and finite element nodes move with time to adapt to the motion of the solid/liquid interface. Because the interface is continuously traced, the thermal and mechanical conditions at the interface can be applied accurately [82]. Assuming the solute is completely mixed, the solid fraction in the mushy region can be expressed explicitly as a function of temperature. However, there is no consideration of fluid flow and mass diffusion during the solidification.

Tszeng et al. [83] and Chen et al. [84] introduced a temperature recovery method for phase-change problems. Chen et al. [56] extended the conduction model, which uses temperature recovery method to solidification of binary alloy problems employing a moving grid methodology. A set of transport equations was used similar to that of Bennon and Incropera [9]. However, there were source terms in the energy and species balance equations. An energy boundary condition was used instead for the solid/liquid interface thus,

$$k_{\mathrm{s}}\frac{\partial T_{\mathrm{s}}}{\partial \mathbf{n}} - k_{\mathrm{l}}\frac{\partial T_{\mathrm{l}}}{\partial \mathbf{n}} = \rho L\frac{\partial x}{\partial t} \text{ on } \varepsilon(t), \tag{10.48}$$

and species boundary conditions,

$$(\nabla f^{\alpha}).\mathbf{n} = 0 \text{ on } \Gamma_{\mathrm{e}}(t), \tag{10.49}$$

$$f^{\alpha} = f_{\mathrm{e}}^{\alpha} \text{ on } \varepsilon(t), \tag{10.50}$$

in which $\mathbf{n}$ represents the outward normal vector. $\Gamma_{\mathrm{s}}(t)$ and $\Gamma_{\mathrm{l}}(t)$ denote the fixed boundary of the solid and non-solid respectively, and $\varepsilon(t)$ denotes solidus line.

It was found that the temperature recovery method was useful in reducing the computational effort without sacrificing the computational accuracy and time. Thus Chen et al. [56], determined the interface location using the temperature recovery method at each time step. After determining the solid zone, the momentum, energy and species transport equations were solved for the non-solid region. Since the governing equations are nonlinear, the results of the previous time step were used for the velocity terms $\mathbf{V}^*$ in the momentum, energy and species balance equations for linearization. In addition, the previous step solutions for the temperature and species distribution were used for decoupling coupled terms in the momentum balance equations.

The numerical solution procedure of Chen et al. [56] can be summarized as follows:

1. At the beginning of each time step, temperature fields were obtained by solving the energy equation, omitting the convective terms. The phase interface zone was determined from the given temperature distributions.
2. A new mesh for the non-solid domain was created since a new interfacial position was obtained from the result of step (1) above.
3. The discretized and linearized momentum, temperature and species balance equations were solved for the non-solid domain in a staggered manner. At the first iteration, the velocity distributions were obtained with the values of temperature and species composition from the previous step. Using these velocity distributions, the energy and species balance equations were solved.

Li [85] studied the transient evolution of the fluid flow, heat transfer and solidification phenomena during continuous casting of aluminum alloys using a computational methodology based on the deformable finite-element concept. In this approach the moving front was modeled by deforming appropriate elements and the accompanying fluid and thermal fields were calculated using a mixed Eulerian–Lagrangian formulation. The effect of solidification on the fluid behavior was modeled by the temperature dependent viscosity and the latent heat release by the Scheils equation. Two types of casting processes were numerically simulated namely, round ingot casting and spreading casting. In the case of round ingot casting, it was found that the temperature distribution changed rapidly at the initial stage but the change slowed down later in the process. Findings for spreading casting showed that fluid flow had a strong effect on the thermal field and was mainly responsible for redistributing thermal energy in the system.

## 10.4 Multi-domain (Two-Region) Methods

In the two-region methods, the total solidification domain is divided into two regions and the interface region, and each region is treated separately [86–88]. Since the volume of each region changes with time, the method is often referred to as the variable domain method. It can also be divided into the following five groups:

- Fixed grid method
- Variable space grid method
- Variable time-step method
- Boundary immobilization method
- Isotherm migration method

In the fixed grid method, a special differencing scheme is applied near the interface, considering the latter is a boundary. Either the grid size or number of grids is adjusted in the variable space grid method such that the interface lies on a grid point. The time step is selected such that the interface moves one grid per time step in a variable time-step method. In the boundary immobilization method, the interface remains fixed by the transformation of coordinates. The isothermal migration method consists of exchanging one of the spatial variables and temperature, making the former a dependent variable.

The prediction of only the temperature distribution during continuous casting of a metal does not usually require detailed modeling of the flow and species transport processes in the melt. For certain purposes, energy-equation based models can be useful to understand and optimize the process. However, such models cannot be used to predict inhomogeneities in the chemical composition of a solidified material (i.e. macrosegregation) caused by uncontrolled species transport due to melt convection.

There are generally two main approaches for handling the mushy region in multi-domain methods. The first approach considers the planar interface between the solid and liquid regions and solves only transport equations for the liquid and solid phases. The interface is tracked explicitly with appropriate boundary conditions [87, 89–92].

In this kind of problems involving solidification from below, the transport equations for solid and liquid phases are;

$$\frac{\partial T}{\partial t} = \alpha_s \frac{\partial^2 T}{\partial z^2} \quad (z < h(t)), \tag{10.51}$$

$$\frac{\partial T}{\partial t} + u\frac{\partial T}{\partial z} = \alpha_l \frac{\partial^2 T}{\partial z^2} \quad (z > h(t)), \tag{10.52}$$

$$\frac{\partial C}{\partial t} + u\frac{\partial C}{\partial z} = DC_{zz}, \tag{10.53}$$

where $\alpha_s$ and $\alpha_l$ are the thermal diffusivities of the solid and liquid phases respectively, $z$ is the axial (vertical) coordinate and $h(t)$ is the evolving height of the solid–liquid interface. The boundary conditions for these equations are (for solidification from below):

$$T = T_B(z = 0), \tag{10.54}$$

$$T = T_h, C = C_h(z = h(t)), \tag{10.55}$$

and

$$T \to T_\infty, C \to Co(z \to \infty \text{ or } t \to 0). \tag{10.56}$$

Two flux conditions are typically applied at the unknown position of the interface $h(t)$. Conservation of heat requires that the latent heat of solidification must be diffused away from the interface thus;

$$L\frac{\partial h}{\partial t} = K_s T_z \,|_{z=h-} - k_l T_z \,. \tag{10.57}$$

Conservation of solute at the interface,

$$r(1-k)C_h \frac{\partial h}{\partial t} = -DC_z \,|_{z=h+}, \tag{10.58}$$

where $r (= \rho_s/\rho_l)$ is the ratio of densities of the solid and liquid phases.

Assuming thermodynamic equilibrium, the temperature and concentration at the interface are related as:

$$T(h,t) = -mC(h_+, t), \tag{10.59}$$

where $m$ is the liquidus slope.

The above system of equations admits a similarity solution in terms of erf function. The growth rate and the position of the interface are usually calculated from the equation obtained from the solution of the transport equations coupled with the boundary conditions. The resulting relation is generally a complex transcendental equation whose eigen values $\lambda$ are related to the interface location as $h(t) = 2\lambda\sqrt{Dt}$.

Coriell et al. [92] solved the transient one-dimensional heat and solute diffusion equation without considering convective effect. The situation considered was similar to a diffusion-controlled problem of Coates and Kirkaldy [93] and Maugis et al. [90], which has multiple similarity solutions. Assuming local equilibrium at the solid/metal interface Coriell et al. [92] obtained the equation for the parabolic growth rate which was solved numerically. Using a lead–tin alloy as an example to examine phase stability, it was found that the diffusion path begins at the composition and temperature of one phase, crosses the two-phase region between the liquidus and solidus lines, and terminates at the composition and temperature of the other phase.

The second approach treats the mushy region as a third region and solves a separate set of transport equations for this region. Hills et al. [94] systematically developed a full set of thermodynamic equations for the mushy region, while solving a much-reduced one-dimensional set to approximately analyze the growth of a binary alloy. In their analysis, the growth of the mush was assumed to advance with a prescribed constant velocity. To improve this growth-constraining assumption, Huppert and Worster [89] formulated a simpler mathematical model on the basis of global conservation relations. The models assume that the solid fraction is constant through the mush, which eliminates the possible interaction between convection and solidification.

Worster [2] improved this model by considering the solid fraction as a function of space and time, leading to a qualitative description of the morphology of the mush. Including a marginal-equilibrium condition at the melt/mush interface closes the model. Unlike the above mathematical models in which no buoyancy-induced convection is considered, Fowler [95], based on the thermodynamic relation developed by Hills et al. [94], employed the Darcy's law to be the momentum conservation in the mush, to successfully analyze the convection stability in the mush. Using Fowler's model in the mush as well as the common model in the bulk melt, several investigations [86, 96, 97] have successfully analyzed the stability characteristics of the convective flow in the directional solidification system in which a dendritic mushy layer underlying a semi-infinite bulk fluid layer is considered. The above models treat the mushy and the bulk fluid regions separately, which usually require explicit tracking of the melt/mush interface of regular shape as well as incorporating empirical relations between the solid fraction and the permeability of the mush.

A typical set of transport equations for the mushy region employed by Chiareli et al. [87] and partly developed by Hills et al. [94], Roberts and Loper [98], Fowler [95] and Worster [99] are:

$$C_m \frac{\partial T}{\partial t} + C_e U \nabla T = \nabla (K_m \nabla T) + J_{-s} \frac{\partial \phi}{\partial t}, \tag{10.60}$$

$$(1-\phi)\frac{\partial C}{\partial t} + U \nabla C = \nabla (D_m \nabla C) + rC(1-k)\phi, \tag{10.61}$$

where, $C_m$, $K_m$, and $D_m$ are the average thermal properties which are obtained using solid fraction $\phi$: for example $C_m$ is expressed thus:

$$C_m = \phi C_s + (1 - \phi)C_l, \tag{10.62}$$

where, $C_s$ and $C_l$ are the specific heat of solid and liquid, respectively.

The velocity is calculated from the following equation:

$$\nabla U = (1 - r)\phi_t, \tag{10.63}$$

which implies that there is only shrinkage-induced flow.

Generally two approaches are used for the boundary condition at the mush–melt interface. The first one is zero solid fraction at the edge of the mush or marginal phase equilibrium at the edge of the liquid phase. For a significant range of parameter values, these two alternatives are essentially equivalent.

An alternative for the extra-mush–melt boundary condition proposed by Hills et al. [94] is that the mass fraction of solid, $\phi$, at the edge of the mush is zero. Fowler [95] followed the lead of Hills et al. [94] in the choice of extra condition, but Langer [100] suggested that the advancing dendrite tips are marginally stable to constitutional super cooling. Worster [2] proposed an alternative extra condition, that of marginal equilibrium at the edge of the melt. Kerr et al. [101], Chen et al. [97], Chiareli et al. [87] followed Worster in applying the marginal equilibrium condition. Subsequently Anderson and Worster [102], and Schultze and Worster [103] reverted to the condition that $\phi = 0$.

The boundary conditions at the mushy–melt interfaces are:

$$LV_n[\phi] = [K_m n \nabla T], \tag{10.64}$$

$$r(1 - k)C_e V_n[\phi] = [D_m n \nabla C], \tag{10.65}$$

$$(r - 1)V_n[\phi] = [nU], \tag{10.66}$$

where parenthesis [] denotes the jump in the enclosed quantity across an interface with normal $n$ moving with normal velocity $V_m$.

The boundary conditions at the bottom of the mushy layer at the cooled surface $z = 0$ are:

$$T = T_B, n.U = 0 \text{ (at } Z = 0). \tag{10.67}$$

Chiareli et al. [87] solved the transport equations coupled with the boundary conditions by introducing a similarity variable

$$\eta = \frac{Z}{2\sqrt{K_e t}}, \tag{10.68}$$

and the position of interface is,

$$h(t) = 2\rho\sqrt{K_e t}. \tag{10.69}$$

The resulting ordinary differential equations are solved numerically by shooting on the fixed computational domain $0 \leq \xi \leq 1$ in which $\xi = \eta/\mu$.

The model is used to predict the flow within the interstices of the mushy layer on the overall growth rate and on the distribution of solid within it. Further, the redistribution of solute is calculated indicating the level of macrosegregation to be expected in a fully solidified system.

Obtaining the boundary conditions using continuity of variables and fluxes at the interfaces leaves the problem underdetermined and an auxiliary condition often must arbitrarily be specified at this interface. Loper [104] showed that the underdeterminacy may be resolved by application of an inequality assuring that the melt is not supercooled together with the condition that the mass fraction of solid has physically realistic values. The lack of super cooling is a natural corollary of the assumption of thermodynamics phase equilibrium in the mushy region. For a wide range of parameter values, this corollary leads directly to the "extra" boundary conditions that have been previously proposed, which are the marginal phase equilibrium and zero solid fractions. Therefore no auxiliary condition needs to be applied.

Worster [99] and Chen et al. [97] employed Darcy's law as the momentum equation in the mushy layer and the Navier–Stokes equations as the momentum equation in the fluid layer. A non-slip boundary condition is prescribed at the mushy/melt interface. Specifically the formulation of this multi-domain approach (model 1) involves mathematically the coupling of the two momentum equations through an appropriate set of matching conditions at the melt/mushy interface. The two momentum equations are of different order of differentiation, necessitating the use of some form of empirical conditions. Worster [99] employed the non-slip condition based on the assumption that the characteristic length scale of the flow on the melt/mushy interface is much larger than the dendrite arm spacing. This in general sense, however, may not be the case since boundary-layer mode (BLM) of convection prevails in the system for most of the cases observed in the experiments. An alternative approach (Model 2) was proposed by Beavers and Joseph [105] who systematically developed with a series of experiments, an empirical condition between the horizontal velocities of the fluid and the porous layers (or a homogenous and isotropic mushy layer). This condition primarily allows a slip between the velocities above and below the melt/mushy interface. The validity of this condition was confirmed by Taylor [106] and Saffman [107]. Another alternative (Model 3) is to use the momentum equation in the mushy region, the Brinkman-extended Darcy equation, which is associated with the continuity of the velocities and the first derivative of velocities across the melt/mushy interface. This model is mathematically equivalent to the model developed with the single-domain approach.

Lu and Chen [108] linearized the equations of the three models by introducing small disturbances to the basic state quantities and substituting their combination into the original equations. After obtaining the linearized equations by neglecting higher-order terms of disturbance quantities and applying the normal mode proportional to $\exp(wt + i\alpha x)$, equations were obtained for small disturbances. The resulting equations and the associated boundary conditions of each model consist of a complex eigenvalue problem,

$$F(R_m, \alpha, w, \theta_x, A^*, C^*, H^*, \varepsilon, \pi) = 0, \tag{10.70}$$

in which a set of ordinary differential equations of thirteen-order of models 1 and 2, and 15-order of model 3, subject to relevant boundary conditions, are to be solved numerically.

In (10.70), $R_m = \frac{g\beta \Delta C \Pi_0 H}{\nu\alpha}$ is the Rayleigh number for mushy region, $\Pi_0$ is the reference permeability, $\alpha$ represents the horizontal wave number, $w$ the normal mode of frequency $\varepsilon = \frac{\alpha}{D}$ is the Lewis number, $Pr$ is the Prandtl number, $C^* = \frac{C_s - C_\infty}{\Delta C}$ is the concentration ratio, $F = L/c\Delta T$ is the Stefan number, and $L$ is the latent heat of fusion.

These eigenvalue problems have been solved with a shooting technique, in which the integration is implemented in a truncated domain [109] and the far-field boundary conditions are imposed at a sufficiently large distance so that the integrations are independent of the value of the truncated distance. Noting that in the mushy layer the basic state is known only implicitly, a change of independent variable proposed by Worster [2] is used to avoid the need to invert the transcendental equations. The final equations are those developed by Keller [110], who imposed orthonormalization to improve accuracy, and also linearization developed by Powell [111] to promote convergence of the eigenvalues $R_m$ and $W$.

From the solution of the above equations, Lu and Chen [108] showed that a liquid plume which carries a cooler and less concentrated fluid is released from the mushy region and penetrates into the fluid layer, forming plume convection. It was shown that two types of convection patterns form in the system. The first is a convection pattern circulating between the mushy layer and the fluid layer with a length scale of about the height of the mushy layer. This pattern is called the mushy layer mode (MLM). The second pattern forms above the melt/mush interface and is called BLM.

Figure 10.7 shows the critical flow patterns obtained with the three models. It is shown that BLM is a convection cell sitting above the mush/melt interface and MLM consists of two separated cells, one in the mushy layer and one in the fluid layer. The critical flow patterns for the three models are similar except that the critical wavelength differs from one another.

## 10.5 Boundary Conditions

The solution of the heat transfer equations is largely controlled by the boundary conditions used to specify the problem. In solidification problems, these boundary conditions depend on the nature of the contact between the freezing material and its container as well as the heat transferred by the container to the external cooling media. The latter problem has been well documented in the heat transfer literature over the years where correlations for convective heat transfer to flowing streams of coolant, natural convection and so on abound. The boundary conditions between the freezing material and mold however, are less well documented.

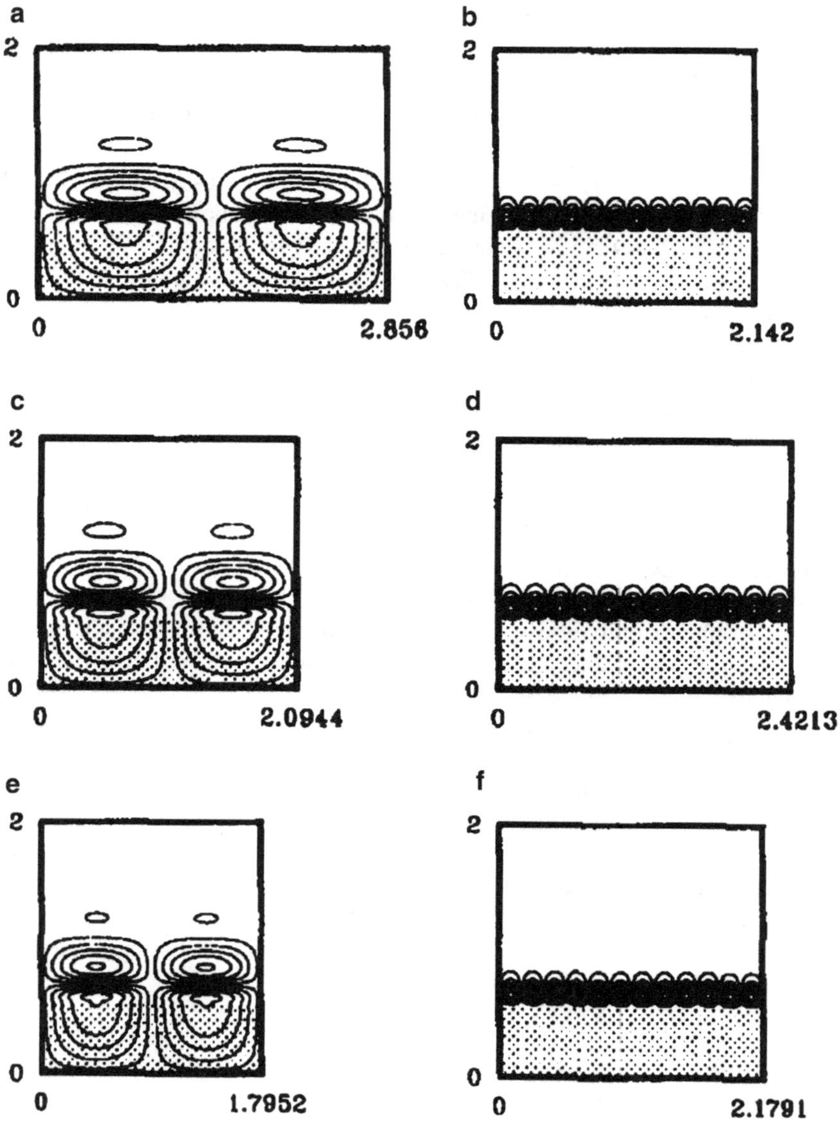


**Fig. 10.7** The critical flow patterns resulted from the three models. (**a**) Model 1, MLM, $\alpha^c = 3.42$, $R_m = 18.896$, (**b**) model 1, BLM, $\alpha^c = 17.6$, $R_m = 9.14$; (**c**) model 2, MLM, $\alpha^c = 3.00$, $R_m = 18.530$ (**d**) model 2, BLM, $\alpha^c = 15.5$, $R_m = 5.653$, (**e**) model 3, MLM, $\alpha^c = 3.50$, $R_m = 18.855$ (**f**) model 3, BLM, $\alpha^c = 17.3$, $R_m = 8.047$. The shadow accounts for the mushy layer [108]

It is frequently the case that the rates of heat transfer between the material and mold changes during the solidification process. When the material is still liquid, but once solidification begins, thermal contractions of the material reduces the heat transfer rate through the formation of an air gap. At the same time, the mold heats and expands contributing further to the formation of the gap. The relative importance of the gap depends on its size and on the thermal conductivities of the material and mold materials.

In solidification modeling, initially, alloy in the mold is generally assumed to be at rest, to be isothermal and of uniform composition, which can be mathematically expressed as:

$$\boldsymbol{u} = 0, T = T_{\rm i}, C_{\rm l} = C_{\rm i},\ f_{\rm l} = 1,\ \text{at}\ t = 0, \tag{10.71}$$

where, $\boldsymbol{u}$ is the velocity vector, and $T_{\rm i}$ and $C_{\rm i}$ are the initial temperature and composition respectively.

The boundary conditions are essentially the same for all the methods reviewed above. However, the multiphase and multi-domain methods need extra boundary conditions due to the additional equations solved. Therefore the general boundary conditions employed in control volume methods and FEMs are discussed together and additional boundary conditions required for other methods are given in subsequent sections.

Boundary conditions for energy equation are usually difficult to assess. This is a result both of the complex heat transfer phenomena with the surroundings and of the possible air gap formation between two different media. The external boundary conditions for energy equation, which can apply to a part $\Gamma_a$ of the domain surface $\Omega_v$ can be of the following types;

$$T(r_i, t) = T(r_i, t)\ \text{for}\ r_i, \varepsilon, \Gamma_\Omega\ \text{(DirichletCondition)}, \tag{10.72}$$

$$-k\frac{\partial T}{\partial n}(r_i, t) = Q_{\rm ext}(r_i, t)\ \text{for}\ r_i, \varepsilon, \Gamma_\Omega\ \text{(Newmann Conduction)}, \tag{10.73}$$

$$-k\frac{\partial T}{\partial n}(r_i, t) = h(r_i, t)[T(r_i, t) - T_a]\ \text{for}\ r_i, \varepsilon, \Gamma_\Omega \text{(Cauchy Conduction)}. \tag{10.74}$$

A prescribed temperature can sometimes decrease at a certain rate along the boundary thus,

$$T = T_{\rm i} - rt, \tag{10.75}$$

where, $T_{\rm i}$ is the initial temperature and $r$ is a prescribed cooling rate. Any combination of the above-cited conditions can be imposed along the boundaries of the domain.

The boundary conditions associated with the momentum equations are no-slip at the solid boundaries, expressed as,

$$u = v = w = 0, \tag{10.76}$$

at the solid boundaries that include completely solidified regions in the domain.

If there is a free surface in the system the shear and surface tension forces balance each other there such that;

$$\mu \frac{\partial u}{\partial y}\bigg|_{y=H} = \sigma_{\mathrm{T}} \frac{\partial T}{\partial x}\bigg|_{y=H} + \sigma_{\mathrm{S}} \frac{\partial f_l}{\partial x}\bigg|_{y=H}, \tag{10.77}$$

where, $\sigma_{\mathrm{T}}$ and $\sigma_{\mathrm{S}}$ are the thermal and solutal surface tension gradients.

If the effect of surface tension can be neglected, then the free surface at the top of the domain is free of normal and tangential stresses. The boundary conditions at the top $(y = H)$ becomes,

$$p = 0, \quad \frac{\partial u}{\partial y} = 0, \quad \frac{\partial T}{\partial y} = 0, \quad \frac{\partial C_l}{\partial y} = 0, \tag{10.78}$$

where, $p$ is the gage pressure.

No transfer of solute mass is allowed at solid boundaries or at an undeformable free surface (if it is assumed at the top of the container) thus,

$$\phi \frac{\partial c_l}{\partial n} = 0, \tag{10.79}$$

along those boundaries. If the assumption of very long container is made, we must require a balance of diffusive and convective transport along that boundary so that

$$D \frac{\partial c_l}{\partial x} + w\,(c_l - c_\infty) = 0, \tag{10.80}$$

at the top boundary.

### 10.5.1 Boundary Conditions in Multiphase Models

The velocity boundary conditions employed in the multiphase models are the no-slip condition for the liquid velocity and zero normal components for solid velocity. However, partial slip may occur for the solid at the wall if the diameter of the solid particles is larger than the length scale of the surface roughness [42]. The tangential velocity of the solid is proportional to its normal gradient at the wall thus:

$$(V_{\mathrm{s}})_t \mathrm{I_w} = -\lambda_{\mathrm{p}} \frac{\partial (V_{\mathrm{s}})_t}{\partial n} \mathrm{I_w}, \tag{10.81}$$

where $\lambda_{\mathrm{p}}$ is the mean distance between particles given by

$$\lambda_{\mathrm{p}} = \frac{d_{\mathrm{e}}}{\varepsilon_{\mathrm{g}}^{1/3}}, \tag{10.82}$$

with $d_{\mathrm{e}}$ and $\varepsilon_{\mathrm{g}}$ being the grain diameter and grain fraction, respectively. Note that for small $d_{\mathrm{e}}$, the slip coefficient $\lambda_{\mathrm{p}}$ approaches zero so that no-slip condition is imposed for the grains.

### 10.5.2 Boundary Conditions for Multi-region Method

Studies employing the multi-domain method usually consider solidification problems cooled from the bottom. There are primarily three approaches for treating boundary conditions between the mush–melt and solid–mush interfaces in the multi-domain approach,. The first model to be considered is that used by Worster [2] and Chen et al. [97] in which the Darcy's equation is employed as the momentum equation in the mushy layer and the Navier–Stokes equation as the momentum equation in the fluid layer and no-slip boundary condition is prescribed at the melt–mushy interface.

Consider a mushy layer that lies above a static solid region and below a semi-infinite fluid region in a binary solution of concentration $C_\infty$, and temperature $T_\infty$, and unidirectional solidification from below. Both the melt/mushy and mushy/solid interfaces move upwards with a constant velocity $V$. The mushy layer extends form $z = 0$ to $z = h(x, y, t)$. The boundary conditions at the $Z \to \infty$ are,

$$\theta \to \theta x, \tag{10.83}$$

$$C^* \to 0, \tag{10.84}$$

$$U \to 0, \tag{10.85}$$

in which

$$\theta = \frac{T - T_{\mathrm{L}}(C_\infty)}{\Delta T}, \quad C^* = \frac{C - C_\infty}{\Delta C},$$

where

$$\Delta T = \Gamma \Delta C = T_{\mathrm{L}}(C_\infty) - T_{\mathrm{E}}, T_{\mathrm{L}}(C_\infty)$$

is the liquidus temperature corresponding to $C_\infty$.

The boundary conditions at the melt/mushy interface $Z = h$ are

$$\theta = C^*, \tag{10.86}$$

$$n\nabla\theta = n\nabla C^*, \tag{10.87}$$

$$[nU] = 0, \tag{10.88}$$

$$[C^*] = 0, \tag{10.89}$$

$$[n\nabla C^*] = 0, \tag{10.90}$$

$$\phi = 1, \tag{10.91}$$

$$[P] = 0, \tag{10.92}$$

$$U - (nU)n = 0, \tag{10.93}$$

where, the square brackets denote the jump of the enclosed quantity across the interface and $n$ is a unit vector normal to the interface. Equation (10.86) is the liquidus relation expressing the marginal equilibrium condition. Equation (10.87) expresses the continuities of temperature and heat flux while (10.88) is an enforced condition assuming the solid at the interface vanishes. Equation (10.92) expresses the continuity of normal stress and (10.93) is the no-slip boundary condition employed by Worster.

At the mushy/solid interface $z = 0$,

$$\theta = -1, \tag{10.94}$$

$$w = 0. \tag{10.95}$$

Since the temperature here is assumed to be the eutectic temperature $T_E$ and due to viscosity the non-penetration condition at the solid boundary is satisfied.

Worster [99] employed the non-slip condition based on the assumption that the characteristic length scale of the flow on the melt/mushy interface is much larger than the space between the arms of the dendrite. This in general sense may not be the case since the BLM convection prevails in the system for most of the cases considered in the experiments. An alternative (the second model) was proposed by Beavers and Joseph [105] who systematically developed an empirical relation between the horizontal velocities of the fluid and the porous layer (or a homogenous and isotropic mushy layer) based on a series of experiments. This condition primarily allows a slip between the velocities above and below the melt/mushy interface. This condition can be expressed as

$$\frac{\partial U_2}{\partial z}\mathrm{I}_{h^+} = \Lambda\sqrt{\frac{\bar{H}}{\Pi(1)}}(U_2\mathrm{I}_{h^+} - U_2\mathrm{I}_{h^-}), \tag{10.96}$$

where $U_2$ accounts for the horizontal velocity vector $(u, v)$, $\Pi(1)$ the permeability at the melt/mushy interface, $\Lambda$ an empirical constant determined experimentally, $\bar{H} = \frac{H^2}{\Pi_0}$ is the inverse of the Darcy number, and $H$ is the characteristic length. This equation means that at the melt/mushy interface a slip between the velocities in the fluid and mushy layer is proportional to the shear rate.

The third model is that the momentum equation in the mushy is the Brinkman-extended Darcy Equation, associated with continuity across the melt/mushy interface. This model is mathematically equivalent to the model developed with the single-domain approach.

## References

1. Poulikakos D (1984) Maximum density effects on natural convection in a porous layer differentially heated in the horizontal direction. Int J Heat Mass Transf 27:2067–2075
2. Worster MG (1986) Solidification of an alloy from a cooled boundary. J Fluid Mech 167: 481–501

3. Sparrow EM, Patankar SV, Ramadhyani S (1977) Analysis of melting in the presence of natural convection in the melt region. J Heat Transf 99:520–526
4. Webb BW, Viskanta R (1986) Analysis of heat transfer during melting of a pure metal from an isothermal vertical wall. Numer Heat Transf 9:539–558
5. Wolff F, Viskanta R (1995) Solidification of a pure metal at a vertical wall in the presence of liquid superheat. Int J Heat Mass Transf 31:1735–1744
6. Kroeger PG, Ostrach S (1974) The solution of a two-dimensional freezing problem including convection effects in the liquid region. Int J Heat Mass Transf 17:1191–1207
7. Sparrow EM, Ramsey JW, Kemink RG (1979) Freezing controlled by natural convection. J Heat Transf 101:578–584
8. Bennon WD, Incropera FP (1987) A continuum model for momentum, heat and species transport in binary solid–liquid phase change system-I. Model formulation. Int J Heat Mass Transf 30(10):2161–2170
9. Bennon WD, Incorpera FP (1987) A continuum model for momentum, heat and species transport in binary solid–liquid phase change system: II. Application to solidification in a rectangular cavity. Int J Heat Mass Transf 30:2171–2187
10. Bennon WD, Incropera FP (1988) Numerical analysis of binary solid–liquid phase change using a continuum model. Numer Heat Transf 13:277–296
11. Voller VR, Prakash C (1987) A fixed grid numerical modeling methodology for convection–diffusion mushy region phase-change problems. Int J Heat Mass Transf 308:1709–1719
12. Voller VR (1989) Development and application of a heat balance integral method for analysis of metallurgical solidification. Appl Math Modell 13:3–11
13. Voller VR, Brent AD, Prakash C (1989) The modeling of heat, mass and solute transport in solidification system. Int J Heat Mass Transf 32:1719–1731
14. Voller VR, Swaminathan CR (1990) Fixed grid techniques for phase change problems: a review. Int J. Numer Meth Eng 30:875–898
15. Patankar SV (1980) Numerical heat transfer and fluid flow. McGraw-Hill, New York
16. Brent AD, Voller VR, Reid KJ (1988) Enthalpy porosity technique for modeling convection–diffusion phase change: application to the melting of a pure metal. Numer Heat Transf 13:297–318
17. Amberg G (1991) Computation of macrosegregation in an iron–carbon cast. Int J Heat Mass Transf 341:217–227
18. Shahani H, Amberg G, Fredriksson H (1992) On the formation of macrosegregations in unidirectionally solidified Sn–Pb and Pb–Sn alloys. Metall Trans A 23A:1992–2301
19. Prakash C, Voller V (1989) On the numerical solution of continuum mixture model equations describing binary solid–liquid phase change. Numer Heat Transf Part B 15:171–189
20. Beckermann C, Viskanta R (1988) .Double-diffusive convection during dendritic solidification of a binary mixture. PCH 10:195–213
21. Christenson MS, Bennon WD, Incropera FP (1989) Solidification of an aqueous ammonium chloride solution in a rectangular cavity-II. Comparison of predicted and measured results. Int J Heat Mass Transf 32(1):69–79
22. Prescott PJ, Incropera FP (1991) Numerical simulation of a solidifying Pb–Sn alloy: the effects of cooling rate on thermosolutal convection and macrosegregation. Metall Mater Trans B 22B:529–540
23. Yoo H, Viskanta R (1992) Effect of anisotropic permeability on the transport process during solidification of a binary mixture. Int J Heat Mass Transf 35(10):2335–2346
24. Chiang KC, Tsai HL (1992) Interaction between shrinkage-induced fluid flow and natural convection during alloy solidification. Int J Heat Mass Transf 35(7):1771–1778
25. Chen JH, Tsai HL (1993) Inverse segregation for a unidirectional solidification of aluminum–copper alloys. Int J Heat Mass Transf 36(12):3069–3075
26. Prescott PJ, Incropera FP, Gaskell DR (1994) Convective transport phenomena and macrosegregation during solidification of a binary metal alloy: II – experiments and comparisons with numerical predictions. Trans ASME 116:742–749
27. Diao QZ, Tsai, HL (1994) Modeling of the formation of under-riser macrosegregation during solidification of binary alloys. Metall Mater Trans A 25A:1051–1062

28. Schneider MC, Beckermann C (1995) A numerical study of the combined effects of microsegregation, mushy zone permeability and flow, caused by volume contraction and thermosolutal convection, on macrosegregation and eutectic formation in binary alloy solidification. Int J Heat Mass Transf 38:3455–3473
29. Ilegbusi OJ, Mat MD (1997) A hybrid model of the mushy region in phase-change problems. J Mater Process Manuf Sci 6:1–15
30. Ilegbusi OJ, Mat MD (1998) Modeling flowability of mushy zone with a hybrid model utilizing coherency solid fraction. Mater Sci Eng A A247:135–141
31. Rady MA, Nada SA (1998) Solidification of hypereutectic binary alloys with buoyancy and surface tension driven natural convection. Heat Mass Transf 34:337–347
32. Vreeman CJ, Krane MJM, Incropera FP (2000) The effect of free-floating dendrites and convection on macrosegregation in direct chill cast aluminum alloys Part I: model development. Int J Heat Mass Transf 43:677–686
33. Mat MD, Ilegbusi OJ (2000) Prediction of macrosegregation in binary alloy solidification using a non-Newtonian mushy model. J. Mater Process Manuf Sci 8:188–217
34. Raw WY, Lee SL (1991) Application of weighting function scheme on convection-conduction phase cage problems. Int J Heat Mass Transf 34:1503–1513
35. Prescott PJ, Incropera FP (1994) Convective transport phenomena and macrosegregation during solidification of a binary metal alloy: I. Numerical predictions. J Heat Transf 116:735–741
36. Krane MJM, Incropera FP (2000) The effect of free-floating dendrites and convection on macrosegregation in direct chill cast aluminum alloys Part II: predictions for Al–Cu and Al–Mg alloys. Int J Heat Mass Transf 43:687–704
37. Ni J, Beckermann C (1991) A volume-averaged two-phase model for transport phenomena during solidification. Metall Mater Trans B 22B:349–361
38. Wang CY, Beckermann C (1993) Single- vs. dual-scale volume averaging for heterogeneous multiphase systems. Int J Multiphase Flow 19:397–407
39. Wang CY, Beckermann C (1993) A multiphase solute diffusion model for dendritic alloy solidification. Metall Trans A 24:2787–2802
40. Wang CY, Beckermann C (1993) A unified solute diffusion model for columnar and equiaxed dendritic alloy solidification. Mater Sci Eng A 171:199–211
41. Wang CY, Beckermann C (1994) Prediction of columnar to equiaxed transition during diffusion-controlled dendritic alloy solidification. Metall Trans A 25:1081–1093
42. Wang CY, Beckermann C (1995) Computer simulations of microstructural development in dendritic alloy solidification with convection. In: Voller VR, Marsh SP, El-Kaddah N (eds) Materials processing in the computer age – II. The Minerals, Materials Society, Warrendale, PA, pp 129–143
43. Reddy AV, Beckermann C (1995) Simulation of the effects of thermosolutal convection, shrinkage induced flow and solid transport on macrosegregation and equiaxed grain size distribution in a Dc continuous aast Al–Cu round. In: Voller VR, Marsh SP, El-Kaddah N (eds) Materials processing in the computer age – II. The Minerals, Materials Society, Warrendale, PA, pp 89–102
44. Reddy AV, Beckermann C (1997) Modeling of macrosegregation due to thermosolutal convection and contraction-driven flow in direct chill continuous casting of an Al–Cu round ingot. Metall Mater Trans B 28:479–489
45. Beckermann C, Ni J (1996) Simulation of sedimentation in globulitic alloy solidification. Int J Heat Mass Transf 23(3):315–324.
46. Spalding DB (1981) Numerical computation of multiphase flow and heat transfer. In: Taylor C, Morgan K (eds) Recent advances in numerical methods in fluids, vol 1. Pineridge Press, Swansea, pp 139–167
47. Rosten H, Spalding DB (1986) PHOENICS beginner's guide and user's manual. CHAM Limited Technical Report TR/100, London, UK
48. Thévoz Ph, Desbiolles JL, Rappaz M (1989) Modeling of equiaxed microstructure formation in casting. Metall Mater Trans A 20A:311–322
49. Voller VR, Beckermann C (1999) A unified model of microsegreegation and coarsening. Metall Mater Trans A 30A:2183–2189

50. Crivelli LA, Idelsohn SR (1986) A temperature-based finite element solution for phase-change problems. Int J Numer Meth Eng 23:99–119
51. Felicelli SD, Heinrich JC, Poirier DR (1991) Simulation of freckles during vertical solidification of binary alloy. Metall Trans B 22:847–859
52. Nandapurkar PJ, Poirier DR, Heinrich JC (1991) Momentum equation for dendritic solidification. Numer Heat Transf 19A:297–311
53. Usmani SA, Lewis WR, Seetharamu NK (1992) Finite element modeling of natural convection-controlled change phase. Int J Numer Meth Fluids 14:1019–1036
54. Felicelli SD, Heinrich JC (1993) Numerical model for dendritic solidification of binary alloys. Numer Heat Transf B 23:461–481
55. Ruan Y, Li BQ, Liu JC (1995) A finite element method for steady-state conduction – advection phase change problems. Finite Elements Anal Des 19:153–168
56. Chen Y, Im YT, Lee J (1995) Finite element simulation of solution with momentum, heat and species transport. J Mater Process Tech 48:571–579
57. Felicelli SD, Heinrich JC, Poirier DR (1998) Three-dimensional simulations of freckles in binary alloys. J Cryst Growth 191:879–888
58. Ravindran K, Brown SGR, Spittle JA (1999) Prediction of the effective thermal conductivity of three-dimensional dendritic regions by the finite element method. Mater Sci Eng 269 A: 90–97
59. Nigro N, Huespe A, Fachiotti (2000) Phasewise numerical integration of finite element method applied to solidification processes. Int J Heat Mass Transf 43:1053–1066
60. Comini G, Del Guidice S, Lewis RW, Zienkiewicz (1974) Finite element solution of non-linear heat conduction problems with special reference to phase change. Int J Numer Meth Eng 8:613–624
61. Chorin JA (1968) Numerical solution of Navier–Stokes equations. Math Comp 22:745–762
62. Ramaswamy B, Jue TC (1992) Some recent trends and developments in finite element analysis for incompressible thermal flows. Int J Numer Meth Eng 35:675–692
63. Ohnaka I (1985) Introduction to analysis of heat transfer and solidification by computer-application for casting process. Maruzem, Tokyo
64. Thomas BG, Samarasekera IV, Brimacombe JK (1984) Comparison of numerical modeling techniques for complex, two-dimensional, transient heat conduction problems. Metall Trans 15B:307–318
65. Dantzig JA (1989) Modeling liquid–solid phase changes with melt convection Int J Numer Meth Eng 28:1769–1785
66. Rolph WDIII, Bathe KJ (1982) An efficient algorithm for analysis of non-linear Heat transfer with phase change. Int J Numer Meth Eng 18:119–134
67. Lewis RW, Roberts PM (1987) Finite element simulation of solidification problems. Appl Sci Res 44:61–92
68. Pham QT (1986) The use of lumped capacitances in the finite-element solution of heat conduction with phase change. Int J Heat Mass Transf 29:285–292
69. Ortega JM, Rheinboldt WC (1980) Iterative solution of nonlinear equations in several variables. Academic Press, New York
70. Lees M (1966) A linear three-level difference scheme for quasi-linear parabolic equations. Math Comput 20:516–622
71. Pardo E, Weckman DC (1990) A fixed grid finite element technique for modeling phase change in steady-state conduction-advection problems. Int J Numer Meth Eng 29(5):969–984
72. Marshall RS, Heinrich JC, Zienkiewicz OC (1978) Natural convection in a square enclosure by a finite element, penalty function method, using primitive fluid variables. Numer Heat Transf 1:315–330
73. Heinrich JC, Marshall RS (1981) Viscous incompressible flow by a penalty function finite element method. Comput Fluids 9:73–83
74. Heinrich JC, Yu CC (1988) Finite element simulation of buoyancy-driven flows with emphasis on natural convection in a horizontal circular cylinder. Comput Meth Appl Mech Eng 69:1–27
75. Ganesan S, Poirier DR (1990) Conservation of mass and momentum for the flow of interdendritic liquid during solidification. Metall Trans B 21B:173–181

76. Steven G (1982) Internally Discontinuous Finite elements for moving interface problems. Int J Numer Methods Eng 18:(4), 569–582
77. Fachinotti VD, Cardono A, Huespe AE (1999) Fast convergent and accurate temperature model for phase-change heat conduction. Int J Numer Meth Eng 44:1863–1884
78. Miller K, Miller RN (1981) Moving finite elements. I: SIAM J Numer Anal 18:1019–1057
79. Lynch DR, O'Neill K (1981) Continuously deforming finite elements for the solution of parabolic problems, with and without phase change. Int J Numer Meth Eng 17:81–96
80. Zabaras N, Ruan Y (1989) A deforming finite element method for analysis of inverse Stefan problems. Int J Numer Meth Eng 28:295–313
81. Zabaras N, Ruan Y (1990) Moving and deforming finite-element simulation of two-dimensional Stefan problems. Commun Appl Numer Meth 6:495–506
82. Ruan Y, Liu JC, Ricmond O (1993) A deforming finite element method for analysis of alloy solidification problems. Finite Elem Anal Des 12:49–63
83. Tszeng TC, Im YT, Kobayashi S (1989) Thermal analysis of solidification by the temperature recovery method. Int J Mach Tools Manuf 29:107
84. Chen YH, Im YT, Lee ZH (1991) Three dimensional finite element analysis with phase change by temperature recovery method. Int J Mach Tools Manuf 31:1
85. Li BQ (1997) Numerical simulation of flow and temperature evolution during the initial phase of steady-state solidification. J Mater Process Technol 71:402–413
86. Worster MG (1991) Natural convection in a mushy layer. J Fluid Mech 167:481–501
87. Chiareli AOP, Huppert HE, Worster MG (1994) Segregation and flow during the solidification of alloy. J Cryst Growth 139:134–146
88. Paradies CJ, Smith RN, Glicksman ME (1997) The influence of convection during solidification on fragmentation of the mushy zone of a model alloy. Metall Mater Trans A 28A:875–883
89. Huppert HE, Worster MG (1985) Dynamic solidification of a binary melt. Nature 314:703–707
90. Maugis P, Hopfe WD, Morral JE, Kirkaldy JS (1996) Degeneracy of diffusion paths in ternary, two-phase diffusion couples. J Appl Phys 79:7592
91. Maugis P, Hopfe WD, Morral JE, Kirkaldy JS (1997) Multiple interface velocity solutions for ternary biphase infinite diffusion couples. Acta Mater 45:1941
92. Coriell SR, McFadden GB, Sekerka RF, Boettinger WJ (1998) Multiple similarity solutions for solidification and melting. J Cryst Growth 191:573–585
93. Coates DE, Kirkaldy JS (1971) Morphological stability of alpha.-b phase interfaces in the copper-zinc-nickel system at 775.deg. Metall Trans 2:3467
94. Hills RN, Loper DE, Roberts PH (1983) A thermodynamically consistent model of a mushy zone. Q J Mech Appl Math 36:505–539
95. Fowler AC (1985) The formation of freckles in binary alloys. IMA J Appl Math 35:159–174
96. Amberg G, Homsy GM (1993) Nonlinear analysis of buoyant convection in binary solidification with application to channel formation. J Fluid Mech 252:79–98
97. Chen F, Yang TY, Lu JW (1993) Influence of convection on solidification of binary solutions cooling from below. J Appl Phys 74:7531
98. Roberts PH, Loper DE (1983) Towards a theory of the structure and evolution of a dendrite layer. In: Soward AM (ed) Stellar and planetary magnetism. Gordon and Breach, New York, pp 329–349
99. Worster MG (1992) Instabilities of the liquid and mushy regions during solidification of alloys. J Fluid Mech 237:649–669
100. Langer JS (1980) Instabilities and pattern formation in crystal growth. Rev Appl Math 52:1
101. Kerr RC, Woods AW, Worster MG, Huppert HE (1990) Solidification of an alloy cooled from above part 2. Non-equilibrium interface kinetic. J Fluid Mech 217:331–348
102. Anderson DM, Worster MG (1995) Weakly nonlinear analysis of convection in mushy layers during the solidification of binary alloys. J Fluid Mech 302:307
103. Schultze TP, Worster MG (1998) A numerical investigation of steady convection in mushy layers during the directional solidification of binary alloys. J Fluid Mech 356:199
104. Loper DE (2001) On the boundary conditions at a mush–melt interface. J Cryst Growth 222:655–666

105. Beavers GS, Joseph DD (1968) Boundary conditions at a naturally permeable wall. J Fluid Mech 30:197
106. Taylor GI (1971) A model for the boundary condition of a porous material Part 1. J Fluid Mech 49:319
107. Saffman PG (1971) On the boundary condition at the interface of a porous medium. St Appl Math 50:93
108. Lu JW, Chen F (1997) Assessment of mathematical models for the flow in directional solidification. J Cryst Growth 171:601–613
109. Coriell SR, Cordes MR, Boettinger WJ, Sekerka RF (1980) Convective and interfacial instabilities during unidirectional solidification of binary alloy. J Cryst Growth 49:13–28
110. Keller HB (1976) Numerical solution of two points boundary value problems. Regional conference series in applied mathematics. SIAM, Philadelphia, PA
111. Powell MJD (1970) A hybrid method for nonlinear equations. In Numerical methods for nonlinear algebraic equations. In: Rabinowitz PH (ed) Gordon and Breach, London

# Chapter 11
# Review of Nanoscale and Microscale Phenomena in Materials Processing

## 11.1 Introduction

Due to recent rapid developments in a variety of nanoscale and microscale devices [1, 2], information on the basic characteristics of multiphase transport phenomena in nanochannels and microchannels has become increasingly important. A characteristic feature of such systems is that a surface force or an interfacial force predominates over the body force. The wettability of solid walls of channels and surface tension of liquid, therefore, play an essential role in analyzing the transport phenomena.

The wettability is quantitatively evaluated in terms of the contact angle, $\theta$, usually defined using the sessile drop method [3], as shown in Fig. 11.1. This method involves placing a droplet on a solid plate for a sufficiently long time. The plate is wetted by the liquid in the droplet for $90° > \theta \geq 0°$, while it is poorly wetted for $180° \geq \theta \geq 90°$. A bubble cannot stay on a wetted solid body, while a bubble can attach to a poorly wetted solid body until the volume of the bubble reaches a certain critical value. This critical value depends on the contact angle, $\theta$, and the surface tension of the liquid, $\sigma$. Consequently, the behavior of dispersed gas phase (bubble and slug) changes drastically depending on the wettability of the channel. The effect of the wettability on the characteristics of gas–liquid two-phase flows becomes more significant as the size of the channel decreases.

This chapter is divided into fundamentals and applications of nanoscale and microscale transport phenomena as follows:

### *11.1.1 Fundamentals*

- Generation methods of nanobubbles and microbubbles
- Removal of nanobubbles and microbubbles from a gas–liquid mixture
- Flow patterns of gas–liquid two-phase flows in microchannels
- Flow characteristics in microchannels
- Heat transfer in microchannels

M. Iguchi and O.J. Ilegbusi, *Modeling Multiphase Materials Processes: Gas-Liquid Systems*, DOI 10.1007/978-1-4419-7479-2_11,

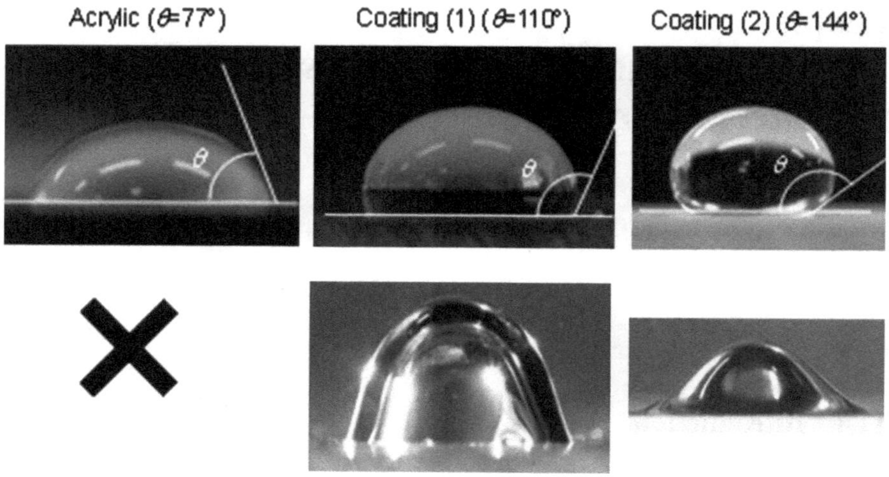


**Fig. 11.1** Water droplet and bubble on rod of different wettability ($\theta$ contact angle, *upper figures*: water droplet, *lower figures*: air bubble)

- Numerical simulation of nanoscale and microscale transport phenomena
- Mixing in microchannels and microreactors
- Measurement methods

### *11.1.2 Applications*

- Enhancement of gas dissolution rate
- Microfluidic devices
- Fuel cell

## 11.2 Definitions and Generation Method of Nanoscale and Microscale

### *11.2.1 Bubbles*

#### 11.2.1.1 Nanobubble and Microbubble

Microbubbles and nanobubbles are not clearly defined. For example, the following classification is commonly used with reference to the bubble diameter [4].

- Microbubble: $d_B = 10\,\mu m$ to $(50–60)\,\mu m$
- Micro-nanobubble: $d_B = (500–600)$ nm to $10\,\mu m$
- Nanobubble: $d_B < (500–600)$ nm

### 11.2.2 Generation Method

Various methods have been proposed for generating nanobubbles and microbubbles to study their basic characteristics. Some commercial generation methods of microbubbles are briefly reviewed in Serizawa et al. [5].

1. *Single-hole nozzle*
   Sano et al. [6] found that when a single-hole nozzle is wetted by a liquid, a bubble growing at the end of the nozzle spreads toward the outer edge. Accordingly, the volume of the nozzle is governed by the outer diameter of the nozzle. On the other hand, the volume of a bubble generated from a wetted nozzle is governed by the inner diameter of the nozzle. Consequently, a wetted nozzle is best suited for generating nanobubbles and microbubbles.
   The diameter of a bubble, $d_B$, generated successively from a wetted nozzle of around 1 mm in inner diameter is given by the relation [7]:

$$d_B = [6Q_g/(1.06\pi)]^{1/3}[\sigma/(\rho_L g^3)]^{1/12}(\rho_L/\rho_g)^{1/15}\left[d_{ni}/\left(Q_g{}^2/g\right)^{1/5}\right]^{5/30} \tag{11.1}$$

   where $Q_g$ is the gas flow rate, $\sigma$ is the surface tension, $\rho_L$ is the density of liquid, g is the acceleration due to gravity, $\rho_g$ is the density of gas, and $d_{ni}$ is the inner diameter of the nozzle. The diameter of a bubble becomes small as the inner diameter of the nozzle decreases. However, it is generally difficult to generate nanobubbles using a single-hole nozzle.
2. *Porous nozzle*
   Nanobubbles and microbubbles can be generated from a porous nozzle as long as the nozzle is wetted by the liquid and the gas flow rate is low. A bubble growing at the exit of one of the pores of a poorly wetted porous nozzle coalesces with bubbles growing at the exits of pores located around it. As a result, only large bubbles equivalent to the diameter of the porous nozzle are successively generated from the poorly wetted porous nozzle.
   Kikuzaki et al. [8] generated monodispersed bubbles with a mean diameter of less than 1 μm from a hydrophilic porous glass membrane with uniform size-controlled pores. The size of the bubbles was controlled by varying the size of the membrane. For example, nanobubbles with a mean diameter of 0.72 μm were generated for a mean pore diameter of 0.084 μm.
3. *Turbulent breakup*
   Daughter bubbles can be generated by introducing a mother bubble into a highly turbulent liquid flow. Martinez-Bazan et al. [9] found that the bubble size probability density function of the daughter bubbles depends not only on the size of the mother bubble, but also on the value of the dissipation rate of turbulent kinetic energy.
   Sadatomi et al. developed a microbubble generator inserting a spherical body in a flowing water channel [10]. Fujikawa et al. generated microbubbles using a rotational porous plate [11].

Breakup of bubbles in a water jet is usually used for generating microbubbles [12, 13]. This is a relatively simple and convenient method.

4. *Ultrasonication*
   Kim et al. [14] generated many nanobubbles by ultrasonication using a palladium electrode. This method uses cavitation of liquid caused by ultrasonic wave. The average size of the bubbles was in the range of 300–500 nm.
5. *Depressurization*
   Microbubbles can be generated from high-pressure liquid with high gas solubility by depressurization. Inaba et al. [15] reported that the growth of an air bubble from its nucleus is influenced by water viscosity, water velocity at the nozzle, and air solubility in pressurized water.
6. *Packed bed*
   Microbubbles can be generated by flowing a liquid–gas mixture through a Rashig ring packed bed, as demonstrated by Miyahara [16].
7. *Cavitation nozzle*
   A cavitation nozzle is suitable for generating many microbubbles [17]. A venturi meter is also very useful for this purpose.
8. *Gas–liquid two-phase swirl flow*
   Ohnari et al. developed a simple and convenient microbubble generator [4]. This generator is regarded as a type of depressurization method.

## 11.3 Removal of Gas from Gas–Liquid Mixture

There exist many metallurgical reactions generating bubbles in a reactor. When the size of a bubble in the reactor is large enough, the buoyancy force being one of body forces is dominant. The bubble is therefore carried up to the bath surface of the reactor and escapes into the atmosphere. This situation cannot be expected when the size of the reactor is very small because the buoyancy force loses its validity. Many kinds of bubble removal methods have been proposed, as listed below.

- Centrifugal force method [18, 19]
- Ultrasonic method [20, 21]
- Surface tension difference method [22–24],
- Electrostatic force method [25]
- Electromagnetic force method [26–28]
- Y-junction of different branch angle [29–32]
- Y-junction of different wettability [3, 33, 34]

The last method [3, 33, 34] is expected to be promising in nanoscale and microscale reactors since surface forces predominate in such situations. This method relies on the fact that gas is likely to attach to a poorly wetted solid body [35–39]. An example will be briefly reviewed below.

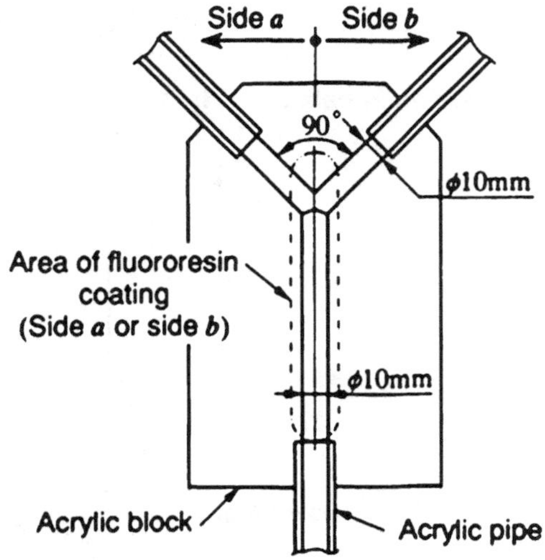


**Fig. 11.2** Y-junction

Figure 11.2 shows a Y-junction made of transparent acrylic resin. One branch was coated with repellent to change its wettability. The original acrylic resin is wetted by water. Figure 11.3 shows a schematic of the experimental apparatus. The Y-junction was placed on a horizontal plane and a gas–liquid two-phase flow was introduced into the main pipe of the Y-junction. Figure 11.4 shows that bubbles move preferably in the poorly wetted branch. Under a certain condition, all the bubbles penetrated into the poorly wetted side. As the size of the Y-junction becomes smaller, the surface force becomes more dominant. Accordingly, more efficient bubble removal can be realized.

## 11.4 Flow Pattern of Gas–Liquid Two-Phase Flow in Microchannels

Information on flow patterns of gas–liquid two-phase flows in nanochannels is currently not available. Therefore, this section focuses on microchannels.

Serizawa et al. [40] observed two-phase flow patterns in horizontal microchannels of 20, 25, and 100 μm in inner diameter using horizontal air–water flows. The test pipe was made of transparent silica, and the test section was approximately 10 mm. Since the surface tension force predominates over the gravitational force, the flow patterns are quite different from those in macropipes. A flow pattern map is discussed here.

The flow patterns of gas–liquid two-phase flows in microchannels are reviewed in Kraus et al. [41] and Akbar et al. [42]. The entire flow regime map in microchannels can be divided into four regions defined by the Weber numbers $We_{LS}$ and $We_{GS}$ as follows [42]:

1. Surface tension-dominated region, including bubbles, plug, and slug (where flow pattern is dominated by large and elongated bubbles); $3.0 \geq We_{LS}$, $0.11\, We_{LS}^{0.315} \geq We_{GS}$ and $We_{LS} > 3,\ 1.0 \geq We_{GS}$

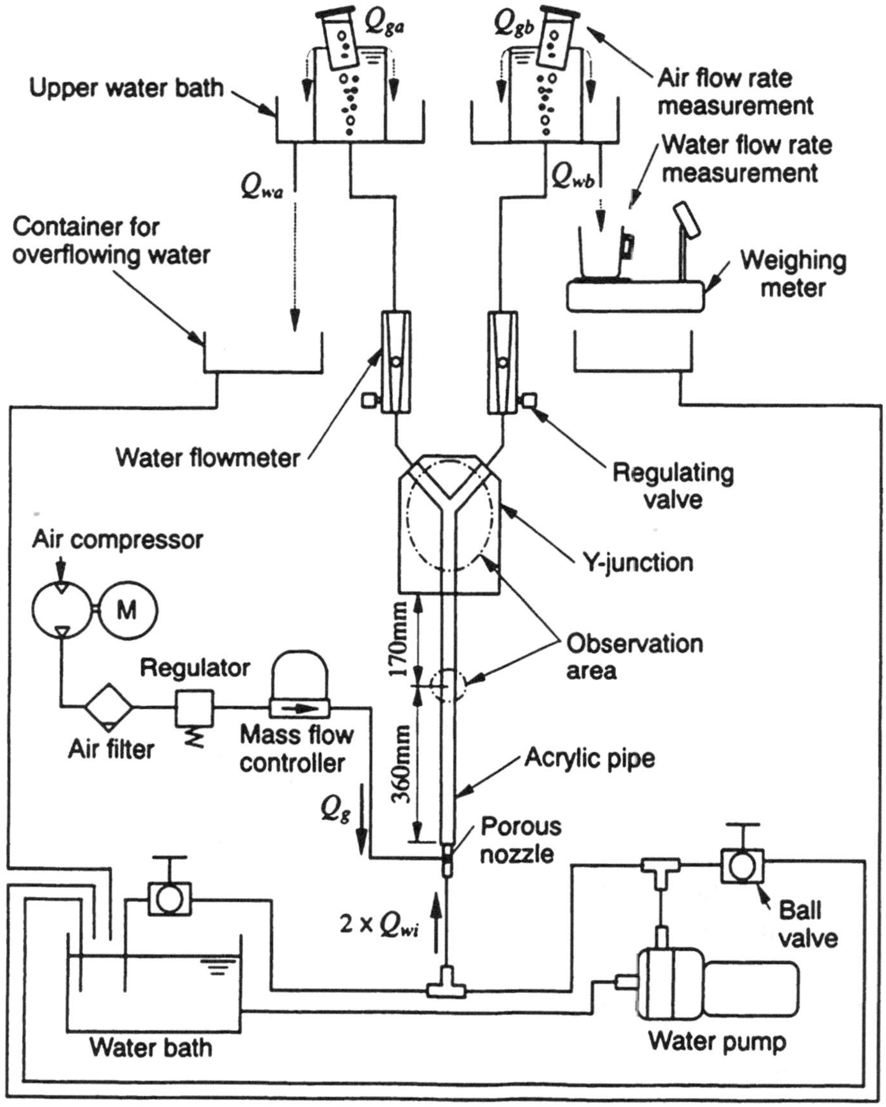


**Fig. 11.3** Schematic of experimental apparatus

2. Inertia-dominated zone 1, including annular and wavy-annular regimes; $3.0 \geq We_{LS}$, $We_{GS} \geq 11.0 We_{LS}^{0.14}$
3. Inertia-dominated zone 2, including the dispersed flow regimes; $We_{LS} > 3.0$, $We_{GS} > 1.0$
4. Transition zone.

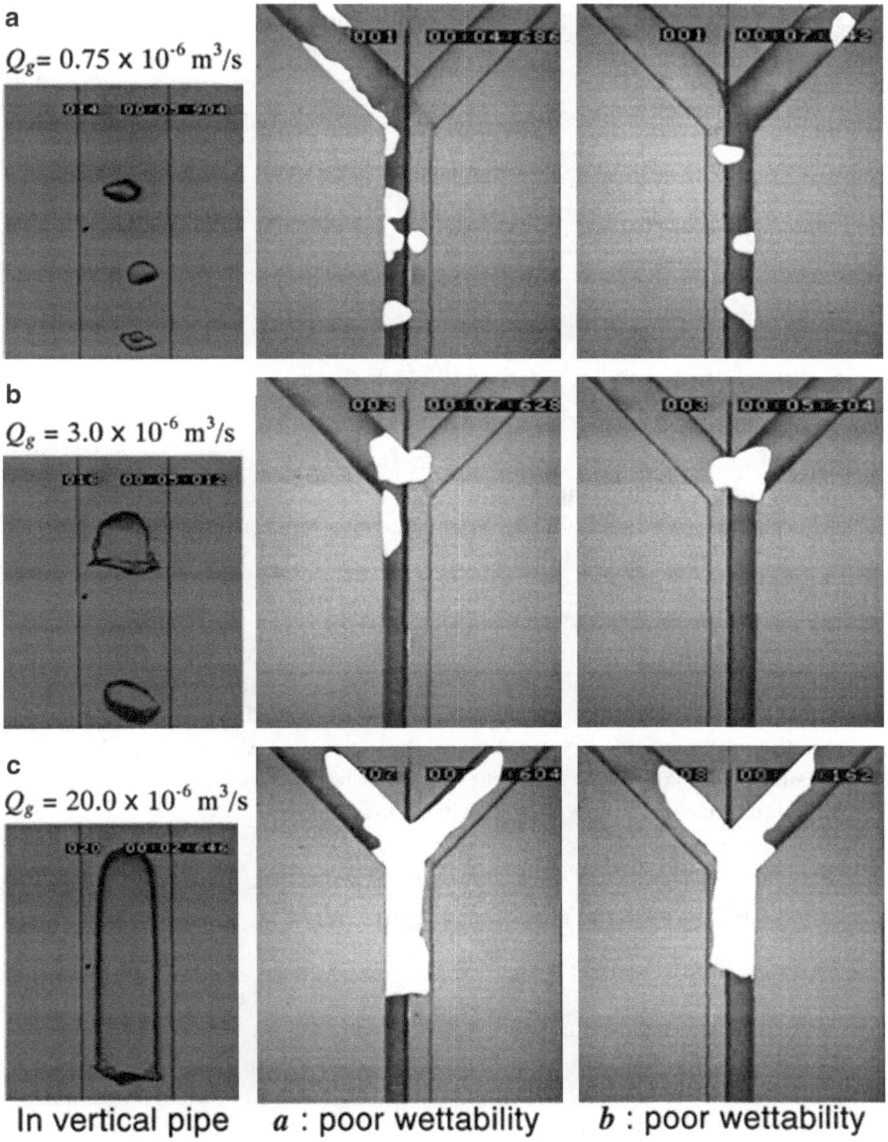


**Fig. 11.4** Bubbles moving near Y-junction (*white part*: bubble)

In the above expressions, $We_{LS}$ and $We_{GS}$ are defined as

$$We_{LS} = U_{LS}{}^2 D_H \rho_L/\sigma \tag{11.2}$$

$$We_{GS} = U_{GS}{}^2 D_H \rho_G/\sigma \tag{11.3}$$

$$U_{LS} = Q_L/A \tag{11.4}$$

$$U_{GS} = Q_G/A \tag{11.5}$$

$$D_H = 4A/S \tag{11.6}$$

where $U_{LS}$ and $U_{GS}$ are the liquid and gas superficial velocities, respectively, $\rho_L$ and $\rho_G$ are the densities of liquid and gas, respectively, $D_H$ is the hydrodynamic channel diameter, $\sigma$ is the surface tension, and $Q_L$ are $Q_G$ are the liquid and gas flow rates, respectively, $A$ is the cross-sectional area, and S is the peripheral length of the pipe.

The two-phase flow patterns in parallel microchannels are investigated by Hetsroni et al. [43] The effects of wettability and channel diameter on gas–liquid two-phase flows in microchannels have also been extensively investigated [44–46].

## 11.5 Flow Characteristics in Microchannels

A variety of microparticle image velocimetry (micro PIV) have been developed for measuring flows in microchannels [47–49]. Westerweel et al. [47] proposed a micro PIV which is particularly suitable for quasi-stationary and periodic flows and carried out velocity measurement in a straight microchannel with a near-rectangular trapezoidal cross-section of about $50 \times 20\,\mu\text{m}^2$ and a hydraulic diameter of $26\,\mu\text{m}$. The total length of the channel was 35 mm. A reasonable result was obtained even very close to the wall although a single-phase flow was considered. Park et al. [49] also developed a micro PIV and measured the mean velocity in microchannels of 99 and $516\,\mu\text{m}$ diameter. The accuracy of the method was confirmed by the good agreement between the measured velocity profiles and the theoretical parabolic profiles.

Transition from laminar to turbulent flow was observed by Bau and Pfahler [50]. Investigation of turbulent flow in microchannels is limited.

## 11.6 Heat Transfer in Microchannels

Peng et al. [51] conducted experiments to measure the bubble nucleation temperatures for four working fluids boiled on a platinum wire confined in capillary tubes. The bubble nucleation temperature markedly depended on the size of the microchannel.

Kroeker et al. [52] investigated the pressure drop and thermal characteristics of heat sinks with circular microchannels using the continuum model consisting of the conventional Navier–Stokes equations and the energy conservation equation.

Steinke et al. carried out experimental investigation to study the control of dissolved gases and their effects on heat transfer and pressure drop during the flow of water in a microchannel [53].

As a fundamental investigation on heat pipes, Tchikanda et al. [54] derived analytical expressions for the mean velocity of a liquid flowing in an open rectangular microchannel. The flow is driven by pressure gradients and shear stress on the liquid/vapor interface. Koizumi et al. [55] developed a microheat pipe with asymmetric cross-section and observed boiling two-phase flow.

The effects of microchannel size, mass flow rate, and heat flux on boiling incipience or bubble cavitation in a microchannel were investigated by Li and Cheng [56]. The effects were also estimated of contact angle, dissolved gas, and the existence of microcavities and corners in the microchannel on bubble nucleation and cavitation temperature.

## 11.7 Numerical Simulation of Transport Phenomena

Nagayama et al. [57] carried out nonequilibrium molecular dynamic simulations to study the effect of interface wettability on the pressure driven flow of a Lennard–Jones fluid in a nanochannel. The velocity profile changed significantly depending on the wettability of the wall. The no-slip boundary condition breaks down for a hydrophobic wall. Siegel et al. [58] developed a two-dimensional computational model for fuel cells.

## 11.8 Mixing in Microchannels and Microreactors

Mixing in microchannels is one of the most important technologies responsible for the efficiency of microreactors [59, 60]. Suzuki et al. developed particle tracking velocimetry (PTV) for the measurement of chaotic mixing in a microreactor [61].

## 11.9 Measurement Method

A promising device for measuring nanoscale and microscale channel flows is micro PIV, as mentioned in the previous Sect. "Flow Characteristics in Microchannels." Santiego et al. [62] developed a micro PIV for measuring the instantaneous and mean liquid velocities around a circular cylinder of a diameter of 30 $\mu$m placed in a microchannel. The diameter of the tracer particles ranged from 100 to 300 nm. Visualization of flow field in microchannels [63,64] and Brownian motion of a tracer particle [65] can also provide useful insight. Micro PIV is also used for observing the motion of blood cell in a capillary [66, 67]. Micro hot-wire anemometer [68] and Doppler velocimetry [69] have also been recently developed for such studies in which bulk carbon nanotube were used as a wire.

## 11.10 Enhancement of Gas Dissolution Rate

The dissolution rate of gas into liquid is highly enhanced by introducing microbubbles [70]. Ohnari et al. used micro-/nanobubbles for enhancing the growth rate of bivalves, such as scallops and oysters [71] and for waste water treatment [72–74].

## 11.11 Microfluidic Devices

Microfluidic devices in liquid handling systems have been investigated [75–78]. Microreactors, microvalves, and micropumps should be referred to review papers by Reyes et al. [1] and Auroux et al. [2].

## 11.12 Fuel Cell

A novel fuel cell concept was proposed by Choban et al. [79] Many studies have discussed useful methods for enhancing the efficiency of the conventional fuel cells [80–87].

## 11.13 Closing Remarks

An understanding of transport phenomena in nanochannels and microchannels is essential for the development of microdevices and microreactors. Precise measurement devices for the transport phenomena are desirable. The flow field in nanoscale and microscale channels is largely driven by the surface tension force. Since the same situation is encountered in microgravity, the information available in this chapter could also be relevant to materials processing in space.

## References

1. Reyes DR, Iossifidis D, Auroux P-A, Manz A (2002) Micro total analysis systems. 1. Introduction, theory, and technology, Anal Chem 74:2623–2636
2. Auroux PA, Iossifidis D, Reyes P-A, Manz A (2002) Micro total analysis systems. 2. Analytical standard operations and applications, Analytical Chemistry 74:2637–2652
3. Inoue T, Iguchi M, Mizuno Y (2000) Separation of gas and liquid based on wettability difference of circular pipes. Proc. 1st Jpn. Soc, Multiphase Flow Conference pp 241–242
4. Ohnari H, Ohnari H, Nakayama T, Shakutui H (2003) Proc. JSME Annual Meeting, Osaka, July 23–25, pp 221–222
5. Serizawa A, Yahiro T (2001) Proc. JSMF Annual Meeting, Kita-kyusyu, July 12–13, pp 139–140
6. Sano M, Fujita Y, Mori K (1976) Formation of bubbles at single nonwetted nozzles in mercury, Metal. Trans. 7B:300–301
7. Iguchi M, Chihara T (1998) Water model study of the frequency of bubble formation under reduced and elevated pressures, Metall Mater Trans B 29B:755–761
8. Kukizaki M, Nakashima T, Song J, Kohama Y (2004) Monodispersed nano-bubbles generated from porous glass membrane and bubble size control, Kagaku Kogaku Ronbunshu 30–5:654–660
9. Martinez-Bazan JL, Montanes, Lasheras JC (1999) On the breakup of an air bubble injected into a fully developed turbulent flow. Part 2. Size PDF of the resulting daughter bubbles, J. Fluid Mech. 401:183–207

10. Sadatomi M, Kawahara A, Kano K, Ohtomo A (2002) Proc. JSMF Annual Meeting, Nagoya, July 29–31, pp 13–14
11. Fujikawa S, Zhang R, Hayama S, Peng G (2003) The control of micro-air-bubble generation by a rotational porous plate, Int. J. Multiphase Flow 29:1221–1236
12. Ito H, Tomodokoro T, Takemura F, Hishida K (2002) Proc. JSMF Annual Meeting, Nagoya, July 29–31, pp 157–158
13. Shakouchi T, Oike T, Nishiyama S (2002) Proc. JSMF Annual Meeting, Nagoya, July 29–31, pp 159–160
14. Kim M, Song G, Kim J.D (2000) Zeta potential of nanobubbles generated by ultrasonication in aqueous alkyl polyglycoside solutions, J. Colloid and Interface Science 223:285–291
15. Inaba H, Horibe A, Haruki N, Nakao T (2004) Micro bubbles generation by depressurization from high pressure water, Proc. JSMF Annual Meeting, Okayama, August 5–7, pp 325–326
16. Miyahara T, Norikane H, Maeda J, Ono Y (2004) Micro-bubble formation by liquid flow through raschig ring packed bed, Proc. JSMF Annual Meeting, Okayama, August 5–7, pp 327–328
17. Inui T, Serizawa A, Yahiro T (2003) Proc. JSMF Annual Meeting, Osaka, July 23–25, pp 165–166
18. Imai R, Yano T (1997) A study of bubble separation utilizing centrifugal force under a reduced gravity condition, Trans. JSME(B) 63–611:2296–2302
19. Shakutsui H, Watanabe K, Onari H, Saga T, Kadowaki H (2000) A311 flow patterns in swirl gas-liquid two-phase flow in a vertical pipe (Phase separation), Proc. of 4th JSME-KSME Thermal Engineering Conference, October 1–6, Kobe, Japan, vol. 3, pp 69–72
20. Kamimura H, Yoshihara S, Azuma H (1993) J Jpn Soc Microgravity Application 10–4:129
21. Mochizuki O (1996) Active control of bubble, Proc. 5th Symposium on Short Period Microgravity Application, pp 27–30
22. Miyazaki Y (1996) Development of high temperature fluid flow technologies in microgravity, NEDO-ITK-9507, pp 159–172
23. Jpn. Soc. Microgravity Application (1996) Frontiers in experiments in space, B-1135:219
24. Azuma H et al (1983) Proc. 27th Symposium on Science and Technology in Space, pp 182–183
25. Imai R, Yano T (1994) Bubble transfer by electrostatic force, preprint of Jpn. Soc. Mech. Eng, No. 930–9(1993), pp 388–390; Trans. JSME(B) 60:3979–3986
26. Okuzawa T, Kojima Y, Tsubouchi K, Takagi Y, Hamamoto T (1992) Fundamental investigation of an electromagnetic degasser, Trans. JSME(C) 58:3543
27. Wakayama N (1996) Development of high temperature fluid flow technologies in microgravity. NEDO-ITK-9507, pp 99–112
28. Okuzawa T et al (1990) Proc. 7th Symposium on Microgravity Application, pp 367–372
29. Lahey RT Jr (1986) Current understanding of phase separation mechanisms in branching conduits, Nuclear Engineering and Design 95:pp 145–161
30. Azzopardia BJ, Purvisand A, Govana AH (1987) Annular two-phase flow split at an impacting T, Int. J. Multiphase Flow 13–5:605–614
31. Hwang ST, Soliman HM, Lahey RT Jr (1989) Phase separation in impacting wyes and tees, Int. J. Multiphase Flow 15–6:965–975
32. Asano H, Fujii T, Takenaka N, Watanabe T, Ogura A, Arakawa T (2000) A309 The phase separation characteristics of a gas-liquid two-phase flow in an impacting Y-junction: Application to microgravity condition (Phase separation), Proc. 4th JSME-KSME Thermal Eng. Conf. 3:3–57-3–62
33. Inoue T, Iguchi M, Mizuno Y (2001) Separation of gas and liquid using wettability difference of T-junction, J Jpn Soc. Multiphase Flow 15–2:158–164
34. Mizuno Y, Iguchi M (2002) Separation of gas and liquid based on wettability difference of a horizontally placed Y-junction, Jpn. Soc. Mech. Eng 68–675:2984–2989
35. Terauchi Y, Iguchi M, Kosaka H, Yokoya S, Hara S (1999) Wettability effect on the flow pattern of air—water two-phase flows in a vertical circular pipe, Tetsu-to- Hagane 85:645–651
36. Mizuno Y, Shimizu T, Sonoyama N, Iguchi M (2000) Attachment of bubbles to a horizontal circular cylinder of poor wettability, Jpn. J. Multiphase Flow 14:166–175

37. Mizuno Y, Iguchi M (2001) Behavior of bubbles attaching to and detaching from solid body of poor wettability. ISIJ Int 41 Suppl, S56–S59
38. Iguchi M, Terauchi Y (2001) Boundaries among bubbly and slug flow regimes in air–water two-phase flows in vertical pipe of poor wettability, Int. J. Multiphase Flow 27:729–735
39. Iguchi M, Terauchi Y (2001) Microgravity effects on the rising velocity of bubbles and slugs in vertical pipes of good and poor wettability, Int. J. Multiphase Flow 27:2189–2198
40. Serizawa A, Feng Z, Kawahara Z (2002) Proc. JSMF Annual Meeting, Nagoya, July 29–31, pp 281–284
41. Kraus T, Gunther A, Mas N, Schmidt MA, Jensen KF (2004) An integrated multiphase flow sensor for microchannels, Exp. In Fluids 36:819–832
42. Akbar MK, Plummer DA, Ghiaasiaan SM (2003) On gas–liquid two-phase flow regimes in microchannels, Int. J. Multiphase Flow 29:855–865
43. Hetsroni G, Mosyak A, Segal Z, Pogrebnyak E (2003) Two-phase flow patterns in parallel micro-channels, Int. J. Multiphase Flow 29:341–360
44. Chung PM-Y, Kawaji M (2004) The effect of channel diameter on adiabatic two-phase flow characteristics in microchannels, Int. J. Multiphase Flow 30:735–761
45. de Gennes PG (1998) The dynamics of reactive wetting on solid surfaces, Physica A 249:196–205
46. Paterson A, Fermigier M, Jenffer P, Limat L (1995) Wetting on heterogeneous surfaces: Experiments in an imperfect Hele-Shaw cell, Physical Review E 51–2:1291–1299
47. Westerweel J, Geelhoed PF, Lindken R (2004) Single-pixel resolution ensemble correlation for micro-PIV applications, Exp. in Fluids 37:375–384
48. Bourdon CJ, Olsen MG, Gorby AD (2004) Power-filter technique for modifying depth of correlation in microPIV experiments, Exp. in Fluids 37:263–271
49. Park JS, Cjoi CK, Kihm KD (2004) Optically sliced micro-PIV using confocal laser scanning microscopy (CLSM), Exp. in Fluids 37:105–119
50. Bau HH, Pfahler JN (2001) Experimental observations of liquid flow in micro conduits, 39th AIAA Aerospace Sciences Meeting and Exhibit (2001–1), AIAA 2001–0722
51. Peng XF, Tien Y, Lee DJ (2001) Bubble nucleation in microchannels: statistical mechanics approach, Int. J. Heat Mass Transfer 44:2957–2964
52. Kroeker CJ, Soliman HM, Ormiston SJ (2004) Three-dimensional thermal analysis of heat sinks with circular cooling micro-channels, Int. J. Heat and Mass Transfer 44:4733–4744
53. Steinke ME, Kandlikar SG (2004) Control and effect of dissolved air in water during flow boiling in microchannels, Int. J. Heat and Mass Transfer 47:1925–1935
54. Tchikanda SW, Nilson RH, Griffiths SK (2004) Modeling of pressure and shear-driven flows in open rectangular microchannels, Int. J. Heat and Mass Transfer 47:527–538
55. Koizumi Y, Ohtake H, Sakurai H (2004) Study on micro heat transport device, Proc. JSMF Annual Meeting, Okayama August 5–7:345–346
56. Li J, Cheng P (2004) Bubble cavitation in a microchanne, Int. J. Heat Mass Transfer 47:2689–2698
57. Nagayama G, Cheng P (2004) Effects of interface wettability on microscale flow by molecular dynamics simulation. Int. J. Heat Mass Transfer 47:501–513
58. Siegel NP, Ellis MW, Nelson DJ, von Spakovsky MR (2004) A two-dimensional computational model of a PEMFC with liquid water transport. J. Power Sources 128:173–184
59. Kuksenok O, Yeomans JM, Balazs AC (2002) Using patterned substrates to promote mixing in microchannels. Physical Review E 65(031502):1–8
60. Sato Y, Irisawa G, Ishizuka M, Hishisa K, Maeda M (2003) Visualization of convective mixing in microchannel by fluorescence imaging. Measurement Sci. Tech. 14:114–121
61. Suzuki H, Nakano M, Kasagi N, Ho C-M (2003)ISMME B22–035:397–402
62. Santiego JG, Wereley ST, Meinhart CD, Beebe DJ, Adrian RJ (1998) A particle image velocimetry system for microfluidics. Exp. Fluids 25:316–319
63. Meinhart CD, Wereley ST, Santiego JG (1999) PIV measurements of a microchannel flow. Exp. Fluids 27:414–419
64. Meinhart CD, Wereley ST, Gray MHB (2000) Volume illumination for two-dimensional particle image velocimetry. Measurement Sci. Tech. 11:809–814

65. Sato Y, Inaba S, Hishisa K, Maeda M (2003) Spatially averaged time-resolved particle-tracking velocimetryin microspace considering Brownian motion of submicron fluorescent particles. Exp. Fluids 35:167–177
66. Smith ML, Long DS, Damiano ER, Ley K (2003) Near-Wall m-PIV reveals a hydrodynamically relevant endothelial surface layer in venules in vivo. Biophysical J. 85:637–645
67. Secom TW, Hsu R, Pries AR (2001) Motion of red blood cells in a capillary with an endothelial surface layer: Effect of flow velocity. Am. J. Physiol. Hear. Circ. Physiol. 281:629–636
68. Wang V, Li W (2003) Proc. 4th ASME-JSME Joint FED Conf FEDSM2003-45067
69. Sethuram S, Samimy M, Lempert W (2002) Pa. Am. Inst. Aeronat. Astronaut. AIAA-2002-0690
70. Koito H, Takemura F, Hihara E, Matsumoto Y (2000) JSMF Annual Meeting, Sendai July 13–14, pp 177–178
71. Ohnari H, Harada N, Ohnari H, Nakayama T (2004) Proc. JSMF Annual Meeting, Okayama August 5–7, pp 335–336
72. Nakata A, Ohnari H, Ohnari H, Nakayama T (2004) Proc. JSMF Annual Meeting, Okayama August 5–7, pp 339–340
73. Matsuo K, yamahara Y, Oda T, Maeda K, Shakutui H, Ohnari H (2004) Proc. JSMF Annual Meeting, Okayama August 5–7, pp 341–342
74. Himuro S, Shakutui H, Matsuo K, Ohnari H (2004) Proc. JSMF Annual Meeting Okayama, August 5–7, pp 343–344
75. Lee J, Moon H, Fowler J, Schorllhammer T, Kim C-J (2002) Electrowetting and electrowetting-on-dielectric for microscale liquid handling. Sensors and Actuators A 95:259–268
76. Karlsson R, Karlsson A, Orwar O (2003) A nanofluidic switching device. J. Am. Chem. Soc. 125:8442–8443
77. Garnier N, Grigoriev RO, Schatz MF (2003) Optical manipulation of microscale fluid flow. Phys Rev Lett 91–5(054501):1–4
78. Shinohara K, Sygii Y, Aota A, Hibara A, Tokeshi M, Kitamori T, Okamoto K (2004) High-speed micro-PIV measurements of transient flow in microfluidic devices. Meas. Sci. Tecjnol. 15:1965–1970
79. Choban ER, Markoski LJ, Wieckowski A, Kenis PJA (2004) Microfluidic fuel cell based on laminar flow. J. Power Sources 128:54–60
80. Melli TR, de Santos JM, Kolb WB, Scriven LE (1990) Cocurrent downflow in networks of passages. Microscale Roots of Macroscale Flow Regimes. Ind. Eng. Chem. Res. 29:2367–2379
81. Hui M-H, Blunt MJ (2000) Effects of wettability on three-phase flow in porous media. J. Phys. Chem. B 104:3833–3845
82. Bewer T, Beckmann T, Dohle H, Mergel J, Stolten D (2004) Novel method for investigation of two-phase flow in liquid feed direct methanol fuel cells using an aqueous H2O2 solution. J. Power Sources 125:1–9
83. Wang ZH, Wang CY, Chen KS (2001) Two-phase flow and transport in the air cathode of proton exchange membrane fuel cells. J. Power Sources 94:40–50
84. Tüeber K, Pocza D, Hebling C (2003) Visualization of water buildup in the cathode of a transparent PEM fuel cell. J. Power Sources 124:403–414
85. Mench MM, Dong OL, Wang CY (2003) In situ water distribution measurements in a polymer electrolyte fuel cell. J. Power Sources 124:90–98
86. Satija R, Jacobson DL, Arif M, Werner SA (2004) In situ neutron imaging technique for evaluation of water management systems in operating PEM fuel cells. J. Power Sources 129:238–245
87. You L, Liu H (2002) A two-phase flow and transport model for the cathode of PEM fuel cells. Int. J. Heat Mass Transfer 45:2277–2287

# Appendix 1

## Dimensionless Numbers Useful in Process Metallurgy

| Group | Formula | Nomenclature | Quantities Represented | Application |
|---|---|---|---|---|
| Bagnold number, Ba | $\frac{d^2\gamma\rho}{\mu}$ | $\rho$ particle density<br>$\gamma$ velocity gradient<br>d particle diameter<br>$\mu$ fluid viscosity | $\frac{\text{particle collissional force}}{\text{fluid viscous force}}$ | Granular flows |
| Bond number, Bo | $\frac{g(\rho-\rho^t)d_p}{\sigma}$ | $d_p$ drop diameter<br>$\rho$ drop density<br>$\rho_1$ density of surrounding fluid<br>$\sigma$ surface tension | $\frac{\text{gravitational force}}{\text{surface tension force}}$ | Spraying |
| Drag Coefficient, $C_d$ | $\frac{g(\rho-\rho^t)}{\rho u^2}$ | L characteristic length dimension of object<br>$\rho$ density of object<br>u relative velocity | $\frac{\text{gravitational force}}{\text{inertial force}}$ | Free settling |
| Friction factor, f | $\frac{nL(\Delta P_1/\rho)}{2u^2 l}$ | L characteristic length dimension of conduit cross section<br>$\Delta P_1/\rho$ friction head length of pipe | $\frac{\text{shear stress}}{\text{velocity head}}$ | Friction drops in conduits |
| Froude number, Fr | $\frac{u^2}{gL}$ | L characteristic length dimension of system | $\frac{\text{inertial force}}{\text{gravitational force}}$ | Wave and surface behavior pouring streams |

(continued)

M. Iguchi and O.J. Ilegbusi, *Modeling Multiphase Materials Processes: Gas-Liquid Systems*, DOI 10.1007/978-1-4419-7479-2,

(continued)

| Group | Formula | Nomenclature | Quantities Represented | Application |
|---|---|---|---|---|
| Froude number, modified, $Fr_m$ | $\frac{\rho_g u^2}{(\rho_l - \rho_g)\, gL}$ | $\rho_g$ gas density<br>$\rho_l$ liquid density | $\frac{\text{inertial force}}{\text{gravitational force}}$ | Fluid behavior of gas-liquid systems |
| Galileo number, Ga | $\frac{gp^2 L^2}{\mu^2}$ | μ viscosity | $\frac{(\text{inertial}) \times (\text{gravitational})}{(\text{viscous force})^2}$ | Flow in baths of viscous liquids |
| Grashof number, Gr | $\frac{g\rho^2 L^2 \beta \Delta T}{\mu L u^2}$ | L characteristic length dimension<br>density of liquid<br>coefficient of thermal volume expansion | $(\text{Re})\left(\frac{\text{buoyancy force}}{\text{viscous force}}\right)$ | Free convention heat transfer |
| Grashof number, $Gr_m$ (mass transfer) | $\frac{g\rho^2 L^2 \beta' \Delta x}{\mu^2}$ | β′ coefficient of density change with concentration<br>Δx concentration of driven force | $(\text{Re})\left(\frac{\text{buoyancy force}}{\text{viscous force}}\right)$ | Free convection mass transfer |
| Hartmann number, Ha | $LB\left(\frac{\sigma_e}{\mu}\right)^{1/2}$ | L characteristic length<br>B characteristic magnetic flux density<br>$\sigma_e$ electric conductivity | $\frac{(\text{electromagnetic force})}{\text{viscous force}}$ | |
| Lewis number, Le | $\frac{\rho C_P D_{A \to B}}{k}$ | density of liquid<br>Cp specific<br>DA→B molecular diffusivity<br>k thermal conductivity | $\frac{\text{molecular diffusivity}}{\text{thermal diffusivity}}$ | Combined mass and heat transfer |

(continued)

(continued)

| Group | Formula | Nomenclature | Quantities Represented | Application |
|---|---|---|---|---|
| Mach number, Ma | $\frac{u}{u_s}$ | u fluid velocity<br>$u_s$ velocity of sound in fluid | | High speed flow |
| Morton number, Mo | $\frac{g\mu L^4}{\rho L(\sigma)^3}$ | $\mu_L$ liquid viscosity | $\frac{\text{(gravitational)} \times \text{(viscous) force}}{\text{surface tension force}}$ | Velocity of bubbles in liquids |
| Peclet number, Pe (mass transfer) | $\frac{Lu}{D'_{A\rightarrow B}}$ | L characteristic length<br>D′A→B characteristic molecular diffusivity | $\frac{\text{bulk mass transport}}{\text{mass transfer by diffusion}}$ | Mass transfer in reactions |
| Peclet number, Pe (heat transfer) | $\frac{Lu\rho C_P}{k} = \frac{Lu}{\alpha}$ | u liquid velocity<br>$C_P$ specific heat<br>α thermal diffusivity | $\frac{\text{heat transfer by bulk motion}}{\text{conductive heat transfer}}$ | Forced convection |
| Prandtl number, Pr | $C_p\mu/k$ | $C_P$ specific heat of liquid<br>μ viscosity<br>*k* thermal conduction | $\frac{\text{momentum diffusivity}}{\text{thermal diffusivity}}$ | Forced and free convection |
| Power number, Pn | $\frac{P}{\rho n^3 L^5}$ | P′ power input to agitator<br>L characteristic dimension of agitator paddle<br>n angular speed of rotation | $\frac{\text{drag force}}{\text{viscous force}}$ | Power consumption in agitated vessels |
| Rayleigh number, Ra | $\frac{\rho^2 L^2 Cp\beta\Delta T}{k}$ | β coefficient of thermal volume expansion<br>ΔT temperature different across film<br>L characteristic length dimension | $\frac{\text{heat transfer by convection}}{\text{heat transfer by conduction}}$ | Free convection |

(continued)

(continued)

| Group | Formula | Nomenclature | Quantities Represented | Application |
|---|---|---|---|---|
| Reynolds number, Re | $\frac{Lu\rho}{\mu}$ | L characteristic length dimension of system<br>ρ fluid density<br>μ fluid viscosity | $\frac{\text{inertial force}}{\text{viscous force}}$ | Fluid flow |
| Stoke's number, St | $\frac{dV\rho}{\mu}$ | ρ particle density<br>D particle diameter<br>V fluid velocity<br>μ fluid viscosity | $\frac{\text{particle nertia}}{\text{fluid inertia}}$ | mesoscopic particulate flow |
| Weber number, We | $\frac{\rho L u^2}{\sigma}$ | σ surface tension | $\frac{\text{inertial force}}{\text{surface tension force}}$ | Bubble formation atomization of liquid jet |

# Appendix 2

## Transport Properties of Air

| Air (0.01 MPa) | | | | Air (0.1 MPa) | | | | Air (1.0 MPa) | | | | Air (10 MPa) | | | |
|---|---|---|---|---|---|---|---|---|---|---|---|---|---|---|---|
| Temp. (K) | Density (kg/m³) | Dynamic viscosity (μPa s) | Kinematic viscosity (mm²/s) | Temp. (K) | Density (kg/m³) | Dynamic viscosity (μPa s) | Kinematic viscosity (mm²/s) | Temp. (K) | Density (kg/m³) | Dynamic viscosity (μPa s) | Kinematic viscosity (mm²/s) | Temp. (K) | Density (kg/m³) | Dynamic viscosity (μPa s) | Kinematic viscosity (mm²/s) |
| 80 | 0.43732 | | | 80 | | | | 80 | | | | 80 | | | |
| 100 | 0.34897 | | | 100 | 3.5648 | | | 100 | | | | 100 | | | |
| 120 | 0.29051 | | | 120 | 2.9392 | | | 120 | 33.605 | | | 120 | | | |
| 140 | 0.24888 | | | 140 | 2.5064 | | | 140 | 27.046 | | | 140 | 573.36 | | |
| 160 | 0.21771 | | | 160 | 2.187 | | | 160 | 22.938 | | | 160 | 396.53 | | |
| 180 | 0.19349 | | | 180 | 1.9408 | | | 180 | 20.025 | | | 180 | 273.18 | | |
| 200 | 0.17412 | | | 200 | 1.7448 | | | 200 | 17.821 | | | 200 | 213.78 | | |
| 220 | 0.15828 | | | 220 | 1.5851 | | | 220 | 16.082 | | | 220 | 179.69 | | |
| 240 | 0.14508 | | | 240 | 1.4523 | | | 240 | 14.669 | | | 240 | 156.91 | | |
| 260 | 0.13392 | | | 260 | 1.3401 | | | 260 | 13.491 | | | 260 | 140.19 | | |
| 273.15 | 0.12747 | 17.22 | 135.1 | 273.15 | 1.2754 | 17.23 | 13.51 | 273.15 | 12.818 | 17.28 | 1.348 | 273.15 | 131.36 | 19.47 | 0.1482 |
| 300 | 0.11606 | 18.62 | 160.4 | 300 | 1.1609 | 18.62 | 16.04 | 300 | 11.638 | 18.67 | 1.604 | 300 | 116.88 | 20.47 | 0.1751 |
| 320 | 0.1088 | 19.68 | 180.9 | 320 | 1.0882 | 19.69 | 18.09 | 320 | 10.895 | 19.73 | 1.811 | 320 | 108.29 | 21.32 | 0.1969 |
| 340 | 0.1024 | 20.63 | 201.5 | 340 | 1.024 | 20.63 | 20.15 | 340 | 10.242 | 20.67 | 2.018 | 340 | 101.03 | 22.09 | 0.2186 |
| 360 | 0.096711 | 21.54 | 222.7 | 360 | 0.96707 | 21.54 | 22.27 | 360 | 9.6649 | 21.58 | 2.233 | 360 | 94.785 | 22.85 | 0.2411 |
| 380 | 0.09162 | 22.42 | 244.7 | 380 | 0.9161 | 22.42 | 24.47 | 380 | 9.15 | 22.45 | 2.454 | 380 | 89.344 | 23.61 | 0.2643 |
| 400 | 0.087039 | 23.27 | 267.4 | 400 | 0.87025 | 23.27 | 26.74 | 400 | 8.6879 | 23.3 | 2.682 | 400 | 84.55 | 24.35 | 0.288 |
| 420 | 0.082893 | 24.09 | 290.6 | 420 | 0.82878 | 24.1 | 29.08 | 420 | 8.2708 | 24.13 | 2.917 | 420 | 80.286 | 25.09 | 0.3125 |

(continued)

(continued)

| Air (0.01 MPa) | | | | Air (0.1 MPa) | | | | Air (1.0 MPa) | | | | Air (10 MPa) | | | |
|---|---|---|---|---|---|---|---|---|---|---|---|---|---|---|---|
| Temp. (K) | Density ($kg/m^3$) | Dynamic viscosity (μPa s) | Kinematic viscosity ($mm^2/s$) | Temp. (K) | Density ($kg/m^3$) | Dynamic viscosity (μPa s) | Kinematic viscosity ($mm^2/s$) | Temp. (K) | Density ($kg/m^3$) | Dynamic viscosity (μPa s) | Kinematic viscosity ($mm^2/s$) | Temp. (K) | Density ($kg/m^3$) | Dynamic viscosity (μPa s) | Kinematic viscosity ($mm^2/s$) |
| 440 | 0.079125 | 24.9 | 314.7 | 440 | 0.79108 | 24.9 | 31.48 | 440 | 7.8922 | 24.93 | 3.159 | 440 | 76.462 | 25.82 | 0.3377 |
| 460 | 0.075685 | 25.69 | 339.4 | 460 | 0.75666 | 25.69 | 33.95 | 460 | 7.5471 | 25.72 | 3.408 | 460 | 73.011 | 26.54 | 0.3635 |
| 480 | 0.072531 | 26.46 | 364.8 | 480 | 0.72512 | 26.46 | 36.49 | 480 | 7.2311 | 26.48 | 3.662 | 480 | 69.877 | 27.25 | 0.39 |
| 500 | 0.06963 | 27.21 | 390.8 | 500 | 0.6961 | 27.21 | 39.09 | 500 | 6.9407 | 27.24 | 3.925 | 500 | 67.016 | 27.95 | 0.4171 |
| 550 | 0.0633 | 29.03 | 458.6 | 550 | 0.6328 | 29.03 | 45.88 | 550 | 6.3081 | 29.06 | 4.607 | 550 | 60.839 | 29.67 | 0.4877 |
| 600 | 0.058025 | 30.78 | 530.5 | 600 | 0.58006 | 30.78 | 53.06 | 600 | 5.7816 | 30.8 | 5.327 | 600 | 55.747 | 31.33 | 0.562 |
| 650 | 0.053561 | 32.47 | 606.2 | 650 | 0.53544 | 32.47 | 60.64 | 650 | 5.3366 | 32.49 | 6.088 | 650 | 51.469 | 32.95 | 0.6402 |
| 700 | 0.049735 | 34.1 | 685.6 | 700 | 0.49719 | 34.1 | 68.59 | 700 | 4.9555 | 34.12 | 6.885 | 700 | 47.819 | 34.53 | 0.7221 |
| 750 | 0.04642 | 35.69 | 768.8 | 750 | 0.46405 | 35.69 | 76.91 | 750 | 4.6253 | 35.7 | 7.718 | 750 | 44.665 | 36.07 | 0.8076 |
| 800 | 0.043518 | 37.23 | 855.5 | 800 | 0.43505 | 37.23 | 85.58 | 800 | 4.3364 | 37.25 | 8.59 | 800 | 41.91 | 37.58 | 0.8967 |
| 850 | 0.040959 | 38.74 | 945.8 | 850 | 0.40946 | 38.74 | 94.61 | 850 | 4.0816 | 38.75 | 9.494 | 850 | 39.4482 | 39.06 | 0.9893 |
| 900 | 0.038683 | 40.22 | 1,040 | 900 | 0.38671 | 40.22 | 104 | 900 | 3.8552 | 40.23 | 10.44 | 900 | 37.325 | 40.5 | 1.085 |
| 1,000 | 0.034815 | 43.08 | 1,237 | 1,000 | 0.34805 | 43.08 | 123.8 | 1,000 | 3.4702 | 43.09 | 12.42 | 1,000 | 33.657 | 43.32 | 1.287 |
| 1,100 | 0.03165 | 45.84 | 1,448 | 1,100 | 0.31641 | 45.84 | 144.9 | 1,100 | 3.1553 | 45.85 | 15.38 | 1,100 | 30.654 | 46.05 | 1.502 |
| 1,200 | 0.029012 | 48.51 | 1,672 | 1,200 | 0.29005 | 48.52 | 167.3 | 1,200 | 2.8928 | 48.52 | 16.77 | 1,200 | 28.148 | 48.7 | 1.73 |
| 1,300 | 0.026781 | 51.11 | 1,908 | 1,300 | 0.26774 | 51.11 | 190.9 | 1,300 | 2.6706 | 51.12 | 19.14 | 1,300 | 26.024 | 51.28 | 1.97 |
| 1,400 | 0.024868 | 53.64 | 2,157 | 1,400 | 0.24862 | 53.64 | 215.8 | 1,400 | 2.4802 | 53.65 | 21.63 | 1,400 | 24.2 | 53.79 | 2.223 |
| 1,500 | 0.02321 | 56.11 | 2,418 | 1,500 | 0.23205 | 56.11 | 241.8 | 1,500 | 2.3152 | 56.12 | 24.24 | 1,500 | 22.617 | 56.25 | 2.487 |

# Transport Properties of Argon

| Argon (0.01 MPa) | | | | Argon (0.1 MPa) | | | | Argon (1.0 MPa) | | | | Argon (10 MPa) | | | |
|---|---|---|---|---|---|---|---|---|---|---|---|---|---|---|---|
| Temp. (K) | Density (kg/m$^3$) | Dynamic viscosity (μPa s) | Kinematic viscosity (mm$^2$/s) | Temp. (K) | Density (kg/m$^3$) | Dynamic viscosity (μPa s) | Kinematic viscosity (mm$^2$/s) | Temp. (K) | Density (kg/m$^3$) | Dynamic viscosity (μPa s) | Kinematic viscosity (mm$^2$/s) | Temp. (K) | Density (kg/m$^3$) | Dynamic viscosity (μPa s) | Kinematic viscosity (mm$^2$/s) |
| 83.78 | 0.57559 | | | 83.78 | 1, 534.70 | | | 83.78 | | | | 83.78 | | | |
| 100 | 0.4815 | | | 100 | 4.9142 | | | 100 | 1,314.10 | | | 100 | 1,347.80 | | |
| 120 | 0.40085 | | | 120 | 4.0575 | | | 120 | 47.185 | | | 120 | 1,219.20 | | |
| 140 | 0.34321 | | | 140 | 3.4614 | | | 140 | 37.743 | | | 140 | 1,062.90 | | |
| 160 | 0.30047 | | | 160 | 3.0204 | | | 160 | 31.932 | | | 160 | 831.62 | | |
| 180 | 0.26693 | | | 180 | 2.6801 | | | 180 | 27.836 | | | 180 | 491.51 | | |
| 200 | 0.2403 | | | 200 | 2.4093 | | | 200 | 24.745 | | | 200 | 337.55 | | |
| 220 | 0.21844 | | | 220 | 2.1886 | | | 220 | 22.31 | | | 220 | 270.87 | | |
| 240 | 0.20023 | | | 240 | 2.0051 | | | 240 | 20.332 | | | 240 | 231.26 | | |
| 260 | 0.18482 | | | 260 | 1.8502 | | | 260 | 18.69 | | | 260 | 203.97 | | |
| 273.15 | 0.17592 | | | 273.15 | 1.7607 | | | 273.15 | 17.752 | | | 273.15 | 189.99 | | |
| 280 | 0.17161 | | | 280 | 1.7174 | | | 280 | 17.301 | | | 280 | 183.6 | | |
| 300 | 0.16017 | 22.7 | 141.7 | 300 | 1.6025 | 22.71 | 14.17 | 300 | 16.11 | 22.91 | 1.422 | 300 | 167.59 | 25.73 | 0.1401 |
| 320 | 0.15015 | 23.92 | 159.3 | 320 | 1.5021 | 23.94 | 15.94 | 320 | 15.077 | 24.12 | 1.6 | 320 | 154.56 | 26.64 | 0.1724 |
| 340 | 0.14132 | 25.1 | 177.6 | 340 | 1.4135 | 25.11 | 17.76 | 340 | 14.17 | 25.28 | 1.784 | 340 | 143.69 | 27.57 | 0.1919 |
| 360 | 0.13347 | 26.64 | 199.6 | 360 | 1.3348 | 26.65 | 19.97 | 360 | 13.369 | 26.77 | 2.002 | 360 | 134.43 | 28.9 | 0.215 |
| 380 | 0.12644 | 27.91 | 220.7 | 380 | 1.2644 | 27.92 | 22.08 | | | | | | | | |
| 400 | 0.12012 | 28.9 | 240.6 | 400 | 1.2011 | 28.91 | 24.07 | 400 | 12.013 | 29.03 | 2.417 | 400 | 119.43 | 30.85 | 0.2583 |
| 420 | 0.1144 | 30 | 262.2 | 420 | 1.1439 | 30.01 | 26.23 | 420 | 11.435 | 30.13 | 2.635 | 420 | 113.23 | 31.83 | 0.2811 |
| 440 | 0.1092 | 31.08 | 284.6 | 440 | 1.0918 | 31.09 | 28.48 | 440 | 10.911 | 31.2 | 2.859 | 440 | 107.7 | 32.79 | 0.3045 |
| 460 | 0.10445 | 32.13 | 307.6 | 460 | 1.0443 | 32.14 | 30.78 | 460 | 10.433 | 32.25 | 3.091 | 460 | 102.73 | 33.74 | 0.3284 |
| 480 | 0.1001 | 33.15 | 331.2 | 480 | 1.0008 | 33.16 | 33.13 | 480 | 9.9951 | 33.27 | 3.324 | 480 | 98.231 | 34.68 | 0.353 |

(continued)

(continued)

| Argon (0.01 MPa) | | | | Argon (0.1 MPa) | | | | Argon (1.0 MPa) | | | | Argon (10 MPa) | | | |
|---|---|---|---|---|---|---|---|---|---|---|---|---|---|---|---|
| Temp. (K) | Density (kg/m$^3$) | Dynamic viscosity (μPa s) | Kinematic viscosity (mm$^2$/s) | Temp. (K) | Density (kg/m$^3$) | Dynamic viscosity (μPa s) | Kinematic viscosity (mm$^2$/s) | Temp. (K) | Density (kg/m$^3$) | Dynamic viscosity (μPa s) | Kinematic viscosity (mm$^2$/s) | Temp. (K) | Density (kg/m$^3$) | Dynamic viscosity (μPa s) | Kinematic viscosity (mm$^2$/s) |
| 500 | 0.096092 | 34.15 | 355.4 | 500 | 0.96075 | 34.16 | 35.56 | 500 | 9.5931 | 34.26 | 3.571 | 500 | 94.135 | 35.6 | 0.3782 |
| 520 | 0.092396 | | | 520 | 0.92379 | | | 520 | 9.2225 | | | 520 | 90.388 | | |
| 540 | 0.088974 | | | 540 | 0.88956 | | | 540 | 8.8796 | | | 540 | 86.945 | | |
| 550 | 0.087356 | 36.54 | 418.3 | 550 | 0.87338 | 36.55 | 41.85 | 550 | 8.7176 | 36.65 | 4.204 | 550 | 85.325 | 37.82 | 0.4432 |
| 560 | 0.085796 | | | 560 | 0.85778 | | | 560 | 8.5615 | | | 560 | 83.768 | | |
| 580 | 0.082838 | | | 580 | 0.8282 | | | 580 | 8.2655 | | | 580 | 80.826 | | |
| 600 | 0.080077 | 38.9 | 485.8 | 600 | 0.80059 | 38.91 | 48.6 | 600 | 7.9894 | 39 | 4.881 | 600 | 78.092 | 40.05 | 0.5129 |
| 620 | 0.077493 | | | 620 | 0.77476 | | | 620 | 7.7312 | | | 620 | 75.545 | | |
| 640 | 0.075072 | | | 640 | 0.75055 | | | 640 | 7.4893 | | | 640 | 73.165 | | |
| 650 | 0.073917 | 41.14 | 556.6 | 650 | 0.739 | 41.15 | 55.68 | 650 | 7.3739 | 41.23 | 5.591 | 650 | 72.033 | 42.19 | 0.5857 |
| 660 | 0.072797 | | | 660 | 0.7278 | | | 660 | 7.2621 | | | 660 | 70.936 | | |
| 680 | 0.070656 | | | 680 | 0.7064 | | | 680 | 7.0483 | | | 680 | 68.842 | | |
| 700 | 0.068637 | 43.3 | 630.9 | 700 | 0.68622 | 43.31 | 63.11 | 700 | 6.8468 | 43.39 | 6.337 | 700 | 66.873 | 44.26 | 0.6619 |
| 750 | 0.064061 | 45.39 | 708.6 | 750 | 0.64047 | 45.4 | 70.89 | 750 | 6.3903 | 45.47 | 7.115 | 750 | 62.42 | 46.28 | 0.7414 |
| 800 | 0.060057 | 47.43 | 789.7 | 800 | 0.60044 | 47.44 | 79.01 | 800 | 5.9909 | 47.51 | 7.93 | 800 | 58.536 | 48.26 | 0.8244 |
| 850 | 0.056525 | 49.42 | 874.3 | 850 | 0.56513 | 49.43 | 87.47 | 850 | 5.6387 | 49.5 | 8.779 | 850 | 55.116 | 50.19 | 0.9106 |
| 900 | 0.053384 | 51.37 | 962.3 | 900 | 0.53373 | 51.38 | 96.27 | 900 | 5.3256 | 51.44 | 9.659 | 900 | 52.08 | 52.09 | 1 |
| 1,000 | 0.048046 | 55.14 | 1,148 | 1,000 | 0.48037 | 55.14 | 114.8 | 1,000 | 4.7935 | 55.2 | 11.52 | 1,000 | 46.926 | 55.77 | 1.188 |
| 1,100 | 0.043678 | 58.77 | 1,346 | 1,100 | 0.4367 | 58.77 | 134.6 | 1,100 | 4.3583 | 58.82 | 13.5 | 1,100 | 42.715 | 59.34 | 1.389 |
| 1,200 | 0.040039 | 62.26 | 1,555 | 1,200 | 0.40032 | 62.27 | 155.6 | 1,200 | 3.9956 | 62.32 | 15.6 | 1,200 | 39.212 | 62.79 | 1.601 |
| 1,300 | 0.036959 | 65.66 | 1,777 | 1,300 | 0.36953 | 65.67 | 177.7 | 1,300 | 3.6888 | 65.71 | 17.81 | 1,300 | 36.255 | 66.14 | 1.824 |
| 1,400 | 0.034319 | | | 1,400 | 0.34314 | | | 1,400 | 3.426 | | | 1,400 | 33.727 | | |

(continued)

(continued)

| Argon (0.01 MPa) | | | | Argon (0.1 MPa) | | | | Argon (1.0 MPa) | | | | Argon (10 MPa) | | | |
|---|---|---|---|---|---|---|---|---|---|---|---|---|---|---|---|
| Temp. (K) | Density (kg/m$^3$) | Dynamic viscosity (μPa s) | Kinematic viscosity (mm$^2$/s) | Temp. (K) | Density (kg/m$^3$) | Dynamic viscosity (μPa s) | Kinematic viscosity (mm$^2$/s) | Temp. (K) | Density (kg/m$^3$) | Dynamic viscosity (μPa s) | Kinematic viscosity (mm$^2$/s) | Temp. (K) | Density (kg/m$^3$) | Dynamic viscosity (μPa s) | Kinematic viscosity (mm$^2$/s) |
| 1,500 | 0.032031 | | | 1,500 | 0.32027 | | | 1,500 | 3.1982 | | | 1,500 | 31.545 | | |
| 1,600 | 0.030029 | | | 1,600 | 0.30026 | | | 1,600 | 2.9991 | | | 1,600 | 29.644 | | |
| 1,700 | 0.028263 | | | 1,700 | 0.2826 | | | 1,700 | 2.8234 | | | 1,700 | 27.978 | | |
| 1,800 | 0.026693 | | | 1,800 | 0.26691 | | | 1,800 | 2.6674 | | | 1,800 | 26.508 | | |
| 1,900 | 0.025288 | | | 1,900 | 0.25287 | | | 1,900 | 2.528 | | | 1,900 | 25.205 | | |
| 2,000 | 0.024024 | | | 2,000 | 0.24024 | | | 2,000 | 2.4026 | | | 2,000 | 24.044 | | |

# Transport Properties of Ethanol

Thermodynamic properties of saturated ethanol ($C_2H_5OH$)

| Temp. $T$(k) | Density $\rho$ (kg/m$^3$) | Surface tension $\sigma$ (mN/m) | Dynamic viscosity $\eta$ (μPa s) | Kinematic viscosity $\nu$ (mm$^2$/s) |
|---|---|---|---|---|
| 160 | | 34.35 | | |
| 170 | | 33.51 | 57,980 | |
| 180 | | 32.66 | 33,230 | |
| 190 | | 31.81 | 20,430 | |
| 200 | | 30.95 | 13,580 | |
| 210 | | 30.09 | 9,556 | |
| 220 | | 29.23 | 6,985 | |
| 230 | | 28.36 | 5,242 | |
| 240 | | 27.49 | 4,004 | |
| 250 | | 26.62 | 3,103 | |
| 260 | | 25.74 | 2,436 | |
| 270 | | 24.86 | 1,937 | |
| 273.15 | 806.2 | 24.58 | 1,807 | 2.241 |
| 280 | 800.5 | 23.97 | 1,558 | 1.946 |
| 290 | 792.1 | 23.08 | 1,269 | 1.602 |
| 300 | 783.5 | 22.19 | 1,045 | 1.334 |
| 310 | 774.8 | 21.28 | 869.8 | 1.123 |
| 320 | 766.1 | 20.38 | 731.2 | 0.9544 |
| 330 | 757.7 | 19.47 | 620.4 | 0.8188 |
| 340 | 748.8 | 18.55 | 530.7 | 0.7087 |
| 350 | 739.3 | 17.62 | 457.1 | 0.6183 |
| 360 | 729.3 | 16.69 | 395.8 | 0.5427 |
| 370 | 718.6 | 15.75 | 344.4 | 0.4793 |
| 380 | 707.3 | 14.8 | 300.7 | 0.4251 |
| 390 | 695.3 | 13.85 | 263.3 | 0.3787 |
| 400 | 682.5 | 12.88 | 230.9 | 0.3383 |
| 410 | 668.8 | 11.9 | 202.7 | 0.3031 |
| 420 | 654.2 | 10.92 | 178.1 | 0.2722 |
| 430 | 638.6 | 9.92 | 156.3 | 0.2448 |
| 440 | 621.9 | 8.9 | 137.1 | 0.2205 |
| 450 | 603.9 | 7.87 | 120.2 | 0.199 |
| 460 | 583 | 6.82 | 105.1 | 0.1803 |
| 470 | 565.5 | 5.74 | 91.7 | 0.1622 |
| 480 | 543.5 | 4.64 | 79.9 | 0.147 |
| 490 | 511.3 | 3.5 | 69.4 | 0.1357 |
| 500 | 458.2 | 2.3 | 60.2 | 0.1314 |
| 510 | 372 | 1 | | |
| 516.3 | 276 | 0 | | |

# Transport Properties of Helium

| Helium (0.01 MPa) | | | | Helium (0.1 MPa) | | | | Helium (1.0 MPa) | | | | Helium (10 MPa) | | | |
|---|---|---|---|---|---|---|---|---|---|---|---|---|---|---|---|
| Temp. (K) | Density ($kg/m^3$) | Dynamic viscosity (μPa s) | Kinematic viscosity ($mm^2/s$) | Temp. (K) | Density ($kg/m^3$) | Dynamic viscosity (μPa s) | Kinematic viscosity ($mm^2/s$) | Temp. (K) | Density ($kg/m^3$) | Dynamic viscosity (μPa s) | Kinematic viscosity ($mm^2/s$) | Temp. (K) | Density ($kg/m^3$) | Dynamic viscosity (μPa s) | Kinematic viscosity ($mm^2/s$) |
| 20 | 0.24074 | | | 20 | 2.4104 | | | 20 | 24.28 | | | 20 | 148.5 | | |
| 40 | 0.12033 | | | 40 | 1.2008 | | | 40 | 11.765 | | | 40 | 92.3 | | |
| 60 | 0.08022 | | | 60 | 0.80071 | | | 60 | 7.8609 | | | 60 | 65.81 | | |
| 80 | 0.060167 | | | 80 | 0.60074 | | | 80 | 5.916 | | | 80 | 51.29 | | |
| 100 | 0.048135 | | | 100 | 0.48073 | | | 100 | 4.7459 | | | 100 | 42.13 | | |
| 120 | 0.040114 | | | 120 | 0.40069 | | | 120 | 3.9633 | | | 120 | 35.8 | | |
| 140 | 0.034384 | | | 140 | 0.34351 | | | 140 | 3.4027 | | | 140 | 31.139 | | |
| 160 | 0.030086 | | | 160 | 0.30061 | | | 160 | 2.9812 | | | 160 | 27.57 | | |
| 180 | 0.026743 | | | 180 | 0.26724 | | | 180 | 2.6527 | | | 180 | 24.739 | | |
| 200 | 0.024069 | | | 200 | 0.24053 | | | 200 | 2.3894 | | | 200 | 22.438 | | |
| 220 | 0.021881 | | | 220 | 0.21868 | | | 220 | 2.1737 | | | 220 | 20.531 | | |
| 240 | 0.020058 | | | 240 | 0.20047 | | | 240 | 1.9938 | | | 240 | 18.923 | | |
| 260 | 0.018515 | | | 260 | 0.18506 | | | 260 | 1.8413 | | | 260 | 17.549 | | |
| 273.15 | 0.017624 | 18.69 | 1,060 | 273.15 | 0.17615 | 18.69 | 106.1 | 273.15 | 1.7532 | 18.69 | 10.66 | 273.15 | 16.75 | 18.76 | 1.12 |
| 280 | 0.017193 | | | 280 | 0.17185 | | | 280 | 1.7105 | | | 280 | 16.362 | | |
| 300 | 0.016047 | 19.93 | 1,242 | 300 | 0.1604 | 19.93 | 124.3 | 300 | 1.5971 | 19.93 | 12.48 | 300 | 15.326 | 19.99 | 1.304 |
| 320 | 0.015044 | 20.83 | 1,385 | 320 | 0.15038 | 20.83 | 138.5 | 320 | 1.4978 | 20.83 | 13.91 | 320 | 14.413 | 20.88 | 1.449 |
| 340 | 0.014159 | 21.71 | 1,533 | 340 | 0.14154 | 21.71 | 153.3 | 340 | 1.4101 | 21.71 | 15.4 | 340 | 13.602 | 21.76 | 1.6 |
| 360 | 0.013372 | 22.58 | 1,689 | 360 | 0.13368 | 22.58 | 168.9 | 360 | 1.3321 | 22.59 | 16.96 | 360 | 12.878 | 22.63 | 1.757 |
| 380 | 0.012669 | 23.44 | 1,850 | 380 | 0.12664 | 23.44 | 185.1 | 380 | 1.2623 | 23.44 | 18.57 | 380 | 12.228 | 23.48 | 1.92 |
| 400 | 0.012035 | 24.29 | 2,019 | 400 | 0.12031 | 24.29 | 201.9 | 400 | 1.1994 | 24.29 | 20.25 | 400 | 11.639 | 24.33 | 2.09 |
| 420 | 0.011462 | 25.12 | 2,192 | 420 | 0.11459 | 25.12 | 219.2 | 420 | 1.1425 | 25.12 | 21.99 | 420 | 11.105 | 25.16 | 2.266 |
| 440 | 0.010941 | 25.95 | 2,372 | 440 | 0.10938 | 25.95 | 237.3 | 440 | 1.0908 | 25.95 | 23.79 | 440 | 10.618 | 25.98 | 2.447 |

(continued)

(continued)

| Helium (0.01 MPa) | | | | Helium (0.1 MPa) | | | | Helium (1.0 MPa) | | | | Helium (10 MPa) | | | |
|---|---|---|---|---|---|---|---|---|---|---|---|---|---|---|---|
| Temp. (K) | Density (kg/m$^3$) | Dynamic viscosity (μPa s) | Kinematic viscosity (mm$^2$/s) | Temp. (K) | Density (kg/m$^3$) | Dynamic viscosity (μPa s) | Kinematic viscosity (mm$^2$/s) | Temp. (K) | Density (kg/m$^3$) | Dynamic viscosity (μPa s) | Kinematic viscosity (mm$^2$/s) | Temp. (K) | Density (kg/m$^3$) | Dynamic viscosity (μPa s) | Kinematic viscosity (mm$^2$/s) |
| 460 | 0.010465 | 26.76 | 2,557 | 460 | 0.10463 | 26.76 | 255.8 | 460 | 1.0435 | 26.76 | 25.64 | 460 | 10.171 | 26.79 | 2.634 |
| 480 | 0.010029 | 27.57 | 2,749 | 480 | 0.10027 | 27.57 | 275 | 480 | 1.0002 | 27.57 | 27.56 | 480 | 9.7607 | 27.59 | 2.827 |
| 500 | 0.0096282 | 28.36 | 2,946 | 500 | 0.096259 | 28.36 | 294.6 | 500 | 0.96028 | 28.36 | 29.53 | 500 | 9.3821 | 28.38 | 3.025 |
| 520 | 0.0092579 | | | 520 | 0.092558 | | | 520 | 0.92346 | | | 520 | 9.0317 | | |
| 540 | 0.008915 | | | 540 | 0.08913 | | | 540 | 0.88936 | | | 540 | 8.7065 | | |
| 550 | 0.0087529 | 30.31 | 3,463 | 550 | 0.08751 | 30.31 | 346.4 | 550 | 0.8723 | 30.31 | 34.71 | 550 | 8.5526 | 30.33 | 3.546 |
| 560 | 0.0085966 | | | 560 | 0.85948 | | | 560 | 0.85768 | | | 560 | 8.4039 | | |
| 580 | 0.0083002 | | | 580 | 0.082985 | | | 580 | 0.82819 | | | 580 | 8.1216 | | |
| 600 | 0.0080235 | 32.21 | 415 | 600 | 0.08022 | 32.21 | 401.5 | 600 | 0.80065 | 32.21 | 40.23 | 600 | 7.8577 | 32.23 | 4.102 |
| 620 | 0.0077647 | | | 620 | 0.077633 | | | 620 | 0.77489 | | | 620 | 7.6103 | | |
| 640 | 0.0075221 | | | 640 | 0.075207 | | | 640 | 0.75073 | | | 640 | 7.3781 | | |
| 650 | 0.0074063 | 34.07 | 4,600 | 650 | 0.07405 | 34.07 | 460.1 | 650 | 0.73921 | 34.07 | 46.09 | 650 | 7.2672 | 34.09 | 4.691 |
| 660 | 0.0072941 | | | 660 | 0.072929 | | | 660 | 0.72804 | | | 660 | 7.1595 | | |
| 680 | 0.0070796 | | | 680 | 0.070784 | | | 680 | 0.70667 | | | 680 | 6.9536 | | |
| 700 | 0.0068773 | 35.89 | 5,219 | 700 | 0.068762 | 35.89 | 522 | 700 | 0.68653 | 35.89 | 52.28 | 700 | 6.7591 | 35.9 | 5.311 |
| 750 | 0.0064189 | 37.68 | 5,870 | 750 | 0.064179 | 37.68 | 587.1 | 750 | 0.64085 | 37.68 | 58.8 | 750 | 6.3174 | 37.69 | 5.966 |
| 800 | 0.0060177 | 39.43 | 6,552 | 800 | 0.060169 | 39.43 | 655.3 | 800 | 0.60087 | 39.43 | 65.62 | 800 | 5.9298 | 39.44 | 6.651 |
| 850 | 0.0056637 | 41.15 | 7,266 | 850 | 0.05663 | 41.15 | 726.7 | 850 | 0.56559 | 41.15 | 72.76 | 850 | 5.587 | 41.16 | 7.367 |
| 900 | 0.0053491 | 42.85 | 8,011 | 900 | 0.05348 | 42.85 | 801.2 | 900 | 0.56559 | 42.85 | 80.21 | 900 | 5.2817 | 42.85 | 8.113 |
| 1,000 | 0.0048142 | | | 1,000 | 0.048137 | 46.16 | 958.9 | 1,000 | 0.48088 | | | 1,000 | 4.7612 | | |
| 1,100 | 0.0043765 | | | 1,100 | 0.043761 | 49.38 | 1,128 | 1,100 | 0.43722 | | | 1,100 | 4.334 | | |
| 1,200 | 0.0040118 | | | 1,200 | 0.040115 | 52.52 | 1,309 | 1,200 | 0.40083 | | | 1,200 | 3.9771 | | |
| 1,300 | 0.0037032 | | | 1,300 | 0.03703 | 55.59 | 1,501 | 1,300 | 0.37003 | | | 1,300 | 3.6744 | | |
| 1,400 | 0.0034387 | | | 1,400 | 0.034385 | 58.6 | 1,704 | 1,400 | 0.34363 | | | 1,400 | 3.4146 | | |
| 1,500 | 0.0032095 | | | 1,500 | 0.032093 | 61.55 | 1,918 | 1,500 | 0.32074 | | | 1,500 | 3.189 | | |

# Transport Properties of Mercury

Thermodynamic properties of saturated mercury ($H_g$)

| Temp. $T$ (K) | Density $\rho'$ (kg/m$^3$) | Surface tensiton $\sigma$ (mN/m) | Temp. $T$(K) | Dynamic viscosity $\eta$ (μPa s) | Kinematic viscosity $\nu$(mm$^2$/ s) |
|---|---|---|---|---|---|
| 234.28 | 13,691.52 | | 640 | 62.03 | 13.5 |
| 253.15 | 13,644.51 | | 660 | 64.22 | 10.3 |
| 273.15 | 13,595.03 | | 680 | 66.43 | 7.97 |
| 293.15 | 13,545.83 | 486.5 | 700 | 68.65 | 6.28 |
| 313.15 | 13,496.89 | 482.4 | 720 | 70.87 | 5.02 |
| 333.15 | 13,448.19 | 478.3 | 740 | 73.09 | 4.07 |
| 353.15 | 13,399.71 | 474.2 | 760 | 75.32 | 3.34 |
| 373.15 | 13,351.42 | 470.1 | 780 | 77.55 | 2.77 |
| 393.15 | 13,303.30 | 466 | 800 | 79.78 | 2.32 |
| 413.15 | 13,255.31 | 461.9 | 850 | 85.35 | 1.56 |
| 433.15 | 13,207.45 | 457.8 | 900 | 90.9 | 1.1 |
| 453.15 | 13,159.68 | 453.7 | 950 | 96.42 | 0.804 |
| 473.15 | 13,111.97 | 449.6 | 1,000 | 101.9 | 0.608 |
| 493.15 | 13,064.31 | | 1,050 | 107.35 | 0.472 |
| 513.15 | 13,016.65 | | 1,073.15 | 109.86 | 0.423 |
| 533.15 | 12,968.98 | | | | |
| 553.15 | 12,921.27 | | | | |
| 573.15 | 12,873.50 | | | | |
| 593.15 | 12,825.60 | | | | |
| 613.15 | 12,777.60 | | | | |
| 633.15 | 12,729.40 | | | | |
| 653.15 | 12,681.10 | | | | |
| 673.15 | 12,632.60 | | | | |
| 693.15 | 12,583.80 | | | | |
| 713.15 | 12,534.80 | | | | |
| 733.15 | 12,485.40 | | | | |
| 753.15 | 12,435.80 | | | | |
| 773.15 | 12,386 | | | | |
| 793.15 | 12,336 | | | | |
| 813.15 | 12,285 | | | | |
| 833.15 | 12,234 | | | | |
| 853.15 | 12,182 | | | | |
| 873.15 | 12,130 | | | | |
| 893.15 | 12,078 | | | | |
| 913.15 | 12,025 | | | | |
| 933.15 | 11,972 | | | | |
| 953.15 | 11,918 | | | | |
| 973.15 | 11,863 | | | | |
| 993.15 | 11,809 | | | | |
| 1,013.15 | 11,753 | | | | |
| 1,033.15 | 11,697 | | | | |
| 1,053.15 | 11,641 | | | | |
| 1,073.15 | 11,584 | | | | |

# Transport Properties of Methanol

| Thermodynamic properties of saturated methanol ($CH_3OH$) | | | | |
|---|---|---|---|---|
| Temp. $T$(K) | Density $\rho$ (kg/m$^3$) | Surface tension $\sigma$ (mN/m) | Dinamic viscosity $\eta$ ($\mu$Pa s) | Kinematic viscosity $\nu$ (mm$^2$/s) |
| 180 | | 30.67 | 10,400 | |
| 190 | | 30.9 | 6,490 | |
| 200 | | 30.12 | 4,390 | |
| 210 | | 29.33 | 3,180 | |
| 220 | | 28.54 | 2,411 | |
| 230 | | 27.75 | 1,889 | |
| 240 | | 26.95 | 1,514 | |
| 250 | | 26.14 | 1,234 | |
| 260 | 822.3 | 25.33 | 1,019 | 1.239 |
| 270 | 813 | 24.52 | 852.8 | 1.049 |
| 273.15 | 810 | 24.26 | 808.1 | 0.9977 |
| 280 | 803.6 | 23.69 | 721.2 | 0.8975 |
| 290 | 794.3 | 22.86 | 617 | 0.7768 |
| 300 | 784.9 | 22.03 | 533 | 0.6791 |
| 310 | 775.6 | 21.18 | 464.8 | 0.5993 |
| 320 | 766.2 | 20.33 | 408.8 | 0.5335 |
| 330 | 756.9 | 19.47 | 362.1 | 0.4784 |
| 340 | 748 | 18.6 | 323 | 0.4318 |
| 350 | 738 | 17.72 | 289.6 | 0.3924 |
| 360 | 727.6 | 16.83 | 260.7 | 0.3583 |
| 370 | 716.8 | 15.93 | 235.5 | 0.3285 |
| 380 | 705.4 | 15.01 | 213.2 | 0.3022 |
| 390 | 693.4 | 14.09 | 193.2 | 0.2786 |
| 400 | 680.6 | 13.15 | 175.2 | 0.2574 |
| 410 | 667 | 12.19 | 159 | 0.2384 |
| 420 | 652.2 | 11.22 | 144 | 0.2208 |
| 430 | 636 | 10.22 | 130 | 0.2044 |
| 440 | 618.2 | 9.21 | 117 | 0.1893 |
| 450 | 598.1 | 8.16 | 106 | 0.1772 |
| 460 | 575 | 7.09 | 94.9 | 0.165 |
| 470 | 548 | 5.97 | 84.9 | 0.1549 |
| 480 | 516 | 4.8 | 75.8 | 0.1469 |
| 490 | 475 | 3.57 | 67.4 | 0.1419 |
| 500 | 423 | 2.22 | 59.7 | 0.1411 |
| 510 | 350 | 0.61 | | |
| 512.58 | 272 | 0 | | |

# Transport Properties of Nitrogen

| Nitrogen (0.01 MPa) | | | | Nitrogen (0.1 MPa) | | | | Nitrogen (1.0 MPa) | | | | Nitrogen (10 MPa) | | | |
|---|---|---|---|---|---|---|---|---|---|---|---|---|---|---|---|
| Temp. (K) | Density (kg/m$^3$) | Dynamic viscosity (μPa s) | Kinematic viscosity (mm$^2$/s) | Temp. (K) | Density (kg/m$^3$) | Dynamic viscosity (μPa s) | Kinematic viscosity (mm$^2$/s) | Temp. (K) | Density (kg/m$^3$) | Dynamic viscosity (μPa s) | Kinematic viscosity (mm$^2$/s) | Temp. (K) | Density (kg/m$^3$) | Dynamic viscosity (μPa s) | Kinematic viscosity (mm$^2$/s) |
| 80 | 0.42271 | | | 80 | 4.3793 | | | 80 | 798.49 | | | 80 | 819.12 | | |
| 100 | 0.33758 | | | 100 | 3.4366 | | | 100 | 690.09 | | | 100 | 733.97 | | |
| 120 | 0.28109 | | | 120 | 2.8402 | | | 120 | 32.081 | | | 120 | 632.06 | | |
| 140 | 0.24084 | | | 140 | 2.424 | | | 140 | 26.002 | | | 140 | 499.72 | | |
| 160 | 0.21068 | | | 160 | 2.1157 | | | 160 | 22.111 | | | 160 | 344.02 | | |
| 180 | 0.18724 | | | 180 | 1.8778 | | | 180 | 19.329 | | | 180 | 248.73 | | |
| 200 | 0.1685 | | | 200 | 1.6883 | | | 200 | 17.213 | | | 200 | 199.37 | | |
| 220 | 0.15317 | | | 220 | 1.5337 | | | 220 | 15.538 | | | 220 | 169.39 | | |
| 240 | 0.14039 | | | 240 | 1.4052 | | | 240 | 14.173 | | | 240 | 148.77 | | |
| 260 | 0.12959 | | | 260 | 1.2967 | | | 260 | 13.037 | | | 260 | 133.44 | | |
| 273.15 | 0.12335 | 16.62 | 134.7 | 273.15 | 1.234 | 16.62 | 13.47 | 273.15 | 12.388 | 16.64 | 1.343 | 273.15 | 125.26 | 18.67 | 0.149 |
| 280 | 0.12033 | | | 280 | 1.2038 | | | 280 | 12.075 | | | 280 | 121.45 | | |
| 300 | 0.11231 | 17.87 | 159.1 | 300 | 1.1233 | 17.87 | 15.91 | 300 | 11.249 | 17.89 | 1.59 | 300 | 111.74 | 19.54 | 0.1749 |
| 320 | 0.10529 | 18.77 | 178.3 | 320 | 1.0529 | 18.77 | 17.83 | 320 | 10.531 | 18.79 | 1.784 | 320 | 103.66 | 20.23 | 0.1952 |
| 340 | 0.0991 | 19.65 | 198.3 | 340 | 0.99089 | 19.65 | 19.83 | 340 | 9.9005 | 19.67 | 1.987 | 340 | 96.801 | 20.94 | 0.2163 |
| 360 | 0.09359 | 20.51 | 219.1 | 360 | 0.93576 | 20.51 | 21.92 | 360 | 9.3427 | 20.52 | 2.196 | 360 | 90.892 | 21.66 | 0.2383 |
| 380 | 0.08866 | 21.35 | 240.8 | 380 | 0.88645 | 21.35 | 24.08 | 380 | 8.8454 | 21.36 | 2.415 | 380 | 85.733 | 22.37 | 0.2609 |
| 400 | 0.08423 | 22.17 | 263.2 | 400 | 0.84209 | 22.17 | 26.33 | 400 | 8.399 | 22.17 | 2.64 | 400 | 81.18 | 23.09 | 0.2844 |
| 420 | 0.08022 | 22.97 | 286.3 | 420 | 0.80196 | 22.97 | 28.64 | 420 | 7.996 | 22.97 | 2.873 | 420 | 77.126 | 23.81 | 0.3087 |
| 440 | 0.07657 | 23.75 | 310.2 | 440 | 0.76548 | 23.75 | 31.03 | 440 | 7.6303 | 23.76 | 3.114 | 440 | 73.488 | 24.52 | 0.3337 |
| 460 | 0.07324 | 24.52 | 334.8 | 460 | 0.73218 | 24.52 | 33.49 | 460 | 7.297 | 24.53 | 3.362 | 460 | 70.202 | 25.23 | 0.3594 |
| 480 | 0.07019 | 25.28 | 360.2 | 480 | 0.70166 | 25.28 | 36.03 | 480 | 6.9917 | 25.28 | 3.616 | 480 | 67.216 | 25.93 | 0.3858 |

(continued)

(continued)

| Nitrogen (0.01 MPa) | | | | Nitrogen (0.1 MPa) | | | | Nitrogen (1.0 MPa) | | | | Nitrogen (10 MPa) | | | |
|---|---|---|---|---|---|---|---|---|---|---|---|---|---|---|---|
| Temp. (K) | Density (kg/m$^3$) | Dynamic viscosity (μPa s) | Kinematic viscosity (mm$^2$/s) | Temp. (K) | Density (kg/m$^3$) | Dynamic viscosity (μPa s) | Kinematic viscosity (mm$^2$/s) | Temp. (K) | Density (kg/m$^3$) | Dynamic viscosity (μPa s) | Kinematic viscosity (mm$^2$/s) | Temp. (K) | Density (kg/m$^3$) | Dynamic viscosity (μPa s) | Kinematic viscosity (mm$^2$/s) |
| 500 | 0.06739 | 26.02 | 386.1 | 500 | 0.67359 | 26.02 | 38.63 | 500 | 6.7113 | 26.02 | 3.877 | 500 | 64.489 | 26.62 | 0.4128 |
| 520 | 0.06479 | | | 520 | 0.64767 | | | 520 | 6.4256 | | | 520 | 61.987 | | |
| 540 | 0.06239 | | | 540 | 0.62368 | | | 540 | 6.2132 | | | 540 | 59.682 | | |
| 550 | 0.06126 | 27.82 | 454.1 | 550 | 0.61234 | 27.82 | 45.43 | 550 | 6.1001 | 27.83 | 4.562 | 550 | 58.596 | 28.32 | 0.4833 |
| 560 | 0.06017 | | | 560 | 0.6014 | | | 560 | 5.9911 | | | 560 | 57.551 | | |
| 580 | 0.05809 | | | 580 | 0.58066 | | | 580 | 5.7844 | | | 580 | 55.574 | | |
| 600 | 0.05615 | 29.55 | 526.2 | 600 | 0.56131 | 29.55 | 52.64 | 600 | 5.5916 | 29.56 | 5.287 | 600 | 53.733 | 29.98 | 0.5579 |
| 620 | 0.05434 | | | 620 | 0.5432 | | | 620 | 5.4112 | | | 620 | 52.016 | | |
| 640 | 0.05264 | | | 640 | 0.52623 | | | 640 | 5.2422 | | | 640 | 50.409 | | |
| 650 | 0.05183 | 31.22 | 602.3 | 650 | 0.51813 | 31.22 | 60.26 | 650 | 5.1616 | 31.23 | 6.05 | 650 | 49.643 | 31.59 | 0.6363 |
| 660 | 0.05105 | | | 660 | 0.51028 | | | 660 | 5.0835 | | | 660 | 48.902 | | |
| 680 | 0.04955 | | | 680 | 0.49528 | | | 680 | 4.9341 | | | 680 | 47.485 | | |
| 700 | 0.04813 | 32.84 | 682.3 | 700 | 0.48113 | 32.84 | 68.23 | 700 | 4.7933 | 32.84 | 6.851 | 700 | 46.151 | 33.15 | 0.7183 |
| 750 | 0.04492 | 34.41 | 766 | 750 | 0.44906 | 34.41 | 76.63 | 750 | 4.4742 | 34.41 | 7.691 | 750 | 43.13 | 34.68 | 0.8041 |
| 800 | 0.04212 | 35.93 | 853.1 | 800 | 0.421 | 35.93 | 85.34 | 800 | 4.1951 | 35.93 | 8.565 | 800 | 40.489 | 36.17 | 0.8933 |
| 850 | 0.03964 | 37.41 | 943.8 | 850 | 0.39624 | 37.41 | 94.41 | 850 | 3.9488 | 37.41 | 9.474 | 850 | 38.159 | 37.63 | 0.9867 |
| 900 | 0.03744 | 38.86 | 1,038 | 900 | 0.37423 | 38.86 | 103.8 | 900 | 3.7299 | 38.86 | 10.42 | 900 | 36.088 | 39.05 | 1.082 |
| 1,000 | 0.03369 | 41.65 | 1,236 | 1,000 | 0.33681 | 41.65 | 123.7 | 1,000 | 3.3578 | 41.65 | 12.4 | 1,000 | 32.563 | 41.81 | 1.284 |
| 1,100 | 0.03063 | 44.33 | 1,447 | 1,100 | 0.3062 | 44.33 | 144.8 | 1,100 | 3.0532 | 44.33 | 14.52 | 1,100 | 29.673 | 44.46 | 1.498 |
| 1,200 | 0.02808 | 46.9 | 1,670 | 1,200 | 0.28069 | 46.9 | 167.1 | 1,200 | 2.7994 | 46.9 | 16.75 | 1,200 | 27.258 | 47.01 | 1.725 |
| 1,300 | 0.02592 | 49.39 | 1,906 | 1,300 | 0.2591 | 49.39 | 190.6 | 1,300 | 2.5846 | 49.39 | 19.11 | 1,300 | 25.21 | 49.48 | 1.963 |
| 1,400 | 0.02407 | 51.8 | 2,152 | 1,400 | 0.2406 | 51.8 | 215.3 | 1,400 | 2.4004 | 51.8 | 21.58 | 1,400 | 23.45 | 51.88 | 2.212 |

(continued)

(continued)

| Nitrogen (0.01 MPa) | | | | Nitrogen (0.1 MPa) | | | | Nitrogen (1.0 MPa) | | | | Nitrogen (10 MPa) | | | |
|---|---|---|---|---|---|---|---|---|---|---|---|---|---|---|---|
| Temp. (K) | Density (kg/m$^3$) | Dynamic viscosity (μPa s) | Kinematic viscosity (mm$^2$/s) | Temp. (K) | Density (kg/m$^3$) | Dynamic viscosity (μPa s) | Kinematic viscosity (mm$^2$/s) | Temp. (K) | Density (kg/m$^3$) | Dynamic viscosity (μPa s) | Kinematic viscosity (mm$^2$/s) | Temp. (K) | Density (kg/m$^3$) | Dynamic viscosity (μPa s) | Kinematic viscosity (mm$^2$/s) |
| 1,500 | 0.02246 | 54.13 | 2,410 | 1,500 | 0.22456 | 54.13 | 241 | 1,500 | 2.2407 | 54.13 | 24.16 | 1,500 | 21.921 | 54.2 | 2.473 |
| 1,600 | 0.02106 | | | 1,600 | 0.21053 | | | 1,600 | 2.1009 | | | 1,600 | 20.579 | | |
| 1,700 | 0.01982 | | | 1,700 | 0.19815 | | | 1,700 | 1.9776 | | | 1,700 | 19.394 | | |
| 1,800 | 0.01872 | | | 1,800 | 0.18714 | | | 1,800 | 1.868 | | | 1,800 | 18.337 | | |
| 1,900 | 0.01773 | | | 1,900 | 0.1773 | | | 1,900 | 1.7698 | | | 1,900 | 17.391 | | |

# Transport Properties of Oxygen

| Oxygen (0.1 MPa) | | | | Oxygen (1.0 MPa) | | | | Oxygen (5.0 MPa) | | | | Oxygen (10 MPa) | | | |
|---|---|---|---|---|---|---|---|---|---|---|---|---|---|---|---|
| Temp. (K) | Density (kg/m$^3$) | Dynamic viscosity (μPa s) | Kinematic viscosity (mm$^2$/s) | Temp. (K) | Density (kg/m$^3$) | Dynamic viscosity (μPa s) | Kinematic viscosity (mm$^2$/s) | Temp. (K) | Density (kg/m$^3$) | Dynamic viscosity (μPa s) | Kinematic viscosity (mm$^2$/s) | Temp. (K) | Density (kg/m$^3$) | Dynamic viscosity (μPa s) | Kinematic viscosity (mm$^2$/s) |
| 75 | | 303.8 | 0.252 | 75 | | 308.1 | 0.2553 | 75 | | 327.9 | 0.2701 | 75 | | 353.9 | 0.289 |
| 80 | 1,190.30 | 253.6 | 0.2143 | 80 | 1,192 | 256.9 | 0.2168 | 80 | 1,198.90 | 272.1 | 0.2282 | 80 | 1,207.20 | 292.1 | 0.2432 |
| 85 | | 216.2 | 0.1863 | 85 | | 218.9 | 0.1883 | 85 | | 230.7 | 0.1972 | 85 | | 233.8 | 0.1983 |
| 90 | | 188.2 | 0.1656 | 90 | | 190.3 | 0.1671 | 90 | | 199.8 | 0.1742 | 90 | | 212.1 | 0.1833 |
| 95 | | 7.21 | 1.728 | 95 | | 168.4 | 0.1512 | 95 | | 176.3 | 0.1569 | 95 | | 186.2 | 0.1642 |
| 100 | 3.9443 | 7.64 | 1.937 | 100 | 1,092.80 | 150.7 | 0.1384 | 100 | 1,103.70 | 157.8 | 0.1436 | 100 | 1,116.10 | 166.6 | 0.15 |
| 105 | | 8.05 | 2.15 | 105 | | 135.5 | 0.1276 | 105 | | 142.3 | 0.1325 | 105 | | 150.4 | 0.1384 |
| 110 | | 8.46 | 2.373 | 110 | | 121.6 | 0.1177 | 110 | | 128.5 | 0.1227 | 110 | | 136.5 | 0.1284 |
| 115 | | 8.86 | 2.605 | 115 | | 108.6 | 0.108 | 115 | | 115.8 | 0.1135 | 115 | | 124 | 0.1196 |
| 120 | 3.2532 | 9.26 | 2.85 | 120 | 38.378 | 9.79 | 0.2554 | 120 | 993.93 | 104 | 0.1051 | 120 | 1,015.10 | 112.3 | 0.1111 |
| 125 | | 9.65 | 3.09 | 125 | | 10.14 | 0.2828 | 125 | | 93.43 | 0.09755 | 125 | | 101.7 | 0.1035 |
| 130 | | 10.03 | 3.35 | 130 | | 10.49 | 0.3104 | 130 | | 84.34 | 0.09141 | 130 | | 92.31 | 0.09688 |
| 135 | | 10.4 | 3.611 | 135 | | 10.83 | 0.3383 | 135 | | 75.78 | 0.08579 | 135 | | 84.32 | 0.09155 |
| 140 | 2.7741 | 10.77 | 3.883 | 140 | 30.422 | 11.18 | 0.3674 | 140 | 845.45 | 67.12 | 0.08009 | 140 | 892.39 | 76.95 | 0.08675 |
| 145 | | 11.13 | 4.16 | 145 | | 11.52 | 0.3966 | 145 | | 58.13 | 0.07418 | 145 | | 69.78 | 0.08207 |
| 150 | | 11.48 | 4.442 | 150 | | 11.85 | 0.4262 | 150 | | 48.43 | 0.06793 | 150 | | 62.77 | 0.07751 |
| 155 | | 11.82 | 4.729 | 155 | | 12.18 | 0.4566 | 155 | | 18.57 | 0.06643 | 155 | | 55.96 | 0.07314 |
| 160 | 2.4202 | 12.16 | 5.186 | 160 | 25.669 | 12.5 | 0.4873 | 160 | 207.5 | 16.53 | 0.07971 | 160 | 716.13 | 49 | 0.06882 |
| 165 | | 12.49 | 5.326 | 165 | | 12.82 | 0.5186 | 165 | | 16.34 | 0.09085 | 165 | | 41.72 | 0.06446 |
| 170 | | 12.82 | 5.634 | 170 | | 13.14 | 0.5509 | 170 | | 15.954 | 0.09817 | 170 | | 35.08 | 0.06171 |
| 180 | 2.1473 | 13.47 | 6.274 | 180 | 22.348 | 13.76 | 0.6165 | 180 | 140.14 | 16.04 | 0.1147 | 180 | 423.39 | 26.08 | 0.06228 |
| 190 | | 14.11 | 6.942 | 190 | | 14.38 | 0.6848 | 190 | | 16.33 | 0.1308 | 190 | | 22.57 | 0.06906 |

(continued)

(continued)

| Oxygen (0.1 MPa) | | | | Oxygen (1.0 MPa) | | | | Oxygen (5.0 MPa) | | | | Oxygen (10 MPa) | | | |
|---|---|---|---|---|---|---|---|---|---|---|---|---|---|---|---|
| Temp. (K) | Density ($kg/m^3$) | Dynamic viscosity (μPa s) | Kinematic viscosity ($mm^2/s$) | Temp. (K) | Density ($kg/m^3$) | Dynamic viscosity (μPa s) | Kinematic viscosity ($mm^2/s$) | Temp. (K) | Density ($kg/m^3$) | Dynamic viscosity (μPa s) | Kinematic viscosity ($mm^2/s$) | Temp. (K) | Density ($kg/m^3$) | Dynamic viscosity (μPa s) | Kinematic viscosity ($mm^2/s$) |
| 200 | 1.9302 | 14.75 | 7.642 | 200 | 19.853 | 15.01 | 0.7569 | 200 | 114.09 | 16.72 | 0.1469 | 200 | 277.61 | 21.28 | 0.07737 |
| 210 | | 15.38 | 8.371 | 210 | | 15.62 | 0.8309 | 210 | | 17.16 | 0.1633 | 210 | | 20.8 | 0.08601 |
| 220 | 1.7533 | 16.01 | 9.132 | 220 | 17.892 | 16.24 | 0.9088 | 220 | 98.155 | 17.64 | 0.1801 | 220 | 219.89 | 20.7 | 0.09487 |
| 230 | | 16.63 | 9.92 | 230 | | 16.85 | 0.9941 | 230 | | 18.14 | 0.1974 | 230 | | 20.79 | 0.1039 |
| 240 | 1.6062 | 17.25 | 10.74 | 240 | 16.302 | 17.46 | 1.072 | 240 | 86.912 | 18.65 | 0.2148 | 240 | 186.72 | 21 | 0.1131 |
| 250 | | 17.86 | 11.59 | 250 | | 18.06 | 1.158 | 250 | | 19.17 | 0.2329 | 250 | | 21.29 | 0.1225 |
| 260 | 1.4821 | 18.45 | 12.45 | 260 | 14.982 | 18.64 | 1.245 | 260 | 78.372 | 19.68 | 0.2513 | 260 | 164.24 | 21.62 | 0.1321 |
| 270 | | 19.02 | 13.33 | 270 | | 19.21 | 1.335 | 270 | | 20.19 | 0.2701 | 270 | | 21.97 | 0.1418 |
| 273.15 | 1.4104 | | | 273.15 | 14.229 | | | 273.15 | 73.749 | | | 273.15 | 152.8 | | |
| 280 | 1.3758 | 19.59 | 14.24 | 280 | 13.867 | 19.77 | 1.426 | 280 | 71.581 | 20.7 | 0.2894 | 280 | 147.59 | 22.35 | 0.1517 |
| 290 | | 20.16 | 15.18 | 290 | | 20.33 | 1.522 | 290 | | 21.21 | | 290 | | 22.75 | 0.1619 |
| 300 | 1.2837 | 20.72 | 16.14 | 300 | 12.912 | 20.88 | 1.619 | 300 | 66.005 | 21.72 | 0.3293 | 300 | 134.58 | 23.16 | 0.1722 |
| 310 | | 21.27 | 17.12 | 310 | | 21.42 | 2.695 | 310 | | 22.23 | 0.3499 | 310 | | 23.58 | 0.1828 |
| 320 | 1.2033 | 21.8 | 18.12 | 320 | 12.082 | 21.95 | 1.817 | 320 | 61.32 | 22.95 | 0.3743 | 320 | 124.03 | 24 | 0.1936 |
| 330 | | 22.32 | 19.13 | 330 | | 22.47 | 1.921 | 330 | | 23.2 | 0.3916 | 330 | | 24.42 | 0.2044 |
| 340 | 1.1323 | 22.84 | 20.17 | 340 | 11.355 | 22.98 | 2.025 | 340 | 57.311 | 23.69 | 0.4134 | 340 | 115.23 | 24.84 | 0.2155 |
| 350 | 1.0999 | 23.35 | 21.23 | 350 | 11.024 | 23.48 | 2.131 | 350 | 55.513 | 24.17 | 0.4353 | 350 | 111.35 | 25.27 | 0.2268 |
| 400 | | 25.82 | 26.83 | 400 | | 25.94 | 2.695 | 400 | | 26.51 | 0.5514 | 400 | | 27.4 | 0.2861 |
| 450 | | 28.14 | 32.9 | 450 | | 28.25 | 3.308 | 450 | | 28.74 | 0.6763 | 450 | | 29.49 | 0.35 |
| 500 | | 30.33 | 39.4 | 500 | | 30.43 | 3.962 | 500 | | 30.86 | 0.8624 | 500 | | 31.51 | 0.4178 |
| 550 | | 32.4 | 46.3 | 550 | | 32.49 | 4.653 | 550 | | 32.88 | 0.9506 | 550 | | 33.45 | 0.4897 |
| 600 | | 34.37 | 53.58 | 600 | | 34.45 | 5.384 | 600 | | 34.8 | 1.099 | 600 | | 35.31 | 0.565 |
| 700 | | 38.08 | 69.27 | 700 | | 38.14 | 6.957 | 700 | | 38.44 | 1.417 | 700 | | 38.86 | 0.7262 |

(continued)

(continued)

| Oxygen (0.1 MPa) | | | | Oxygen (1.0 MPa) | | | | Oxygen (5.0 MPa) | | | | Oxygen (10 MPa) | | | |
|---|---|---|---|---|---|---|---|---|---|---|---|---|---|---|---|
| Temp. (K) | Density ($kg/m^3$) | Dynamic viscosity (μPa s) | Kinematic viscosity ($mm^2/s$) | Temp. (K) | Density ($kg/m^3$) | Dynamic viscosity (μPa s) | Kinematic viscosity ($mm^2/s$) | Temp. (K) | Density ($kg/m^3$) | Dynamic viscosity (μPa s) | Kinematic viscosity ($mm^2/s$) | Temp. (K) | Density ($kg/m^3$) | Dynamic viscosity (μPa s) | Kinematic viscosity ($mm^2/s$) |
| 800 | | 41.52 | 86.32 | 800 | | 41.58 | 8.666 | 800 | | 41.9 | 1.766 | 800 | | 42.19 | 0.9008 |
| 900 | | 44.72 | 104.6 | 900 | | 44.77 | 10.5 | 900 | | 44.99 | 2.113 | 900 | | 45.31 | 1.087 |
| 1,000 | | 47.7 | 124 | 1,000 | | 47.75 | 12.44 | 1,000 | | 47.95 | 2.523 | 1,000 | | 48.22 | 1.284 |
| 1,100 | | 50.55 | 144.5 | 1,100 | | 50.59 | 14.5 | 1,100 | | 50.77 | 2.937 | 1,100 | | 51.02 | 1.493 |
| 1,200 | | 53.25 | 166.1 | 1,200 | | 53.3 | 16.66 | 1,200 | | 53.41 | 3.368 | 1,200 | | 53.69 | 1.712 |
| 1,300 | | 55.84 | 188.6 | 1,300 | | 55.88 | 18.92 | 1,300 | | 56.03 | 3.825 | 1,300 | | 56.23 | 1.939 |

# Transport Properties of Silicone Oil

Silicone oil

Property of 298 K

| Products name (Shin-Etsu Chemical Co., Ltd) | Density ($kg/m^3$) | Kinematic viscosity ($mm^2/s$) | Surface tension (mN/m) |
|---|---|---|---|
| KF96L-1 | 818 | 1 | 16.9 |
| KF96L-2 | 873 | 2 | 18.3 |
| KF96L-5 | 915 | 5 | 19.7 |
| KF96L-10 | 935 | 10 | 20.1 |
| KF96L-50 | 960 | 50 | 20.8 |
| KF96L-100 | 965 | 100 | 20.9 |
| KF96L-350 | 970 | 350 | 21.1 |
| KF96L-1000 | 970 | 1,000 | 21.2 |

Kinematic viscosity ($mm^2/s$)

| Temp. $T$ (K) | 218.15 | 223.15 | 228.15 | 238.15 | 248.15 | 273.15 | 298.15 | 323.15 | 373.15 | 423.15 | 473.15 |
|---|---|---|---|---|---|---|---|---|---|---|---|
| KF96-50 | 586 | 437 | 327 | 220 | 164 | 88 | 50 | 32.5 | 15.9 | 9.13 | 5.81 |
| KF96-100 | 1,223 | 909 | 700 | 480 | 324 | 171 | 100 | 64.6 | 31.3 | 17.9 | 11.4 |
| KF96-350 | 4,570 | 3,350 | 2,530 | 1,700 | 1,190 | 598 | 350 | 220 | 107 | 59.1 | 37.1 |
| KF96-1000 | 16,100 | 10,000 | 7,350 | 4,860 | 3,400 | 1,720 | 1,000 | 633 | 302 | 165 | 104 |

Kinematic viscosity ratio

| Temp. $T$(K) | 218.15 | 223.15 | 228.15 | 238.15 | 248.15 | 273.15 | 298.15 | 323.15 | 373.15 | 423.15 | 473.15 |
|---|---|---|---|---|---|---|---|---|---|---|---|
| KF96-50 | 11.7 | 8.74 | 6.54 | 4.4 | 3.28 | 1.76 | 1 | 0.65 | 0.318 | 0.183 | 0.116 |
| KF96-100 | 12.2 | 9.09 | 7.00 | 4.8 | 3.24 | 1.71 | 1 | 0.646 | 0.313 | 0.179 | 0.114 |
| KF96-350 | 13.1 | 9.57 | 7.23 | 4.86 | 3.40 | 1.71 | 1 | 0.629 | 0.306 | 0.169 | 0.106 |
| KF96-1000 | 16.1 | 10.0 | 7.35 | 4.86 | 3.4 | 1.72 | 1 | 0.633 | 0.302 | 0.165 | 0.104 |

# Transport Properties of Water

$H_2O$

| Temp. (K) | Density (kg/m$^3$) | Dynamic viscosity (μPa s) | Kinematic viscosity (mm$^2$/s) | Surface tension (mN/m) |
|---|---|---|---|---|
| 273.15 | 999.78 | 1,791.40 | 1.792 | 75.65 |
| 280 | 999.93 | 1,435.40 | 1.435 | 74.68 |
| 290 | 998.87 | 1,085.30 | 1.087 | 73.21 |
| 300 | 996.62 | 854.4 | 0.8573 | 71.69 |
| 310 | 993.42 | 693.7 | 0.6983 | 70.11 |
| 320 | 989.43 | 577.2 | 0.5834 | 68.47 |
| 330 | 984.75 | 489.9 | 0.4974 | 66.79 |
| 340 | 979.44 | 422.5 | 0.4314 | 65.04 |
| 350 | 973.59 | 369.4 | 0.3794 | 63.25 |
| 360 | 967.21 | 326.7 | 0.3378 | 61.41 |
| 370 | 960.37 | 291.8 | 0.3039 | 59.52 |
| 380 | 953.08 | 263 | 0.2759 | 57.59 |
| 390 | 945.36 | 238.8 | 0.2527 | 55.61 |
| 400 | 937.22 | 218.5 | 0.2331 | 53.58 |
| 410 | 928.67 | 201.1 | 0.2165 | 51.52 |
| 420 | 919.7 | 186.2 | 0.2024 | 49.42 |
| 430 | 910.32 | 173.3 | 0.1903 | 47.28 |
| 440 | 900.51 | 162 | 0.1799 | 45.11 |
| 450 | 890.26 | 152.2 | 0.171 | 42.9 |
| 460 | 879.55 | 143.5 | 0.1632 | 40.66 |
| 470 | 868.36 | 135.9 | 0.1564 | 38.4 |
| 480 | 856.66 | 129 | 0.1506 | 36.11 |
| 490 | 844.41 | 122.8 | 0.1454 | 33.81 |
| 500 | 831.57 | 117.2 | 0.1409 | 31.48 |
| 520 | 803.9 | 107.3 | 0.1335 | 26.79 |
| 540 | 773.06 | 98.7 | 0.1276 | 22.09 |
| 560 | 738.18 | 90.8 | 0.1230 | 17.41 |
| 580 | 697.79 | 83.3 | 0.1193 | 12.81 |
| 600 | 649.3 | 75.6 | 0.1165 | 8.39 |
| 620 | 586.47 | 67.2 | 0.1145 | 4.28 |
| 640 | 481.96 | 55 | 0.1142 | 0.82 |

# Index

Kolibri 360 GmbH
Ferdinand-Jühlke-Straße 7
99095 Erfurt, Deutschland
produktsicherheit@kolibri360.de